b

Human
Osteology

SECOND EDITION

Human Osteology

SECOND EDITION

Text by

Tim D. White

Department of Integrative Biology and
Laboratory for Human Evolutionary Studies
Museum of Vertebrate Zoology
University of California
Berkeley, California

Images by

Pieter Arend Folkens

Alaska Whale Foundation
Benicia, California

ACADEMIC PRESS

An Elsevier Science Imprint

San Diego   San Francisco   New York   Boston   London   Sydney   Tokyo

Front cover photograph: Male adult skull. For more information, see Figure 17.10

This book is printed on acid-free paper.

Academic Press
An Imprint of Elsevier Science
525 B Street, Suite 1900, San Diego, California 92101-4495, USA
http://www.academicpress.com

Academic Press
Harcourt Place, 32 Jamestown Road, London NW1 7BY, UK
http://www.academicpress.com

Library of Congress Catalog Card Number: 99-61961

International Standard Book Number: 0-12-746612-6

PRINTED IN THE UNITED STATES OF AMERICA
03   04   05   06   MM   9   8   7   6   5   4

Contents

CHAPTER 5
Dentition

CHAPTER 6
Hyoid and Vertebrae

CHAPTER 7
Thorax: Sternum and Ribs

CHAPTER 8
Shoulder Girdle: Clavicle and Scapula

CHAPTER 9
Arm: Humerus, Radius, and Ulna

CHAPTER 10
Hand: Carpals, Metacarpals, and Phalanges

CHAPTER 11
Pelvic Girdle: Sacrum, Coccyx, and Os Coxae

CHAPTER 12
Leg: Femur, Patella, Tibia, and Fibula

CHAPTER 18

Osteological and Dental Pathology

CHAPTER 19

Postmortem Skeletal Modification

CHAPTER 20
The Biology of Skeletal Populations: Discrete Traits, Distance, Diet, Disease, and Demography

CHAPTER 21
Molecular Osteology

CHAPTER 22
Forensic Case Study: Homicide: "We Have the Witnesses but No Body"

CHAPTER 23
Forensic Case Study: Child Abuse, the Skeletal Perspective

CHAPTER 24
Archeological Case Study: The Bioarcheology of the Stillwater Marsh, Nevada

CHAPTER 25
Archeological Case Study: Anasazi Remains from Cottonwood Canyon

Preface to the Second Edition

With nearly a decade of advances in osteological research—and the positive response to the first edition—it was time to revise *Human Osteology*. This revision was driven by colleagues and students who found the first volume valuable and called for "more and better." We have strengthened and updated each chapter, added a host of new figures, tables, and features, and incorporated new standards. Among the advances are a new glossary and new sections on morphogenesis, bone modification, and disease and demography. The chapter on assessing age, sex, stature, ancestry, and identity has been greatly strengthened, and an occupation section has been added to the paleopathology chapter. A new chapter on molecular osteology and four new case studies have been added.

Many of our colleagues contributed excellent suggestions for revision. We have tried to incorporate as many as were feasible. In particular, we thank those authors who wrote published reviews, as well as Susan Ánton, Donna Boyd, Kristian Carlson, Mark Fleishman, David Frayer, Marie Geise, Haskel Greenfield, Mark Griffin, Rebecca Keith, Murray Marks, Debra Martin, David Mills, Mary Ellen Morbeck, Robert Paine, John Verano, and Richard Wilkinson. We obviously couldn't add all the things that all the users and reviewers requested, but we have done our best to honor all the good advice from these colleagues.

The most important contributor to the completion of the second edition was David DeGusta. His research and writing skills are apparent throughout, and he contributed much of the new chapter on molecular osteology. As with the first edition, Lyman Jellema was tireless in tracking down the bones to illustrate the new growth sections, and we sincerely appreciate his professionalism, kind assistance, and attention to detail (Lyman even sent cat toys to prevent the felid Lubaka from chewing on specimens). Susan Chin helped to construct the glossary and the guide to electronic resources in osteology. Clark Larsen, Phil Walker, and Juan Luis Arsuaga contributed background and photographs of their work featured in the new case studies, and Robert Paine contributed new photographs in Chapter 2. Gene Hammel helped with demographic questions, Henry Gilbert helped with figures, and Jose Miguel Carretero provided critical observations on the hand skeleton. Alan Shabel was a skilled and tireless proofreader. Thanks again go to the students in Berkeley's "Osteo U" for all their critical observations and helpful suggestions that made this a better book.

Tim D. White
Pieter Arend Folkens

Preface to the First Edition

Anatomists, forensic scientists, osteologists, paleontologists, and archeologists frequently encounter human remains in their work. These remains are used by such researchers to investigate both the recent and the ancient past. For these scholars, an illustrated human osteology book represents an important tool for identification and analysis in both field and laboratory.

In the educational setting, memorizing the name of each bone in the human skeleton is easily accomplished. Each year thousands of students prove this in introductory anatomy and physical anthropology courses. This book is intended to serve as a text for students who wish to advance their osteological skills beyond this level—to be able to accurately identify isolated and fragmentary skeletal remains, and to use these remains to learn something about the individuals represented by only bones and teeth.

Recent professional literature suggests that some students of physical anthropology have misinterpreted the deemphasis of "traditional" osteometric methods—that they have mistakenly concluded that basic identification no longer has an important role in the field. This book puts great emphasis on the identification of bones and teeth. Such identification is fundamental to students and professionals in paleontological, archeological, and forensic contexts. Basic identification of element, side, and taxon is a prerequisite to further interpretive analysis in all three settings. In this text, we devote attention to these basics while attempting to provide introduction and access to the wide range of modern work on the human skeleton.

I (T. W.) do not consider myself a human osteologist in the traditional sense of the word. My interests are more archeological and paleontological. I wrote this book after teaching a decade's worth of introductory human osteology classes at the University of California at Berkeley. For many of those years I supplemented Bass's text (1987 and earlier editions) with a variety of other books and papers. With this background, I concluded that my notes, handouts, and experience could be added to information available in a number of other sources and distilled into a single text for use in a one-semester college course on human osteology.

It is essential that instructors secure skeletal remains to use in teaching osteology. The two-dimensional images in this textbook should prove valuable in instruction and learning. They have been carefully crafted to show a maximum amount of anatomical information. These photographs cannot, however, substitute for work with the actual bones and teeth. For students to derive the most knowledge from this book they should have access to a minimum of one articulated adult human skeleton, one complete disarticulated adult skeleton, one child's skeleton, and a variety of fragmentary human bones. It should be understood that this is a bare minimum, and it is recommended that students be introduced to a much wider range of skeletal materials during their study. If archeological material is to be used in instruction, it is essential that *every* piece be individually labeled and that the bone be solid enough to withstand heavy handling.

Whatever success Pieter Folkens and I have achieved in developing a useful book we owe largely to the approximately 175 students who met the challenge of weekly laboratory examinations and exhaustive midterms in what one class described as "Osteo U." Gary Richards was a graduate of the first class in osteology that I taught. He has served as a teaching associate in most of the others. Thanks are extended to him for the hard work, patience, and talent he devoted to the class.

A number of colleagues have made this book possible. William Woodcock at Academic Press first suggested the text. Academic Press editor John Thomas made it readable, and Linda Shapiro, Kerry Pinchbeck, Lisa Herider, Nancy Olsen, and Chuck Arthur saw it through to completion. Gene Hammel and M. J. Tyler of the Department of Anthropology provided the initial Macintosh that made it possible, and Apple Computer's generous grant to the Institute of Human Origins assisted in the preparation of the text. Bruce Latimer and Lyman Jellema of the Cleveland Museum of Natural History helped us to obtain the Hamann-Todd postcranial skeleton illustrated here. Mitchell Day assisted in providing specimens for illustrations from the University of the Pacific School of Dentistry's Atkinson Cranio-Osteological Collection of the A. W. Ward Museum, and Jane Becker and Gary Thodas of that institution provided assistance in extending this loan. Susan Anton and Gary Richards provided invaluable assistance in obtaining material from the University of California Lowie Museum of Anthropology collections. Their curatorial work on these collections, as well as support of this work by the National Science Foundation, is gratefully acknowledged. Kent Lightfoot of Berkeley's Archaeological Research Facility provided assistance with records and photographs (14.3–14.6). Betty Clark of Berkeley's Cowell Hospital assisted in providing the radiographs used in the text. Thanks go to Bob Jones of U.C. Berkeley's Museum of Vertebrate Zoology for removing the grease from the illustrated skeleton. Andrew Mackenzie, Mike Black, and Richard May assisted in choosing Lowie Museum specimens used to illustrate variation. Larissa Smith of the Institute of Human Origins helped with translations.

Clark Larsen provided helpful critical comments on an early draft of the manuscript. Yoel Rak and Bill Kimbel critically reviewed Chapter 4, improving it considerably. Owen Lovejoy, Bruce Latimer, and Scott Simpson thoroughly reviewed later drafts of the book, providing innumerable corrections, clarifications, and valuable suggestions. Walter Hartwig read the page proofs. The students in U.C. Berkeley's Introduction to Human Osteology, Fall 1986, 1987, 1988, and 1990 classes, deserve special mention for commenting extensively on drafts of this text. Special thanks go to Jeni McKeighen for editorial and indexing assistance. We thank all of these reviewers for their contributions to the book and take full responsibility for any errors of fact that made it past this set of dedicated readers.

Ray Wood and Thomas Holland of the University of Missouri provided advice on specimen photography. Bill Pack generously offered his photographic studio and his expertise, and LuAnn Taylor and Todd Telander gave valued assistance during photographic sessions. Film processing and printing were accomplished by the good people at GAMMA Photographic Labs of San Francisco. Larry Zimmerman, Douglas Ubelaker, George Gill, and Owen Lovejoy kindly granted permission to reprint their photographs and drawings, and Scott Simpson provided much help in obtaining illustrative material.

Thanks go to the Harry Frank Guggenheim Foundation for support of research conducted by T. W. in parallel with this book.

Generous assistance with the case studies was provided by D. C. Johanson, Bill Kimbel, Berhane Asfaw, Gen Suwa, and Gerald Eck (Olduvai); by Jerry Fetterman (Cottonwood Canyon); and by Owen Lovejoy and Scott Simpson (Cleveland Homicide). These workers took valuable time from their ongoing research efforts in order to round out the book by sharing their unique photographs, insights, and experience.

Tim D. White
Pieter A. Folkens

CHAPTER **1**

Introduction

Bones have been entombed for millions of years in sediments left by ancient lakes, swamps, and rivers that once dotted Ethiopia's Afar Depression. Today, as erosion cuts into these ancient layers, the fossils reach the barren Afar surface, often shattering into small, glistening, multicolored fragments of bone and teeth. Thinly scattered among the osseous and dental remains of thousands of crocodile, turtle, hippopotamus, giraffe, carnivore, baboon, pig, horse, antelope, and other animals found in the surface fossil assemblages are the remains of primitive human ancestors. While paleontologists search Afar outcrops for osteological clues about prehistoric human form, archeologists uncover the osteological remains of the inhabitants of Herculaneum who perished as Mount Vesuvius erupted, burying them with their possessions. On the other side of the globe, anthropologists probe into a recent grave containing skeletal parts that may be those of a Nazi war criminal.

1.1
Human Osteology

A thread that binds these and thousands of other investigations is **human osteology,** the study of human bones. The scientists performing the investigations employ their knowledge of the human skeleton in recovering and interpreting the bones. Outside of anatomical and medical science, there are three main areas in which knowledge of human osteology is often applied. First, osteological work is often aimed at identification of the relatively recently deceased and is usually done in a legal context. This work, which pertains to the public forum, most often a court of law, is called **forensic osteology,** a division of forensic anthropology. The other two contexts in which human osteological knowledge is commonly applied are historical. The context can be ancient and purely **paleontological,** as for the Pliocene precultural hominids of Africa. Alternatively, the context can be relatively recent, part of an **archeological** record. For example, human bones recovered from the Aztec centers in Mexico were chronicled just a few hundred years ago by the Spanish during the "con-

1

Figure 1.1 Osteology and associated scientific disciplines.

quest" of Mexico. Osteological analysis of materials from such cultural contexts is routinely undertaken as part of archeological research. Archeologists concentrate on cultural residues of former human occupations, but they stand to gain a great deal of valuable information from the skeletal remains of the ancient inhabitants. It has recently become fashionable to refer to the study of human remains from archeological contexts as **bioarcheology.**

The information that human or protohuman bones can provide makes the recovery of bones a critically important activity. Skeletal anatomy (including teeth) reflects the combined action of genes and environment. The skeleton forms the framework for the body, whereas the teeth form a direct interface between the organism and its environment. Bones carry in their shape, or **morphology,** the signature of soft tissues with which they were embedded during life—tissues including muscles, ligaments, tendons, arteries, nerves, veins, and organs. These soft tissues usually disappear soon after death. The skeleton, however, often preserves evidence of the former existence and nature of many of these other body parts.

Because the bones and teeth of the skeleton are resistant to many kinds of decay, they often form the most lasting record of an individual's existence. It is possible to estimate an individual's age, sex, and stature from the bones and teeth. Study of the skeleton often makes it possible to discern a variety of pathologies from which the individual may have suffered. Analysis of groups of individuals may offer insights into prehistoric population structure, biological affinities, cultural behaviors, and patterns of disease. The evolutionary history of humanity itself may be read from the fossil record—a record comprising mostly teeth and bones. Figure 1.1 illustrates the place of osteology in relation to other scientific disciplines.

1.2
A Guide to the Text

The goal of **forensic** osteology often involves the identification of an unknown individual. The process of personally identifying the remains of the recently dead individual is called **individuation.** To narrow the possibilities, the forensic osteologist first ascertains whether the remains are human and then begins to explore the individual characteristics such as age, sex, and stature, comparing these variables in the hopes of obtaining an exclusive match with what is known about the missing individual. The human osteologist working in an **archeological** context usually cannot perform such personal identification. Instead, bioarcheological concerns are with characteristics of the individual and with the insights that skeletal remains of many individuals, representing biological populations, can provide on diet, health, biological affinity, and population history. The osteologist working in a **paleontological** context (and note that many fossils of Pliocene and Pleistocene age are found in archeological contexts) is also interested in using the normally rare fossil remains of human ancestors and other relatives to learn all these things and also to discern evolutionary relationships.

Skeletal remains can provide meaningful clues about the recent and the ancient past to all these investigators. To use these clues, one must master some fundamentals. In a real sense, fundamentals are what this text is about.

The most difficult part of writing this book has been choosing among the many things that could have been included in a single volume. The "perfect" osteology book would be a gigantic volume illustrating all stages of skeletal growth, all variations in the adult skeleton, and all skeletal elements of all mammals that might be confused with a human. It would cover bone histology, skeletal embryology and morphogenesis, and biomechanics. The volume would contain bibliographic reference to all papers published in osteology and would include full accounts of every skeletal measurement and identification technique published and all applicable tables. But given the constraints of format, production, and cost, our goal was to produce a single text that is simultaneously accessible to the college student and useful to the practicing professional osteologist.

A book such as this cannot possibly do justice to the large body of professional literature on human osteology available in other books and in journals such as the *American Journal of Physical Anthropology, Journal of Human Evolution, Journal of Dental Research, Human Biology, Paleopathology Newsletter,* and *Ossa.* Rather than reprint the data found in these primary sources, and in secondary compilations such as Krogman and İşcan (1986) and Bennett (1993), we have built pathways to them. The **Suggested Further Readings** sections briefly introduce samples of books and professional papers that the reader seeking further details on each chapter topic might consult. All of these references, as well as those cited in the chapter texts, are included in the **Bibliography** at the end of the book. Neither the suggested further readings nor the bibliography attempts to be comprehensive. Instead, these sources were compiled to provide the professional with an orientation to the primary literature and to give the student a set of sound departure points for further study in human osteology and related topics. This approach is intended to encourage all users of the book to directly consult the original literature in the acquisition and application of osteological knowledge. Through this direct approach, students will achieve a better understanding of the nature of original research in human osteology and a firsthand familiarity with the results of this research.

In this book we focus on the first two basic questions that any human osteologist must answer about a bone or collection of bones whether in a forensic, archeological, or paleontological context:

- Are the bones human?
- How many individuals are present?

Whether the context is forensic, archeological, or paleontological, these questions must be answered before further analysis is possible. This text, a guide to human osteology, emphasizes the anatomy of the human skeleton. The skeletal remains of some other animals, particularly when fragmentary, are often difficult to distinguish from human bones and teeth. Although there are no general differences that ensure effective sorting between human and nonhuman bones, the first step in answering the question of whether the bones are human is to acquire familiarity with the human skeleton in all of its many variations in shape and size. Ubelaker (1989, Fig. 63) illustrates whole bones of mammals most often confused with humans, but this confusion rarely involves whole bones. Fragmentary nonhuman remains are more problematic and more frequently encountered. Once a familiarity with the range, or envelope, of variation characterizing the modern human species is achieved, further work in comparative osteology of both extinct and extant mammals becomes a much easier task. With further comparative work comes more experience, and with that experience the osteologist is better able to make the basic identifications required.

Information on determining the age, sex, stature, and identity of skeletalized individuals is provided in later chapters of the text. These second-level questions and many others, including those about biomechanical capability, phylogenetic relationships, and geographic affinities, however, can be accurately answered only after the elements and individuals have been correctly identified. Too often the first, basic identifications are overlooked or hastily performed, and the succeeding analyses are thus built on weak foundations.

We conclude this introductory chapter by considering some advice on methods and techniques for studying, learning, and teaching human osteology. Chapter 2 is an introduction to bone as a tissue. Skeletal growth is introduced here, along with a presentation of the major internal structure and functions of bones, teeth, and joints. The critical topic of intraspecific variation (variation within a species) is discussed and its various sources are identified and illustrated. In Chapter 3 we introduce anatomical terminology—the vocabulary of osteology that is essential in the scientific study of bony and dental tissues.

Chapters 4–13 form the core of the book. In these chapters we consider one anatomical region at a time, beginning with the bones of the skull. A separate chapter and format are employed for the dentition. We introduce Chapters 4–13 with brief accounts of the phylogenetic history of the body segment(s) and osteological elements described in the chapter. This approach sets the osteological elements in a broad evolutionary framework. These introductory statements are mostly drawn from the excellent functional human anatomy text by Cartmill, Hylander, and Shafland (1987), and the interested reader may pursue further details there and in other comparative texts such as Jarvik (1980).

In the descriptive chapters (4–13), each bone is shown individually, in various views, by means of photographs. For ease of comparison, the scale for all individual bones is natural size; the teeth are shown twice natural size. For paired bones, only the right side is shown unless otherwise indicated. Orientations for the articulated crania are standard, and other bones are illustrated in orientations showing the most anatomy. For example, the "anterior view" of the frontal is a photograph of the anterior surface of that bone. Because our focus is on external morphology, and because the dimension of depth is sacrificed when depicting bones and teeth on the printed page, we have developed new methods to illustrate osteological form. Our goal is to accurately portray external morphology while minimizing the confusing stains, translucency, and shine found in natural bones and teeth. Details of the preparation and photography of these specimens are given in this volume's appendix.

Human osteology texts often depict human skeletons and their elements as if they were interchangeable. Sets of illustrations that provide little or no visual information on the relative sizes of different parts of the skeleton are standard. Furthermore, bones are often unscaled and/or derived from several individuals in such illustrations. This sacrifices additional information on proportionality and fosters typological thought. To remedy this situation, we chose the skeleton of a single individual to illustrate all the elements of the postcranial skeleton. All postcranial elements shown in the descriptive chapters of this text are from a single modern human individual, Hamann-Todd specimen number 857, a 24-year-old black male who stood 5' 7" tall (170 cm) and weighed 138 lb (62.6 kg). This skeleton was chosen for its relatively few unusual features (noted in the captions for each element), completeness, and excellent preservation. Articulated cranial and all dental specimens illustrated in Chapters 4 and 5 are of recent Mesoamerican origin, and the disarticulated bones of the cranium are all from a single modern individual from Southern Asia. Note that these disarticulated cranial bones are from a 16-year-old individual in which the bony elements could be disarticulated and shown intact. This young individual had not fully developed all the markers of cranial robusticity illustrated by the articulated skull. Other illustrated specimens are identified in figure legends or in the appendix, where a full listing is provided.

Each descriptive chapter is organized systematically. The element is first named and its articulations identified. Under the **Anatomy** section for each element, the major parts and osteologically significant features of each element are identified in **bold** typeface and described. These parts and features are identified in the accompanying photographs with the same letters used in the text. This format encourages students to use the text in a self-examination mode and keeps the osteological images uncluttered for comparative use.

The function and soft tissue relations for most structures are identified in *italic* typeface. It is not our intention to give a complete listing of muscle origins and insertions. Instead, we provide the data on soft tissue to make functional sense of bony features while constantly reinforcing the reality of bone as an integral part of the musculoskeletal system. Readings cited at the end of Chapter 4 include human anatomy texts that the student may wish to obtain as companions to this one. Students interested in pursuing soft tissue anatomy are urged to consult these books.

Ossification of each element is briefly considered in the descriptive chapters under a section called **Growth.** Further details on development of various elements are presented in Chapter 17. The **Possible Confusion** and **Siding** sections provide information to complement the illustrations and allow effective identification of isolated and fragmentary skeletal elements.

In Chapter 14 we consider the discovery and recovery of osteological material. This chapter also covers transport of the remains to the laboratory, primary cleaning of the material, and restoration. In Chapter 15 we trace skeletal material through a variety of analytical techniques, including measurement and photography, and we conclude with a section on the reporting of human osteological remains.

In Chapter 16 we consider ethics in osteology. Chapter 17 is a guide to the assessment of an individual's age, sex, race, and stature from skeletal remains. In Chapter 18 we consider some of the most common pathologies encountered in human skeletal remains. Chapter 19 is a discussion of **taphonomy,** the study of processes that affect skeletal remains as they move along the often tortuous path between death and curation. In Chapter 20 we address the subject of how the biology of now-dead human populations, particularly the diet, demography, and affinities, might be studied. Chapter 21 covers the rapidly growing field of molecular osteology.

The text then considers six case studies to show how the fundamentals outlined in the first twenty-one chapters have been applied to very different investigations involving hominid skeletal material. There are two studies each from forensics, bioarchaeology, and human paleontology.

Measurements in this text are expressed in the metric system as is standard in osteology. Carter (1980) provides a good history of the English and metric systems for students unfamiliar with metric terms, or for those who remain unconvinced of the metric system's utility in modern scientific investigation.

1.3
Teaching Osteology

Instruction in human osteology should begin at the undergraduate level, whether in biology or anthropology. I (TW) have found that an intensive, one-semester, upper division course in osteology provides a good foundation for undergraduate and graduate students interested in forensic anthropology, bioarcheology, and hominid paleontology. Indeed, courses in these specialty fields prove far more meaningful to students with a good foundation in human osteology. I have found that students learn best when challenged by frequent examination, when able to access a wide series of comparative specimens during their studies, and when kept to a rigorous schedule of weekly quizzes and frequent midterm exams (with timed identification stations).

This book is topically organized for ease of access by student and professional. Instructors will have their own preferences for the order of presentation, and the book is organized to allow this. My personal preference for teaching Berkeley's introductory human osteology course is to begin with Chapters 1 and 3, and then proceed directly to Chapter 4 on the skull. By covering the skull and dentition chapters early in the semester, the students are continually challenged with the most difficult parts of the skeleton. They work on this for the remainder of the course, thereby learning and retaining this information best. During the students' laboratory work on the skull and teeth, I cover Chapters 3 (bone biology) and 16 (ethics) and proceed to lecture on recovery and analysis (Chapters 14 and

15), covering the cranial half of Chapter 17 (age, sex, etc.). Case studies are introduced right through the course. At the midsemester mark we turn to the postcranial skeleton, studying Chapters 6–13 and exploring case studies, population biology (Chapter 20), and molecular osteology (Chapter 21). After the students have mastered basic identification of all elements in the skeleton, the semester finishes up with a look at how paleopathology (Chapter 18) and taphonomy (Chapter 19) extend the morphological envelopes of human skeletal remains. The book and the lectures provide the vocabulary (see the Glossary), basic concepts, and references necessary for the student to approach, use, learn, and eventually master the primary professional research literature on human osteology.

1.4
Resources for the Osteologist

The most important single resource for the student, instructor, and acting professional is a collection of skeletal remains. Ideally, the laboratory should have a growth series of skeletons of single individuals of known age to accompany mounted skeletons of several individuals, as well as element collections in which many individuals of known sex, age, occupation, and pathology are represented. A collection of skeletal remains from a variety of modern nonhuman animals is also very important, as is a cast collection of fossil hominoids. Access to a human anatomy laboratory with cadavers to dissect is also desirable. A library with every published resource on forensics, bioarcheology, and human paleontology is essential. Of course, like the ideal textbook, the ideal laboratory can only be approached, never realized.

Beyond the physical plant of the laboratory, the key resources for the human osteologist have traditionally been equipment, specimens, and publications. The advent of electronic publishing and educational software and the explosion of resources on the World Wide Web has dramatically extended resources available to the osteologist. These media are rapidly growing, evolving, and volatile, but it is important that the osteologist be aware of them. Table 1.1 is an attempt to identify and assess the most important and useful web sites pertaining to human osteology.

1.5
Studying Osteology

The gulf between knowing the names of elements in an adult skeleton and correctly identifying the taxon, element, and side of an isolated, fragmentary bone or tooth is a formidable one. It must often seem that an instructor is performing magic in correctly identifying, for example, a human left upper third premolar. It is, however, far from magic. The ability to identify skeletal material is a skill that can be acquired only through intensive study of actual specimens. This study is best reinforced by frequent formal and informal quizzes and examinations. It is not enough to be able to side and identify intact elements, because intact elements are

Table 1.1
Electronic Resources in Osteology[a]

STARTING POINTS

http://www.nitehawk.com/alleycat/anth-faq.html
  Extensive links to anthropology resources on the internet, including mailing lists, USENET discussion groups, FTP files, and World Wide Web sites.

http://www.sscf.ucsb.edu/anth/alpha.gifs/alphnet.html
  UC Santa Barbara's anthropology site with an alphabetical listing of good anthropology resources on the web.

http://www.geocities.com/CapeCanaveral/Lab/9893
  Good resource for links to other sites covering osteology, NAGPRA, forensics, paleopathology and anatomy. New links added regularly.

http://www.medstat.med.utah.edu/kw/osteo
  Overview of human osteology, including forensic anthropology and paleopathology, with photos.

http://www.tamu.edu/anthropology/news.html
  Contains breaking news about anthropology and archaeology from ABC, CNN, *USA Today, Washington Post,* Nando, Archaeology, university press releases, and other sources.

CAREERS

http://www.museum.state.il.us/ismdepts/anthro/dlcfaq.html
  Discusses jobs available for archaeologists, the training and education required, colleges or universities recommended, general introductory books on archaeology, archaeological excavation opportunities, and further links.

http://www.usd.edu/anth/handbook/hbjob.html
  Considers what one can do with a BA/BS degree in anthropology.

BONE BIOLOGY

http://www.lumen.luc.edu/lumen/MedEd/Histo/frames/h_fram10.html
  Color histology slides and descriptions of bone development, muscles, connective tissue, and neural tissue.

http://www.kumc.edu/instruction/medicine/anatomy/histoweb
  Color histology slides show bone formation in a developing skull, epiphyseal plates, long bone development, and decalcified bone.

http://edcenter.med.cornell.edu/CUMC_PathNotes/Skeletal/Bone_TOC.html
  Cornell University Medical College's site with lecture notes on bone biology and pathology. Radiographs and micro- and macroscopic color images of bone biology and pathologies.

SKELETAL ANATOMY

http://www.gwc.maricopa.edu/class/bio201/index.html
  Online tutorials for the skull, hand, and wrist and major superficial muscles with illustrated and photographic images that include point-and-click anatomical identifications. Useful for reviewing anatomy online.

http://anatomy.uams.edu/HTMLpages/anatomyhtml/medcharts
  Tables describe the details of gross anatomy structures such as bones, muscles, nerves, and joints. Good for review.

http://www.medmedia.com/med.html
  Wheeless' *Textbook of Orthopaedics*—a complete online textbook. Contains good sections on skeletal anatomy, bone biology, joints, muscles, and nerves.

SOFT TISSUE ANATOMY

http://info.med.yale.edu/caim/edu_resources/anatomy/head_anatomy/head_anatomy.html
  Yale School of Medicine's site on Head Anatomy. Anatomical illustrations of the head in anterior, lateral, and basal views; includes labeled details of associated soft tissues: arteries, veins, glands, muscles, and osteological landmarks. Anatomically labeled radiographs of the skull.

Table 1.1
Electronic Resources in Osteology[a] *(continued)*

http://www.ptcentral.com/muscles

Good reference of skeletal muscles of the human body, including each muscle's origin, insertion, action, blood supply, and innervation. Muscles are grouped by body regions as well as alphabetically. Links to other anatomy-related sites.

http://www.neuromus.ucsd.edu/MusIntro/Jump.html

Indexed links to topics in muscle physiology including muscle structure, movement, and muscle joint interactions. Text and illustrations.

http://www.nlm.nih.gov/research/visible/visible_human.html

The National Library of Medicine's Visible Human project has created a digital atlas of the human body.

DENTAL ANTHROPOLOGY

http://www.sscf.ucsb.edu/~walker/sites.html

The Dental Anthropology Association's site with a visual database of the human dentition. Each tooth can be rotated 360 degrees to provide comprehensive views of tooth morphology. Has good links to other dental anthropology sites.

FIELDWORK

http://www.cincpac.com/afs/testpit.html

Archaeological fieldwork opportunities worldwide, listed by geographic region.

FORENSICS

http://medstat.med.utah.edu/kw/osteo/resources/resources.html

A reference guide to graduate programs in forensic anthropology.

http://www.uncwil.edu/people/albertm/career.html

Discusses the difference between forensic anthropologists and forensic pathologists, and provides information on education and careers in forensic anthropology.

http://www.tncrimlaw.com/forensic

Provides extensive links to other forensic anthropology sites.

LAB TECHNIQUES

http://www.palmer.edu/Foundation/Archives/Page3~2.html

Describes skeletal curation techniques.

http://www.duke.edu/~mtb3/castingmanual/titlepage.html

Contains a good detailed explanation of the cast-making process, with step-by-step instructions for making a mold and casting techniques.

http://www.flmnh.ufl.edu/natsci/vertpaleo/resources/prep.html

Discusses fossil preparation and conservation techniques, including preparing fossils with consolidants, adhesives, hand and mechanical tools, chemicals, treatment of subfossil bone, restoring crushed skulls, mounting fossil specimens, and conserving artifacts.

http://www-museum.unl.edu/research/vertpaleo/musnote2.html

Description of steps taken to preserve vertebrate fossils. Discussion of museum curatorial techniques and the nature of fossilization.

OSTEOMETRICS

http://konig.la.utk.edu/howells.html

W. W. Howells' (1973) craniometric data can be downloaded from this site.

http://www.cmnh.org/research/physanth

The Cleveland Museum's Hamann-Todd skeletal database available for online searches by specimen number, age, height, weight, sex, or ethnic origin. Casts of human and nonhuman primate crania and postcrania available for sale.

Table 1.1
Electronic Resources in Osteology*a* *(continued)*

RADIOGRAPHY

http://web.wn.net/~usr/ricter/web/medradhome.html
   Good links to other radiography sites and anatomy sites. Includes sections on radiology resources, radiologic anatomy, and gross anatomy and a reference desk. Good place to start searching for radiography online.

http://www.ptcentral.com/radiology
   Radiographs of the upper and lower extremities and of the spine. Includes pathologies and links to other related sites.

http://www.rad.washington.edu/AnatomyModuleList.html
   University of Washington's Department of Radiology site featuring radiographic anatomy of the entire skeleton, divided into regions. Also features an online muscle atlas with color illustrations of the lower extremities regions, and gives the origin, insertion, action, innervation, and arterial supply of each muscle.

http://www.rad.washington.edu/AnatCaseList.html
   Over 80 radiograph cases with brief description of some form of paleopathology such as an avulsion fracture or a hip dislocation. Good source for paleopathological radiography.

NAGPRA

http://www.uiowa.edu/~anthro/reburial/repat.html
   Contains the text of NAGPRA, links to repatriation notices, and other information relating to the law and regulations. Case studies, ethics codes, policy statements, text of selected state laws, organizations, bibliographies, and articles on NAGPRA-related issues are available.

http://www.cast.uark.edu/other/nps/nagpra.html
   National Archaeological Database site contains NAGPRA-related documents. Document searches available by state, county, keyword, etc.

http://www.cr.nps.gov/aad/nagpra.htm
   National Park Service NAGPRA site. Laws pertaining to NAGPRA and other cultural resource issues are available in full text.

http://www.nmnh.si.edu/anthro/repatriation/intro.html
   Covers history and background of NAGPRA, including on-going work in repatriating human remains and cultural objects in association with Native American groups. Description of the documentation process for repatriating human remains. Table lists all repatriated remains as of 1996.

PALEOPATHOLOGY

http://www.usd.edu/~archlab/paleo.html
   Paleopathology image collection (with text descriptions) of skeletons from the Crow Creek massacre and from the University of South Dakota collection.

http://www.medlib.med.utah.edu/WebPath/BONEHTML/BONEIDX.html
   Bone and Joint Pathology Index site, featuring bone and joint diseases with views at gross, x-ray, MRI, and microscopic levels. Osteoporosis tutorial.

http://www.kumc.edu/instruction/medicine/pathology/ed/ch_26/ch_26_f.html
   Bone and joint pathology site with description of bone formation and development and an extensive list of paleopathologies with color illustrations, x-rays, micro- and macroscopic images.

http://www.merck.com/!!upkcn11x3upkcn11x3/pubs/mmanual/html/lgoidlei.html
   Merck Manual's online site covers osteoarthritis, bone and cartilage disorders, infections of bones and joints, arthritis associated with spondylitis.

http://www.mic.ki.se/Diseases/c5.html
   Extensive links to sites covering musculoskeletal diseases with color photographs or illustrations.

Table 1.1
Electronic Resources in Osteology[a] *(continued)*

http://www.sbu.ac.uk/~dirt/museum/skeleton.html
Paleopathology site covers pathologies of the chest, hand, wrist, upper extremities, pelvis, hip and thigh, and lower extremities. Covers a wide range of pathologies in radiographs. Brief description of symptoms and pathologic changes.

HUMAN PALEONTOLOGY
http://www.talkorigins.org/faq/homs/specimen.html
Describes all fossil hominid species (type specimen, when and where they were found, age, etc.). Focuses on refuting creationist arguments against evolution; has links to other sites.

http://www.sew.csuohio.edu/public/sew/gallery/paleontology/index.html
Color photo gallery of prominent hominid fossils.

http://www.amnh.org/enews/iskulls.html
American Museum of Natural History site. A "family tree" of hominid fossil crania that can be viewed in 3-D and compared to modern human crania. Color illustrations.

http://www.digitaldarwins.sarc.msstate.edu
An online museum and lab featuring three-dimensional images of skulls, dentition, and endocranial casts.

SOCIETIES
http://www.utexas.edu/cons/aapa
American Association of Physical Anthropologists (AAPA) site contains membership information, and discusses careers in physical anthropology, graduate programs, publications, the annual meeting, and links to associated societies.

http://www.ameranthassn.org/index.html
The American Anthropological Association (AAA) site includes sections on ethics, publications, career placement issues, and links to other anthropology sites.

http://www.aafs.org
Site of the American Academy of Forensic Sciences, containing statement of organization's purpose, membership application, career information, and links to other related sites.

MAILING LISTS
PaleoAnthro mailing list: majordomo@list.pitt.edu
The mailing list discusses paleoanthropology, physical anthropology, prehistoric archaeology, human and nonhuman primate evolution. Also comes in digest form. Mainly nonacademic participants.

---

[a] These electronic resources are found via the World Wide Web. Unlike printed books and journals, the "shelf life" of these resources is often short, and the electronic medium is therefore considered more volatile. In other words, by the time you attempt to use the addresses here, some of them will have changed, or the web site will have been discontinued. Others, however, are being added all the time. Many of these sites are good complements to this book.

rarely found in field paleontological situations and only sometimes found in archeological and forensic contexts. The many hominid specimens "recovered" from faunal collections but misidentified as nonhominid are ample testimony to the superficial knowledge of the hominid skeleton that even many specialists have. Learning to identify bones and teeth can be slow, painful, and frustrating, but the rewards make the effort worthwhile. The loss of scientific data and the professional embarrassment caused by a misidentification make the effort essential.

Students may find several techniques useful in learning the skeleton. First, remember that the osteologist always has an intact comparative skeleton close at hand, even in remote field locations. That skeleton is embedded in his or her own body. It is useful to visualize and even **palpate** (feel your own bones through the skin) the way an isolated skeletal element might "plug into" your own body. This is particularly true for identifying and siding teeth, which are conveniently exposed in the osteologist's mouth. Never ridicule an osteologist who holds a radius against the right forearm and then shifts it to the left forearm before identifying it; that osteologist will probably correctly side the bone. When identifying the side of any identified skeletal element, all you have to do is establish three axes in space: top to bottom, side to side, and front to back. Anatomical features of the bone will assist in this and thereby provide the clues necessary for correct identification and orientation.

A second tip for learning to identify skeletal remains involves hierarchies of decisions. Begin identifying a bone by deciding which elements it cannot be. For example, a radius fragment cannot be a cranial bone or tooth (which excludes hundreds of possibilities), a tibia or humerus (too small), a metatarsal (too big), or even a fibula (wrong shape). You will be surprised how soon you can exclude all but the correct choice if you approach identification in this way.

No matter how often you misidentify a bone or tooth, keep trying. Do not constantly use the articulated skeleton or intact skull as a crutch. There is great osteological truth to the idea that "with every mistake we must surely be learning." Try to learn from your mistakes. Try to identify the kinds of mistakes you are making. Do your misidentifications most often involve immature specimens? If so, set out a growth series for each skeletal element and see how the shape of the bone changes with growth. Are you having trouble with nonhuman skeletal parts? If so, look at a range of human variation to get a good idea of how much variation to expect. Are you simply confusing one part of the skeleton for another? If so, look at all the elements in the skeleton that might mimic each other when fragmentary and check the Possible Confusion sections in the descriptive chapters of this text. In the field, keep a copy of this book nearby. It is much easier to carry than a skeleton, and the lifesize photographs in many views should facilitate comparisons and identifications.

1.6
Working with Human Bones

Because the results of human osteology have an impact on so many disciplines, there are a variety of career options that involve human skeletal remains. Most of these choices involve the academic setting. As a result, most professional human osteologists work in colleges, universities, and museums. Even in forensic anthropology, the most "applied" of human osteological endeavors, coroners, medical examiners, and law enforcement agencies most often turn to local or national specialists employed in higher education (Galloway and Simmons, 1997). Most bioarcheologists and hominid paleontologists are also employed as teachers and researchers in academic settings, often teaching in departments of anthropology or biology or in medical school departments of anatomy. Some bioarcheologists are employed by cultural resource management firms, but these positions are often short-term and project oriented.

Wherever employed, the human osteologist is involved with specimens in collections. As you study human bones and teeth, always respect them as objects of scientific inquiry. In some ways, skeletal resources are like books in a library. Bones and teeth have the potential, if read correctly, to inform us about the living, breathing people they once belonged to. Treat these objects with care; some of them are fragile and all of them are irreplaceable.

Respect any system of organization in which you find skeletal material. Never mix bones and teeth of different individuals, even for a short time or with the best of intentions. Remember that mixing of bones results in a loss of contextual information—an action that is potentially even more devastating than physical breakage of an element. In the library, history books shelved incorrectly in the biology section become unavailable to any historian who wishes to consult them. Bones reshelved in the wrong place are almost impossible to retrieve.

Finally, respect the people who came before you by treating their bones with care. Respect the generations of students and professionals who will follow you by keeping the bones and their provenance intact.

Suggested Further Readings

There are several introductory osteology textbooks. These books are highly variable in their content, the quality of their illustrations, and their coverage. In addition to these, the human anatomy books identified at the end of Chapter 4 are useful supplements to the study of human osteology.

Anderson, J. E. (1962) *The Human Skeleton: A Manual for Archaeologists.* Ottawa: Dept. of Northern Affairs and National Resources. 164 pp.
>An introductory manual with line drawings.

Bass, W. M. (1995) *Human Osteology: A Laboratory and Field Manual* (3rd Edition). Columbia, Missouri: Missouri Archaeological Society. 361 pp.
>A laboratory and field manual that emphasizes identification.

Bennett, K. A. (1993) *A Field Identification Guide for Human Skeletal Identification* (2nd Edition). Springfield, Illinois: C. C. Thomas. 113 pp.
>A vital reference for forensic work with tables and techniques for estimating sex, age, statue, and race.

Brothwell, D. R. (1981) *Digging Up Bones* (3rd Edition). Ithaca, New York: Cornell University Press. 208 pp.
>A beginner's guide to recovery and analysis of skeletal remains.

Buikstra, J. E., and Ubelaker, D. H. (1994) *Standards for Data Collection from Human Skeletal Remains.* Fayetteville, Arkansas: Arkansas Archeological Survey Report Number 44. 206pp.
>The essential standards volume in North America.

Goldberg, K. E. (1985) *The Skeleton: Fantastic Framework.* New York: Torstar Books. 165 pp.
>An enjoyable, readable guide to the skeleton for the layperson. Excellent color photographs.

Schwartz, J. H. (1995) *Skeleton Keys.* New York: Oxford University Press. 362 pp.

A textbook of human osteology.

Shipman, P., Walker, A., and Bichell, D. (1985) *The Human Skeleton.* Cambridge, Massachusetts: Harvard University Press. 343 pp.

A guide that stresses functional aspects of the skeleton at many levels.

Snow, C. C. (1982) Forensic anthropology. *Annual Review of Anthropology* 11:97–131.

A concise history and introduction to the field of forensic anthropology, especially osteology.

Steele, D. G., and Bramblett, C. A. (1988) *The Anatomy and Biology of the Human Skeleton.* College Station, Texas: Texas A&M University Press. 291 pp.

An atlas dedicated to the identification and biology of the human skeleton.

Tekiner, R. (1971) *Human Skeletal Morphology: A Laboratory Manual for Anthropology.* New York: Continental House. 95 pp.

A very simple beginner's manual focusing on identification.

Ubelaker, D. H. (1989) *Human Skeletal Remains: Excavation, Analysis, Interpretation* (2nd Edition). Washington, D.C.: Taraxacum. 172 pp.

A guide to human osteology that emphasizes recovery and interpretation. Excellent photographs of excavations. The text features comparative photographs of skeletal elements of large mammals common in North America.

Ubelaker, D. H. (1996) Skeletons testify: Anthropology in forensic science. *Yearbook of Physical Anthropology* 39: 229–244.

An overview of osteology's role in forensic science.

CHAPTER 2

Bone Biology

Bone as connective tissue and bones as elements may be studied on several hierarchical levels. Information derived from many skeletons may be used in reconstructing **population** biology (Chapter 20). Study of the various elements of a single skeleton may be used to elucidate the biological aspects of an **individual** (Chapters 17 and 18). Such assessments are built on a foundation that emphasizes the identification of the bony **elements** that constitute each human skeleton (Chapters 4–13). Before embarking on a systematic consideration of human skeletal elements, it is useful to consider bone biology. This chapter should therefore be considered an essential stepping stone to the descriptive and interpretive chapters to follow.

It is important to note the multiple functions of bone as a tissue and of bones as organs. Bones act as essential mechanical components of the **musculoskeletal system.** They serve to protect and support soft tissues; to anchor muscles, tendons, and ligaments; and as the rigid levers that muscles operate to produce movement. Bones also function as physiologically critical centers for the production of blood cells, as storage facilities for fat, and as reservoirs of important elements such as calcium (essential in blood clotting and muscle contraction). Bone as tissue is adapted to these functions. The varied mechanical and physiological functions of bones as organs are intimately related to the gross and microscopic (including molecular) structure of bone tissue, which we review in this chapter.

Bone is a dynamic tissue that allows for growth during **ontogeny** (development) of the individual. It is shaped and reshaped by cells that reside within it. Because of this, the gross shape, or **morphology,** of bones can be altered during life. The shape and size of bones and teeth can also vary dramatically between different individuals. Before introducing bone biology at the level of the molecule, the cell, and the gross element, it is critical that we examine a property of all biological structure, the property of **variation.** Understanding and appreciating variation in bony and dental gross anatomy is critically important in any work with the human skeleton.

2.1
Variation

For a random living sample of fifty male and fifty female individuals from various human populations, it would be easy to establish physical characteristics that would allow each person to be individually recognized. Variation would be employed in sorting the population, and only in the rare instance of identical twins would there be much difficulty in distinguishing different people. This is because the human species, like other species, exhibits variation. This variation extends to the teeth and bones of the skeleton.

In identifying our friends and acquaintances, we make use of our ability to recognize variation. Without variation in physical features it would be impossible for us to identify one another at the group and individual level. In fact, our use of soft tissue variation seems so natural that we take it for granted. Oddly enough, however, the amount of variation in the hard tissues of the body is often not anticipated by students of osteology. Shape and size in human bones and teeth vary widely, and analysis of this variation makes human osteology simultaneously challenging for the beginner and useful for the professional.

There are four major factors leading to variation in human skeletal anatomy. One source of this variation is **ontogeny,** or growth. A great deal of skeletal variation in size and shape is observed along the continuum of growth between fetus and adult. This variation can be used by the osteologist in determining the age at death from skeletal remains. In Chapter 17 we discuss how such analysis is conducted.

A second source of skeletal variation in humans is the **sex** of the individual. Humans are moderately **sexually dimorphic** in body size, and in any given skeletal population this dimorphism is manifested in the smaller relative size of female bones and teeth. This size variation is accompanied by shape variation, which allows certain skeletal elements to be used in determining sexual identity of prehistoric remains. In Chapter 17 we discuss sex determination of osteological material.

A third type of variation is **geographic,** or **population based.** Different human groups vary in many skeletal and dental characteristics. This geographic variation can be employed to assess the geographic (sometimes called racial) affinity of skeletal remains. In Chapters 17 and 20 we consider the use of this kind of variation in the study of past and present human populations.

Finally, even individuals of the same age, sex, and population differ in anatomy; apart from identical twins, no two people are identical in their external size and shape. Skeletal elements are not exceptions to this rule. Normal variation between different individuals of the same age, sex, and population is called **individual,** or **idiosyncratic,** variation. This variation can be substantial, but it is too often overlooked. Figure 2.1 illustrates the influences of ontogeny, sex, and idiosyncrasy on variation in the talus, one of the bones of the ankle.

A profusion of classifications of fossils has been created by the failure to appreciate normal skeletal variation in modern species and the failure to understand the principles and goals of taxonomy. In the resulting forest of family trees, ill-conceived species and genera are hung like ornaments. Normal variation within closely related **extant** (modern) species must

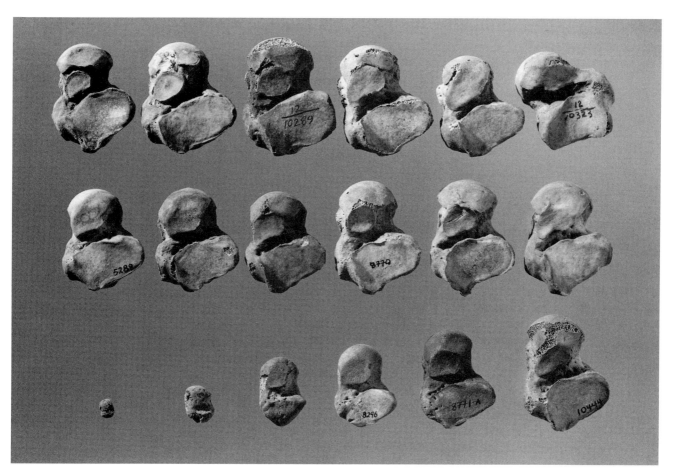

Figure 2.1 Normal variation in a bone of the ankle, the talus (viewed from the inferior, or lower, aspect; the anterior, or front, surface of each bone is toward the top of the page). All of the tali shown here were selected from a single-site skeletal sample of prehistoric Californians to illustrate skeletal variation attributable to age, sex, and idiosyncrasy. All specimens are from the left side except for the specimen in the upper right corner. One-half natural size.

(*top*) **Idiosyncratic variation.** Six adult male tali chosen from a sex-balanced sample of fifty adult individuals. Variation in this series is seen in the size and shape of the overall bone outline as well as in the proportions of the parts of the bone and in the topography of the various surfaces. Such variation is common in human skeletal remains.

(*middle*) **Sex variation.** Three adult female (*left*) and three adult male (*right*) specimens chosen at random from a sex-balanced sample of fifty adult skeletons. *Homo sapiens* is a primate species whose sexual dimorphism in body size is moderate by primate standards, less than the gorilla, but more than the common chimpanzee. In Chapter 17 we consider sexual dimorphism in the human skeleton in depth.

(*bottom*) **Ontogenetic variation.** The specimen at the far left is a talus from a newborn child. Tali from individuals at ages 1.5, 6, 10, 12, and 18 years show ontogenetic changes in size and shape of this skeletal element.

guide our expectations of variation in species whose members lived in the past. For this reason, osteologists unfamiliar with normal variation in the present are inclined to misinterpret similar variation in the past as indicating multiple species. To help avoid such misinterpretation, the

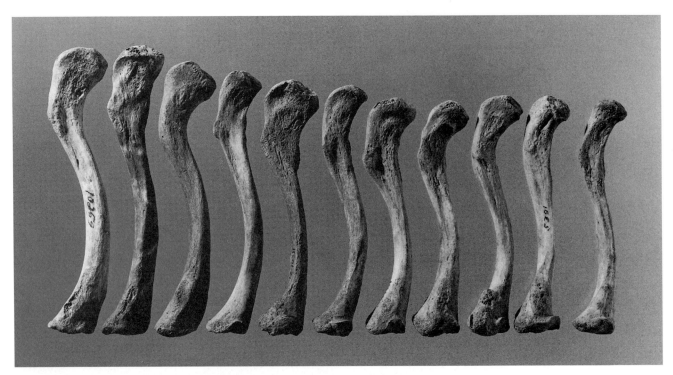

Figure 2.2 Adult human clavicles (collarbones) selected to illustrate the total range of variation in size, outline, and topography in a single-site, sex-balanced sample of eighty normal prehistoric Californians. Note the variation in overall size, in the shape of the bone, and in the topography of the surface. This kind and amount of variation should be expected in any normal sample of similar age and sex composition. These right-side specimens are shown in inferior view, with the lateral (arm end) toward the top of the page. One-half natural size.

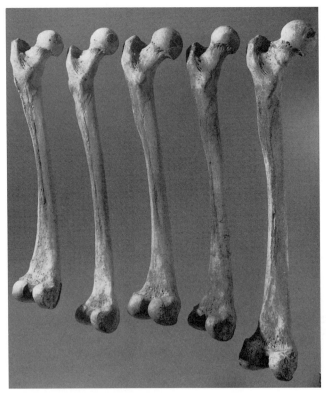

Figure 2.3 Adult human femora (upper leg bones) selected to illustrate the size and shape variation encountered in a single-site, sex-balanced sample of 100 normal prehistoric Californians. There is considerable variation between specimens shown here in size, robusticity, markings of muscle attachments, and proportions and angles of the different parts of each femur. These left-side specimens are shown in posterior view, with the superior (top) end of the bone toward the top of the page. One-fourth natural size.

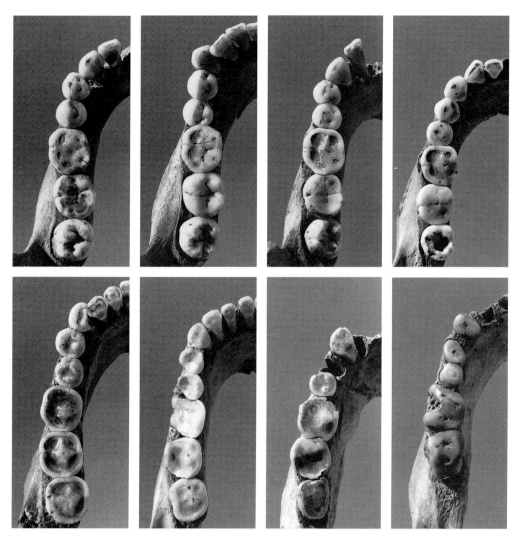

Figure 2.4 Dental variation due to age (reflected in wear), sex, and idiosyncrasy. The tooth rows shown here were selected from a single-site, sex-balanced skeletal sample of sixty prehistoric (Late Horizon) Californians. Note the differences in wear, size, and shape of the teeth as well as the variation in tooth row curvature and length.

paleontologist must become familiar with variation in modern humans and their closest relatives, the great apes, by studying large skeletal collections. In assessing any skeletal element, note ontogenetic changes in shape and size by studying different individuals who died at different ages. To assess normal variation in the adult skeleton arising from the other sources identified previously, try to examine a large, mixed-sex sample of individuals. Figures 2.2, 2.3, and 2.4 illustrate variation in size and shape in a single-site, balanced-sex sample from a prehistoric Californian skeletal population.

The reality of dimensional and morphological variation in the hard tissues of individuals makes **typology,** the practice of choosing one individual to characterize a species, a particularly unsuitable approach to the study and understanding of human osteology and evolution. Yet, to illustrate basic points of anatomy and identification it is necessary to start somewhere. In using this book's chapters on identification (Chapters 4–13), the reader should think of the skeletal elements chosen for illustration as representative, but never as typical. There is no "typical." We illustrate variation in skeletal size and shape in this book to reinforce the fact that such variation in biological structures is normal and to be expected.

2.2
A Few Facts about Bone

Bone is one of the strongest biological materials in existence and is the main supporting tissue of the body. During running, a human loads the bones at the knee joint with a force in excess of five times the weight of the entire body. Yet, despite its great strength, bone is a very lightweight material. The skeleton itself constitutes less than 20% of the weight of the entire body, whereas a framework of steel bars performing the same mechanical functions as the human skeleton would weigh four to five times more. Unlike steel, bone is a **composite** material, formed of protein (collagen) and mineral (hydroxyapatite). Bone differs from steel in yet another sense: it is a living tissue that can repair and reshape itself in response to external stresses. More detailed reviews of the physical, geometric, and mechanical properties of bone as a tissue and bones as organs are provided by Burr (1980) and Currey (1984).

Bone as a material in the skeleton is routinely subjected to compression, tension, shear, bending, and torsion during the lifetime of an individual. In 1869 the German orthopedic surgeon Julius Wolff formulated a physiological law that today bears his name. Wolff recognized that bones are remodeled during life to fit their mechanical functions. Simply put, **Wolff's law,** stated as the "law of bone transformation" in 1884, holds that bone is laid down where needed and resorbed where not needed. Before considering how bones perform this feat at the molecular, cellular, and gross levels, it is important that we understand the role of bones in the musculoskeletal system.

The study of skeletal **biomechanics** uses the laws of physics and the principles of engineering to consider the motions of the skeleton and the forces acting on parts of the skeleton. By studying the topography of joint surfaces, proportions, and the internal structure of bones it is possible to learn about lifeways and capabilities of ancient ancestors as well as more recent prehistoric and historic populations. For example, biomechanical studies of the long bones of Pliocene fossil specimens such as "Lucy" from Hadar imply that strong physical demands were placed on the bones of the early hominids.

2.3
Bones as Elements of the Musculoskeletal System

In the most basic terms, the musculoskeletal system is a system of bony levers operated by **muscles.** Any connection between different skeletal elements is called a **joint.** Bones in the skeleton **articulate** (intersect) at joints and are connected to one another by means of **ligaments** and **cartilage.** Cartilage is a tough and dense but elastic and compressible connective tissue. Bones are moved by muscles acting directly on the bones or via **tendons,** closely packed parallel bundles of collagen fibers. Movement at the joints is controlled and limited by the shapes of the articular surfaces and by ligaments that bind the joints together and prevent dislocation (Figure 2.5).

The hip, elbow, knee, and thumb joints are all examples of freely moving joints called **synovial joints.** The surfaces of the bones participating in synovial joints are coated with a thin (usually 1–5 mm) layer of slick, articular cartilage called **hyaline cartilage.** The area between the adjacent bones is the **joint cavity,** a space lined by a membrane that secretes a lubricant called **synovial fluid,** which resembles egg white in consistency. This fluid nourishes cartilage cells of the joint and is confined to the joint by the fibrous **joint capsule,** a sac made of connective tissue and reinforced by ligaments connecting to the periosteum of the articulating bones (see section 2.4). The combination of hyaline cartilage coating the bone surfaces and synovial fluid lubricating these surfaces gives synovial joints durability with smooth movement and low friction.

Synovial joints are often classified according to the geometric properties of the articulation. The hip joint is a **spheroidal,** or **ball-and-socket** joint, with the hemispherical femur head fitting into the acetabulum, a cavity in the pelvis. This joint structure allows movement in many directions. The elbow and knee joints are called **hinge** joints because they allow a hingelike movement limited mostly to one plane. The joint at the base of the thumb is called a **saddle-shaped,** or **sellar,** joint because of its shape. It allows movement in two basic directions. **Planar** joints allow two bones to slide across one another.

Because of their mobility, synovial joints are the most obvious joints in the musculoskeletal system, but there are two other important joint types in the body: cartilaginous joints and fibrous joints. In **cartilaginous joints** (**synchondroses**) the articulating bones are united by means of cartilage, and very little movement is allowed. The temporary joints between growth centers (described below) in a single growing bone are cartilaginous. Some of these joints persist in adulthood, such as the cartilaginous connections between the ribs and the breastbone (sternum). A **symphysis** is a variety of cartilaginous joint in which the fibrocartilage between the bone surfaces is covered by a thin layer of hyaline cartilage. **Syndesmoses** are tight, inflexible fibrous joints between bones that are united by bands of dense fibrous tissue in the form of membranes or ligaments; an example is the joint at which the two lower leg bones (tibia and fibula) articulate above the ankle (the distal tibiofibular articulation). **Cranial sutures** are fibrous joints of the skull; these are interlocking, usually tortuous joints in which the bones are close together and the fibrous tissue between them is thin. A **gomphosis** is the joint between the roots of the teeth and the bone of the jaws. When any two bony elements fuse together, the result is called a **synostosis.**

Movement of the skeleton takes place, for the most part, at synovial joints. This movement is caused by the muscles, which work by contracting across joints between bones. Muscles usually attach to two different bones, but they may attach to several. Most muscles are connected to bones via **tendons. Ligaments** are cords, bands, or sheets of collagenous

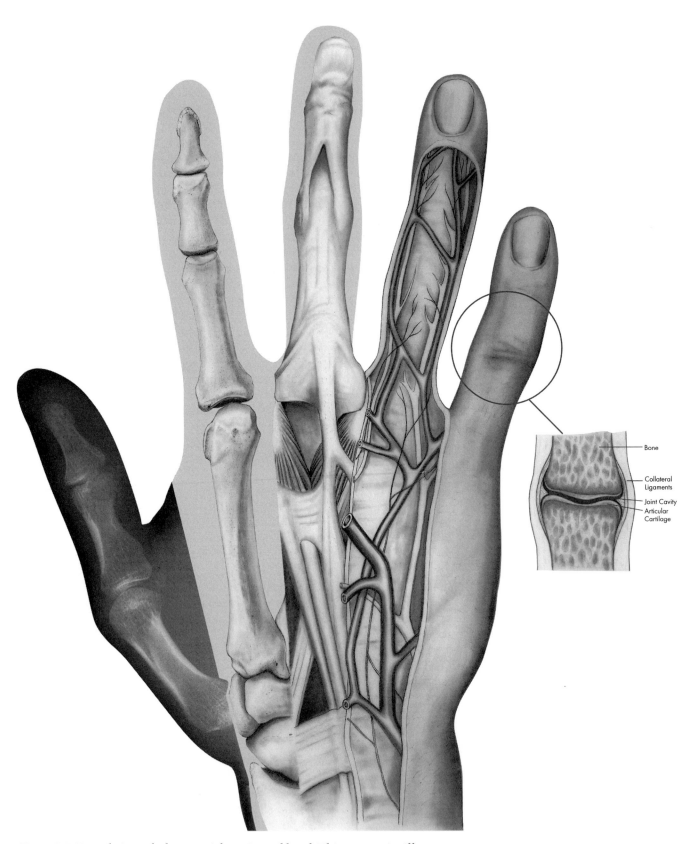

Bone

Collateral
Ligaments

Joint Cavity
Articular
Cartilage

Figure 2.5 Dorsal view of a human right wrist and hand. This composite illustration combines the techniques of cross-sectioning (little finger), dissection (ring and middle fingers), and radiography (thumb). The bones are embedded in a matrix of soft tissues including skin, nerves, arteries, veins, muscles, ligaments, tendons, and joint cartilage. In the assessment of external bone morphology, neither the soft tissue matrix nor the internal structure should be forgotten.

bundles that extend between the bones forming a joint. Ligaments resist tension, thereby strengthening the joint and permitting only movements compatible with the function of the joint.

Muscle attachment sites are conventionally identified in relative terms. The site that stays relatively stable during contraction of the muscle is called the **origin.** For the appendages, this is usually the attachment site closest to the trunk. The site that is moved by the contraction of a muscle is termed the **insertion.** For example, the muscles that flex the fingers originate in the anterior compartment of the forearm and insert on the fingerprint side of the finger bones (phalanges). Actions caused by muscles are usually reciprocal. At the elbow joint, different muscles cause opposite motions such as **extension** (straightening the arm) and **flexion** (bending the arm). Such muscles are called **antagonists.** Muscles are often identified by the primary action that their contraction causes. In Chapters 4–13 we introduce some of the major muscles that move the human skeleton and leave traces of their origins and insertions on the bones. For now, we can easily illustrate several basics of the musculoskeletal system with the human hand and arm. Muscles in the forearm are easily palpated as the hand is clenched and opened. These muscles act via tendons which become very visible across the front and back of the wrist when it is flexed and extended. For example, the *extensor digitorum* muscle, a resident of the forearm, functions in extending the four fingers as it operates via four tendons that cross the wrist.

2.4
Gross Anatomy of Bones

The wide range of element shapes in the human skeleton seems to defy classification. The bones in the body may, however, be partitioned into a few basic but overlapping shapes. The limb bones and many of the hand and foot bones, usually called long bones, are tubular in shape, with expanded ends (Figure 2.6). The bones of the cranial vault, shoulder, pelvis, and rib cage tend to be flat and tabular. The bones of the ankle, wrist, and spine are blocky and irregular. Despite this variety of external form, the makeup of bones at the gross and microscopic levels is remarkably constant.

At the gross level, all of the bones in the adult skeleton have two basic structural components: **compact** and **spongy** bone. The solid, dense bone that is found in the walls of bone shafts and on external bone surfaces is called compact, or **cortical,** bone. At joints, compact bone covered by cartilage during life is called **subchondral bone.** This is recognized as smoother and shinier than nonarticular compact bone, and it lacks the haversian systems described below.

The second kind of bone has a more spongy, porous, lightweight, honeycomb structure. This bone is found under protuberances where tendons attach, in the vertebral bodies, in the ends of long bones, in short bones, and sandwiched within flat bones. This **cancellous,** or **trabecular,** bone is named after the thin bony spicules (**trabeculae**) that form it. The molecular and cellular compositions of compact and trabecular bone tissue are identical; it is only the difference in porosity that separates these gross anatomical bone types.

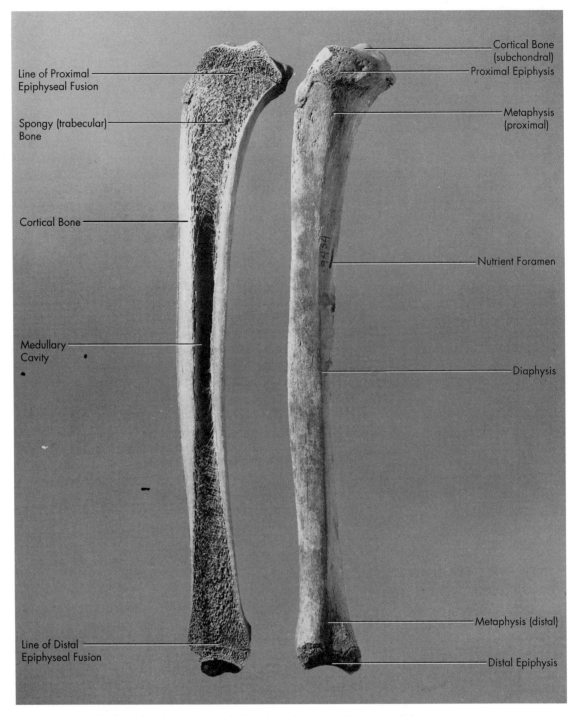

Line of Proximal Epiphyseal Fusion

Spongy (trabecular) Bone

Cortical Bone

Medullary Cavity

Line of Distal Epiphyseal Fusion

Cortical Bone (subchondral)

Proximal Epiphysis

Metaphysis (proximal)

Nutrient Foramen

Diaphysis

Metaphysis (distal)

Distal Epiphysis

Figure 2.6 A left tibia (shin bone) sectioned (cut) to show key elements of the gross anatomy of a human long bone. Note the disposition of the compact and spongy bone. One-half natural size.

Areas of trabecular bone in the growing skeleton constitute sites of the **red marrow,** a blood-forming, or **hemopoietic,** tissue that produces red and white blood cells and platelets. The **yellow marrow,** mainly a reserve of fat cells found in the **medullary cavity** (hollow inside the shaft) of tubular bones, is surrounded by compact bone. During growth, the red marrow is progressively replaced by yellow marrow in most of the long bones. As noted previously, in addition to their role in blood cell production and fat storage, bones function as organs in yet another way: bone tissue represents a calcium reservoir for the body.

Parts of tubular, or long, bones are often described according to the centers of ossification (see section 2.7) that appear during the growth process. The ends of long bones are called the **epiphyses** because they develop from secondary ossification centers of the bone (the articular surfaces of the epiphyses are parts of joints). The shaft of a long bone is called its **diaphysis** because it is the result of the bone's primary ossification center. The expanded, flared ends of the shaft are called **metaphyses.** A good example of these parts is the knee, where the epiphysis at the knee end of the femur fuses to the metaphysis of the shaft when growth is complete.

During life, the outer surface of bones is usually covered with a thin tissue called the **periosteum.** This tissue is missing in dry bones, but in life it coats all bone surfaces not covered by cartilage. The periosteum is a tough, vascularized membrane that nourishes bone. Some of the periosteum's thin fibers penetrate the surface of bone, and others intertwine with tendons to anchor muscles to the bone. The inner surface of bones is lined with an ill-defined and largely cellular membrane called the **endosteum.** Both periosteum and endosteum are **osteogenic** tissues—they contain bone-forming cells that are numerous and active during youth. These cells are reduced in number, but remain potentially active, in adulthood. They may be stimulated to deposit bone when the periosteum is traumatized.

2.5
Molecular Structure of Bone

We now turn to an assessment of bones at more basic molecular and cellular levels. No matter what shape a bone takes at the molecular level, its tissue is basically the same in all mammals. Bone tissue, like fiberglass, is a composite of two kinds of materials. The first component is a large protein molecule known as **collagen,** which constitutes about 90% of bone's organic content. Collagen is the most common protein in the body. Collagen molecules intertwine to form flexible, slightly elastic fibers in bone. The collagen of mature bones is stiffened by a dense inorganic filling of **hydroxyapatite,** the second component. In bone, crystals of this mineral, a form of calcium phosphate, impregnate the collagen matrix. This weave of protein and minerals gives bone its amazing properties. The value of the combination of materials is illustrated by two simple experiments. The mineral component gives bone its hardness and rigidity. When soaked in acid to dissolve these minerals, a bone becomes a rubberlike, flexible structure. On the other hand, when a bone is heated to combust the organic collagen, or leached out in some archeological contexts, it becomes extremely brittle and crumbles.

Characterizations of bone at the molecular level give some clues about its physical properties, but it is important to consider that bone as a tissue must be made and maintained by cells, it must be responsive to stress, and it must be capable of growth. A look at the structure of bone above the level of the collagen fibril and associated mineral provides insight into these dimensions of bone function.

2.6
Histology and Metabolism of Bone

Histology is the study of tissues, usually at the microscopic level. There are two histological types of mammalian bone, **immature** and **mature.** Immature bone (**coarsely bundled bone** and **woven bone**) is the first kind of bone to develop in prenatal life. Its existence is usually temporary, as it is replaced with mature bone as growth continues. Immature bone is usually formed rapidly and characterizes the embryonic skeleton, sites of fracture repair, and a variety of bone tumors. It has a relatively higher proportion of osteocytes (see below) than mature bone. Woven bone is the more phylogenetically primitive bone type in evolutionary terms. It is coarse and fibrous in microscopic appearance, with bundles of collagen fibers arranged in a non-oriented, random pattern.

Both compact and trabecular portions of adult bones are made of **mature,** or **lamellar, bone** tissue, named for the orderly, organized structure produced by the repeated addition of uniform lamellae to bone surfaces during appositional growth. Compact bone is composed of dense bone that cannot be nourished by diffusion from surface blood vessels. **Haversian systems,** with their canals and canaliculi (described in this section), are the solution to this problem. In contrast, more porous trabecular bone receives nutrition from blood vessels in surrounding marrow spaces and lacks haversian systems. Normal adult bone, both compact and trabecular, is histologically lamellar bone. Lamellar bone is usually laid down more slowly than woven bone, which it usually replaces.

Microscopic examination of a transverse section of compact bone in, for example, the tibial shaft reveals the internal structure of haversian bone (Figure 2.7). Such a section resembles an endview of a pile of sawed-off tree trunks. The cross section of each "trunk" often shows approximately four to eight concentric rings, known as **haversian lamellae.** A close examination of each lamella would reveal a bed of parallel collagen fibers. Fibers in successive lamellae, however, are oriented in different directions. This alternation of fiber direction adds strength to the structure.

Each "trunk" in the cross section of compact lamellar bone is known as a **haversian system,** or **secondary osteon.** See Figure 2.8 for illustrations and descriptions of osteons. These haversian systems measure about 300 μm (0.3 mm) in diameter and are about 3–5 mm in length. They represent the basic structural unit of compact bone, and their long axes parallel that of the long bone of which they are part. Passing through the core of each haversian system is a hollow **haversian canal,** through which blood, lymph, and nerve fibers pass. Additional smaller canals, the **Volkmann's canals,** pierce the bone tissue obliquely and at right angles from the periosteal and endosteal surfaces to link the haversian canals, creating a network that supplies blood and lymph to the cells of long bones.

Small cavities found within each lamella are called **lacunae.** Each lacuna harbors an **osteocyte,** a living bone cell. Nutrients are transported to these cells through **canaliculi,** minute fluid-filled channels that radiate

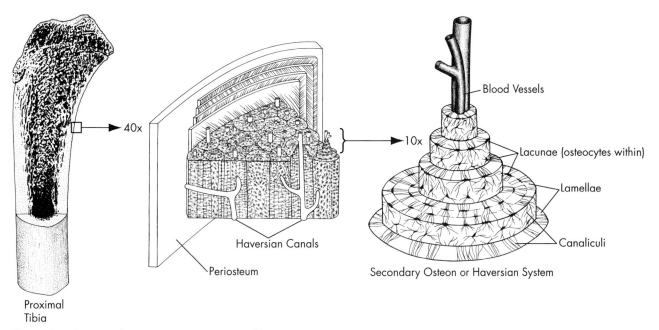

40x

Blood Vessels

Lacunae (osteocytes within)

Lamellae

Canaliculi

10x

Haversian Canals

Periosteum

Proximal
Tibia

Secondary Osteon or Haversian System

Figure 2.7 Gross and microscopic structure of bone.

from the centrally placed haversian canal to lacunae in succeeding lamellae, or from one lacuna to others. These channels in bone tissue enable the living cells to survive in a heavily mineralized environment.

Three primary cell types are involved in forming and maintaining bone tissue. **Osteoblasts** are bone-forming cells responsible for synthesizing and depositing bone material. Osteoblasts are often concentrated just beneath the periosteum. They make large quantities of a material known as **osteoid** (pre-bone tissue), an uncalcified organic matrix rich in collagen. Calcification of bone takes place as crystals of hydroxyapatite, the inorganic component of bone, are deposited into the osteoid matrix. Once surrounded by bony matrix, the osteoblasts are called **osteocytes,** cells which reside in lacunae and are responsible for maintaining bone tissue. **Osteoclasts** are responsible for the **resorption** (removal) of bone tissue. All skeletal elements change dramatically during ontogeny and continue to be capable of change in adulthood. Bone formation takes place throughout life. The reshaping, or **remodeling,** of bone takes place at the cellular level as osteoclasts remove bone tissue and osteoblasts build bone tissue. The opposing processes of bone formation and resorption allow bones to maintain or change their shape and size during growth. Some osteologists distinguish between "modeling" as bone sculpting during growth, and "remodeling" as the process of continuous removal and replacement of bone during life.

2.7
Bone Growth

The histological situation described in section 2.6 accounts for the metabolism of bone and the malleability of bone in the adult. During ontogeny,

Figure 2.8 Bone histology. (A) A comparison of primary and secondary osteons (from Paine and Godfrey, 1997). Polarized lighting, 200×. The primary osteon is an island with lamellar bone streaming around it; the larger secondary osteon intersects the lamellae of primary cortical bone. A primary osteon is composed of a vascular canal without a cement line (because it does not replace preexisting bone). The cement line (sheath), and lamellar bone organized around the central canal characterize the secondary osteon which fills a space left by the disappearance of preexisting bone. (B) An intact secondary osteon with several fragmentary osteons. Secondary osteons are products of bone remodeling. Polarized lighting, 200×. (C) A crowded field of secondary osteons. The large haversian canals indicate incomplete formation of several osteons. Polarized lighting, 100×. See Chapter 17 for a discussion of how analysis of these microscopic histological structures in bone can be employed in individual age assessment. Courtesy of Robert Paine.

however, the skeleton undergoes tremendous growth. Osteocytes do not divide, and, because bone matrix calcifies soon after being produced, the tissue cannot undergo further expansion from within. As a consequence, all bone growth is the result of bone deposition on a preexisting surface. Indeed, bone always develops by replacement of a preexisting connective tissue. Embryologically, bone development (**osteogenesis,** or **ossification**) occurs in two general sites. In **intramembranous ossification,** bones, particularly the frontal and parietal bones of the cranial vault, ossify by apposition on tissue within an embryonic connective tissue membrane. Most bones in the skeleton, however, grow through a process known as **endochondral ossification,** in which bones are preceded by cartilage precursors called cartilage models. Early in its development, *in utero,* the skeleton is flexible, but ossification is initiated before birth. The early skeleton's visible elements are mostly composed of cartilage, a material that is good for rapid growth in a medium where the functions of support are not yet necessary. Cartilage is composed mostly of collagen and, unlike bone, it is flexible and lacks blood vessels in the adult. The only difference between the two distinct mechanisms of ossification is the environment in which ossification occurs. There is no difference between the kind of bone produced.

Fetal ribs, vertebrae, the cranial base, and limb bones begin as cartilage models. Ossification occurs within the cartilage model as it is penetrated by blood vessels. Growth radiates from the location of the initial penetration, which becomes the **nutrient foramen.** A thin membrane called the **perichondrium** surrounds the cartilage model of the long bone. Osteoblasts just beneath the perichondrium in the fetal long bone begin to deposit bone around the outside of the cartilage shaft. Once this occurs, this membrane is called the **periosteum,** a fibrous connective tissue, which in turn deposits more bone layer by layer. As the diameter of the growing long bone shaft increases, osteoclasts on the endosteal surface remove bone and osteoblasts in the periosteum deposit bone. Thus, **appositional growth** allows shaft diameters to enlarge during development. The compact bone of an adult limb bone shaft is periosteal in origin, the original immature shaft having been removed by osteoclasts to form an enlarged medullary cavity. Slow subperiosteal apposition continues throughout life after an adolescent "growth spurt" (Garn, 1972).

Meanwhile, the developing long bone must also grow in length. During growth, the roughened, porous, usually irregular end of an immature long bone's metaphysis marks the region at which most longitudinal growth occurs. Sandwiched between the metaphysis (the primary center of ossification) and the epiphysis (the secondary center of ossification) during development is a cartilaginous center known as the **growth plate (epiphyseal plate),** a tissue layer responsible for bone formation. This plate, a layer of cartilage, "grows" away from the shaft center. The growing cartilage is replaced by bone on the diaphyseal side of the plate. As the individual grows, the epiphyseal plate is pushed farther from the bone's primary growth center (the shaft), lengthening the bone. Ossification and growth of the bone come to a halt when cells at the growth plate stop dividing, and the epiphysis fuses with the metaphysis of the shaft. Because the ends of the long bone flare, substantial remodeling occurs as the bone lengthens during this process (Figure 2.9).

At eleven weeks before birth there are usually about 800 ossification centers, the "bony pieces" of the skeleton. At birth there are about 450 centers. As a rule, "primary" centers appear before birth, and "secondary" centers after birth. The secondary center at the lower end of the femur (upper leg bone) and the one atop the tibia (lower leg bone) begin to appear just prior to birth. Most long bones develop two secondary centers in addition to the primary centers of ossification. A few long bones develop a secondary center at one end only, and typical wrist and ankle bones ossify entirely from their primary centers. By adulthood, all of the primary and secondary centers have fused to yield the average adult human complement of 206 elements, the bones of the adult human skeleton. These bones are listed in Figure 3.1.

2.8
Morphogenesis

Developmental biology is currently a hyperactive field due to the application of molecular techniques to the age-old problem of how form is produced. Discoveries in molecular biology, embryonic limb development,

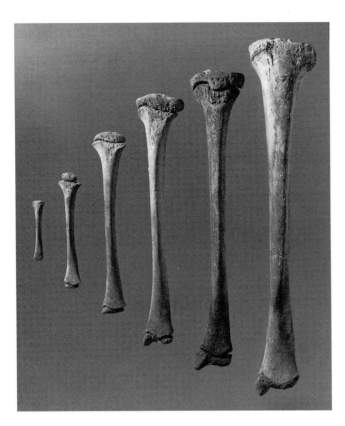

Figure 2.9 Growth series for the left human tibia (lower leg bone). The tibia on the far left is that of a newborn child (B). Larger specimens to the right are from individuals of ages 1.6, 6, 10, 12, and 18 years. Specimens are shown in anterior view, with the proximal (top) end of the bone toward the top of the page. One-fourth natural size.

amphibian limb regeneration, cell–cell communication, and the structure and expression of **morphogens,** growth factors, and **homeobox-**containing genes are rapidly advancing knowledge about how form is shaped during ontogeny.

As Müller points out, self-construction and self-organization are terms that convey the essential properties of development. All humans start with a single, apparently unstructured cell, the fertilized egg. Embryogenesis follows, and then birth and further development. At the end of the process, the fully developed human is an organism with an intricately wired brain that contains over a hundred trillion synaptic contacts that help make it possible for us to ask how the complex shapes and sizes of the human skeleton are encoded and how that code is translated by cells during development. Cells differentiate, communicate, and interact morphologically and functionally. Together they construct multicellular structures such as bones.

The shaping of form, or **morphogenesis,** from simple, seemingly amorphous, generative, starting cells has puzzled biologists and philosophers alike. A big step toward solving the problem came when it was demonstrated that DNA acted as a genetic code. The genome contains information about how to make distinct proteins, rRNA, and tRNA and how to replicate itself. The genome contains elements of a spatiotemporal program that controls the order and pattern of gene expression. As Müller notes, we do not yet understand in detail how a developing human is created on the basis of such minimal information (approximately 100,000 genes in a human). We know that different combinations of genes become effective in time and space in different cells, organs, and body regions. We know that cells interact, influencing each other. For example, a fracture of the adult humerus will stimulate production of bone and healing, but there was no way that the fertilized egg could "know" that any individual's humerus would be fractured at a particular time in adult life.

Basic events in animal development include cell proliferation (recurring cell division), cell differentiation (which occurs in a defined spatial order), and **pattern formation** (the spatiotemporal ordering of molecules, cells, or tissues to form a pattern, which can then develop at different scales). This is the process whereby spatial organization of cell differentiation is controlled. Cells obtain positional information by virtue of their location within a tissue. Cells move and migrate and die according to genetically determined schedules. All of these events are important in morphogenesis, a process tightly choreographed by highly conserved genes and gene arrays.

Bone develops originally from embryonic mesenchymal cells that have a very broad range of development potentialities, giving rise to fat, muscle, and other cells. Along the road to their differentiation into bone-producing cells, a population of cells with more limited potential is formed. These are only able to proliferate into chondroblasts or osteoblasts. These osteoprogenitor cells persist throughout postnatal life and are found in the endosteum and periosteum. They are most active during bone growth but can be reactivated in adult life when fracture repair is initiated (see section 2.9).

During embryogenesis, an **anlage,** or aggregation of cells indicating the first trace of an organ, forms. Recent work in limb development has begun to unravel the process through which an integrated system of sequentially expressed genes and/or gene arrays guides development of the limb by assigning positional address by morphogens (molecules that influence morphogenesis), growth factors, signaling molecules, and homeoboxes (a family of highly conserved base pair sequences of the DNA that encode small proteins that activate specific genes). The homeobox sequence is preserved with only minor modifications in a wide variety of animals, and is very similar in fruit flies, birds, and mammals.

It is already clear that development of a limb is guided by morphologic data sequestered in highly systematic gene arrays and implemented by stereotyped and largely universal cellular response regimens. For example, it was recently shown that implantation of a single acrylic bead soaked in the protein fibroblast growth factor and placed in the flank of an early chick embryo can trigger the formation of an entire new limb. It now appears that most sculpting of the skeletal frame occurs during the earliest phases of embryogenesis. Once the anlage is formed, further skeletal development appears to be directed primarily by the influence of stress history on gene expression by what may be called "assembly rules" which guide the behavior of each connective tissue cell during this process. For the functional morphologist, these insights into morphogenesis have fundamental implications (Lovejoy et al., 1999). What this means to the practicing osteologist is that much individual skeletal variation is the product of the interaction of the environment with these assembly rules rather than a direct readout of some gene(s). For example, the expression of an intertrochanteric line on the femur (see section 12.1.1e) represents an individual variation rather than a species-specific, genetically encoded trait.

2.9
Bone Repair

Bones occasionally break, or **fracture,** when subjected to abnormal stresses or when bone is pathologically weakened. The process of repair begins as soon as the fracture occurs. Blood vessels in the haversian canals, the periosteum, and the marrow are usually ruptured by a fracture. Blood flows into the fracture zone and normally forms a **hematoma** (bloody mass) that coagulates as the blood vessels are sealed off. The periosteum is usually torn at the fracture site and pulled away from the broken bone's ends. This stimulates the osteogenic layer of the periosteum to begin forming a **callus,** fracture repair tissue that forms a sort of natural splint. The callus first consists of fibrous connective tissue that bridges the broken bone surfaces, tying them together. Within two days the osteoblasts respond, and the callus is subsequently mineralized to form woven bone, the **primary bony callus.** The primary bony callus takes about six weeks to develop. Later, this woven bone callus is converted to lamellar bone. If orientation of the broken bone ends is close to the original, and if subsequent movement at the fracture site is limited (especially by immobilizing the bone), the callus may become so remodeled that evidence of fracture is eventually present only in radiographs. Further remodeling may completely eliminate any evidence of the fracture. In Chapter 18 we illustrate some effects of fracture in bone.

Recent work in the clinic has shown that proteins known as bone morphogenetic proteins (BMPs) can be combined with a matrix composed partly of demineralized collagen and applied to serious fractures to speed healing (Alper, 1994). In a fascinating intersection of applied and basic research, it turns out that these proteins are produced by genes belonging to a very ancient family—genes homologous to those in fruit flies. The nightmarish disease called fibrodysplasia ossificans progressiva (FOP) is a heritable disorder of connective tissue characterized by congenital malformation of the large toes and progressive, disabling endochondral osteogenesis in predictable anatomical patterns. Disease progression brings fusion of adjacent bones of the spine, limbs, thorax, and skull, leading to immobilization (Figure 2.10). Disease flare-ups can occur spontaneously or can be induced by minor trauma such as intramuscular drug injections. It was recently shown that this abnormal bone buildup occurs because the FOP patient's white blood cells erroneously manufacture BMP-4, triggering inappropriate heterotopic ("other" + "place") bone growth at sites of injury (Shafritz *et al.,* 1996). The regulatory regions of the BMP-4 gene (or some other gene whose product controls BMP-4 production) are probably implicated in the disease, and further research may identify drugs that would block some part of the BMP-4 pathway in FOP patients. Such remarkable findings at the intersection of basic research in molecular genetics and applied research in the medical clinic are now commonplace in biology. They hold out the promise to allow a fuller understanding of how skeletal form develops and how broken skeletons can be healed.

A basic understanding of bone biology forms part of the foundation necessary to begin identifying various elements of the human skeleton. The rest of the foundation is built in the next chapter, in which we introduce the anatomical terminology essential to every human osteologist.

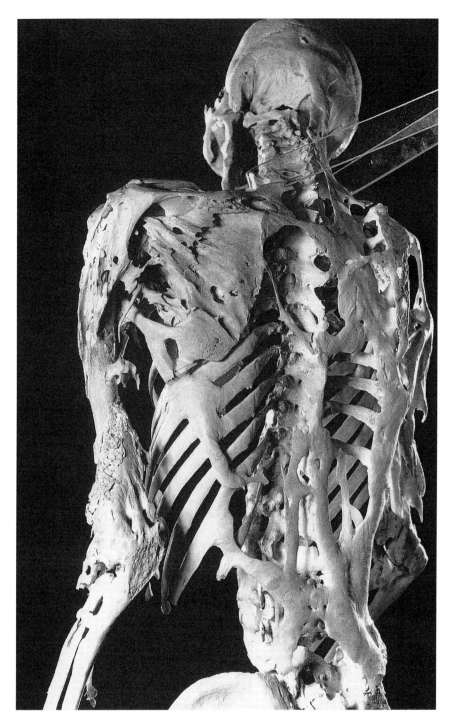

Figure 2.10 Advanced bony manifestations of fibrodysplasia ossificans progressiva in a 39-year-old man. See section 2.9 for details. Courtesy of Fred Kaplan, Mütter Museum, College of Physicians of Philadelphia (Shafritz *et al.*, 1996).

Suggested Further Readings

Active research into bone biology at all levels renders older texts obsolete. The sources below provide supplementation to the discussion presented above.

Bloom, W., and Fawcett, D. W. (1994) *A Textbook of Histology* (12th Edition). New York: Chapman and Hall. 720 pp.
> A comprehensive textbook with chapters on cartilage, bone, and teeth, with fine illustrations.

Currey, J. (1984) *The Mechanical Adaptations of Bones.* Princeton, New Jersey: Princeton University Press. 294 pp.
> An excellent advanced text on the biological and mechanical properties of bone.

Müller, W. A. (1997) *Developmental Biology.* New York: Springer. 382 pp.
> A succinct but thorough introduction to contemporary developmental biology.

Ogden, J. A. (1990) Histogenesis of the musculoskeletal system. In: D. J. Simmons (Ed.) *Nutrition and Bone Development.* New York: Oxford University Press. pp 3–36.
> An excellent review of bone development.

Ortner, D. J., and Putschar, W. G. (1981) *Identification of Pathological Conditions in Human Skeletal Remains.* Washington, D.C.: Smithsonian Institution Press. Smithsonian Contributions to Anthropology 28. 479 pp.
> This illustrated text includes a good chapter on the biology of skeletal tissues.

Shipman, P., Walker, A., and Bichell, D. (1985) *The Human Skeleton.* Cambridge, Massachusetts: Harvard University Press. 343 pp.
> Part I (Chapters 1–6) provides a fairly comprehensive survey of the topics covered in this volume, complete with a variety of illustrations.

Steinbock, R. T. (1976) *Paleopathological Diagnosis and Interpretation.* Springfield, Illinois: C. C. Thomas. 423 pp.
> Chapter 1 provides a concise guide to the basics of bone biology.

CHAPTER 3

Anatomical Terminology

Human anatomists and anthropologists throughout the world use a specific vocabulary to describe the human body. Anatomical nomenclature is both concise and precise, allowing unambiguous communication among all researchers who study skeletal material. Indeed, it is virtually impossible to follow even basic descriptions or interpretations in paleontology, physical anthropology, medicine, human anatomy, and a variety of allied disciplines without a command of the basic, general anatomical terms introduced in this chapter. Here we define the essential planes of reference, directional terms, body motions, and bony parts necessary for the detailed study of human osteology. Anatomical nomenclature has evolved from a classical foundation, and many of the names used to describe bones and their parts are derived from Latin and Greek roots. A working knowledge of these prefixes may facilitate learning the names of bones and their parts.

Anatomical terminology for hominids refers to the body in what is called **standard anatomical position** (Figure 3.1). Standard anatomical position is that of a human standing, feet together and pointing forward, looking forward, with none of the long bones crossed from the viewer's perspective. To prevent the crossing of bones, the palms of the hands must face forward, with the thumbs pointing away from the body. The terms **left** and **right** refer to the sides of the subject being observed, not to the observer's own right or left sides. **Cranial** skeletal anatomy refers to the skull; the remainder of the skeleton is called the **postcranial** anatomy. The **axial** skeleton refers to the bones of the trunk, including the vertebrae, pelvis, ribs, and sternum. The **appendicular** skeleton refers to the bones of the limbs, including the shoulder girdle.

3.1
Planes of Reference

Three basic reference planes are used in human osteology. The **sagittal** (**midsaggital, median,** or **midline**) plane divides the body into symmetrical right and left halves. Any planar slice through the body which parallels

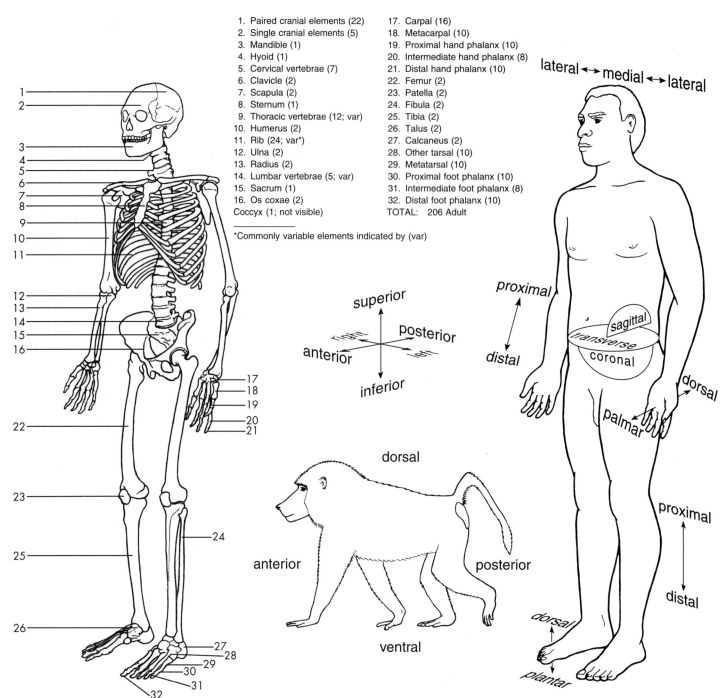

1. Paired cranial elements (22)
2. Single cranial elements (5)
3. Mandible (1)
4. Hyoid (1)
5. Cervical vertebrae (7)
6. Clavicle (2)
7. Scapula (2)
8. Sternum (1)
9. Thoracic vertebrae (12; var)
10. Humerus (2)
11. Rib (24; var*)
12. Ulna (2)
13. Radius (2)
14. Lumbar vertebrae (5; var)
15. Sacrum (1)
16. Os coxae (2)
Coccyx (1; not visible)

17. Carpal (16)
18. Metacarpal (10)
19. Proximal hand phalanx (10)
20. Intermediate hand phalanx (8)
21. Distal hand phalanx (10)
22. Femur (2)
23. Patella (2)
24. Fibula (2)
25. Tibia (2)
26. Talus (2)
27. Calcaneus (2)
28. Other tarsal (10)
29. Metatarsal (10)
30. Proximal foot phalanx (10)
31. Intermediate foot phalanx (8)
32. Distal foot phalanx (10)
TOTAL: 206 Adult

*Commonly variable elements indicated by (var)

Figure 3.1 Directional terms and planes for a human and a quadrupedal mammal. Knowledge of the directional terms and planes of reference is necessary for any work in human anatomy. In human osteology, these terms are essential in the study and comparison of skeletal elements.

the sagittal plane is called a **parasagittal section.** A **coronal (frontal)** plane divides the body into anterior and posterior halves and is placed at right angles to the sagittal plane. A **transverse (horizontal)** plane slices through the body at any height but always passes perpendicular to the sagittal and frontal planes.

3.2
Directional Terms

In osteology it is useful to refer to directions of motion or placement of various skeletal parts. All of the directional terms used here refer to the human body in standard anatomical position, but it is important to note that most of these terms are applicable to all mammals. A few terms may occasionally cause confusion when hominid (the zoological family of primates to which humans and their immediate ancestors and near-relatives belong) and nonhominid bones are being compared, because humans are **orthograde** (trunk upright) bipeds and most other mammals are **pronograde** (trunk horizontal) quadrupeds.

3.2.1 General

a. **Superior:** toward the head end of the hominid body. The *superior* boundary of the human parietal bone is the sagittal suture. **Cephalic** and **cranial** are synonymous terms that may be used homologously for bipeds and quadrupeds.

b. **Inferior:** opposite of superior; for hominids, body parts away from the head. The *inferior* surface of the calcaneus, or heel bone, is the part of the bone that rests nearest to, or lies in contact with, the ground. **Caudal,** toward the tail, is often used in the description of quadrupedal anatomy.

c. **Anterior:** toward the front of the hominid body. The breastbone, or sternum, is located *anterior* to the backbone, or vertebral column. **Ventral,** toward the belly, may be used homologously for bipeds and quadrupeds.

d. **Posterior:** opposite of anterior; for hominids, toward the back of the individual. The occipital bone is on the *posterior* and inferior end of the skull. **Dorsal** is often used for homologous parts of the quadruped anatomy.

e. **Medial:** toward the midline. The left side of the tongue is *medial* to the left half of the mandible.

f. **Lateral:** opposite of medial; away from the midline. The thumb occupies a *lateral* position relative to the little finger in standard anatomical position.

g. **Proximal:** nearest the axial skeleton, usually used for limb bones. The *proximal* end of the upper arm bone, the humerus, is the end toward the shoulder.

h. **Distal:** opposite of proximal; farthest from the axial skeleton. The *distal* end of the terminal foot phalanx fits into the front end of a shoe.

i. **External:** outer. The cranial vault is an *external* covering of the brain.

j. **Internal:** opposite of external; inside. The *internal* surface of the parietal is marked by a set of grooves made by blood vessels that lie external to the brain.

k. **Endocranial:** inner surface of the cranial vault. The brain fills the *endocranial* cavity.

l. **Ectocranial:** outer surface of the cranial vault. The temporal line is on the *ectocranial* surface of the parietal.

m. **Superficial:** close to the surface. The ribs are *superficial* to the heart.

n. **Deep:** opposite of superficial; far from the surface. The dentine core of a tooth is *deep* to the enamel.

o. **Subcutaneous:** just below the skin. The anteromedial surface of the tibia is *subcutaneous.*

3.2.2 Teeth

a. **Mesial:** toward the point on the midline where the central incisors contact each other. The anterior portion of molar and premolar crowns and the medial parts of canines and incisors are the mesial parts of these teeth. The *mesial* surface of the canine touches the incisor next to it, and the mesial surface of the first molar touches the premolar next to it.

b. **Distal:** opposite of mesial. The *distal* half of a premolar is the posterior half of the tooth.

c. **Lingual:** toward the tongue. The *lingual* surfaces of tooth crowns are usually hidden from view when a person smiles.

d. **Labial:** opposite of lingual; toward the lips; usually reserved for incisors and canines. The *labial* surfaces of incisors are observed when a person smiles.

e. **Buccal:** opposite of lingual; toward the cheeks; usually reserved for premolars and molars. A wad of chewing tobacco is often wedged between the cheek and the *buccal* surfaces of the molars of American baseball players.

f. **Interproximal:** in contact with adjacent teeth in the same jaw. Dental floss often gets stuck in *interproximal* areas.

g. **Occlusal:** facing the opposing dental arch, usually the chewing surface of each tooth. Caries ("cavities") are often found on the irregular *occlusal* surfaces of the molar teeth.

h. **Incisal:** the biting, or occlusal, edge of the incisors. The *incisal* edges of the central incisors are used to bite into an apple.

i. **Mesiodistal:** axis running from mesial to distal. The *mesiodistal* dimension of a molar may be reduced by interproximal wear.

j. **Buccolingual** and **labiolingual:** axis running from buccal or labial to lingual. The incisors of Neanderthals often have large *labiolingual* dimensions.

3.2.3 Hands and Feet

a. **Palmar:** palm side of the hand. The *palmar* surface of the digits bears fingerprints.

b. **Plantar:** sole of the foot. The *plantar* surface of the foot is analogous to the palmar surface of the hand.

c. **Dorsal:** top of the foot or the back of the hand. The *dorsal* surfaces of hands and feet often bear hair, whereas the palmar and plantar surfaces do not.

3.3
Motions of the Body

Movement of the body is accomplished by muscles acting directly or via tendons on bones. The less mobile attachment point that anchors a muscle is called the **origin** of the muscle, and the **insertion** is the site of muscle attachment with relatively more movement than the origin.

3.3.1 General

a. **Flexion:** bending movement that decreases the angle between body parts. When a hand is clenched into a fist there is strong *flexion* of the phalanges on the metacarpal heads. By convention, flexion at the shoulder or hip joint is a ventral (forward) movement of the limb.

b. **Extension:** opposite of flexion; a straightening movement that increases the angle between body parts. The classic karate chop is made by a rigid hand in which the fingers are *extended.* By convention, extension at the shoulder or hip joint is a dorsal (backward) swing of the limb.

c. **Abduction:** movement of a body part, usually an appendage, away from the sagittal plane. When the arm is raised to the side from standard anatomical position, *abduction* of the arm occurs. For the special case of fingers and toes, abduction is movement of the digit away from the midline of the hand or foot (spreading the digits).

d. **Adduction:** opposite of abduction; movement of a body part, usually an appendage, toward the sagittal plane. Clicking the heels together, as in a soldier's salute, is accomplished by one or both legs being *adducted.* For the special case of fingers and toes, adduction is movement of the digit toward the midline of the hand or foot (closing the digits).

e. **Circumduction:** a combination of abduction and adduction, as well as flexion and extension, which results in an appendage being moved in a cone-shaped path. When the driver of a slow vehicle signals someone behind to pass, this "waving on" is often done by *circumducting* the arm.

f. **Rotation:** motion that occurs as one body part turns on an axis. The head of the radius *rotates* on the distal humerus.

g. **Opposition:** motion in which body parts are brought together. *Opposition* of the thumb and finger tips allows us to grasp small objects.

3.3.2 Hands and Feet

a. **Pronation:** rotary motion of the forearm that turns the palm from anteriorly facing (thumb lateral) to posteriorly facing (thumb medial). Typewriters are used with the hand in *pronation.*

b. **Supination:** opposite of pronation; rotary motion of the forearm that turns the palm to a position in which the thumb is lateral. When chimpanzees beg for food the hand is often held in *supination.*

c. **Dorsiflexion:** flexion of the entire foot away from the ground. When a mime walks on her heels, her feet are *dorsiflexed.*

d. **Plantarflexion (volarflexion):** opposite of dorsiflexion; flexing of the entire foot inferiorly, toward the ground, at the ankle. Action in both dorsiflexion and plantarflexion occurs at the ankle. When a ballet dancer walks on his toes, his feet are strongly *plantarflexed.*

e. **Eversion:** turning the sole of the foot outward so that it faces away from the midline of the body; also known as pronation of the foot.

f. **Inversion:** turning the sole of the foot inward so that it faces toward the midline of the body; also known as supination of the foot.

3.4
General Bone Features

Whereas the directions and motions described in sections 3.2 and 3.3 have very precise meanings, the series of general terms applied to bony parts are more ambiguous and cross-cutting. The question of when, for example, a tubercle is big enough to be called a tuberosity or a trochanter is rarely faced by the osteologist, who uses the conventional labels for various bones and bone parts. This is because specific terms for nearly all bones and bone parts are already established, and have been for a long time. The "greater trochanter" of a femur, for example, identifies a particular, unique structure for all human osteologists. Recognize that the following terms are often ambiguous by themselves but very particular when coupled with element-specific names introduced in Chapters 4–13 of this book.

3.4.1 Projections and Parts

a. **Process:** a bony prominence. The mastoid *process* forms the prominence behind the ear.

b. **Eminence:** a bony projection; usually not as prominent as a process. The articular *eminence* of the temporal bone is the rounded area with which the mandibular condyle articulates during chewing.

c. **Spine:** generally a longer, thinner, sharper process than an eminence. The vertebral *spines* are used in the identification of various vertebrae.

d. **Tuberosity:** a large, usually rugose (roughened) eminence of variable shape; often a site of tendon or ligament attachment. The deltoid *tuberosity* marks the shaft of the humerus.

e. **Tubercle:** a small, usually rugose eminence; often a site of tendon or ligament attachment. The conoid *tubercle* is found along the inferior edge of the clavicle.

f. **Trochanters:** two large, prominent, blunt, rugose processes found on the femur. The larger of these is called the **greater trochanter;** the smaller is the **lesser trochanter.**

g. **Malleolus:** a rounded protuberance adjacent to the ankle joint. It is easy to palpate (examine by touch) both the lateral and medial *malleoli.*

h. **Boss:** a smooth, round, broad eminence. Female skulls tend to show more *bossing* of the frontal bone than those of males.

i. **Articulation:** an area in which adjacent bones are in contact (via cartilage or fibrous tissue) at a joint. The most proximal surface of the tibia is said to *articulate* with the distal end of the femur.

j. **Condyle:** a rounded articular process. The occipital *condyles* lie on the base of the cranium and articulate with the uppermost vertebra, the atlas.

k. **Epicondyle:** a nonarticular projection adjacent to a condyle. The lateral *epicondyle* of the humerus is located just proximal to the elbow, adjacent to the lateral condylar surface.

l. **Head:** a large, rounded, usually articular end of a bone. The *head* of the humerus is the superior (proximal) end of the bone.

m. **Shaft,** or **diaphysis:** the long, straight section between the ends of a long bone. The femoral *shaft* is roughly circular in cross section.

n. **Epiphysis:** in general usage, usually the end portion or extremity of a long bone which is expanded for articulation. The proximal *epiphysis* of the tibia is the expanded end of the bone which articulates with the femur. See Chapter 2 for more precise definitions of the diaphysis, epiphysis, and metaphysis.

o. **Neck:** the section of a bone between the head and the shaft. The *neck* of the femur is long relative to the size of the femoral head in some early hominids.

p. **Torus:** a bony thickening. The supraorbital *torus* on some *Homo erectus* frontal bones is very thick.

q. **Ridge:** a linear bony elevation, often roughened. The lateral supracondylar *ridge* of the humerus borders the bone above the lateral epicondyle.

r. **Crest:** a prominent, usually sharp and thin ridge of bone; often formed between adjacent muscle masses. The sagittal *crest* is a structure that forms during the development of large temporalis muscles in the gorilla.

s. **Line:** a raised linear surface, not as thick as a torus or as sharp as a crest. The inferior temporal *lines* mark the superior extent of the temporalis muscles.

t. **Hamulus:** a hook-shaped projection. The *hamulus* of the wrist's hamate bone gives the bone its name.

u. **Facet:** a small articular surface, or tooth contact. Bodies of the thoracic vertebrae have *facets* for articulation with the heads of ribs. Occlusal *facets* form on the chewing surfaces of the teeth shortly after crown eruption.

3.4.2 Depressions

a. **Fossa:** a depressed area; usually broad and shallow. The olecranon *fossa* is located on the posterior surface of the distal humerus, where it receives the proximal ulna during full extension of the arm.

b. **Fovea:** a pitlike, depressed area; usually smaller than a fossa. The anterior *fovea* of an unworn molar is seen in occlusal view.

c. **Groove:** a long pit or furrow. The intertubercular *groove* passes between two tubercles on the humerus.

d. **Sulcus:** a long, wide groove. A strong supratoral *sulcus* is present on African ape crania but is weak or absent on *Australopithecus* crania.

e. **Fontanelle:** a space between cranial bones of an infant. The soft spot atop a baby's head indicates the presence of a *fontanelle*.

f. **Suture:** where adjacent bones of the skull meet (articulate). The lambdoidal *suture* is between the occipital and parietal bones.

g. **Foramen:** an opening through a bone, usually a passage for blood vessels and nerves. The mental *foramen* is an opening on the lateral surface of the mandible.

h. **Canal:** a tunnel-like, extended foramen. The carotid *canal* is found at the base of the skull.

i. **Meatus:** a short canal. The external auditory *meatus* is the canal that connects the middle and outer ear.

j. **Sinus:** a cavity within a cranial bone. The frontal *sinus* is well developed in advanced *Homo erectus* crania.

k. **Alveolus:** a tooth socket. The canine *alveolus* in the mandible is deeper than the incisor alveolus.

Suggested Further Readings

Virtually all texts in human anatomy provide guides to anatomical terminology.

Bass, W. M. (1995) *Human Osteology: A Laboratory and Field Manual* (3rd Edition). Columbia, Missouri: Missouri Archaeological Society. 327 pp.
> Appendices for the osteology student are concise sources of information on bone nomenclature.

O'Rahilly, R. (1989) Anatomical terminology, then and now. *Acta Anatomica* 134:291–300.
> A good history of the first twenty-five centuries of anatomical nomenclature.

Twelfth International Congress of Anatomists. (1989) *International Anatomical Nomenclature Committee: Nomina Anatomica* (6th Edition). New York: Churchill-Livingstone.
> This work sets the international standards for anatomical nomenclature.

CHAPTER 4

Skull

The bones of the human skull may be grouped into three sets. Most bones of the skull base lie underneath the brain and are at least partly preformed in cartilage. These bones represent the primitive vertebrate braincase. The facial bones and bones of the roof and sides of the skull are dermal bones, formed in sheets of connective tissue under the skin. Dermal bone served as an armor plating for the primitive fishes, especially as a protective covering for the vital head parts. The rest of the bones of the skull are derived from the gill arches of primitive fish ancestors. Primitive vertebrates had no jaws. The first jaws were derived from the first, or **mandibular,** gill arch. Bones of this structure moved forward under the braincase to become the upper and lower jaws. These primitive jaws became sheathed in, and were eventually replaced by, dermal bones. Meanwhile, original gill arch bones migrated to the middle ear region, where they became two of the three tiny bones associated with hearing.

The skull is the most complex portion of the skeleton and is of major importance for physical anthropology. It is one of the keys to aging, sexing, and understanding the evolutionary history of hominids. The complexity of the human skull can best be understood by recognizing the widely differing functions it performs. It forms the bony foundation for the senses of sight, smell, taste, and hearing. It houses and protects the brain. In addition, the skull forms the framework of the chewing apparatus. Given these varied functions, it is no wonder that the skull is a complex structure.

Before moving to a detailed consideration of the individual bones of the skull, it is useful to consider this part of the anatomy as a unit (Figures 4.1–4.6b). A few convenient landmarks are useful in this analysis. The eye sockets are the **orbits.** The hole between and below the orbits, the nose hole, is the **anterior nasal aperture,** or **piriform aperture.** The ear holes are the **external auditory meati** (the singular is **meatus**), and the large oval hole in the base of the skull is the **foramen magnum.** The thin bony bridges at the sides of the skull are the **zygomatic arches.** The teeth are part of the skull, but because of their importance and peculiar anatomy an entire chapter is devoted to them (see Chapter 5).

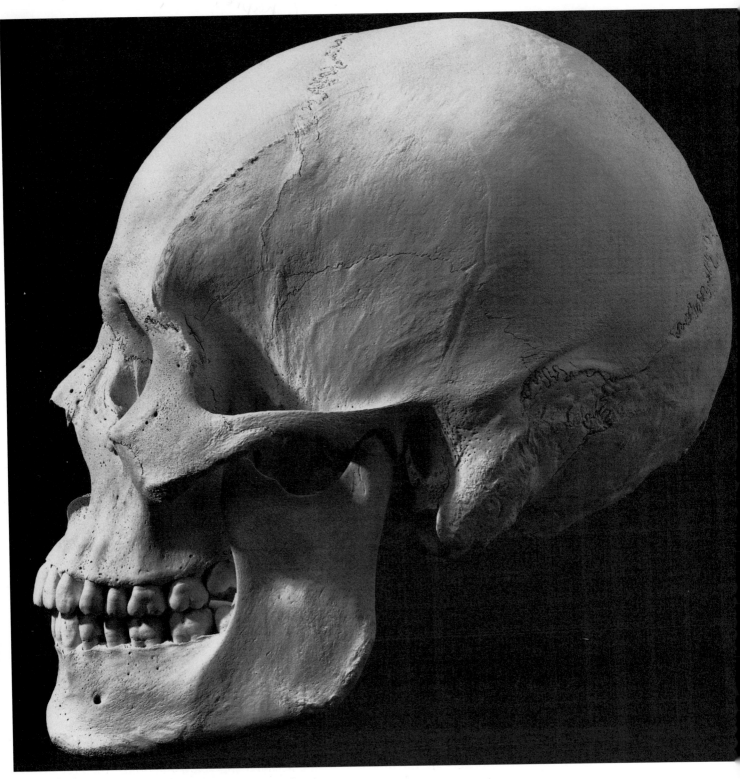

Figure 4.1 **Adult male skull, lateral.** Natural size.

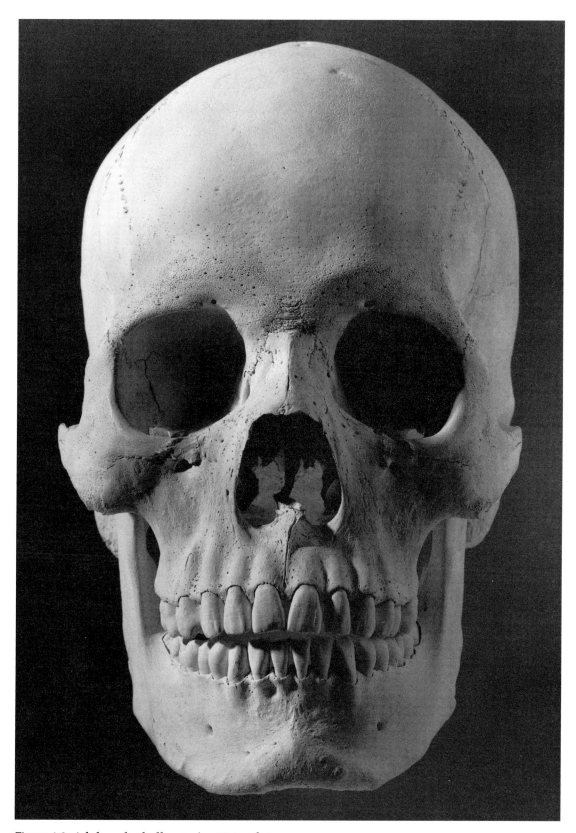

Figure 4.2 **Adult male skull, anterior.** Natural size.

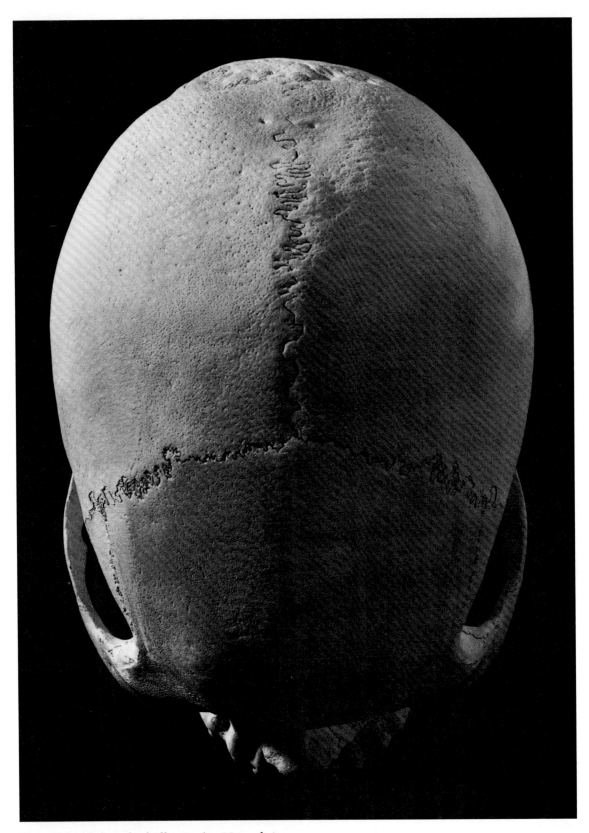

Figure 4.3 **Adult male skull, superior.** Natural size.

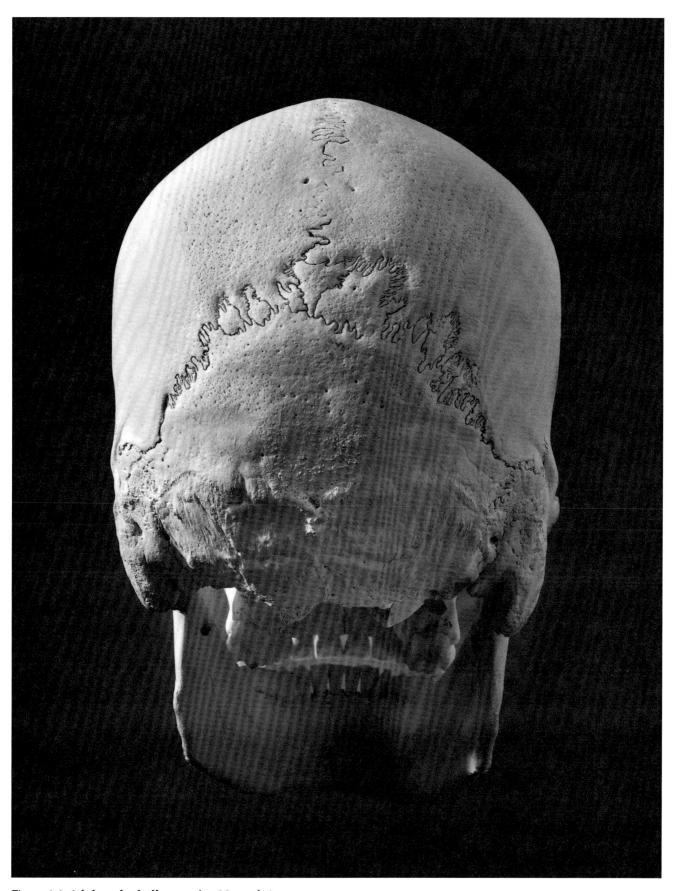

Figure 4.4 **Adult male skull, posterior.** Natural size.

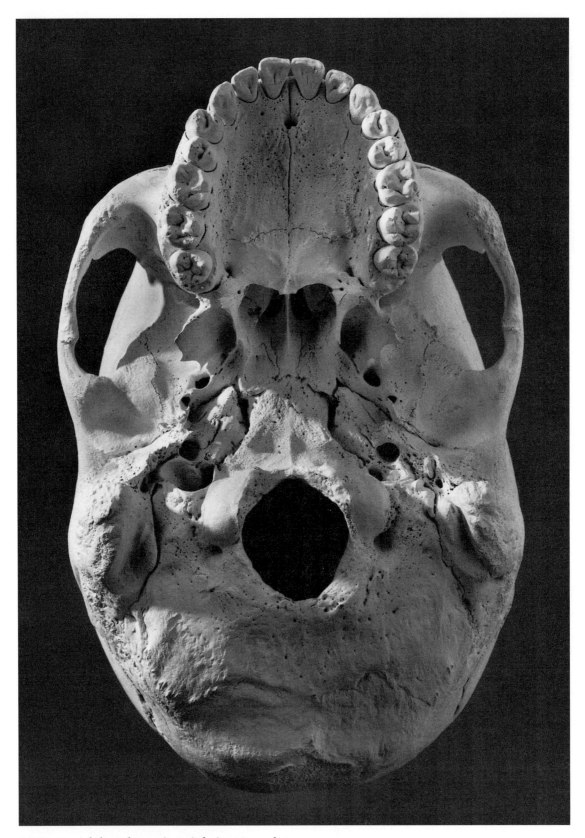

Figure 4.5 **Adult male cranium, inferior.** Natural size.

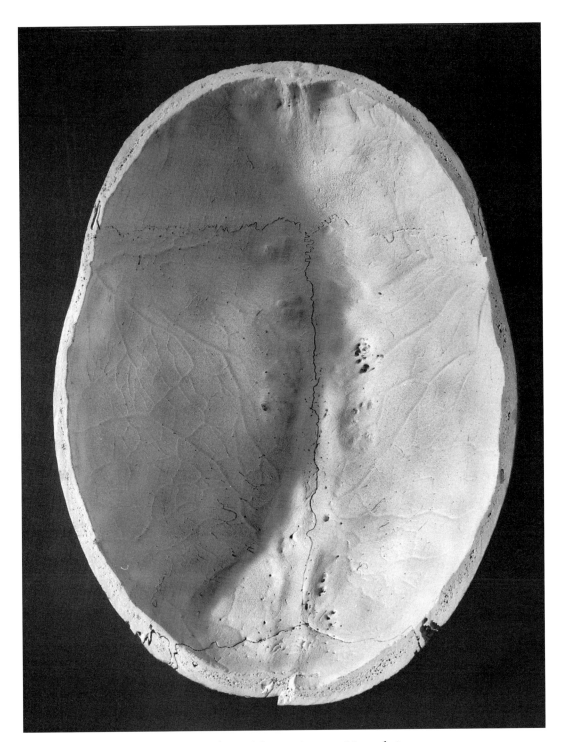

Figure 4.6a **Adult male cranium, endocranial,** superior part. Natural size.

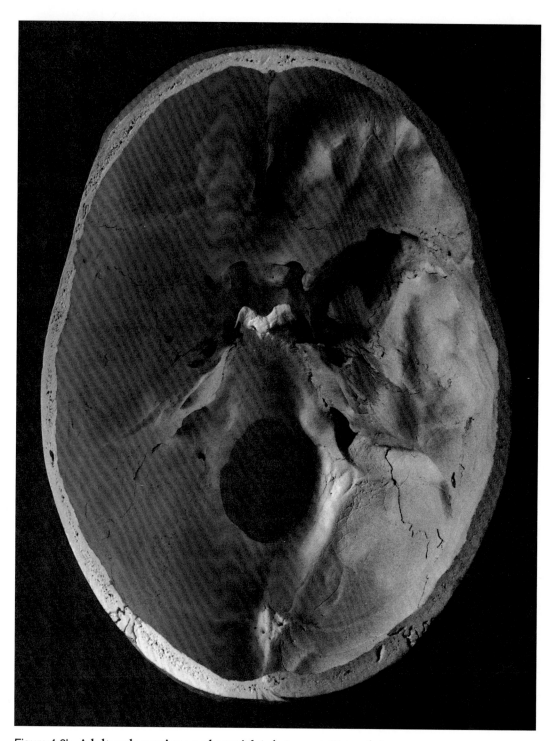

Figure 4.6b **Adult male cranium, endocranial,** inferior part. Natural size.

4.1
Handling the Skull

In addition to being one of the most complex parts of the skeleton, some of the skull's elements are the most delicate. During study, the skull should be handled above a padded surface and stabilized against rolling on the surface by means of sandbags or cloth rings designed for this purpose. It is rarely necessary to store or manipulate the lower jaw in its natural anatomical position relative to the upper jaw. When this position is called for, however, note that the colliding upper and lower teeth are fragile and susceptible to chipping. Care should be exercised when occluding the upper and lower jaws in this fashion, and padding should always be placed between the teeth if the skull is stored in this position.

In handling the skull, common sense and both hands should always be used. A finger or thumb placed in and behind the foramen magnum will not damage the bone, but other openings, such as the orbits or zygomatic arches, are more fragile and should never be used as gripping surfaces. In addition to the thin bones within the orbits, delicate parts which are very susceptible to damage during cranial manipulation include the edges and insides of the nasal aperture and the thin, projecting pterygoid plates and styloid processes at the base of the skull. If teeth become dislodged during study, be sure to replace them immediately in the correct sockets.

The temptation to test the mechanical properties of dry bone by probing, twisting, poking, stabbing, and scraping should always be resisted. However, in the course of handling osteological material, there will be breakage. Fortunately, it is normally a simple matter to glue the bone back together, and this should be done promptly by the laboratory supervisor so that the broken pieces are not permanently lost (see Chapter 14).

4.2
Elements of the Skull

The term "skull" suffers from common misuse. It is worthwhile to review the proper use of cranial terminology.

- The **skull** is the entire bony framework of the head, including the lower jaw.
- The **mandible** is the lower jaw.
- The **cranium** is the skull without the mandible.
- The **calvaria** is the cranium without the face.
- The **calotte** is the calvaria without the base.
- The **splanchnocranium** is the facial skeleton.
- The **neurocranium** is the braincase.
- The three basic divisions of the **endocranial** surface at the base of the neurocranium correspond to the topography of the brain's base. These **anterior, middle,** and **posterior cranial fossae** are respectively occupied by the frontal lobes, temporal lobes, and cerebellum of the brain.

When the ear ossicles (three pairs of tiny bones associated with hearing) are included and the hyoid excluded, there are usually twenty-eight bones in the adult human skull. Distinguishing these bones is occasionally made difficult by the fact that some of them fuse together during adult life. For this reason, it is advisable to begin study with young adult specimens, in which the bones are most readily recognizable. The bones of the skull are listed in Figure 3.1 (p. 36) according to whether they are single (unpaired) or paired in the adult skull. In addition to the bones listed in Figure 3.1, there are often **sutural bones** (also called **Wormian bones,** or **extrasutural bones**), irregular ossicles occurring along sutures. The large, triangular **inca bone** is occasionally found at the rear of human crania.

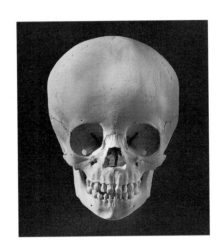
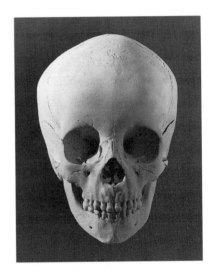

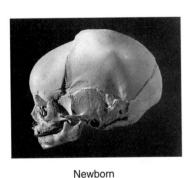
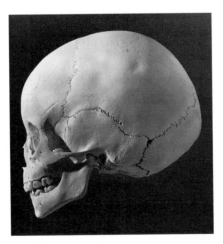
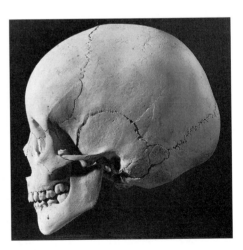

Newborn

3 years

6 years

Figure 4.7 Growth in the human skull. Note the change in proportions of face and vault through the series. All specimens are shown in facial and lateral views. One-third natural size.

4.3
Growth and Architecture, Sutures and Sinuses

The human skull is a fascinating structure when viewed from a phylogenetic perspective. In early vertebrates, two kinds of bone evolved, **dermal bone** and **cartilage-replacement bone.** The modern human skull is derived from both kinds of bone. Dermal bones form the sides and roof of the skull and make up the facial skeleton. The bones of the cranial base, known collectively as the **basicranium,** are mostly preformed in cartilage: the **ethmoid bone** surrounds the olfactory apparatus and nerves, the **sphenoid bone** surrounds the optic nerves, the two **temporal bones** surround the

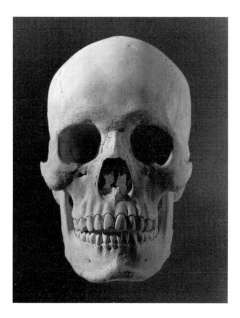

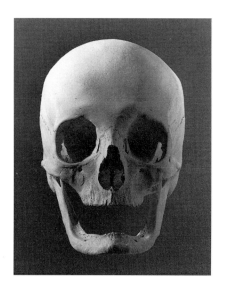

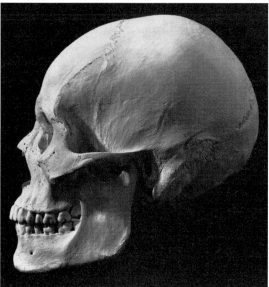

Young adult

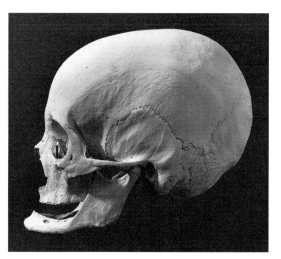

Elderly adult

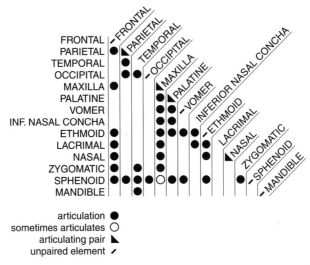

Figure 4.8 Articulation of bones in the adult human skull.

auditory system, and the **occipital bone** surrounds the spinal cord. Some bones, like the occipital, sphenoid, and temporal, combine dermal and cartilage-replacement portions. The remaining bones in the skull (mandible, sphenoid greater wings, ear ossicles, hyoid) are derived from the gill arches of primitive vertebrates.

At birth the skull is made up of forty-five separate elements and is large relative to other parts of the body. The facial part of the newborn skull, however, is relatively small, reflecting the dominance of brain development at this stage of maturation. The face "catches up" to the neurocranium as development, particularly in the mandible and maxilla, proceeds. Important stages in the skull's development include the emergence of the first set of teeth (between the ages of 6 and 24 months), the emergence of the permanent teeth (beginning at about 6 years), and puberty. Figure 4.7 illustrates growth of the skull.

At birth the skull contains intervals of dense connective tissue between plates of bone. These "soft spots," or **fontanelles,** are cartilaginous membranes that eventually harden and turn to bone. In the adult the skull bones contact along joints with interlocking, sawtooth, or zipperlike articulations called **sutures.** Cranial articulations in the adult human skull are summarized in Figure 4.8. Many of these sutures derive their name directly from the two bones that contact across them (Figures 4.9–4.12). For example, the **zygomaticomaxillary sutures** are the sutures between the zygomatics and maxillae, and the **frontonasal sutures** are the short sutures between the frontal and nasals. Some of the sutures in the skull are known by special names. The **sagittal suture** passes down the midline between the parietal bones. The **metopic suture** passes between unfused frontal halves and only rarely persists into adulthood. The **coronal suture** lies between the frontal and parietals. The **lambdoidal suture** passes between the two parietals and the occipital. The **squamosal sutures** are unusual, scalelike, beveled sutures between the temporal and parietal bones. The **sphenooccipital,** or **basilar, suture** (actually a synchondrosis) lies between the sphenoid and the occipital. The **parietomastoid sutures** pass between the parietals and the temporals, constituting posterior extensions of the squamosal suture. The **occipitomastoid sutures** pass between the occipital and temporals on either side of the vault.

Sinuses are void chambers in the cranial bones that enlarge with the growth of the face. There are four basic sets of sinuses, one each in the maxillae, frontal, ethmoid, and sphenoid. These sinuses are linked to the nasal cavity and, in life, irritation of their mucous membranes may cause swelling, draining, and headache-related discomfort.

4.4
Skull Orientation

The most useful and informative comparisons between skulls of different individuals are usually comparisons made when both skulls are in the same orientation. This fact gained early recognition. The French physical anthropologist Paul Broca first defined a plane of orientation using the alveolar area and the condyles of the skull in the 1800s. This proved to be unsatisfactory, because the condyles are often hidden in lateral aspect, and although other anatomists defined a plane through the center of the external auditory meatus and the base of the nasal aperture, difficulties in reproducing this orientation continued. Today the convention used in orienting the skull is the **Frankfurt Horizontal** (**FH**), named for the city in which the convention was established in 1884. The Frankfurt Horizontal is a plane defined by three points—the right and left **porion** points and the left **orbitale.** These points, at the top of each external auditory meatus and the bottom of the left orbit, are defined in section 4.5.

Skulls are normally viewed from six standard perspectives, all in Frankfurt Horizontal (as illustrated in Figures 4.1–4.5). Viewed from above, the skull is seen in **norma verticalis.** When viewed from either side, the skull is seen in **norma lateralis. Norma occipitalis** is the posterior view of the skull. Viewed from the front the skull is seen in **norma frontalis,** and viewed from the base it is seen in **norma basilaris.** All of these views are perpendicular or parallel to the Frankfurt Horizontal.

4.5
Cranial Osteometric Points

Because the skull has been the focus of much physical anthropological investigation, an extensive network of osteometric, or bone-measuring, points has been developed to allow researchers to take comparable measurements on skulls. Early in the 20th century the main focus in physical anthropology was on measuring skulls and comparing these measurements. Today there is a return to an appreciation of the anatomy between the measuring points. Still, it is necessary to have a set of conventions to ensure unambiguous reporting and comparison of osteological material. In addition, even nonmetric descriptions often use a terminology of the skull which makes reference to these measuring points. Indeed, the vocabulary of the skull's measuring points is vital for researchers in osteology and paleontology. In Chapter 15 we provide information on how to measure osteological materials. Before going into a more detailed discussion of cranial anatomy, we introduce the landmarks most frequently used in measurement and description of the human skull. These points are defined for anatomically modern humans. Because of this, it may not be possible or useful to use them when measuring other extant or extinct primate species.

The osteometric points are best considered in two basic sets: unpaired osteometric points are located in the midsagittal plane, and paired points lie on either side of this plane. Figures 4.9–4.12 indicate the major cranial osteometric points defined here.

Three basic kinds of cranial measurements are most often used. The first is the familiar cranial **capacity,** which describes the volume within the neurocranium. Most measurements taken on the skull are **linear** measurements between two points, the line between the two points sometimes being referred to as a **chord.** The third kind of measurement is a curvilinear measurement called an **arc.** The observer should always take care to see that the landmarks being used for measuring points are unbroken and clearly visible. Comparisons should always be made between individuals of similar age. Exceptions to these cautions should always be noted in any osteology report. The list of landmarks given here includes most points in use today. For a slightly extended list, see Martin and Saller (1957), the source of all abbreviations used here.

4.5.1 Unpaired (Midline) Cranial Osteometric Points

Note that the points defined here are arranged in order from the upper incisors around the vault's midline to the lower incisors.

inc. **Incision** is the point at the occlusal surface where the upper central incisors meet.

ids. **Alveolare (infradentale superius)** is the midline point at the inferior tip of the bony septum between the upper central incisors.

pr. **Prosthion** is the midline point at the most anterior point on the alveolar process of the maxillae.

ns. **Nasospinale** is the point where a line tangent to the inferiormost points of the two curves of the inferior nasal aperture margin (edge, rim) crosses the midline.

rhi. **Rhinion** is the midline point at the inferior free end of the internasal suture.

n. **Nasion** is the midline point where the two nasal bones and the frontal intersect.

g. **Glabella** is the most anterior midline point on the frontal bone, usually above the frontonasal suture.

m. **Metopion** is an instrumentally determined, ectocranial midline point on the frontal where the frontal's elevation above the chord from nasion to bregma is greatest.

b. **Bregma** is the ectocranial point where the coronal and sagittal sutures intersect.

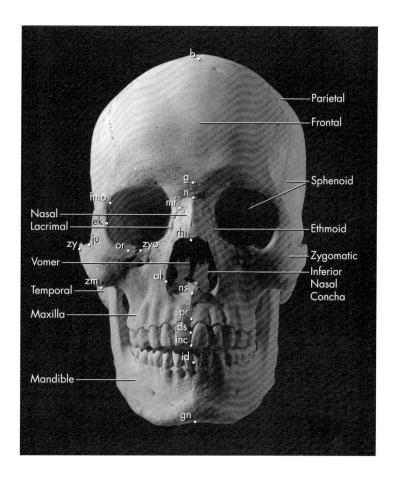

Figure 4.9 Bones and osteometric points of the human skull, anterior. One-half natural size.

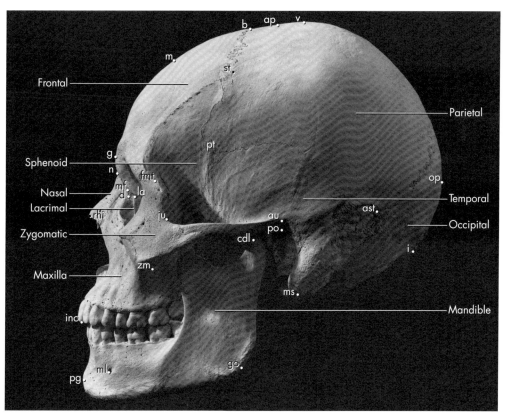

Figure 4.10 Bones and osteometric points of the human skull, lateral. One-half natural size.

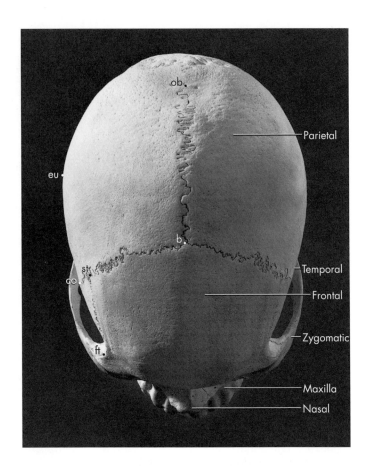

Figure 4.11 Bones and osteometric points of the human skull, superior. One-half natural size.

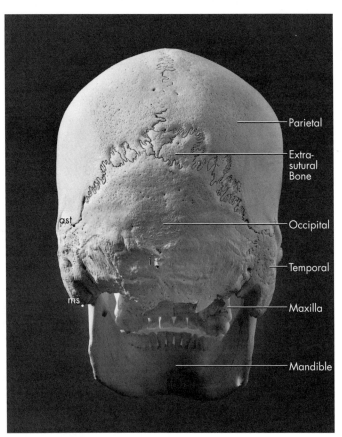

Figure 4.12 Bones and osteometric points of the human skull, posterior. One-half natural size.

v. **Vertex** is determined instrumentally when the skull is in Frankfurt Horizontal. It is the highest ectocranial point on the skull's midline.

ap. **Apex** is an instrumentally determined, ectocranial midline point placed where a coronal plane through the right and left poria (see section 4.5.2) intersects the midsagittal skull outline.

ob. **Obelion** is an ectocranial midline point where a line connecting the parietal foramina (when present) intersects the midline.

l. **Lambda** is the ectocranial midline point where the sagittal and lambdoidal sutures intersect. In cases such as the one illustrated in Figure 4.12, extrasutural bones make placement of this point difficult. When in doubt, choose the point where the lateral halves of the lambdoidal suture and the lower end of the sagittal suture would be projected to meet.

op. **Opisthocranion** is an instrumentally determined point at the rear of the cranium. It is defined as the midline ectocranial point at the farthest chord length from glabella.

i. **Inion** is an ectocranial midline point at the base of the external occipital protuberance. The bony anatomy in this region is highly variable, with crests, lumps, or points of bone possible. Normally, inion is defined as the point at which the superior nuchal lines merge in the external occipital protuberance.

o. **Opisthion** is the midline point at the posterior margin of the foramen magnum.

ba. **Basion** is the midline point on the anterior margin of the foramen magnum. For cranial height measurements, the point is placed on the anteroinferior portion of the foramen's rim. For basinasal and basiprosthion measures, the point is located on the most posterior point of the foramen's anterior rim and is sometimes distinguished as **endobasion.**

sphba. **Sphenobasion** is the point where the midsagittal plane intersects the sphenooccipital suture. This point has been obliterated by synchrondrosis on the specimen illustrated in Figure 4.13.

ho. **Hormion** is the most posterior midline point on the vomer.

alv. **Alveolon** is the point on the interpalatal suture where a line drawn between the posterior ends of the alveolar ridges crosses the midline.

sta. **Staphylion** is the point on the interpalatal suture where a line drawn between the deepest parts of the notches (free edges) at the rear of the palate crosses the midline.

ol. **Orale** is the midline point on the hard palate where a line drawn tangent to the posterior margins of the central incisor alveoli crosses the midline.

gn. **Gnathion** is the most inferior midline point on the mandible.

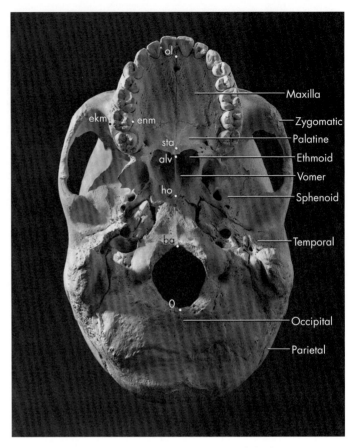

Figure 4.13 Bones and osteometric points of the human cranium, inferior. One-half natural size.

pg. **Pogonion** is the most anterior midline point on the chin of the mandible.

id. **Infradentale** is the midline point at the superior tip of the septum between the mandibular central incisors.

4.5.2 Paired Cranial Osteometric Points

zm. **Zygomaxillare** is the most inferior point on the zygomaticomaxillary suture.

al. **Alare** is instrumentally determined as the most lateral point on the margin of the nasal aperture.

or. **Orbitale** is the lowest point on the orbital margin.

zy. **Zygion** is the instrumentally determined point of maximum lateral extent of the lateral surface of the zygomatic arch.

ju. **Jugale** is the point in the depth of the notch between the temporal and frontal processes of the zygomatic.

ek. **Ectoconchion** is instrumentally determined as the most lateral point on the orbital margin.

mf. **Maxillofrontale** is the point where the anterior lacrimal crest of the maxilla meets the frontomaxillary suture.

la. **Lacrimale** is the point where the posterior lacrimal crest meets the frontolacrimal suture.

d. **Dacryon** is the point where the lacrimomaxillary suture meets the frontal bone.

zyo. **Zygoorbitale** is the point where the orbital rim intersects the zygomaticomaxillary suture.

fmo. **Frontomalare orbitale** is the point where the frontozygomatic suture crosses the inner orbital rim.

fmt. **Frontomalare temporale** is the point where the frontozygomatic suture crosses the temporal line (or outer orbital rim).

ft. **Frontotemporale** is the point where the temporal line reaches its most anteromedial position on the frontal.

st. **Stephanion** is the point where the coronal suture crosses the temporal line.

pt. **Pterion** is a region, rather than a point, where the frontal, temporal, parietal, and sphenoid meet on the side of the vault. The sutural contact pattern in this area is highly variable.

co. **Coronale** is the point on the coronal suture where the breadth of the frontal bone is greatest.

eu. **Euryon** is the instrumentally determined ectocranial point of greatest cranial breadth.

po. **Porion** is the uppermost point on the margin of the external auditory meatus.

au. **Auriculare** is a point vertically above the center of the external auditory meatus at the root of the zygomatic process, a few millimeters above porion.

ast. **Asterion** is the point where the lambdoidal, parietomastoid, and occipitomastoid sutures meet.

ms. **Mastoidale** is the most inferior point on the mastoid process.

ekm. **Ectomolare** is the most lateral point on the outer surface of the alveolar margins of the maxilla, often at the second molar position.

enm. **Endomolare** is the most medial point on the inner surface of the alveolar margin opposite the center of the M^2 crown.

cdl. **Condylion laterale** is the most lateral point on the mandibular condyle.

cdm. **Condylion mediale** is the most medial point on the mandibular condyle.

cr. **Coronion** is the point at the tip of the coronoid process of the mandible.

go. **Gonion** is a point along the rounded posteroinferior corner of the mandible between the ramus and the body. To determine the point, imagine extending the posterior ramus border and the inferior corpus border to form an obtuse angle. The line bisecting this angle meets the curved gonial edge at gonion.

ml. **Mentale** is the most inferior point on the margin of the mandibular mental foramen.

4.6
Learning Cranial Skeletal Anatomy

To learn cranial skeletal anatomy, approach the skull systematically. First, study the skull of a young adult, observing all sutures between the bones. Then, study a growth series, noting how each bone and suture changes during ontogeny. Finally, use the descriptions in sections 4.7–4.21 to learn the features of each bone of the skull. These features are the keys to diagnosing various bones of the skull by element and side. Intact cranial bones have rather unique morphologies. You will rarely have difficulty identifying them.

Because the various bones of the skull are often found in a disarticulated or fragmentary state, each bone or pair of bones must be given individual consideration. To identify and side fragments of the cranium, follow these steps: Determine whether the piece is cranial vault or face. Note any blood vessel impressions, sutures, foramina, surface textures, bone thickness changes, muscle attachments, sinus walls, or tooth roots or sockets. Note the thickness of the piece and its cross-sectional anatomy at the break, including sinus development. Carefully note the morphology of any visible sutures.

4.7
Frontal (Figures 4.14–4.17)

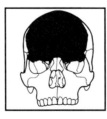 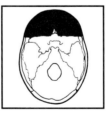

4.7.1 Anatomy

The frontal is located at the front of the neurocranium. It articulates with the parietals, nasals, maxillae, sphenoid, ethmoid, lacrimals, and zygomatics. The frontal is one of the largest and most robust cranial bones. It consists of two general parts, one vertical and one horizontal.

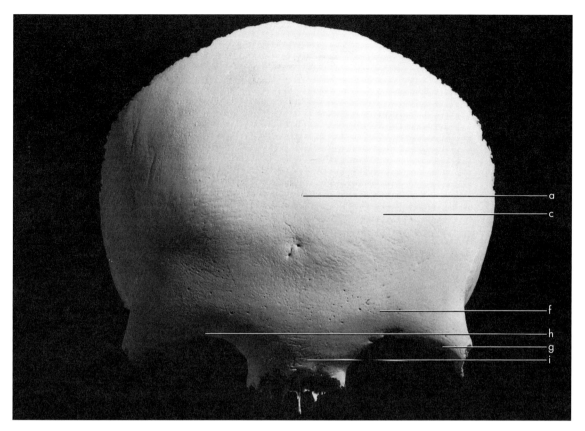

Figure 4.14 **Frontal, anterior** (ectocranial). Natural size. Key: a, frontal squama; c, frontal eminence (frontal tuber, or boss); f, superciliary arch; g, supraorbital margin; h, supraorbital notch (foramen); i, metopic suture (frontal suture).

a. The vertical **frontal squama** forms the forehead.

b. The **horizontal portion** acts to roof the orbits and to floor the frontal lobes of the brain.

c. The **frontal eminences** (**frontal tubers,** or **bosses**) dominate the ectocranial surface. These paired frontal bosses mark the location of the original centers of ossification of this bone.

d. The **temporal lines** on the lateral ectocranial surface mark the attachment of the *temporalis muscle,* a major elevator of the mandible, and its covering, the *temporal fascia,* a fascial sheet that covers the temporalis. The temporal line defines the superior edge of the temporal surface (and fossa). This line becomes a crest in its anterior, lateral extent (on the zygomatic process of the frontal). It often divides into superior and inferior lines as it sweeps posteriorly.

e. The **zygomatic processes** form the most lateral and anterior corners of the frontal.

f. The **superciliary arches** (brow ridges) are the bony tori over the orbits. They are most prominent in males and are sometimes joined by a prominent glabellar region.

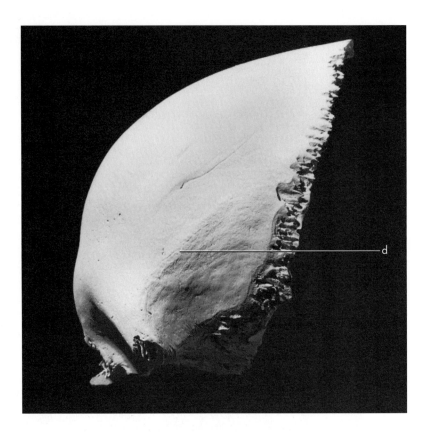

Figure 4.15 **Frontal, left lateral** (ectocranial). Natural size. Key: d, temporal line.

g. The **supraorbital margins** are the upper orbital edges. These are notched or pierced by the supraorbital notch or foramen.

h. The **supraorbital notches** (or **foramina,** if the notches are bridged) are set along the medial half of the superior orbital rim. They transmit the *supraorbital vessels* and *supraorbital nerve* as they pass superiorly to the forehead region.

i. The **metopic suture (frontal suture)** is a vertical suture between right and left frontal halves. Its persistence is variable, but only occasionally does it last into adulthood. Traces of it are most often observed in the glabellar region in adults.

j. **Meningeal grooves** for the *middle meningeal arteries* are present on both sides of the concave endocranial surface of the frontal squama. The *brain* is covered with a tough outer protective membrane, the *dura mater,* whose blood supply comes from the *meningeal arteries*.

k. The **sagittal sulcus** is a vertical groove that runs down the midline of the endocranial surface. It lodges the *superior sagittal sinus,* a large vessel that drains blood from the brain.

l. The **frontal crest** is a midline crest confluent with the anterior end of the sagittal sulcus. This crest gives attachment to the *falx cerebri,* a

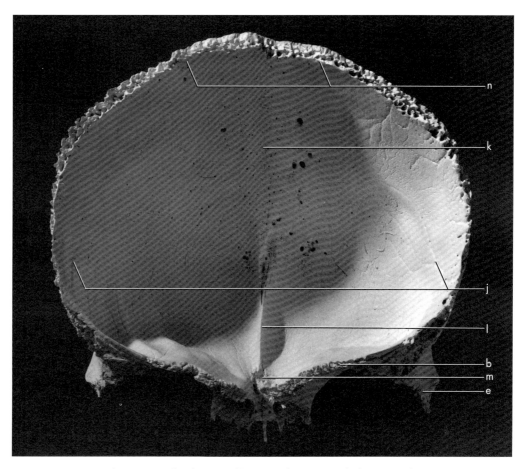

Figure 4.16 **Frontal, posterior** (endocranial). Natural size. Key: b, horizontal portion; e, zygomatic process; j, meningeal grooves; k, sagittal sulcus; l, frontal crest; m, foramen cecum; n, arachnoid foreae (granular foveae).

Figure 4.17 **Frontal, inferior.** Natural size. Key: o, pars orbitalis or orbital plate; p, lacrimal fossa; q, ethmoidal notch; r, frontal sinuses.

strong membrane between the two cerebral hemispheres of the brain.

m. The **foramen cecum,** a foramen of varying size, is found at the root, or base, of the frontal crest and transmits a small vein from the frontal sinus to the superior sagittal sinus.

n. The **arachnoid foveae (granular foveae)** are especially apparent near the coronal suture along the endocranial midline. They are features associated with another covering layer of the brain, the *arachnoid,* which is a delicate, avascular membrane lying beneath the *dura mater.* Tufts of arachnoid, the *arachnoid granulations,* push outward against the dura, causing resorption of the bone and the formation of foveae on the endocranial surface. On both sides of the midline the frontal's endocranial surface bears depressions for convolutions of the frontal lobes of the brain.

o. The **pars orbitalis,** or **orbital plates,** are the horizontal portions of the frontal. Their endocranial surfaces are undulating (bumpy), conforming to the inferior surface of the frontal lobes. Their inferior surfaces (orbital surfaces) are smoother and concave.

p. The **lacrimal fossae,** for the *lacrimal glands,* are found at the lateral, inferior parts of the frontal's orbital (inferior) surfaces.

q. The **ethmoidal notch** is the gap separating the two orbital plates of the frontal. The ethmoid bone fills this notch in the articulated cranium.

r. The **frontal sinuses,** anterior to the ethmoid notch, extend for a variable distance between the outer and inner bone tables of the frontal and sometimes penetrate the orbital plates. Personal identification in forensic cases has been accomplished by employing radiographs and using distinctive patterns of the frontal sinuses for individuation.

4.7.2 Growth

The frontal ossifies on membrane from two primary centers. At birth these centers are separate. They usually fuse along the metopic suture (midline) during the second year of childhood.

4.7.3 Possible Confusion

When fragmentary, the frontal is most often confused with the parietals.

• The meningeal impressions are larger and more dense on the parietals, and the endocranial surfaces of the parietals are less undulating than those of the frontal.

• The frontal is the only major vault bone with a substantial sinus and adjacent orbital rims.

4.7.4 Siding

Isolated fragments of frontal squama may be difficult to side. Siding the frontal, or any other bone or tooth, whether fragmentary or intact, is often simplified by holding the element in its correct orientation adjacent to that region of your own skull. In other words, attempt to imagine the fragment fitting into your own anatomy.

- The coronal suture is posterior and courses anterolaterally, toward the face, from bregma. This means that the sagittal and coronal sutures do not meet at right angles. This fact can be very useful in siding fragments of frontal squama.
- The anteromedially placed frontal sinus is often exposed in broken pieces.
- The ectocranially placed temporal lines swing medially and weaken posteriorly.

4.8 Parietals (Figures 4.18–4.19)

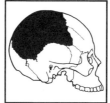

4.8.1 Anatomy

The parietals form the sides and roof of the cranial vault. Each parietal articulates with the opposite parietal and with the frontal, temporal, occipital, and sphenoid. Parietals are basically square and are the largest bones of the vault, with a fairly uniform thickness.

a. The **frontal angle** is located at bregma.

b. The **sphenoidal angle** is located at pterion.

c. The **occipital angle** is located at lambda.

d. The **mastoid angle** is located at asterion.

e. The **parietal tuber** (**boss, eminence**) is the large, rounded eminence centered on the ectocranial surface of the parietal. It marks the center of ossification of the bone.

f. The **temporal lines** dominate the ectocranial surface, passing anteroposteriorly.

g. The **superior temporal line** anchors the *temporal fascia.*

h. The **inferior temporal line** indicates the most superior reach of the *temporalis muscle.*

i. When present, the **parietal foramen** is located close to the sagittal suture near lambda. It transmits a small vein through the parietal to the superior sagittal sinus.

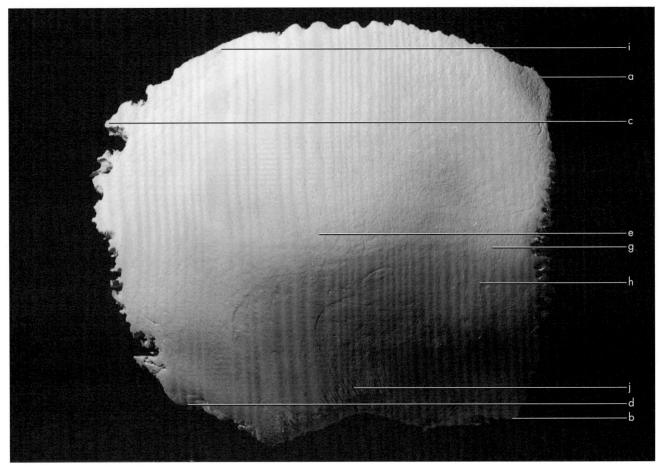

Figure 4.18 **Right parietal, lateral** (ectocranial). Anterior is toward the right, superior is up. Natural size. See Figure 4.1 for more strongly expressed temporal lines. Key: a, frontal angle; b, sphenoidal angle; c, occipital angle; d, mastoid angle; e, parietal tuber (boss, eminence); g, superior temporal line; h, inferior temporal line; i, parietal foramen; j, parietal striae.

j. The **parietal striae** are striations, or "rays," that pass posterosuperiorly for some distance on the ectocranial surface of the parietal from their origin on its beveled squamosal edge.

k. The **meningeal grooves** for *middle meningeal arteries* dominate the endocranial surface of the parietal. These arteries supply the *dura mater.* The most anterior branch parallels the coronal edge of the parietal, and most of the branches traverse the bone toward its occipital angle.

l. The **sagittal sulcus** is made when the parietals are articulated and the shallow grooves along the sagittal edge of each parietal combine along the endocranial midline. This sulcus is a posterior continuation of the same feature on the frontal.

m. The **arachnoid foveae (granular foveae)** are concentrated endocranially along the anterior extent of the sagittal edge of each parietal. They are functionally equivalent to structures of the same name described for the frontal (section 4.7.1n).

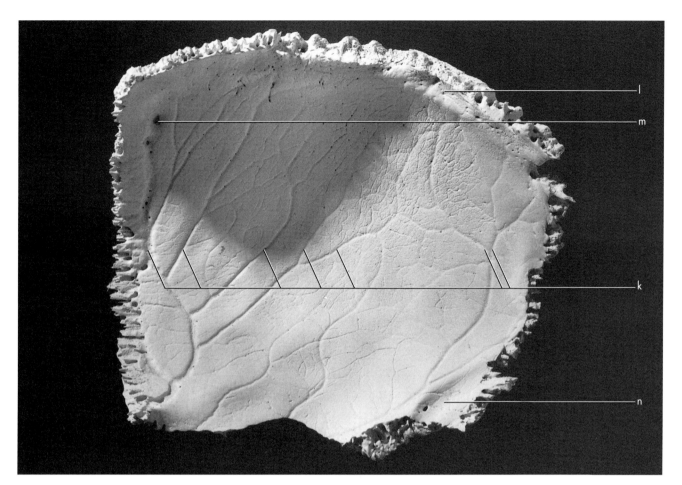

Figure 4.19 **Right parietal, medial** (endocranial). Anterior is toward the left, superior is up. Natural size. Key: k, meningeal grooves; l, sagittal sulcus; m, arachnoid foveae (granular foveae); n, sigmoid (transverse) sulcus.

n. The **sigmoid (transverse) sulcus** crosses the mastoid angle of the parietal, cutting a groove on the endocranial surface. It marks the course of the *transverse (sigmoid) sinus,* a vessel that drains blood from the brain.

4.8.2 Growth

The parietal bone ossifies in membrane, with ossification extending radially from a combined center near the parietal boss. The four corners of the bone are not ossified at birth, and the remaining spaces are the fontanelles.

4.8.3 Possible Confusion

When fragmentary, the parietal is most often confused with the frontal, occipital, or temporal.

- Neither the frontal, occipital, nor temporal has parietal foramina, as many meningeal grooves, or ectocranial striae associated with an externally facing beveled suture (the squamosal).
- The cross section of a parietal is more regular (thickness does not vary so much) than that of the other vault bones.
- The temporal line is a constant feature across the length of the parietal.
- The endocranial surface of this bone is not as undulating and irregular as that of the frontal or occipital.

4.8.4 Siding

Siding is difficult only when the parietal bone is very fragmentary.

- The meningeal grooves are oriented vertically along the coronal suture and more horizontally near the squamosal suture.
- The coronal suture, unlike the sagittal suture, is an interfingering rather than an interlocking, zipperlike, or jigsawlike articulation. The large *anterior middle meningeal vessel* makes an impression along this suture endocranially.
- The thickest corners are the occipital and mastoid angles.
- The mastoid angle bears a sulcus endocranially.
- The squamosal suture is lateral and inferior and the parietal striae angle posterosuperiorly.

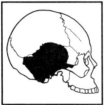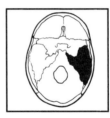

4.9
Temporals (Figures 4.20–4.23)

4.9.1 Anatomy

The temporals form the transition between cranial wall and base, house the delicate organs of hearing, and form the upper surface of the jaw joints. The highly irregular shape of the temporal is related to the bone's varying functions. The temporal articulates with the parietal, occipital, sphenoid, zygomatic, and mandible. The jaw joint, or temporomandibular joint, is often abbreviated TMJ. Parts of the temporal bone are very robust and for this reason are often more resistant to destruction than other parts of the cranial vault.

a. The thin, platelike **squama** rises almost vertically to form the cranial walls and articulate with the parietals along the squamosal suture.

b. The **petrous pyramid** is the massive, dense bony part that dominates the endocranial aspect of the temporal. The sharp superior edge of the endocranial petrous surface angles anteromedially, separating the *temporal* and *occipital lobes* of the brain and housing the internal ear. The petrous is wedged between the occipital and sphenoid. The

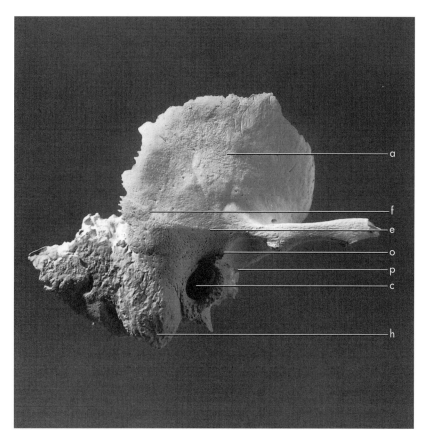

Figure 4.20 **Right temporal, lateral** (ectocranial). Anterior is toward the right, superior is up. Natural size. Key: a, squama; c, external acoustic meatus; e, suprameatal crest; f, supramastoid crest; h, mastoid process; o, postglenoid process; p, entoglenoid process.

end-on view (view from anteromedial) is into the carotid canal (see section 4.9.lv). This petrous part of the bone houses the delicate organs of hearing and equilibrium, including the tiny movable malleus, incus, and stapes bones.

c. The **external acoustic (auditory) meatus (EAM)** is the external opening of the ear canal which passes anteromedially for about 2 cm. The inner end of the canal is closed by the *tympanic membrane* (eardrum) in the living individual.

d. The **zygomatic process** of the temporal is a thin projection of bone which forms the posterior half of the zygomatic arch. Its anterior edge is the serrated zygomaticotemporal suture, its superior edge is an attachment for the *temporal fascia,* and its inferior edge anchors fibers of the *masseter muscle.*

e. The **suprameatal crest** is the superior root of the zygomatic process. It runs horizontally above the EAM where the osteometric point auriculare is located.

f. The **supramastoid crest** is the posterior extension of the suprameatal crest. The continuous raised edge of these crests marks the limit of the *temporalis muscle* and *temporal fascia* attachment.

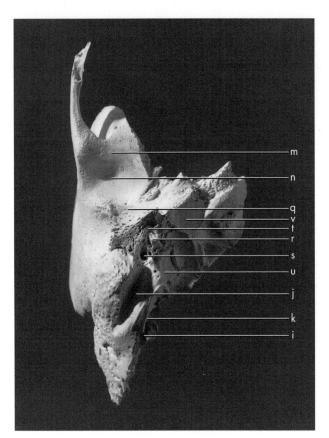

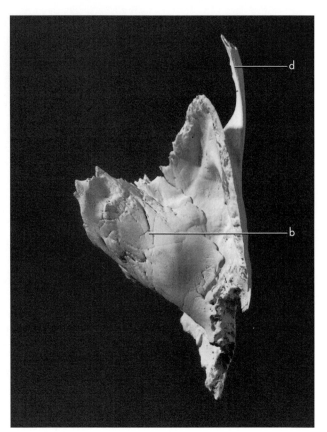

Figure 4.21 **Right temporal, inferior.** Anterior is up, medial is toward the right. Natural size. Key: i, mastoid foramen; j, mastoid notch (digastric groove); k, occipital groove (sulcus); m, articular eminence; n, mandibular fossa (glenoid fossa); q, tympanic; r, styloid process; s, stylomastoid foramen; t, vaginal process; u, jugular fossa; v, carotid canal.

Figure 4.22 **Right temporal, superior.** Anterior is up, medial is toward the left. Natural size. Key: b, petrous pyramid; d, zygomatic process.

g. The **parietal notch** forms on the posterosuperior border of the temporal where the squamosal and parietomastoid sutures meet.

h. The **mastoid process** bears an external surface that is roughened for the attachment of several muscles including the following: *sternocleidomastoideus, splenius capitis*, and *longissimus capitis*. These muscles function in extension and rotation of the head. The *temporalis* muscle may also attach in this region when the supramastoid crest is present on the mastoid area, as in some humans and many fossil hominids. Internally, the thin-walled mastoid process is occupied by a number of variably developed voids known as **mastoid cells.**

i. The **mastoid foramen** (occasionally multiple) is located near the posterior edge of the mastoid process along the occipitomastoid suture. It transmits a small branch of the *occipital artery*, which supplies the *dura mater*, the diploë (spongy bone sandwiched between inner and outer bone tables of cranial vault bones), and the mastoid air cells.

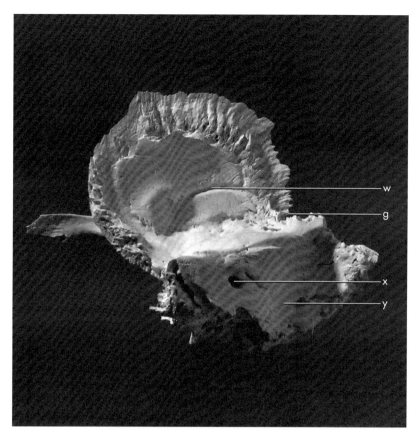

Figure 4.23 **Right temporal, medial** (endocranial). Anterior is toward the left, superior is up. Natural size. Key: g, parietal notch; w, middle meningeal grooves; x, internal acoustic (auditory) meatus; y, sigmoid sulcus.

j. The **mastoid notch (digastric groove)** for the attachment of the *digastric muscle* is the vertically oriented furrow medial to the mastoid process.

k. The **occipital groove (sulcus)** lies just medial to the mastoid notch. It is a shallow furrow that lodges the *occipital artery*.

l. The **temporomandibular articular surface** is the smooth, articular surface inferior to the root of the zygomatic process. There is considerable topographic relief to this inferiorly facing surface.

m. The **articular eminence** forms the anterior portion of the temporomandibular articular surface.

n. The **mandibular fossa (glenoid fossa)** lies posterosuperior to the articular eminence. The eminence and the fossa itself are bounded medially by the sphenosquamosal suture. In chewing, the condyle of the mandible moves anteriorly onto the eminence and posteriorly into the fossa as well as from side to side in actions at the TMJ. In life there is a fibrocartilaginous *articular disk* interposed between the mandibular condyle and the fossa.

o. The **postglenoid process** is a projection that lies just anterosuperior to the EAM, interposed between the **tympanic part** of the bone (which forms most of the rim of the EAM) and the mandibular fossa. This rim is roughened for the attachment of the cartilaginous part of the EAM.

p. The **entoglenoid process** is the inferior projection of the articular surface at the medial edge of the articular eminence.

q. The **tympanic** part of the temporal lies posterior to the TMJ. Its anterior surface, forming the rear wall of the mandibular fossa, is nonarticular.

r. The **styloid process** is a thin, pointed bony rod that points anteroinferiorly from the base of the temporal bone. It is a slender projection of variable length and is fragile and often broken or missing (as on the illustrated specimen, where its distal end has snapped off). It anchors the *stylohyoid* (sometimes partly ossified) *ligament* and several small muscles.

s. The **stylomastoid foramen,** located immediately posterior to the base of the styloid process, is for the exit of the *facial nerve* and the entrance of the *stylomastoid artery.*

t. The **vaginal process** ensheaths the base of the styloid process.

u. The **jugular fossa** is located just medial to the base of the styloid process. This deep fossa houses the *bulb of the internal jugular vein,* a vessel that drains blood from the head and neck.

v. The **carotid canal** is a large circular canal that transmits the *internal carotid artery,* a major source of blood for the head, and the *carotid plexus* of nerves. It is situated medial to the styloid process at the level of the sphenosquamosal suture, just anterior to the jugular fossa.

w. The **middle meningeal grooves** mark the endocranial surface of the temporal. Undulations on this surface are related to convolutions of the *temporal lobe* of the brain.

x. The **internal acoustic (auditory) meatus** is located about midway along the posterior surface of the petrous pyramid and transmits the *facial* and *acoustic nerves* (cranial nerves 7 and 8, respectively) as well as the *internal auditory artery.*

y. The **sigmoid sulcus** is the large, curving groove set at the posterior base of the petrous pyramid on the endocranial surface of the mastoid part of the temporal bone. This sulcus houses the *sigmoid sinus,* an anteroinferior extension of the *transverse sinus,* which is a major vessel draining blood from the brain into the *jugular vein.* Note the continuation of this sulcus onto the posteroinferior corner of the parietal.

4.9.2 Growth

Growth of the temporal is complex, with both membranous and endochondral ossification. It ossifies from eight centers during fetal development, not counting those of the middle ear and tympanic ring. As birth approaches, only three main centers remain: the squama, the petromastoid, and the tympanic ring.

4.9.3 Possible Confusion

Even when fragmentary, it is difficult to confuse the temporal with other bones because of its unique morphology.

- The broken elements that may present some trouble are the squama and zygomatic process.
- The temporal squama overlaps the parietal and is thinner than the parietal or frontal.
- Fragmentary temporal processes of the zygomatic bone are not as thin and long as the zygomatic process of the temporal.

4.9.4 Siding

- In isolated mastoid sections, the mastoid tip points inferiorly and the entire mastoid angles anteriorly.
- The digastric groove is posterior and medial.
- For isolated petrous pyramids, the internal acoustic foramen is posterior and the pyramid tapers anteromedially.
- For isolated fragments of squama, the squamosal suture surface overlaps the parietal's ectocranial surface. Grooves for the middle meningeals branch posteriorly and slightly superiorly.
- For broken zygomatic processes, the articular eminence is posterior, the superior edge of the arch is thinnest, and the zygomaticotemporal suture runs from posteroinferior to anterosuperior.

4.10
Auditory Ossicles (Figure 4.24)

The tiny ear ossicles, the **malleus** (hammer), **incus** (anvil), and **stapes** (stirrup), are housed in the tympanic cavity of the temporal. The first is connected to the *tympanic membrane*, or *eardrum*, and the others are located more medially. These bones are so small that they are best observed under

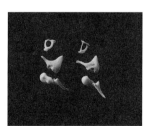

Figure 4.24 **Auditory ossicles** from both right and left sides: *top*, stapes; *middle*, incus; *bottom*, malleus. Natural size.

magnification. They are often lost during skeletal recovery and are seldom studied.

4.11
Occipital (Figures 4.25–4.26)

4.11.1 Anatomy

The occipital bone is set at the rear of the cranium and articulates with the temporals, sphenoid, parietals, and the uppermost vertebra, the atlas.

a. The **foramen magnum** is the large hole in the occipital through which the *brainstem* passes inferiorly into the vertebral canal.

b. The **squamous** portion of the occipital bone is by far the largest, constituting the large plate of bone posterior and superior to the foramen magnum.

c. The **occipital planum** (c-1) is that part of the occipital squama which lies above the superior nuchal lines (see 4.11.1e). The section of squama inferior to the lines is the **nuchal planum** (c-2).

d. The **external occipital protuberance** lies on the ectocranial midline where the occipital and nuchal planes meet. It is highly variable in appearance and heavier and more prominent in male individuals.

e. The **superior nuchal lines** lie to either side of the midline on the ectocranial surface of the squamous portion. The nuchal plane and occipital planes merge at these superiorly convex lines. Several *nuchal muscles* attach to and below these lines and function to extend and rotate the head.

f. The **inferior nuchal lines** parallel the superior lines but are located about midway on the ectocranial nuchal plane. Fascia separating nuchal muscles attach to the line, whereas additional nuchal muscles attach inferior to this line.

g. The **external occipital crest (median nuchal line)** is a highly variable median line or crest that passes between the right and left *nuchal musculature*. It stretches from the external occipital protuberance to the rear of the foramen magnum, anchoring the *nuchal ligament*.

h. The **basilar** part is the thick, square projection anterior to the foramen magnum. This part articulates with the petrous portions of both temporals and with the sphenoid via the sphenooccipital suture.

i. The **lateral,** or **condylar,** parts of the occipital lie to either side of the foramen magnum, articulating with the temporals.

j. The **occipital condyles** are raised oval structures on either side of the foramen magnum. Their inferior surfaces are convex. The articular

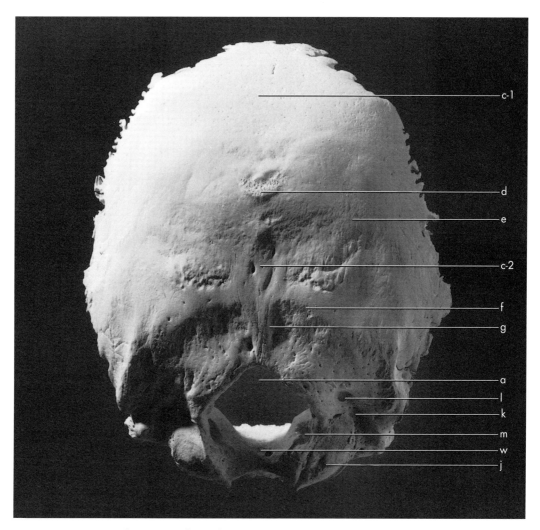

Figure 4.25 **Occipital, posteroinferior** (ectocranial). Superior is up. Natural size. Key: a, foramen magnum; c-1, occipital planum; c-2, nuchal planum; d, external occipital protuberance; e, superior nuchal line; f, inferior nuchal line; g, external occipital crest (median nuchal line); j, occipital condyle; k, condylar fossa; l, condylar foramen (canal); m, hypoglossal canal; w, groove for the medulla oblongata.

surfaces of these condyles fit into the concave facets of the atlas vertebra.

k. The **condylar fossae** are the ectocranial depressions immediately posterior to the condyles. These fossae receive the posterior margin of the superior facet of the atlas vertebra when the head is bent backward.

l. **Condylar foramina** (**canals**) perforate the occipital at the depth of the condylar fossae, where each transmits an *emissary vein.*

m. The **hypoglossal canals** are tunnels through the anterior part of the base (therefore superior in placement) of each condyle. These canals give exit to the *hypoglossal nerves* (cranial nerve 12) and entrance to arteries.

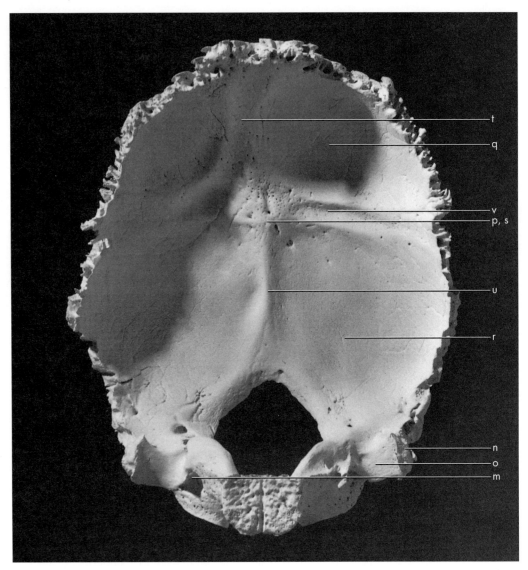

Figure 4.26 **Occipital, anterior** (endocranial). Superior is up. Natural size. Key: m, hypoglossal canal; n, jugular process; o, jugular notch; p, cruciform eminence; q, cerebral fossa; r, cerebellar fossa; s, internal occipital protuberance; t, occipital (sagittal) sulcus; u, internal occipital crest; v, transverse sulcus.

n. The **jugular processes** are laterally directed corners of the bone placed lateral to the condyles. The tips of these processes lie at the anteriormost point along the occipitomastoid suture.

o. The **jugular notch** is excavated into the anterior surface of the jugular process. This notch forms the posterior half of the jugular foramen in the articulated cranium, the anterior half being contributed by the temporal bone (section 4.9.1u).

p. The **cruciform eminence** divides the endocranial surface of the occipital squama into four fossae. It is so named because it is cross-shaped.

q. The **cerebral fossae** are triangular depressions below the lambdoidal suture on the endocranial surface of the occipital. They house the *occipital lobes* of the brain's *cerebrum.*

r. The **cerebellar fossae** occupy the inferior part of the endocranial surface of the occipital squama. Therein rest the *cerebellar lobes* of the brain.

s. The **internal occipital protuberance** lies at the center of the cruciform eminence.

t. The **occipital (sagittal) sulcus** passes superiorly from the internal occipital protuberance. It is a deep endocranial groove marking the posterior extension of the *sagittal sinus,* a major blood drainage pathway from the brain (see sections 4.7.1k and 4.8.1l).

u. The **internal occipital crest** is the inferior arm of the cruciform eminence. Sometimes it bears a sulcus which continues on one or both sides of the foramen magnum. Such a sulcus, called an **occipitomarginal sulcus,** represents an alternative pathway for blood to drain from the brain.

v. The **transverse sulci** form the transverse arms of the cruciform eminence. They house the *transverse sinuses.* The one on the right is usually larger and directly communicates with the sagittal sulcus. However, variations in the soft tissue and bony manifestations of this cranial venous drainage system are common and sometimes pronounced. The transverse sulcus of the occipital connects with the sigmoid sulcus of the temporal and endocranial jugular process, often via the transverse (sigmoid) sulcus on the mastoid corner of the parietal.

w. The **groove for the medulla oblongata** is the hollowing on the endocranial surface of the basilar part of the occipital, the clivus.

4.11.2 Growth

The occipital is another bone with both membranous and endochondral ossification. At birth the occipital consists of four parts, the squama, the lateral parts that bear the condyles, and the basilar part. The squama and lateral portions unite at about age 4, and by age 6 the basilar part attaches to these. The synostosis (fusion) between the occipital and sphenoid (across the sphenooccipital synchondrosis) normally takes place between 18 and 25 years of age.

4.11.3 Possible Confusion

Even when fragmentary, the occipital is difficult to confuse with other bones of the vault.

• There is wide variation in thickness across the occipital squama that is not found in the parietals or frontal.

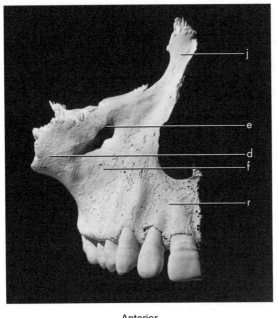

Anterior

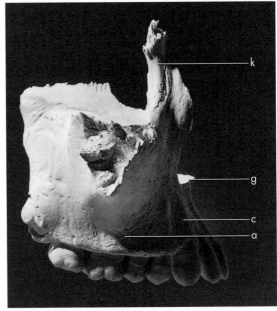

Posterior

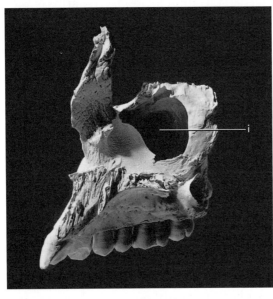

Medial

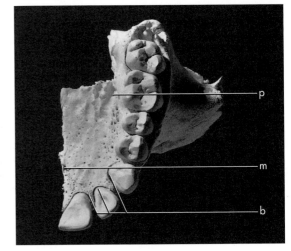

Lateral

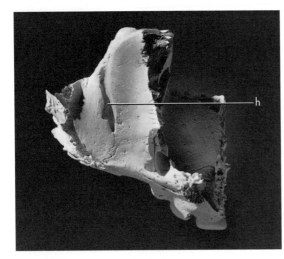

Superior

Inferior

- The occipital lacks meningeal grooves endocranially and has much more ectocranial rugosity than seen on the parietal or frontal.

- Other possible confusion may come when extrasutural bones are encountered. These bones are sometimes quite large, particularly along the lambdoidal suture. One even has a name, the **inca bone.** This is a large, triangular, symmetrical bone placed at the top of the occipital, just below lambda.

4.11.4 Siding

Isolated fragments of occipital are easily sided by locating the lambdoidal suture.

- For isolated condyles, the edge of the foramen magnum is medial and somewhat posterior to the condylar body centers.

- The condylar fossa is posterior, and the hypoglossal canals tunnel from anterolateral to posteromedial.

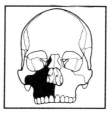
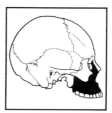

4.12
Maxillae (Figure 4.27)

4.12.1 Anatomy

The maxillae are a pair of bones that form the dominant portion of the face. Functionally, the maxillae hold the tooth roots and form most of the nasal aperture and floor, most of the hard palate, and the floors of the orbits. Most of the maxillary bone is light and fragile, the exception being the portion that holds the teeth. The maxillae comprise four basic processes and articulate with each other and with the frontal, nasals, lacrimals, ethmoid, inferior nasal conchae, palatines, vomer, zygomatics, and sphenoid.

a. The **alveolar process** is the horizontal portion of the maxilla that holds the tooth roots.

b. **Alveoli** for the tooth roots are present all along the alveolar process, except where these have been resorbed following the loss of teeth.

c. The **canine jugum** is a bony eminence over the maxillary canine root on the facial surface of the maxilla.

d. The **zygomatic process** forms much of the cheek.

e. The **infraorbital foramen** is located below the inferior orbital rim on the facial surface and transmits the *infraorbital nerve* and *vessels* to the face.

Figure 4.27 **Right maxilla.** Natural size. Key: a, alveolar process; b, alveoli; c, canine jugum; d, zygomatic process; e, infraorbital foramen; f, canine fossa; g, anterior nasal spine; h, infraorbital sulcus (groove); i, maxillary sinus; j, frontal process; k, anterior lacrimal crest; l, palatine process; m, incisive foramen; p, greater palatine groove; r, nasoalveolar clivus.

f. The **canine fossa** is a hollow of variable extent located on the facial surface just below the infraorbital foramen, where the zygomatic, frontal, and alveolar processes of the maxilla come together.

g. The **anterior nasal spine** is the thin projection of bone on the midline at the inferior margin of the nasal aperture.

h. The **infraorbital sulcus (groove)** is centered on the posterior half of the orbital floor and opens posterosuperiorly. It connects anteroinferiorly with the infraorbital foramen via the **infraorbital canal.**

i. The **maxillary sinus** is the large void superior to the alveolar process and inferior to the orbital floor.

j. The **frontal process** rises to articulate with the frontal, nasals, lacrimal, and ethmoid.

k. The **anterior lacrimal crest** is a vertical crest located on the lateral aspect of the frontal process of the maxilla. The maxilla combines with the lacrimal bone to form the **lacrimal groove** and **canal.** This canal houses the *nasolacrimal duct*, which drains inferiorly into the nasal cavity.

l. The **palatine process** forms the anterior two-thirds of the hard palate and floor of the nasal cavity.

m. The **incisive foramen** perforates the anterior hard palate at the midline.

n. The **incisive canal** is bilobate, opening via the incisive foramen, with each lobe enclosed by one of the maxillae. Each lobe of the canal transmits the *terminal branch of the greater palatine artery* and the *nasopalatine nerve*.

o. The **premaxillary suture** is sometimes seen in the wall of the incisive canal and on the adjacent palatal surface, particularly in young individuals.

p. The **greater palatine groove** at the rear of the hard palate marks the junction of the palatine and alveolar processes. This groove is for the *greater palatine vessels* and *nerve*.

q. The **maxillary tuber** is the rugose surface at the posterior end of the alveolar process. It is variable in expression, articulating with the pyramidal process of the palatine and sometimes with the lateral pterygoid plate of the sphenoid.

r. The **nasoalveolar clivus** is the surface between the canine jugae, the base of the piriform aperture, and the alveolar margin.

4.12.2 Growth

Each maxilla ossifies from two main combined centers, one for the maxilla proper and one for the premaxilla. These fuse early in human development, about the ninth week *in utero*, but the suture between them may persist into adulthood in the region adjacent to the incisive canal.

4.12.3 Possible Confusion

Small fragments of maxilla might be confused with other cranial bones. Because the bone is complex, it is helpful to note the diagnostic features useful in identifying it. These include the alveolar region, the sharp edges of the nasal aperture, the edge of the lacrimal canal, the large maxillary sinus, and the unique, serrated intermaxillary suture.

4.12.4 Siding

Fragments of maxilla may prove difficult to side, and the use of comparative specimens may prove necessary.

- For a broken frontal process, the thinner edge is anterior and medial, the medial surface is vascularized (perforated by blood vessels), and the anterior lacrimal crest is lateral.

- For any segment with alveolar bone preserved, the tooth roots can be used as a guide to medial, lateral, anterior, and posterior.

4.13
Palatines (Figures 4.28–4.29)

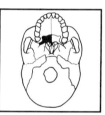

4.13.1 Anatomy

The small, delicate, L-shaped palatine bones form the rear of the hard palate and part of the wall and floor of the nasal cavity. Individual palatine bones are almost never found in an isolated, intact state; they generally accompany the maxillae and sphenoid, to which they are tightly bound. In addition to these two, the palatines articulate with the vomer, inferior nasal conchae, and ethmoid, and with each other.

a. The **horizontal plate** of the palatine forms the posterior third of the hard palate.

b. The **greater palatine foramen (canal)** perforates the rear corner of the hard palate and is formed as the alveolar process of the maxilla meets the horizontal plate of the palatine.

c. The two halves of the **pterygopalatine canal** become visible, sweeping posterosuperiorly, when the maxilla and perpendicular plate of the palatine are disarticulated. This canal transmits the *greater palatine vessels* and *nerve.*

d. The **posterior nasal spine** is located on the superior surface of the horizontal plate. The superior, or nasal-cavity, surface of the plate is smoother and more regular than the palatal surface.

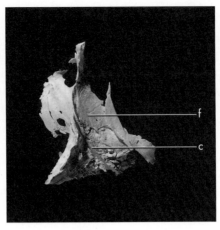
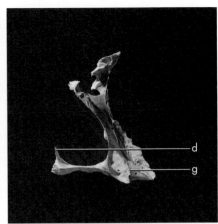

Medial, anterior is toward the left

Lateral, anterior is toward the right

Posterior, lateral is toward the right

Figure 4.28 **Right palatine.** Superior is up. Natural size. Key: c, pterygopalatine canal; d, posterior nasal spine; f, perpendicular (vertical) plate; g, pyramidal process; h, conchal crest.

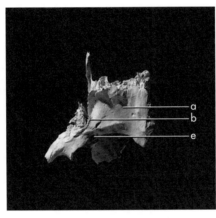

Superior, lateral is toward the right

Inferior, lateral is toward the left

Figure 4.29 **Right palatine.** Anterior is up. Natural size.

e. The **lesser palatine foramina,** for the transmission of the *lesser palatine nerves,* are located on the posterolateral corner of the hard palate posterior to the greater palatine foramina, near the junction of the perpendicular and horizontal plates.

f. The **perpendicular (vertical) plate** is appressed tightly to the posteromedial wall of the maxilla opposite the maxillary sinus, between the pterygoid plates of the sphenoid and the posterior margin of the alveolar process of the maxilla.

g. The posterior border of the perpendicular plate is the thickest border. It bears a serrated groove that articulates with the medial pterygoid plate of the sphenoid. This area of the bone is called the **pyramidal process.**

h. The **conchal crest** is a subhorizontally oriented crest placed not quite halfway up the perpendicular plate on the medial surface of the plate. This crest is for articulation with the inferior nasal concha.

4.13.2 Growth

Palatine bones ossify in membrane from single centers.

4.13.3 Possible Confusion

Because the palatines are almost always attached to the maxillae and sphenoid, identification is not usually difficult. When small, isolated fragments of palatine are encountered, note the free posterior edge of the horizontal plate and the smooth, even concavity on the nasal surface of this plate.

4.13.4 Siding

Because isolated fragments of palatine most often preserve the horizontal plate, note that the superior surface is smooth, that the inferior (palatal) surface is rough, that the posterior edge is nonarticular, and that the greater and lesser palatine foramina are posterolateral.

4.14
Vomer (Figure 4.30)

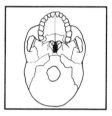

4.14.1 Anatomy

The vomer is a small, thin, plow-shaped, midline bone that occupies and divides the nasal cavity. It articulates inferiorly on the midline with the maxillary halves and palatines, superiorly with the sphenoid via its wings,

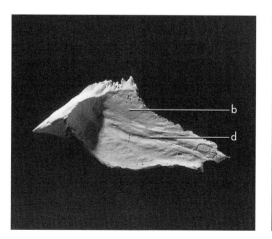
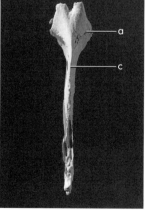

Right lateral, anterior is toward the right

Posterior

Figure 4.30 **Vomer.** Superior is up. Natural size. Key: a, ala or wing; b, perpendicular plate; c, posterior border; d, nasopalatine groove.

and anterosuperiorly with the ethmoid. Thus the bone forms the inferior and posterior part of the nasal septum, which divides the nasal cavity.

a. The **alae,** or **wings,** of the vomer are located on either side of a deep midline furrow on the superior surface of the vomer. This part of the bone is the thickest and sturdiest and is tightly appressed to the sphenoid.

b. The **perpendicular plate** of the vomer is a thin vertical sheet of bone on the midline below the wings.

c. The nonarticular **posterior border** of the vomer divides the posterior nasal opening into two halves.

d. The **nasopalatine grooves** lodge the *nasopalatine nerves* and *vessels,* marking both sides of the perpendicular plate, where they run anteroinferiorly from the alae.

4.14.2 Growth

Another bone with both endochondral and membranous ossification, the vomer ossifies from two plates (laminae) on either side of a median plate of cartilage. By puberty, the lamellae are virtually united, but the bilaminar origin of the bone is discernible in the cleft between the alae.

4.14.3 Possible Confusion

Because of the midline placement of the vomer, symmetry is the best guide to identification. Isolated vomers are rarely found and almost never recovered intact. To avoid confusion with other thin bones such as the sphenoid, note that the vomer has alae and that the perpendicular plate is symmetrical, with a free posterior edge.

4.14.4 Siding

The non-midline portions of the vomer are so small that siding criteria are unnecessary for this bone.

4.15
Inferior Nasal Conchae (Figure 4.31)

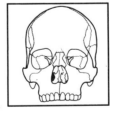

4.15.1 Anatomy

The inferior nasal conchae extend horizontally along the lateral walls of the nasal cavity, articulating with the medial wall of the maxillae and with the palatines. They also articulate with the ethmoid and lacrimals

| Lateral, anterior is toward the right | Medial, anterior is toward the left | Anterior, lateral is toward the left |

Figure 4.31 **Right inferior nasal concha.** Superior is up. Lit from the upper right for detail. Natural size.

superiorly. The bones are rarely found isolated because they are so fragile. Their shape is variable, with the anterior and posterior extremities tapered to a point, and the inferior surface free, thickened, and vascularized. The inferior nasal conchae function in olfaction and in moistening inhaled air.

4.15.2 Growth

The inferior nasal conchae ossify from single centers.

4.15.3 Possible Confusion

Since the inferior nasal conchae are so fragile, they are virtually never found intact as isolated specimens. Small fragments of them might be mistaken for ethmoid, sphenoid, or lacrimal. Note, however, that the surface texture is highly perforated by numerous tiny apertures in the nasal conchae, giving them a fragile and lightweight aspect.

4.15.4 Siding

There is little use for a knowledge of siding inferior nasal conchae.

4.16
Ethmoid (Figures 4.32–4.33)

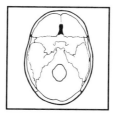

4.16.1 Anatomy

The ethmoid bone is exceedingly light and spongy. It is roughly the size and shape of an ice cube, but only a fraction as heavy. It is located between the orbits, centered on the midline. It articulates with thirteen bones: the

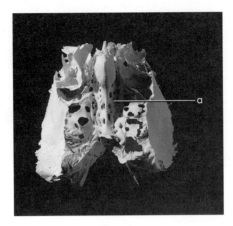

Superior

Figure 4.32 **Ethmoid.** Anterior is up. Natural size. Key: a, cribriform plate.

frontal, sphenoid, nasals, maxillae, lacrimals, palatines, inferior nasal conchae, and vomer. The ethmoid is virtually never found as a unit because of its fragility. It is best viewed in a specially disarticulated skull, where its complexity can be appreciated.

a. The **cribriform plate** is best observed endocranially, where the ethmoid can be seen to fill the ethmoid notch of the frontal. The cribriform plate roofs the nasal cavities, and because it is perforated by many tiny foramina it looks like a sieve. The *olfactory nerves* perforate this plate as they pass up to the brain from the mucous lining of the nose.

b. The **crista galli** is a perpendicular projection of the ethmoid's cribriform plate into the endocranial cavity. It is interposed between *olfactory bulbs,* and its posterior surface anchors the *falx cerebri,* a fold of the *dura mater* extending into the longitudinal fissure of the brain between the two *cerebral hemispheres.*

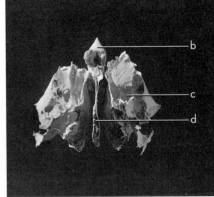

Left lateral, anterior is toward the left   Anterior

Figure 4.33 **Ethmoid.** Superior is up. Natural size. Key: b, crista galli; c, labyrinth or lateral mass; d, perpendicular plate.

c. The **labyrinths,** or **lateral masses,** of the ethmoid lie to either side of the midline and consist of a series of thin-walled ethmoidal cells. The lateral plates of the ethmoid labyrinths form most of the medial orbital walls, and the medial plates form the upper walls of the nasal cavity.

d. The **perpendicular plate** of the ethmoid is a flattened lamina placed at the midline between the lateral masses. It forms part of the nasal septum and articulates inferiorly with the vomer.

4.16.2 Growth

The only basicranial bone that is entirely preformed in cartilage, the ethmoid ossifies from three centers, one for each labyrinth and one for the perpendicular plate. During the first year after birth, the perpendicular plate and crista galli begin to ossify. They are joined in the second year to the labyrinths.

4.16.3 Possible Confusion

The thin plates of the ethmoid might be difficult to identify when found isolated, but this bone is rarely found by itself. More often, pieces of it adhere to the other bones it articulates with, most commonly the frontal or sphenoid. The perpendicular plate might be confused with the vomer, but the vomer has a nonarticular posterior edge and a thickened edge where the alae join.

4.16.4 Siding

Portions of ethmoid that are usually found isolated are the midline crista galli and perpendicular plate.

4.17
Lacrimals (Figure 4.34)

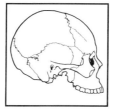

4.17.1 Anatomy

The lacrimals are very small, thin, fragile bones of rectangular shape. The lacrimals make up part of the medial walls of the orbits anterior to the ethmoid. They articulate with the frontal, maxillae, ethmoid, and inferior nasal conchae. They are virtually never found alone but are often attached to facial fragments. The lacrimals have orbital and nasal (lateral and medial) surfaces and four borders.

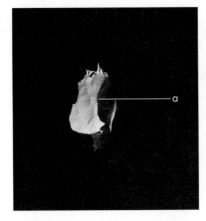

Lateral, anterior is toward the right   Medial, anterior is toward the left

Figure 4.34 **Right lacrimal.** Superior is up. Natural size. Key: a, posterior lacrimal crest.

a. The **posterior lacrimal crest** is a vertical crest on the medial orbital wall that bounds the posterior half of the lacrimal groove (see section 4.12.1k).

4.17.2 Growth

The lacrimal ossifies in membrane from a single center.

4.17.3 Possible Confusion

The lacrimal crest is diagnostic, and although the lacrimal is virtually never found alone, it can often help to identify adjacent bones.

4.17.4 Siding

The lacrimal crest is oriented vertically, and the lacrimal groove is anterior to the crest. The base of the crest sweeps anteriorly to become a margin for the lacrimal canal.

4.18
Nasals (Figure 4.35)

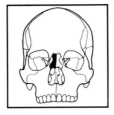

4.18.1 Anatomy

The nasals are small, thin, rectangular bones placed on either side of the midline below the glabellar region of the frontal. Their free inferior ends form the top margin of the nasal aperture. The nasals articulate with the

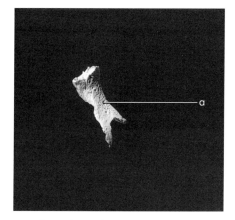
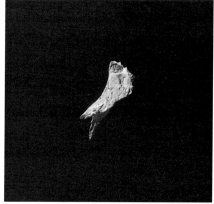

| Lateral, anterior is toward the right | Medial, anterior is toward the left |

Figure 4.35 **Right nasal.** Superior is up. Natural size. Key: a, nasal foramen.

frontal superiorly, with each other medially, and with the frontal processes of the maxillae laterally. They articulate posteriorly with the ethmoid.

a. The **nasal foramen** perforates the facial surface and transmits a vein.

4.18.2 Growth

Each nasal ossifies in membrane from a single center.

4.18.3 Possible Confusion

It is difficult to confuse the nasal with other bones because of its diagnostic internasal suture, the external foramen and internal groove, the smooth outer and rough inner surface, and the nonarticular, free inferior edge.

4.18.4 Siding

Use the criteria mentioned in section 4.18.1 to side nasal bones.

• The free edge is inferior, the thickest articular edge is medial, and the frontal suture is interdigitating and superior.

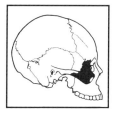

4.19
Zygomatics (Figure 4.36)

4.19.1 Anatomy

The zygomatics form the prominent corners (cheeks) of the face. The edges are easily identifiable, with the rounded orbital rim, the sharp area

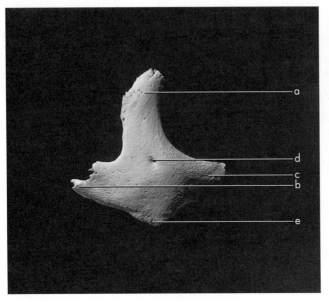 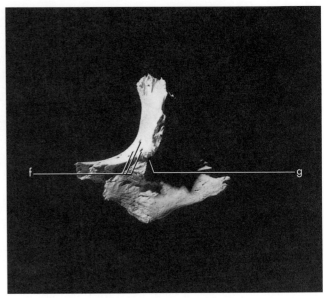

Lateral, anterior is toward the right

Medial, anterior is toward the left

Figure 4.36 **Right zygomatic.** Superior is up. Natural size. Key: a, frontal process; b, temporal process; c, maxillary process; d, zygomaticofacial foramen; e, masseteric origin; f, zygomaticoorbital foramina; g, zygomaticotemporal foramen.

around jugale adjacent to the temporal fossa, and the roughened inferior border. Each zygomatic bone articulates, via its three main processes, with the frontal, sphenoid, temporal, and maxilla.

a. The **frontal process** rises vertically and separates the orbit from the temporal fossa.

b. The **temporal process** extends posteriorly, joining the zygomatic process of the temporal bone to form the zygomatic arch.

c. The **maxillary process** extends toward the midline, forming the inferolateral orbital margin.

d. The **zygomaticofacial foramen** perforates the convex lateral surface of the zygomatic. It is often multiple, allowing the passage of the *zygomaticofacial nerve* and *vessels.*

e. The **masseteric origin,** the roughened, expanded inferior edge of the bone, extends from the zygomaticomaxillary to the temporozygomatic suture. This is the main attachment point for the *masseter muscle,* a major elevator of the mandible.

f. The **zygomaticoorbital foramina** perforate the inferolateral corner of the orbital cavity for the passage of the *zygomaticotemporal* and *zygomaticofacial nerves.*

g. The **zygomaticotemporal foramen** is centered in the temporal surface of the zygomatic. It transmits the *zygomaticotemporal nerve.*

4.19.2 Growth

The zygomatic bone ossifies from three centers that fuse into a single combined center during fetal development.

4.19.3 Possible Confusion

The zygomatic bone's three diagnostic borders make identification easy. The most frequent confusion comes in mistaking the zygomatic process of the temporal for the zygomatic bone.

- None of the processes of the zygomatic bone proper are as thin or extended as the zygomatic process of the temporal.

4.19.4 Siding

To side isolated or fragmentary zygomatic bones, remember the relations of this bone to the orbit, the masseter muscle, and the temporal fossa.

- The masseteric attachment is inferior, and the convex surface is anterior and lateral, perforated by foramina.
- The orbital rim is blunter than the jugal area posterior to it.

4.20 Sphenoid (Figures 4.37–4.40)

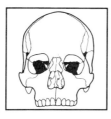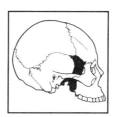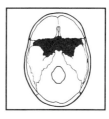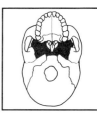

4.20.1 Anatomy

The sphenoid is the most complex bone of the cranium. Although its name means "wedgelike," its shape is far more elaborate. It is very difficult to visualize this bone when working with an articulated cranium, because it has surfaces that face many directions—endocranially, inferiorly, laterally, and anteriorly. The sphenoid is situated between the bones of the cranial vault and those of the face. For this reason, and because many parts of the bone are thin, the sphenoid bone is virtually never found intact in broken crania. Instead, portions of it adhere to other cranial pieces.

The many articulations of the sphenoid were noted as each of the twelve bones it touches were introduced. These are reviewed in the descriptions below. The articulating midline bones are the vomer, ethmoid, frontal, and occipital. The sphenoid also articulates with the paired parietals, temporals, zygomatics, and palatines (and sometimes, the maxillae) lateral to the midline. Examination of the sphenoid is simplified by dealing with four basic parts of the bone, the body, greater and lesser wings, and the pterygoid plates. For an overall perspective, view the sphenoid

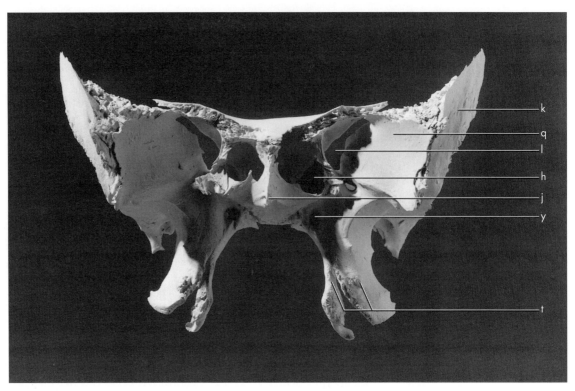

Figure 4.37 **Sphenoid, anterior.** Superior is up. Natural size. Key: h, sphenoidal sinus; j, sphenoidal crest; k, greater wing; l, superior orbital fissure; q, orbital surface; t, pterygoid process; y, pterygoid canal.

from behind, visualizing it as a flying animal with a central body, two pairs of wings, and dangling talons (the pterygoid plates).

a. The **body** is the only part of the sphenoid that lies on and immediately adjacent to the midline. This is the most robust part of the bone. Its anterior surface forms the superoposterior wall of the nasal cavity and articulates with the cribriform and perpendicular plates of the ethmoid. Posteriorly, the body articulates with the occipital across the sphenooccipital suture, and anteroinferiorly it articulates with the vomer.

b. The **optic canals** are seen to either side of the body. They pass anteroinferior to the lesser wings, just medial and superior to the superior orbital fissure. The *optic nerve* and *ophthalmic artery* pass through these canals on their way to the eyeballs.

c. The **sella turcica** ("Turkish saddle") is located endocranially, posterior and inferior to the optic canals, atop the body of the sphenoid, decorated by the four clinoid processes.

d. The **hypophyseal (pituitary) fossa** is the deepest depression of the sella. It holds the *pituitary gland*, manufacturer of growth hormones.

e. The **dorsum sellae** is the square plate of bone that forms the posterior boundary of the sella turcica.

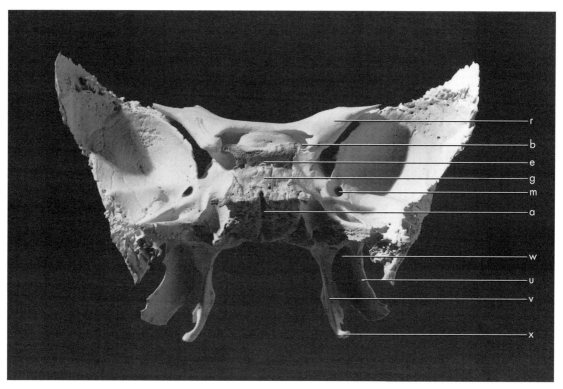

Figure 4.38 **Sphenoid, posterior.** Superior is up. Natural size. Key: a, body; b, optic canal; e, dorsum sellae; g, clivus; m, foramen rotundum; r, lesser wing; u, lateral pterygoid plate (lamina); v, medial pterygoid plate (lamina); w, pterygoid fossa; x, pterygoid hamulus.

f. The **posterior clinoid processes** are the two highly variable tubercles located at the superolateral corners of the dorsum sellae.

g. The **clivus** is the slight endocranial hollow that slopes posteriorly from the dorsum sellae toward the sphenooccipital suture.

h. The **sphenoidal sinuses** are large, paired hollows within the body of the sphenoid.

i. The **sphenoidal rostrum** is a midline bony projection on the anteroinferior surface of the body of the sphenoid. It fits into the fissure between the alae of the vomer.

j. The **sphenoidal crest** is continuous with the rostrum, extending superiorly from it on the anterior surface of the body of the sphenoid. This midline sphenoidal crest articulates with the perpendicular plate of the ethmoid and forms part of the septum of the nose.

k. The **greater wings** of the sphenoid (right and left) are attached to the body. They are the segments that extend the farthest laterally from the body, forming most of the middle cranial fossae endocranially and much of the temporal fossae ectocranially. The greater wings articulate with the temporals, parietals, frontal, zygomatics, and maxillae.

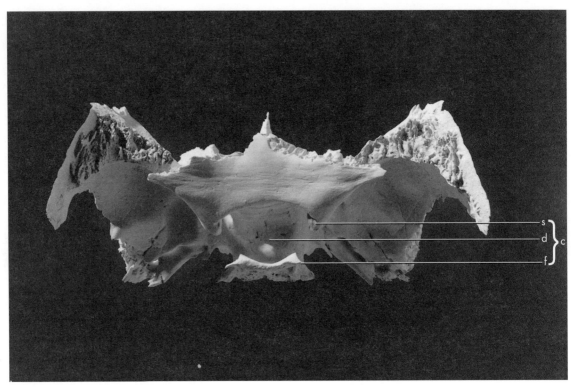

Figure 4.39 **Sphenoid, superior.** Anterior is up. Natural size. Key: c, sella turcica; d, hypophyseal (pituitary) fossa; f, posterior clinoid process; s, anterior clinoid process.

l. The **superior orbital fissures** are the open spaces (gaps) between the inferior surfaces of the lesser wings and the anterior surfaces of the greater wings. The fissures are visible at the back of the orbits in an anterior view of the cranium. The superior orbital fissure and three foramina identified below are best seen on the endocranial surface of the sphenoid. These openings are arranged in the form of an arc that sweeps posterolaterally from the midline in the area where the greater wing and body merge. The arc is sometimes called the "crescent of foramina."

m. The **foramen rotundum** is situated in the most anterior and medial part of both right and left middle cranial fossae at the junction of the greater wings and the body. These foramina transmit the *maxillary nerves* that run just inferior to the superior orbital fissures.

n. The **foramen ovale** is located posterior to the foramen rotundum on each side, approximately in line with the dorsum sellae in endocranial view. These foramina transmit the *mandibular nerves* and *accessory meningeal arteries.*

o. The **foramen spinosum** is located on each greater wing just posterolateral to the foramen ovale. The foramina spinosa are set in the posteroinferior spines of the sphenoid, very close to the temporal bones. They transmit the *middle meningeal vessels* and branches from the *mandibular nerves.*

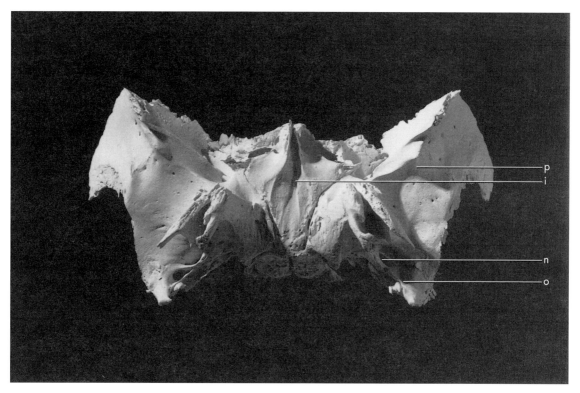

Figure 4.40 **Sphenoid, inferior.** Anterior is up. Natural size. Key: i, sphenoidal rostrum; n, foramen ovale; o, foramen spinosum; p, infratemporal crest.

p. The **infratemporal crests** mark the ectocranial surfaces of the greater wings. They form the base of the temporal fossae at about the level of the zygomatic arches.

q. The **orbital surfaces** of the greater wings of the sphenoid, which form the lateral wall of each orbit, are very smooth and flat in comparison to the endocranial surfaces.

r. The **lesser wings,** which are much smaller than the greater, are thin, wing-shaped posterior projections of the endocranial surface. These partially floor the right and left frontal lobes of the brain. They arise from the superior surface of the body and articulate with the horizontal orbital plates of the frontal.

s. The **anterior clinoid processes** are the posteriormost projections of the lesser wings. These give attachment to the *tentorium cerebelli*, a segment of *dura mater* that separates the cerebellum from the occipital part of the *cerebral hemispheres* of the brain.

t. The **pterygoid processes** of the sphenoid are seen only from below or to the side of the cranium. The pterygoid processes are each divided into two thin plates.

u. The **lateral pterygoid plate (lamina)** is a thin vertical plate of bone seen in lateral view of the cranium.

v. The **medial pterygoid plate (lamina)** is a thin vertical plate of bone that roughly parallels the lateral plate in orientation but is set closer to the midline. Each pair of pterygoid plates articulates anteriorly with the palatines. These four thin projections provide attachment for the *medial pterygoid muscles,* mandibular elevators.

w. The **pterygoid fossae** are the rough-floored hollows between the medial and lateral pterygoid plates.

x. The **pterygoid hamulus** is the hooklike process forming the posterolateral, basal corner of each medial pterygoid plate.

y. The **pterygoid canals** perforate the bone above the pterygoid plates and run along the base of these plates.

4.20.2 Growth

The sphenoid is mostly formed in cartilage; only the pterygoid plates form dermally. Growth is complex, with a number of ossific centers involved, but this bone is recognizable by the time of birth.

4.20.3 Possible Confusion

Because it has so many parts, broken fragments of sphenoid are often difficult to identify. Only an intensive study of the various anatomical parts introduced above will allow confident identification of isolated fragments. Fortunately, fragments of sphenoid are usually attached to other cranial bones, and identification is aided by this fact. The parts most often found isolated are the greater wings and the body.

The nature of the suture between the temporal, parietal, and sphenoid helps in the identification and siding of sphenoid fragments. The sphenoid overlaps the parietal superiorly and underlaps the temporal posteriorly. It abuts the temporal in basicranial aspect.

4.20.4 Siding

- For the pterygoid plates, the pterygoid fossa faces posteriorly, and the plates have nonarticular, sharp posterior and inferior edges.
- For the greater wing, the smooth, flat orbital surfaces must face anteriorly, and the concave temporal surface must face laterally.
- The endocranial surface is posterosuperior, and its base is marked by the foramina rotunda, ovale, and spinosa. The latter foramen is at the spine of the greater wing, the most posterolateral extent of the wing, often on the sphenosquamosal suture.
- For a fragmentary lesser wing, the free nonarticular end is posterior, the wing tips face laterally, and parts of the frontal often adhere anteriorly. The anterior clinoid process points posteriorly.

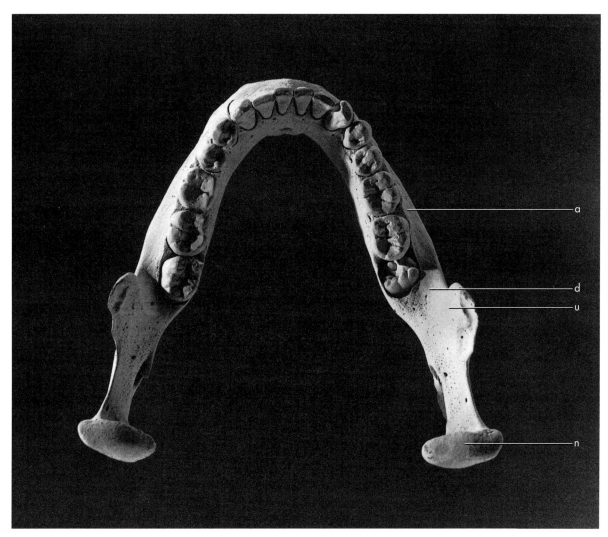

Figure 4.41 **Mandible, superior.** Natural size. Key: a, body (corpus or horizontal ramus); d, extramolar sulcus; n, mandibular condyle; u, endocoronoid ridge or buttress.

4.21
Mandible (Figures 4.41–4.43)

4.21.1 Anatomy

The mandible, or lower jaw, articulates through its condyles (via an articular disk) with the temporal bones at the temporomandibular joint. The primary function of this bone is in **mastication** (chewing). The mandible holds the lower teeth and provides insertion surfaces for the muscles of mastication. These two functions are performed by the two basic parts of the mandible, the body (corpus) and the ascending ramus.

  a. The **body** (**corpus**, or **horizontal ramus**) is the thick bony part of the mandible that anchors the teeth. With its emplanted teeth, the cor-

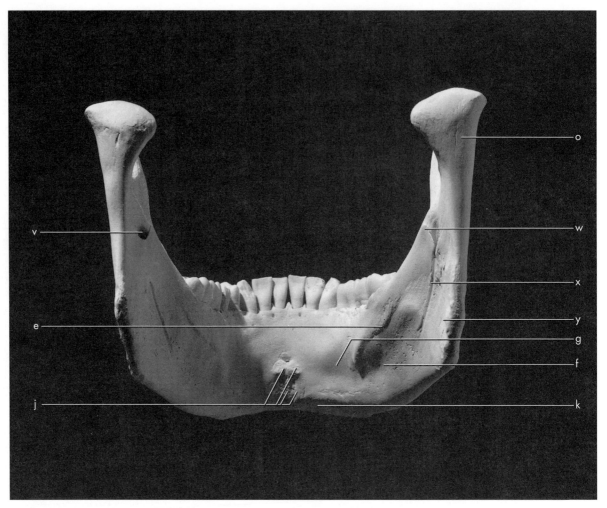

Figure 4.42 **Mandible, posterior.** Natural size. Key: e, mylohyoid line; f, submandibular fossa; g, sublingual fossa; j, mental spines; k, digastric fossae; o, condylar neck; v, mandibular foramen; w, lingula; x, mylohyoid groove (sulcus); y, pterygoid tuberosities.

pus of the mandible is very hard, dense, and resistant to destruction. For this reason, mandibular corpora outlast other body parts in bone assemblages that have been ravaged by carnivores or subjected to physical degradation.

b. The **mental foramen** is the large, sometimes multiple foramen located on the lateral corpus surface, near midcorpus, below the premolar region. This foramen transmits the *mental vessels* and *nerve.*

c. The **oblique line** is a weak eminence that passes from the root of the ramus to the area at the rear of the mental foramen.

d. The **extramolar sulcus** is the gutter between the root of the anterior edge of the ramus and the lateral alveolar margin of the last molar. This area gives rise to the *buccinator muscle,* the muscle of the cheek.

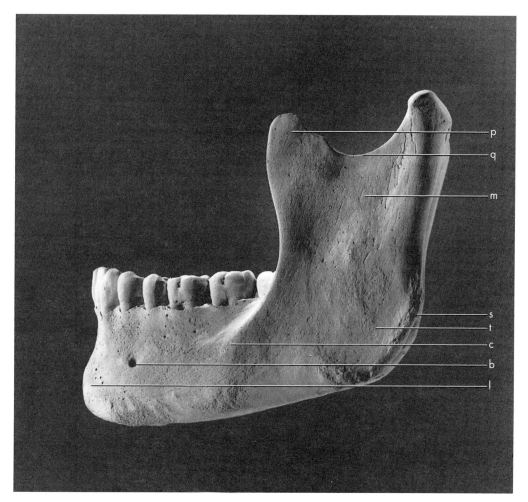

Figure 4.43 **Mandible, lateral.** Natural size. Key: b, mental foramen; c, oblique line; l, mental protuberance (mental eminence); m, ramus (ascending ramus); p, coronoid process; q, mandibular notch (incisura); s, masseteric tuberosity; t, masseteric fossa.

e. The **mylohyoid line** crosses the medial corpus surface obliquely, beginning near the alveolar margin at the last molar position and diminishing as it runs anteriorly and inferiorly. It marks an attachment site for the *mylohyoid muscle*, a muscle that forms the muscular floor of the mouth cavity and acts to raise the *tongue* and hyoid bone.

f. The **submandibular fossa** is the hollow beneath the alveolar portion that runs along the medial corpus, inferior to the mylohyoid line. In life, the *submandibular gland*, one of the salivary glands, rests in this fossa.

g. The **sublingual fossa** is the hollowing beneath the alveolar region, superior to the mylohyoid line in the premolar region. The *sublingual gland*, another salivary gland, rests in this fossa.

h. A **mandibular torus** is the variably developed thickening of the alveolar margin just lingual to the cheek teeth. This feature takes on a billowed appearance in its most extreme manifestations but is often imperceptible.

i. The **mandibular symphysis** technically refers only to the midline surfaces of unfused right and left mandibular halves in individuals less than 1 year of age. Nontechnically, it is often used to refer to the anterior region of the mandible between the canines.

j. The **mental spines** lie near the inferior margin of the inner (posterior) surface of the anterior corpus. They are variable in prominence and anchor the *genioglossal* and *geniohyoid muscles,* muscles of the tongue.

k. The **digastric fossae** are the pair of roughened depressions at the base of the corpus adjacent to the posterior midline. They face posteroinferiorly and are attachment sites for the *digastric muscles,* depressors of the mandible.

l. The **mental protuberance (mental eminence)** is the triangular eminence, or bony chin, at the base of the corpus in the anterior symphyseal region. It is separated from the incisor alveolar region by a pronounced incurvation in modern humans.

m. The **ramus** (or **ascending ramus**) is considerably thinner than the corpus. This vertical part of the mandible rises above the level of the teeth and articulates with the cranial base.

n. The **mandibular condyle** is the large, rounded, articular prominence on the posterosuperior corner of the ramus. It articulates at the temporomandibular joint.

o. The **condylar neck** is the area just anteroinferior to the condyle. A head of the *lateral pterygoid muscle* attaches to the anteromedial surface of the neck just below the articular surface of the condyle, in the **pterygoid fovea.** This muscle acts to depress and stabilize the mandibular condyle during chewing.

p. The **coronoid process** of the ramus is thin and triangular, varying widely in shape and robusticity. Its anterior border is thickened and convex, and its posterior edge is concave and thinner. Both medial and lateral surfaces of this process receive the insertion of the *temporalis muscle.*

q. The **mandibular notch (incisura)** is the notch between the condyle and coronoid process.

r. The **angle** (or **gonial angle**) is the rounded posteroinferior corner of the mandible. The *masseter muscle* attachment is centered on the lateral ramus surface, all along the angle.

s. The **masseteric tuberosity** is the raised, roughened area at the lateral edge of the gonial angle at which the *masseter muscle* attaches. This area is often joined by oblique ridges raised by masseter attachment. When the edge of the gonial angle projects far laterally from the rest of the ramus, the gonial area is said to be strongly **everted.**

t. The **masseteric fossa** is a variably expressed hollowing on the lateral surface of the gonial angle.

u. The **endocoronoid ridge,** or **buttress,** is the vertical ridge extending inferiorly from the coronoid tip on the inner (medial) aspect of the ramus.

v. The **mandibular foramen** enters the bone obliquely, centered in the medial surface of the ramus. The *alveolar vessels* and *inferior alveolar nerve* enter the bone through this opening, running through the mandible via the **mandibular canal.**

w. The **lingula** is a sharp, variably shaped projection at the edge of the mandibular foramen. It is the attachment point for the *sphenomandibular ligament.*

x. The **mylohyoid groove (sulcus)** crosses the medial ramus surface, running anteroinferiorly from the edge of the mandibular foramen. It lodges the *mylohyoid vessels* and *nerve.*

y. The **pterygoid tuberosities** interrupt the medial surface of the gonial angle posteroinferior to the mylohyoid groove. They mark the insertion of the *medial pterygoid muscle,* an elevator of the mandible.

4.21.2 Growth

The mandibular halves are separate at birth; they join during the first year at the symphysis. At birth the mandible holds unerupted deciduous teeth in crypts below the surface. The eruption of these teeth and their permanent counterparts effects dramatic changes on the mandible during ontogeny. Loss of the permanent teeth results in resorption of the alveolar portion of the mandible.

4.21.3 Possible Confusion

Only small fragments of mandible can be confused with other bones.

• Where tooth sockets are present, the bone must be maxilla or mandible. The former has a sinus above the molar roots.

• The mandible corpus has a much thicker cortex than the maxilla, as well as a basal contour.

• The coronoid process might be mistaken for thin cranial bones such as the sphenoid or zygomatic. Note, however, that the coronoid does not articulate, and its edges are therefore nonsutural.

4.21.4 Siding

To side fragments of mandible, remember that the incisors are anterior and closer to the midline than the molars, and that the ramus is posterior, with greatest relief on its medial surface.

• For isolated condyles, the border of the mandibular notch is continuous with the lateral side of the condyle (most of the condyle itself lies medial to the plane of the ramus).

- For isolated coronoids, the notch surface is posterior, the tip superior, and the endocoronoid buttress medial.

- For isolated gonial angles, the tuberosities for the medial pterygoid are medial and anterosuperiorly directed.

4.22 Mastication

Rather than studying bones as inert objects with strange names and processes, they are best appreciated as the living foundations of the organism. Our introduction to the skull revealed bony structures that house a variety of organs such as the brain and the eyes. Much of the anatomy of the skull is devoted to its function in chewing. Study of the masticatory system has given physical anthropologists insight into the diet of extant and extinct primates, including human ancestors. It is useful to conclude our examination of the skull with an analysis of the musculoskeletal system behind human mastication. This analysis provides an excellent reminder that the external and internal architecture of bones is strongly related to function.

Chewing takes place through the coordinated action of the musculoskeletal system of the skull. Abundant evidence of the soft-tissue components of this system has been noted in the form of muscular and ligamentous attachments. Foramina and grooves for blood vessels and nerves on the mandible, zygomatics, frontal, parietals, temporals, and other bones are more evidence of soft tissues.

The masticatory system is devoted to generating forces across the opposing mandibular and maxillary teeth. There are elevators of the mandible (muscles that pull the lower jaw and its teeth up and against the maxillary teeth) and depressors of the mandible. The primary muscles that elevate the mandible during chewing (and their major attachments) are as follows: The **temporalis muscle** originates on the side of the cranial vault inferior to the superior temporal line and inserts on the sides, apex, and anterior surface of the coronoid process of the mandible. The **masseter muscle** originates on the inferior surface of the zygomatic arch and inserts on the lateral surface of the mandibular ramus and the gonial angle of the mandible. The **medial pterygoid muscle** originates on the medial surface of the lateral pterygoid plate of the sphenoid and inserts on the medial surface of the mandibular gonial angle.

When the teeth are forcefully clenched, it is easy to palpate the active masseter and temporalis on either side of the jaw and temple. The act of clenching involves stimulation of fibers in each of these muscles. This stimulation comes from nerves which can be traced back to the brain. The muscle fibers contract, and this contraction brings the attachment points of the muscles closer together, forcefully elevating the mandibular teeth against their maxillary counterparts. Any food between the teeth is reduced by this activity to a smaller size and then passed farther down into the digestive system. All of this coordinated, complex working of the masticatory system takes place thousands of times each day without our paying much attention to it.

Suggested Further Readings

The descriptions of cranial anatomy in this chapter, and the descriptions of postcranial anatomy in Chapters 6–13, may be supplemented by any of several osteology and anatomy texts.

Bass, W. M. (1995) *Human Osteology: A Laboratory and Field Manual* (4th Edition). Columbia, Missouri: Missouri Archaeological Society. 361 pp.
> A laboratory and field manual emphasizing identification. The format is similar to that used in this book.

Breathnach, A. S. (Ed.) (1965) *Frazer's Anatomy of the Human Skeleton* (6th Edition). London: J. and A. Churchill. 253 pp.
> An anatomy text with a skeletal orientation. Good drawings, with an emphasis on function.

Cartmill, M., Hylander, W., and Shafland, J. (1987) *Human Structure.* Cambridge, Massachusetts: Harvard University Press. 448 pp.
> A functionally, evolutionarily oriented introductory anatomy textbook.

Fazekas, G., and Kosa, F. (1978) *Forensic Fetal Osteology.* Budapest: Akadémiai Kiadó. 414 pp.
> Excellent illustrations of fetal skeletons at various stages of development.

Grant, J. C. B. (1972) *An Atlas of Anatomy* (6th Edition). Baltimore, Maryland: Williams and Wilkins. 1448 pp.
> A large format atlas, excellent for examining soft- and hard-tissue relationships.

Krogman, W. M., and İşcan, M. Y. (1986) *The Human Skeleton in Forensic Medicine* (2nd Edition). Springfield, Illinois: C. C. Thomas. 551 pp.
> An essential source book; the descriptions and references to work on skeletal growth make good supplements to this book.

McMinn, R. M. H., and Hutchings, R. T. (1977) *Color Atlas of Human Anatomy.* Chicago: Year Book Medical Publishers. 352 pp.
> Excellent photographic illustrations in a large, well-organized format.

McMinn, R. M. H., Hutchings, R. T., and Logan, B. M. (1987) *The Human Skeleton: A Photographic Manual.* Chicago, Illinois: Year Book Medical Publishers.
> A large-format manual that includes full-size foldouts of an articulated skeleton. Many color views of all individual bones of the skeleton.

Rohen, J. W., and Yokochi, C. (1988) *Color Atlas of Anatomy* (2nd Edition). New York: Igaku-Shoin. 469 pp.
> Superb color plates of hard- and soft-tissue human anatomy.

Shipman, P., Walker, A., and Bichell, D. (1985) *The Human Skeleton.* Cambridge, Massachusetts: Harvard University Press. 343 pp.
> Part II presents a guide to bones of the skeleton from a functional perspective. The narrative style may prove useful to students reviewing the material presented in Chapters 4–13 of this volume.

Steele, D. G., and Bramblett, C. A. (1988) *The Anatomy and Biology of the Human Skeleton.* College Station, Texas: Texas A&M University Press. 291 pp.

An atlas dedicated to the identification and biology of the human skeleton.

Warwick, R., and Williams, P. L. (Eds.) (1973) *Gray's Anatomy* (35th British Edition). Philadelphia: W. B. Saunders. 1471 pp.
The classic work in human anatomy.

CHAPTER 5

Dentition

The upper and lower jaws are dermal bones with a fascinating evolutionary history. Teeth in the upper and lower jaws seem to have evolved from fish scales. The living core of a mammalian tooth is a dense bonelike material called **dentin,** or **dentine,** a special type of calcified but slightly resilient connective tissue that extends into a socket in the jaw. The crown of the tooth is covered with a layer of extremely hard, brittle material known as **enamel.** These basic tissues have been molded by evolution into an impressive variety of shapes and sizes among the vertebrates.

The dentition is one of the most important parts of human anatomy for the osteologist. Teeth owe their importance in paleontology and anthropology to a variety of factors. Of all the skeletal elements, teeth are the most resistant to chemical and physical destruction. They are therefore overrepresented relative to other parts of the skeleton in almost all archeological and paleontological assemblages. In addition to being abundant, teeth constitute a focus of anthropological and paleontological interest because they are so informative about the individual possessing them. Teeth provide information on the age, sex, health, diet, and evolutionary position of extant and extinct mammals, hominids included.

Teeth are formed within the jaws and then erupt. Unlike the changing shapes of other skeletal elements, tooth crown morphology can only be altered by **attrition** (tooth wear), breakage, or demineralization once the crown erupts. Tooth morphology can be used to effectively differentiate between populations within a species, species within a genus, and so forth. The stability and adaptive significance of tooth form establish the dentition as a centerpiece in many comparative populational and evolutionary studies. Finally, teeth are the only hard tissues of the body that are directly observable without dissection or radiography.

Dental measurement and terminology are discussed in Chapter 3. Use of the dentition in sexing and aging is discussed in Chapter 17. Dental pathology is reviewed in Chapter 18. Use of the dentition in estimating population distance is covered in Chapter 20. The principles of identification discussed in the present chapter form the very foundation on which all other aspects of dental studies are based.

5.1
Dental Form and Function

Teeth constitute the part of the skeleton that directly interfaces with the environment, acting to seize and masticate (chew) food material, which is subsequently passed farther along the digestive system. The internal composition and external morphology of teeth are adapted to this function in considerable detail among mammals. In adult hominids (Figure 5.1) **incisors** are the eight spatulate teeth in the front of the upper and lower jaws (four in each jaw, two on the right and two on the left). Unworn incisors display sharp, thin cutting edges. Modern human **canines** function primarily as posterior extensions of the incisor rows, but they retain a more

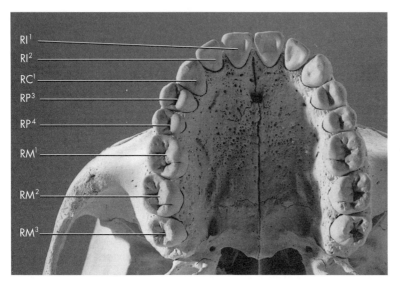

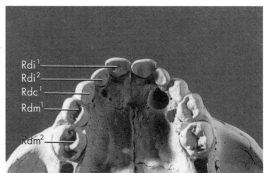

Maxilla

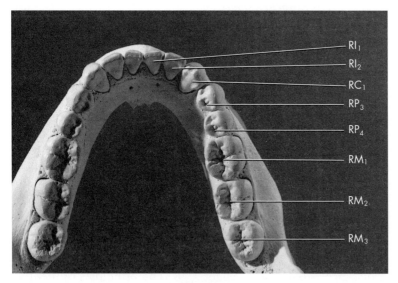

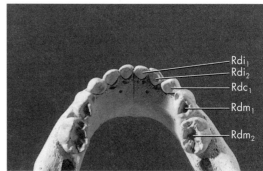

Mandible

Figure 5.1 Permanent (*left*) and deciduous (*right*) dentition. Natural size.

conical shape than the incisors. Human **premolars** are termed "bicuspids" by dentists. In humans there are four premolars (two pairs) in the upper jaw and four in the lower jaw. In primitive mammals there were eight in each jaw, but the mesial (anterior) two pairs of premolars have been lost during the course of primate evolution. For that reason, the first premolar behind the canine in the human dentition is identified as the third premolar since it occupies the third of the original four premolar positions. **Molars** make up the rest of the human dentition. These are the largest teeth; their extensive chewing surfaces emphasize crushing and grinding rather than shearing of food material. There are usually six molars (two sets of three teeth) in both upper and lower adult human jaws.

Deciduous (**primary**) teeth are the first to form, erupt, and function in the first years of life. These teeth are systematically shed and replaced by their **permanent** (**secondary**) counterparts throughout childhood and adolescence (see section 5.4).

5.2
Dental Terminology

In section 3.2.2 we identify some directional terms specific to the dentition. Because work on dental anatomy makes extensive use of these terms, a brief review is essential here. In dental anatomy, the **mesial** portion of the tooth is closest to the point where the central incisors contact each other, and **distal** is the opposite of mesial. The **lingual** part of the tooth crown is toward the tongue. **Labial** is the opposite of lingual but is usually reserved for the incisors and canines. **Buccal** is also the opposite of lingual but is usually reserved for the premolars and molars, where the term refers to that part of the tooth which lies toward the cheeks. **Interproximal** tooth surfaces contact adjacent teeth, and the chewing surface of the tooth is the **occlusal** surface. The tooth roots are suspended in **sockets** (**alveoli**) in the mandible and maxillae by a periodontal ligament.

Study of the dentition is considerably simplified through the use of a shorthand that unambiguously identifies each tooth. In this shorthand, the capital letters **I, C, P,** and **M** identify the permanent incisors, canines, premolars, and molars, respectively. When these are shown in lower case, preceded by the letter **d,** the **i, c,** and **m** denote deciduous incisors, canines, and molars. Because the deciduous molars are replaced by permanent premolars, paleontologists sometimes refer to these teeth as deciduous premolars. Thus the paleontological abbreviation **dp** is the equivalent of the anthropological **dm.**

The positions of all teeth are indicated by numbers, referring to the position that the tooth holds in the tooth row. Thus incisors can be **1**s or **2**s (**centrals** or **laterals**). Human canines are all **1**s. Premolars can be **3**s or **4**s (or 1s or 2s for the non-paleontologically inclined), and molars can be **1**s, **2**s, or **3**s. Right and left teeth are designated by the letters **R** and **L,** respectively. Upper and lower teeth are indicated by subscripting or superscripting the position numbers (placing a line above or below them avoids confusion in handwritten designations). Thus, the left deciduous second mandibular incisor is designated **Ldi$_2$,** whereas the right permanent first maxillary molar is designated **RM^1.** (In an alternative system of nomenclature, human canine position is indicated by a bar representing the occlusal plane; i.e., $\overline{\text{C}}$ for lower canine, $\underline{\text{C}}$ for upper.)

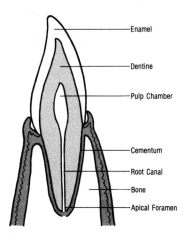

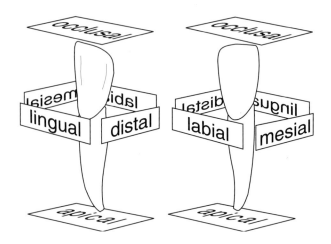

Figure 5.2 Anatomy of a tooth.

5.3
Anatomy of a Tooth

Figure 5.2 shows the various elements of a human tooth. There are eighteen features so commonly found in teeth of all kinds that they are worth listing separately:

1. The **crown** is the part of the tooth covered by enamel.

2. The **root** is the part of the tooth that anchors the tooth in the alveolus of the mandible or maxilla.

3. The **neck** (**cervix**) is the constricted part of the tooth at the junction of the crown and root.

4. **Enamel,** the specialized hard tissue that covers the crown, is both avascular and acellular. It is about 97% mineralized, essentially fossilized once it is formed.

5. The **cervicoenamel line** (or **junction, CEJ**) is the line encircling the crown which is the most rootward extent of the enamel.

6. The **dentinoenamel junction** (**DEJ**) is the boundary between the enamel cap and the underlying dentin.

7. **Dentin** (or **dentine**) is the tissue that forms the core of the tooth. This tissue has no vascular supply but is supported by the vascular system in the pulp and is lined on the inner surface (the walls of the pulp cavity) by odontoblasts, dentin-producing cells. These cells have the same relationship to dentin that osteoblasts have to bone. Dentin underlies the enamel of the crown and encapsulates the pulp cavity, the central soft tissue space within a tooth. Occlusal wear may expose dentin, and because dentin is softer than enamel the resulting exposures are usually occlusally concave.

8. The **pulp chamber** is the expanded part of the pulp cavity at the crown end of the tooth.

9. The **root canal** is the narrow end of the pulp cavity at the root end of the tooth.

10. **Cementum** is a bonelike tissue that covers the external surface of tooth roots.

11. **Calculus** is a calcified deposit commonly found on the sides of tooth crowns. The origins of calculus lie with plaque, colonies of microorganisms that establish themselves on the teeth.

12. The **pulp** is the soft tissue within the pulp chamber. This includes nerves and blood vessels.

13. The **apical foramen** is the opening at each root **apex,** or tip, through which nerve fibers and vessels pass from the alveolar region to the pulp cavity.

14. A **cusp** is an occlusal projection of the crown. Major cusps on hominid molars are individually named (Figure 5.3). Knowledge of these cusps, their relative sizes and wear, is often valuable in identifying isolated teeth. Cusps of the upper teeth end with the suffix **-cone,** whereas cusps on the lower teeth end with the suffix **-conid.** The tip of a cusp is the **apex.** Ridges that descend from cusp apices are **crests.**

   a. The **protocone** is the mesiolingual cusp on an upper molar. Cusplets, grooves, or other topographic features on its mesiolingual surface are called **Carabelli's effects.**

   b. The **hypocone** is the distolingual cusp on an upper molar.

   c. The **paracone** is the mesiobuccal cusp on an upper molar.

   d. The **metacone** is the distobuccal cusp on an upper molar.

   e. The **protoconid** is the mesiobuccal cusp on a lower molar. Cusplets, grooves, or other forms on the protocone's mesiobuccal surface are called **protostylid effects.**

   f. The **hypoconid** is the distobuccal cusp on a lower molar.

   g. The **metaconid** is the mesiolingual cusp on a lower molar.

   h. The **entoconid** is the distolingual cusp on a lower molar.

   i. The **hypoconulid** is the fifth, distalmost cusp on a lower molar.

   j. Cusplets on the incisal edges of unworn incisors are called **mammelons.**

15. A **fissure** is a cleft on the occlusal surface between cusps. Fissures divide the cusps into patterns. The most widely acknowledged of these is the **Y-5 pattern,** a pattern in which the five lower molar cusps are arranged in a Y pattern (for a review, see Johanson, 1979).

16. The primitive mammalian cusp pattern was a triangle of cusps in both upper and lower molars. From this pattern a remarkable variety of forms has arisen through evolution, ranging from the tall

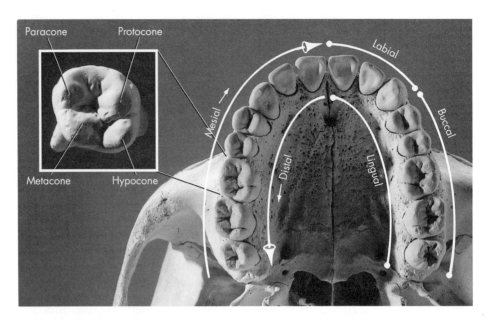

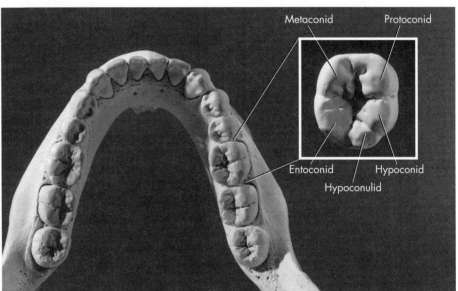

Figure 5.3 Anatomy and directional terms for the dentition. Arcades shown natural size; first molars shown two times natural size.

columnar molars of warthogs to the bladelike molars of some carnivores. In humans, as in most other primates, the mesial (anterior) part of the molar is the **trigon** (**trigonid** in lower molars). The distal (posterior) part of the primate molar, added to the modified original triangle of cusps, is the **talon** (or **talonid** in lower molars).

17. **Interproximal contact facets** (**IPCF**s) are facets formed between adjacent teeth in the same jaw; **occlusal contact facets** result from contact of mandibular and maxillary teeth during chewing.

18. A **cingulum** is a ridge of enamel that partly or completely encircles the sides of a tooth crown and is usually not present in human molars and premolars.

In addition to these anatomical parts of teeth, a few additional terms used to describe and interpret teeth are also useful. **Supernumerary** teeth are teeth that exceed the expected number of teeth in any given category. For example, supernumerary molars (fourth molars) are rare in humans but more common in apes. **Agenesis** is the lack of tooth formation at a given position. **Hypoplasia** (hypomineralization) is a disturbance of enamel formation which often manifests itself in transverse lines, pits, or other irregularities on the enamel surface. **Hypercementosis** is a condition in which an excess of cementum forms on the root. **Taurodontism** refers to the condition in which the pulp chamber is inflated relative to the normal condition. **Caries** is a disease process resulting in the demineralization of dental tissues. **Shovel-shaped** incisors have strongly developed mesial and distal lingual marginal ridges, imparting a "shovel" appearance to the tooth. Several of these features are given further consideration in Chapters 18 and 20, where the uses of teeth in paleopathology and the study of populational affiliation are discussed.

5.4
Dental Development

Even before birth, germs of the deciduous teeth have formed within the jaws. When formation of each deciduous and permanent crown is complete and some root formation has occurred, it is erupted. When developing within the jaw, the **tooth buds,** or **germs,** reside in hollows in the alveolar bone called **crypts.** Within the crypt, calcification of the enamel cap of a tooth crown begins at the cusp apices and proceeds rootward. Crown formation, including enamel calcification, is completed before eruption and before the roots are completely formed. The last parts of a tooth to develop, after eruption, are the root apices. Before replacement by their adult counterparts, the roots of the deciduous teeth are **resorbed** (eaten away by osteoclasts) prior to the shedding of each tooth. In Chapter 17 we describe the timing of these processes and discuss their use in aging subadult human dentitions.

Enamel is formed by cells called **ameloblasts** through a process known as **amelogenesis.** Unlike bone, dental tissues are usually not remodeled during life. Once formation of the enamel is complete, only changes through physical (wear) or chemical (decay) means are possible. Dentin is formed by cells called **odontoblasts** through a process known as **odontogenesis. Primary dentin** is laid down during tooth formation and **secondary dentin** is laid down during the stage of root maturation.

The processes of tooth genesis are similar in all mammals. The striking differences between mammalian tooth shape, size, and structure come as a result of the differing activity of the two cell types, and this activity is in turn regulated by the DNA. Thus, because of the stability of posteruptive form, teeth have a better possibility of more directly reflecting the genes than other parts of the skeleton. For these reasons, teeth are widely used in the assessment of biological distance between human populations (Chapter 20). Insults to the organism during the developmental span of any tooth, however, can directly affect the morphology of the tooth. An example of this is the phenomenon of enamel hypoplasia.

Once erupted, teeth begin to wear away as they are used in mastication. Wear is usually most pronounced on the lingual occlusal surfaces of maxillary premolars and molars and on the buccal occlusal surfaces of mandibular premolars and molars. For the anterior teeth, the wear pattern is more variable because these teeth are often used in functions besides simple chewing. Conditions of overbite, underbite, and edge-to-edge occlusion are some of the variants seen in humans.

5.5
Tooth Identification

Because disassociated teeth are relatively abundant, and because of their importance in osteological work in forensics, archeology, and paleontology, it is necessary to be able to fully and accurately identify isolated teeth. A full and exact identification of each of the twenty deciduous and thirty-two permanent human teeth seems like a formidable task, but a little organization and some help from our closest relatives, the great apes, simplifies the job considerably.

Any modern dentition reflects millions of years of evolution. As a general rule, the variation between different teeth of the human dentition has decreased, or become "homogenized," whereas morphological variation at each tooth position has tended to increase. Thus, compared to ape dentition, human teeth are more difficult to individually identify. For those interested in the range of variation commonly encountered in human teeth, we recommend Taylor (1978) and Woelfel and Scheid (1997). See Figure 2.4 for an example.

In both apes and humans, it is fairly easy to determine whether teeth are incisors, canines, premolars, or molars. Most problems of identification come in distinguishing between left and right, and between different incisor, premolar, and molar positions in the upper and lower jaws. A few hints and an organizational structure greatly facilitate the study and recognition of individual teeth. When learning to distinguish between teeth, it is easiest to begin with the apes; the criteria used to discriminate between ape teeth are, for the most part, the same ones used to identify human teeth.

The steps toward identification of isolated teeth are outlined here in a logical order. It should prove possible to identify virtually all unworn and most worn human teeth by following this order. In using these identification guidelines, even for unworn teeth, always keep variation in mind. The criteria outlined here are the least variable. Identify each tooth on the basis of its crown morphology, check the identification on the basis of root number and shape, and double-check the identification by observing wear, including the presence, placement, and shape of interproximal contact facets. It is best to use several criteria for each identification. Check each one independently and make a decision based on the majority when there is conflict.

All of the identification steps outlined here assume that the tooth is a human tooth with little or no wear. In identifying worn teeth, attempt to mentally reconstruct the original, unworn crown morphology. Only through experience with a range of human teeth does the researcher become familiar with the normal, expected variation within the human species. Thus, when it is not initially obvious whether the tooth is human, working through the steps in detail may make it possible to identify and eliminate nonhuman teeth. Worn bear and pig teeth superficially resemble human teeth, but moderate experience with small samples of worn and unworn human teeth is almost always adequate to allow accurate diagnosis.

Remember that most of the criteria presented here for diagnosing modern human teeth to set, arch, class, or type are relative and hence depend on comparing one observation, index, or size against another. Also, note that the osteologist conveniently carries a handy full or partial comparative dentition in the mouth at all times. For identification of tooth category, arch, and side, the osteologist may find it useful to imagine properly placing the unidentified tooth into his or her own dentition. In identifying isolated teeth, or parts of teeth, proceed according to the sequence outlined in sections 5.6–5.10, with reference to Figure 5.3.

The teeth in Figures 5.4–5.19 are all reproduced twice natural size. Root orientation in mesial, buccal, labial, and lingual views is anatomical (maxillary roots point up, toward the top of the page). Occlusal views are shown with the mesial crown edge toward the top of the page. All "left" side teeth are photographic negative reversals of actual right side teeth. Note that the enamel of the illustrated teeth is not shiny. Enamel is translucent and shiny, but these characteristics combine to obscure topography during photography. To solve this problem, we coated the teeth illustrated here with opaque, thin pigment prior to photography.

5.6
To Which Tooth Category (Class) Does the Specimen Belong?
(Figures 5.4–5.5)

5.6.1 Incisors

- Incisor crowns are flat and bladelike.
- The outline of the incisor's occlusal dentine patch exposed by wear is rectangular or square.

5.6.2 Canines

- Canine crowns are conical and tusklike.
- The outline of the canine's occlusal dentine patch exposed by wear is diamond-shaped.
- Canine roots are longer than other roots in the same dentition.
- Some canines may be mistaken for incisors. Note, however, the longer, larger canine root relative to crown height and the oval canine crown cross section.

5.6.3 Premolars

- Premolar crowns are round, shorter than canine crowns, and smaller than molar crowns. They usually have two cusps.

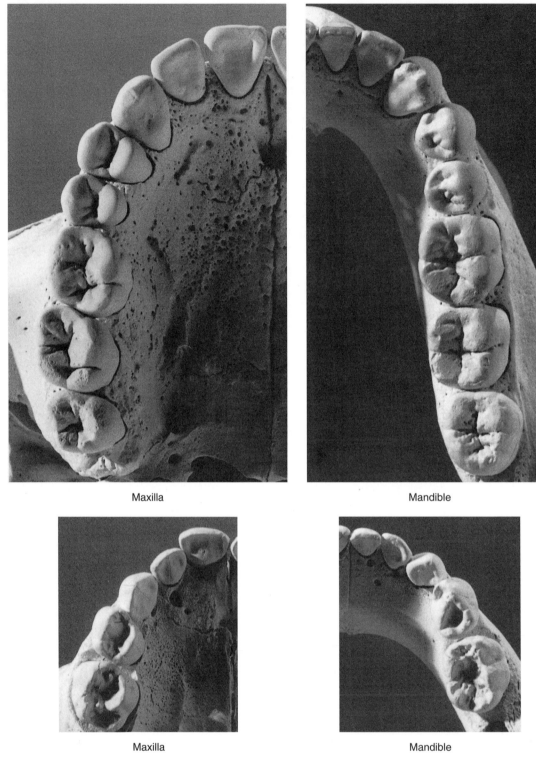

Maxilla Mandible

Maxilla Mandible

Figure 5.4 A comparison of maxillary and mandibular half arches for the permanent (*top*) and deciduous (*bottom*) dentitions. Occlusal, right side, two times natural size.

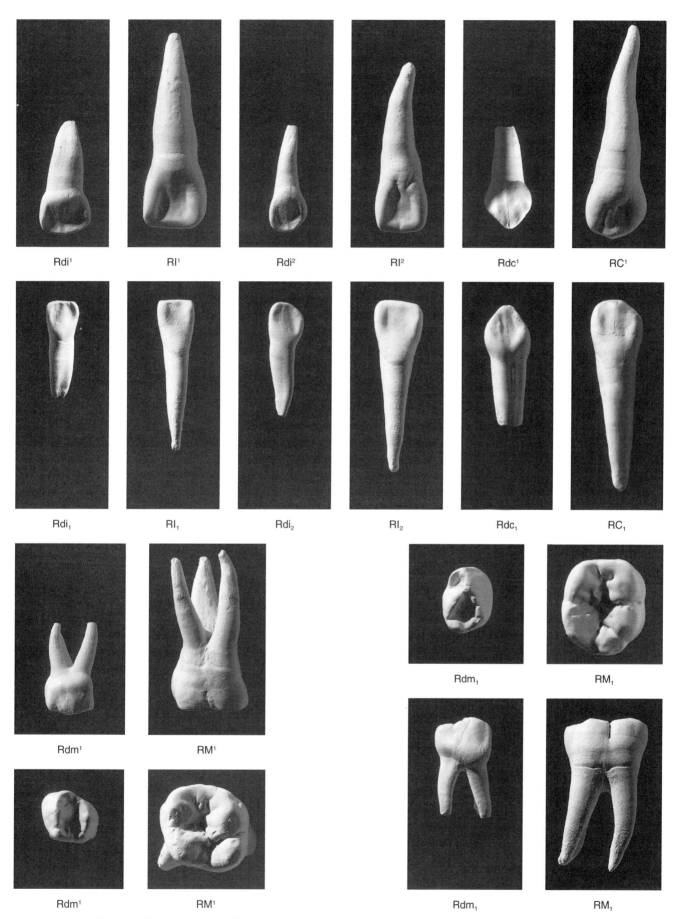

Rdi¹   RI¹   Rdi²   RI²   Rdc¹   RC¹

Rdi₁   RI₁   Rdi₂   RI₂   Rdc₁   RC₁

Rdm¹   RM¹   Rdm₁   RM₁

Rdm¹   RM¹   Rdm₁   RM₁

Figure 5.5 Deciduous and permanent teeth compared. Incisors and canines in lingual view; molars in buccal and occlusal views. Right side, two times natural size.

- Premolars are usually single-rooted.
- Some lower third premolars may be mistaken for canines. Note, however, the smaller crown height and shorter root of the premolar.

5.6.4 Molars

- Molar crowns are larger, squarer, and bear more cusps than other teeth.
- Molars usually have multiple roots.
- Reduced third molars are sometimes mistaken for premolars. To avoid this, note the relationship of root length to crown height, the round or oval outline of the premolars in occlusal aspect, and the comparatively regular cusp pattern on the premolar crowns.

5.7
Is the Tooth Permanent or Deciduous? (Figures 5.4–5.5)

5.7.1 Diagnostic Criteria

- Deciduous teeth are smaller in size within any tooth category.
- Deciduous crowns have enamel that is thinner relative to crown size.
- Deciduous tooth crowns are more bulbous in shape, with the enamel along the crown walls often bulging out above the enamel line more prominently than in permanent teeth.
- Deciduous tooth roots are thinner and shorter. The deciduous molar roots are more divergent.
- Deciduous tooth roots are often partly resorbed, particularly below the crown center of deciduous molars.

5.7.2 Special Cases

- Deciduous first upper and lower molar crowns have peculiar shapes. The upper is triangular in outline, with a strongly projecting buccal paracone surface. The lower has a low talonid and an extensive buccal protoconid surface.

5.8
Is the Tooth an Upper or a Lower?

To identify deciduous teeth, adult criteria outlined both above and below are applicable.

For incisors and canines, view the crown lingually. Gauge the maximum mesiodistal length and maximum crown height (correcting the latter for wear if necessary). When the height dimension is twice the length, it is probably a lower (i.e., tall narrow crowns).

For molars, determine the mesiodistal crown axis by observing the placement of the protocone or protoconid (mesial and lingual, and mesial and buccal, respectively; the largest and most heavily worn cusp) and the disposition of the interproximal contact facets (IPCFs), which must be mesial and distal.

5.8.1 Upper versus Lower Incisors (Figure 5.6)

- Upper incisor crowns are broad (mesiodistally elongate) relative to their height. Lower incisor crowns are narrow compared to their height.

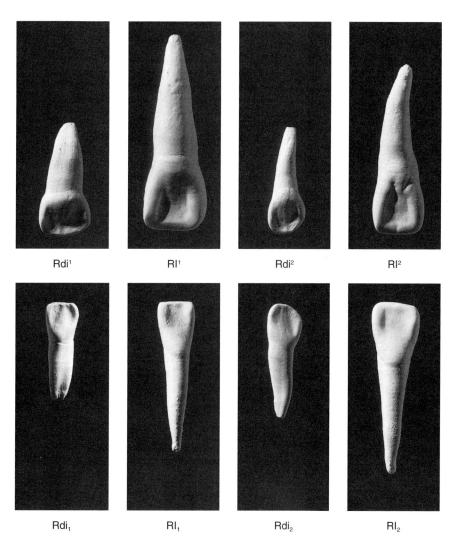

Rdi¹   RI¹   Rdi²   RI²

Rdi₁   RI₁   Rdi₂   RI₂

Figure 5.6 Upper and lower deciduous and permanent incisors compared. Right side, lingual view, two times natural size.

- Upper incisor crowns have much lingual relief. Lower incisor crowns have comparatively little lingual topography.
- Upper incisor roots are usually more circular in cross section. Lower incisor roots are usually more mesiodistally compressed in cross section.

5.8.2 Upper versus Lower Canines (Figure 5.7)

- Upper canine crowns are broad (mesiodistally elongate) relative to their height. Lower canine crowns are narrow relative to their height.
- Upper canine crowns have much lingual relief. Lower canines have comparatively little lingual relief.
- Upper canine crowns have apical wear that is mostly lingual. Lower canines have apical occlusal wear that is mostly labial.

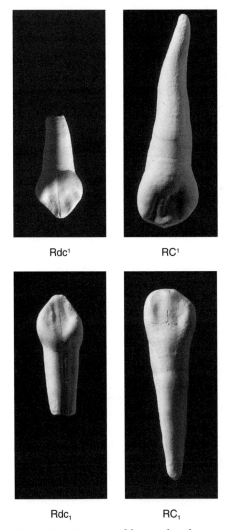

Rdc¹   RC¹

Rdc₁   RC₁

Figure 5.7 Upper and lower deciduous and permanent canines compared. Right side, lingual view, two times natural size.

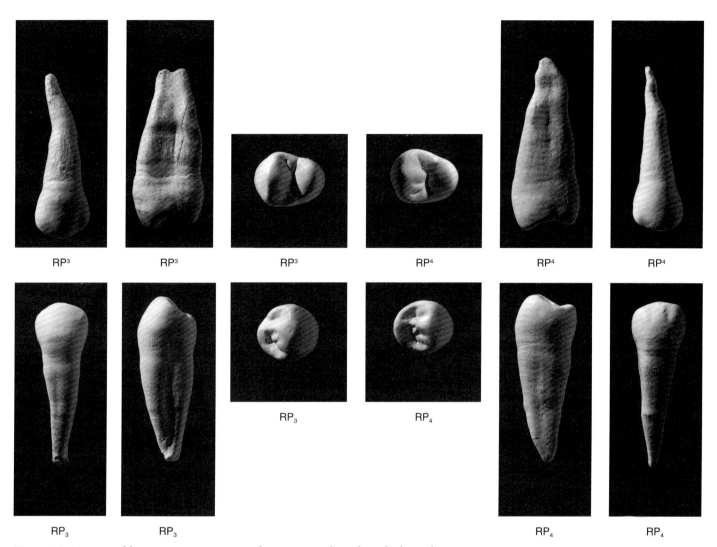

Figure 5.8 Upper and lower permanent premolars compared. Right side, buccal, mesial, and occlusal views, two times natural size.

5.8.3 Upper versus Lower Premolars (Figure 5.8)

- Upper premolar crowns have two cusps of nearly equal size. Lower premolar crowns show comparatively high disparity in buccal and lingual cusp size, the buccal cusp dominating the lingual in height and area.

- Upper premolar crowns have strong occlusal grooves oriented mesiodistally (median grooves) between the major cusps. Lower premolar crowns have comparatively weak median grooves.

- Upper premolar crowns are more oval in occlusal outline. Lower premolar crowns are more circular in occlusal outline.

5.8.4 Upper versus Lower Molars (Figure 5.9)

- Upper molar crowns usually have three or four major cusps. Lower molar crowns usually have four or five major cusps.

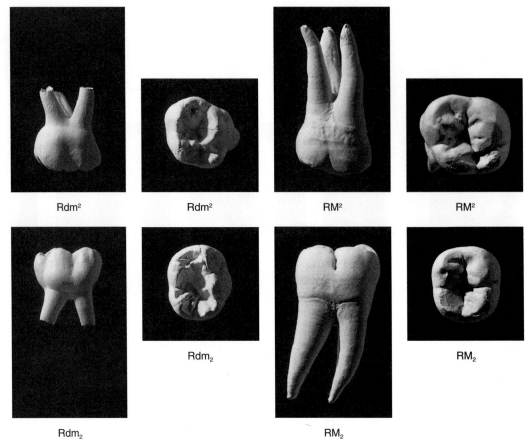

Rdm²   Rdm²   RM²   RM²

Rdm₂   RM₂

Rdm₂   RM₂

Figure 5.9 Upper and lower deciduous and permanent molars compared. Right side, buccal and occlusal views, two times natural size.

- Upper molar crowns have outlines in the shape of a rhombus (skewed rectangle) in occlusal view. Lower molar crowns have square, rectangular, or oblong outlines.

- Upper molar crowns have cusps placed asymmetrically relative to the mesiodistal crown axis. Lower molar crowns have cusps placed symmetrically about the crown midline.

- Upper molars usually have three major roots that are variably fused. Lower molars usually have two major roots but occasionally have three.

5.9
Where in the Arch Is the Tooth Located?

5.9.1 Upper Incisors: I¹ versus I² (Figure 5.10)

- Upper central incisor (I¹) crowns are larger than upper lateral incisor (I²) crowns.

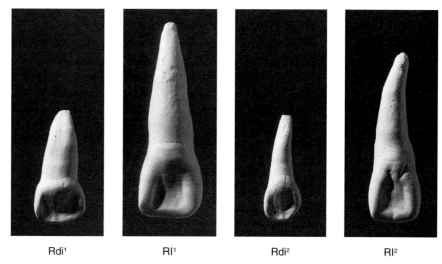

Rdi¹   RI¹   Rdi²   RI²

Figure 5.10 Upper deciduous and permanent incisors compared. Right side, lingual view, two times natural size.

- Upper central incisor (I¹) crowns have a greater mesiodistal length: height ratio than upper lateral incisor (I²) crowns in labial view.
- Upper central incisor (I¹) crowns are more symmetrical in labial view than upper lateral incisor (I²) crowns.
- Upper central incisor (I¹) roots are shorter and stouter relative to crown size than upper lateral incisor (I²) roots.

5.9.2 Lower Incisors: I₁ versus I₂ (Figure 5.11)

- Lower central incisor (I₁) crowns are slightly smaller than lower lateral incisor (I₂) crowns.

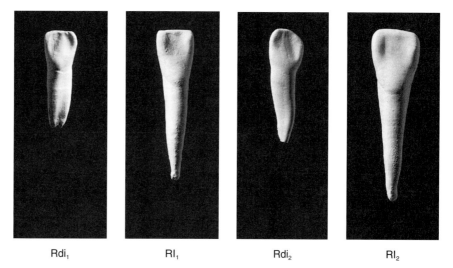

Rdi₁   RI₁   Rdi₂   RI₂

Figure 5.11 Lower deciduous and permanent incisors compared. Right side, lingual view, two times natural size.

- Lower central incisor (I_1) crowns have a smaller mesiodistal length: height ratio than lower lateral incisor (I_2) crowns in labial view.
- Lower central incisor (I_1) crowns are slightly more symmetrical in labial view than lower lateral incisor (I_2) crowns; the distal I_2 crown edges flare distally in this view.
- Lower central incisor (I_1) roots are shorter, both relative to crown height and absolutely, than I_2 roots.

5.9.3 Upper Premolars: P³ versus P⁴ (Figure 5.12)

- Upper third premolar (P³) crowns have major lingual cusps that are small compared to the major buccal cusps. Upper fourth premolar (P⁴) crowns have major buccal and lingual cusps of more equivalent size in occlusal view.
- Upper third premolar (P³) crowns have less symmetric, more triangular outlines in occlusal view than upper fourth premolar (P⁴) crowns. The latter are rounder because the relative mesiodistal length of the major buccal cusp is not as great as in the P³ crown (the lingual cusp of the P⁴ crown is relatively larger in both length and area).
- Upper third premolar (P³) crowns have more concave mesial surfaces and more deeply indented mesial occlusal outlines than upper fourth premolar (P⁴) crowns.
- Upper third premolar (P³) crowns show more mesiobuccal projection of the cervicoenamel line than upper fourth premolar (P⁴) crowns.
- Upper third premolar (P³) crowns contact mesially with the canine. The resulting small mesial IPCF is distinctive, usually curved, and vertically elongate. The mesial IPCF on the upper fourth premolar (P⁴) is usually more symmetrical and buccolingually elongate.
- Upper third premolar (P³) roots are usually double, bilobed, or apically bifurcated. Upper fourth premolar roots (P⁴) are usually single.

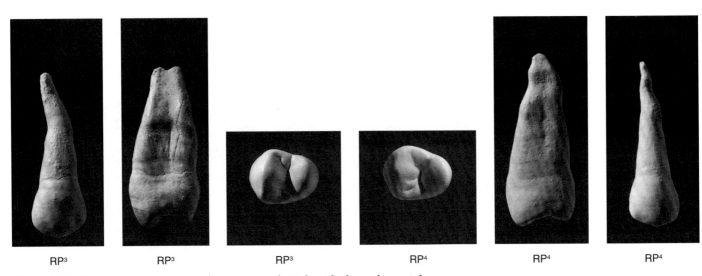

RP³   RP³   RP³   RP⁴   RP⁴   RP⁴

Figure 5.12 Upper permanent premolars compared. Right side, buccal, mesial, and occlusal views, two times natural size.

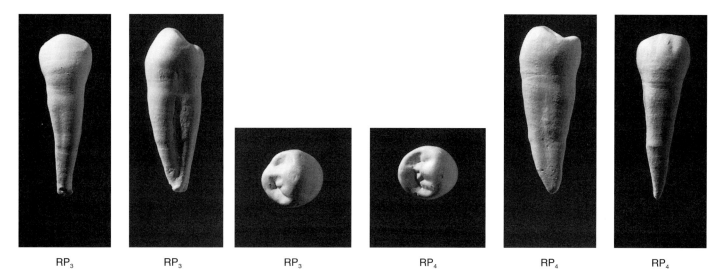

| RP₃ | RP₃ | RP₃ | RP₄ | RP₄ | RP₄ |

Figure 5.13 Lower permanent premolars compared. Right side, buccal, mesial, and occlusal views, two times natural size.

5.9.4 Lower Premolars: P₃ versus P₄ (Figure 5.13)

- Lower third premolar (P₃) crowns have a major lingual cusp that is small, relative to the dominant major buccal cusp, in both occlusal area and height. The major lingual cusp is often expressed merely as a small lingual ridge. Lower fourth premolar (P₄) crowns have major buccal and lingual cusps of more equivalent size, and the major buccal cusp is less pointed than on a P₃ crown.

- Lower third premolar (P₃) crowns have a mesial fovea placed very mesially, close to the mesial occlusal edge in occlusal view. Lower fourth premolar crowns have more distally placed mesial foveae.

- Lower third premolar (P₃) crowns have less symmetry of occlusal outline than lower fourth premolar (P₄) crowns.

- Lower third premolar (P₃) crowns have much smaller talonids than lower fourth premolar (P₄) crowns.

- Lower third premolar (P₃) crowns bear mesial (canine) IPCFs analogous to those discussed above for the upper counterparts.

5.9.5 Upper Molars: M¹ versus M² versus M³ (Figure 5.14)

- Upper first molar (M¹) crowns have four well-developed cusps arranged in a rhombic shape. Upper third molars (M³) tend to be smaller and more crenulate (furrowed on the occlusal surface) than first molars, with more irregular cusp positioning relative to the major crown axes. Upper third molars (M³) usually lack a hypocone in humans. Upper second molar (M²) crowns are intermediate to the first and third molar crowns in morphological attributes.

- Upper first molars (M¹) have three long, separate, and divergent roots. Upper third molars (M³) tend to have fused roots and lack distal IPCFs. Upper second molar (M²) roots are intermediate.

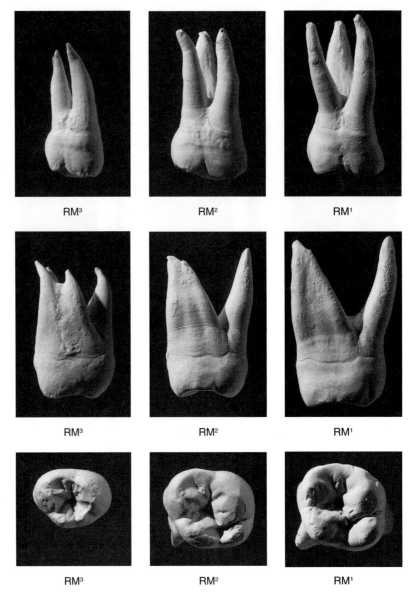

RM³   RM²   RM¹

RM³   RM²   RM¹

RM³   RM²   RM¹

Figure 5.14 Upper permanent molars compared. Right side, buccal, mesial, and occlusal views, two times natural size.

5.9.6 Lower Molars: M_1 versus M_2 versus M_3 (Figure 5.15)

- Lower first molar (M_1) crowns have five well-developed cusps usually arranged in the classic Y-5 pattern. Lower third molars (M_3) usually have four or fewer cusps with more variable arrangement. Lower third molars (M_3) tend to be smaller and more crenulate than first molars, with more irregular cusp positioning relative to the major crown axes. Lower second molar (M_2) crowns are intermediate to the first and third molar crowns in morphological attributes.

- Lower first molars (M_1) have two long, separate, and divergent roots. Lower third molars (M_3) tend to have fused roots and lack distal IPCFs. When the mesial and distal roots remain separate, the distal one is

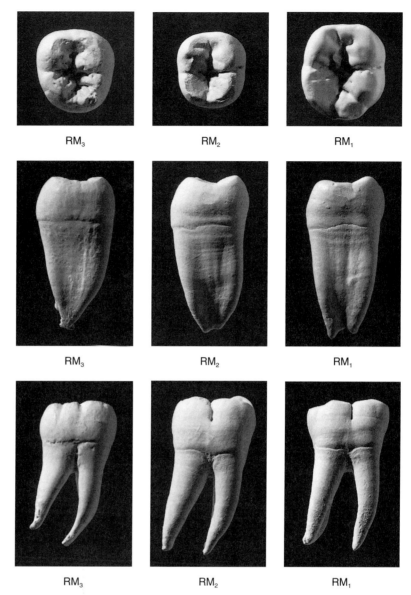

Figure 5.15 Lower permanent molars compared. Right side, occlusal, mesial, and buccal views, two times natural size.

postlike, set below a rounded posterior crown profile. Lower second molar (M_2) roots are intermediate.

5.10
Is the Tooth from the Right or the Left Side?

5.10.1 Upper Incisors (Figure 5.16)

Make the first two observations in labial view, orienting the specimen as it would rest in the dentition, with the occlusal surface horizontal. The root axis angles posterolaterally in the maxilla.

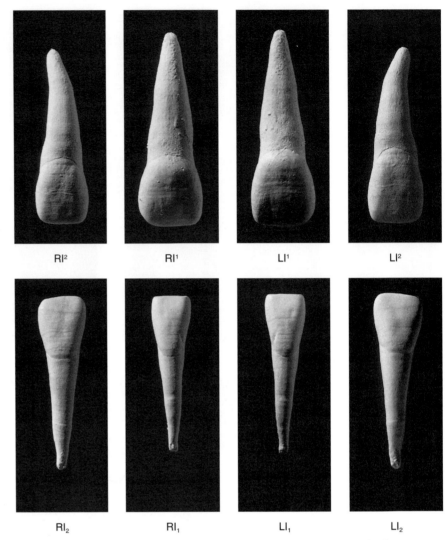

RI²   RI¹   LI¹   LI²

RI₂   RI₁   LI₁   LI₂

Figure 5.16 Right and left upper and lower permanent incisors. Labial view, two times natural size.

- The distal occlusal corner is more rounded than the mesial.
- The long axis of the root angles distally relative to the vertical (cervicoincisal) axis of the crown, the root tip usually leaning distally.
- The I^1/I^1 IPCF is more planar (flatter), wider, and more symmetrically placed than the more irregular, vertically elongate I^1/I^2 IPCF.
- The distal root surface is more deeply grooved than the mesial root surface.

5.10.2 Lower Incisors (Figure 5.16)

Make the first, third, and fourth observations in labial view, orienting the specimen as it would rest in the dentition, with the occlusal surface horizontal. The root axis angles posterolaterally.

- The distal occlusal corner is more rounded than the mesial.
- The I_1/I_1 IPCF is more planar (flatter) and symmetrically placed than the I_1/I_2 IPCF.
- The occlusal wear most often angles distally and inferiorly relative to the vertical (cervicoincisal) axis of the crown.
- The long axis of the root angles distally relative to the vertical (cervicoincisal) axis of the crown, the root tip usually leaning distally.

5.10.3 Upper Canines (Figure 5.17)

Make the first two observations in labial view with the tooth oriented as it would rest in the dentition.

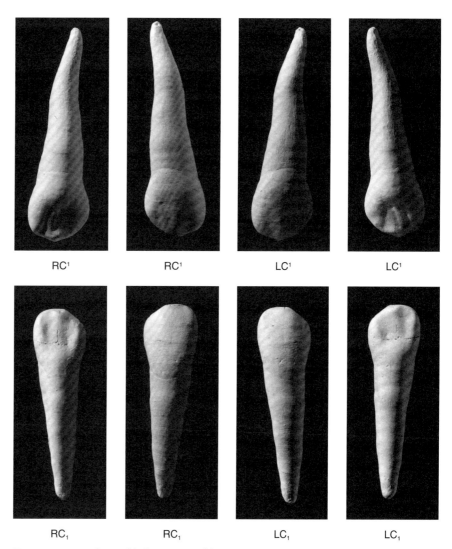

| RC¹ | RC¹ | LC¹ | LC¹ |

| RC₁ | RC₁ | LC₁ | LC₁ |

Figure 5.17 Right and left upper and lower permanent canines. Lingual and labial views, two times natural size.

- The mesial occlusal edge (ridge joining crown shoulder with apex) is usually shorter than the distal occlusal edge.
- The long axis of the root angles distally relative to the vertical (cervicoincisal) axis of the crown.
- The distal IPCF (for P³) is usually larger and especially broader than the mesial (lateral incisor) IPCF.
- The distal root surface is more deeply grooved than the mesial root surface.
- The mesial enamel line has a higher crownward arch.

5.10.4 Lower Canines (Figure 5.17)

All of the criteria used to side upper canines are applicable to lower canines.

5.10.5 Upper Premolars (Figure 5.18)

The IPCFs (when present) are mesial and distal, and the major median groove between cusps is oriented mesiodistally.

- The major lingual cusp is centered mesially relative to the major buccal cusp. Note for worn teeth that the center of the dentine exposure usually corresponds to the placement of the original cusp apex.
- The major lingual cusp is smaller, less occlusally prominent, and usually more heavily worn than the major buccal cusp.
- The long axis (axes) of the root(s) angle(s) distally relative to the vertical (cervicoincisal) axis of the crown.

5.10.6 Lower Premolars (Figure 5.18)

The IPCFs (when present) are mesial and distal.

- The major buccal cusp is larger, more occlusally prominent, and usually more heavily worn than the major lingual cusp.
- The major lingual cusp is centered mesially relative to the main buccolingual axis of the crown in occlusal view.
- The talonid, if present, is distal.
- The long axis (axes) of the root(s) angle(s) distally relative to the vertical (cervicoincisal) axis of the crown.

5.10.7 Upper Molars (Figure 5.19)

The IPCFs (when present) are located on the mesial and distal crown faces.

- The protocone is the largest, most heavily worn cusp. It occupies the mesiolingual crown corner.

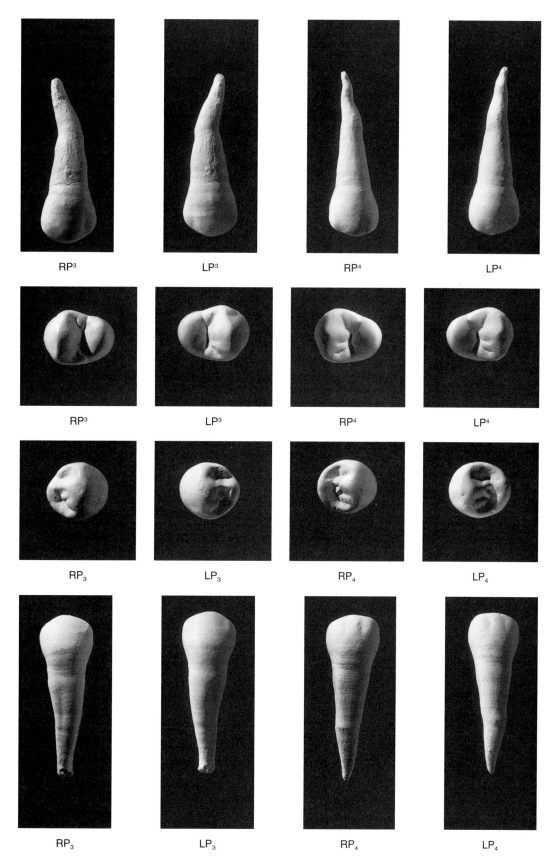

RP³   LP³   RP⁴   LP⁴

RP³   LP³   RP⁴   LP⁴

RP₃   LP₃   RP₄   LP₄

RP₃   LP₃   RP₄   LP₄

Figure 5.18 Right and left upper and lower permanent premolars. Buccal and occlusal views, two times natural size.

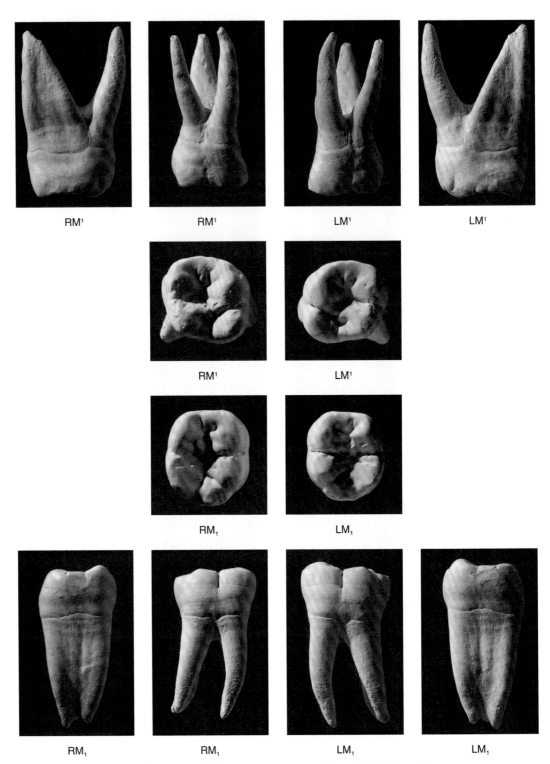

RM¹ RM¹ LM¹ LM¹

RM¹ LM¹

RM₁ LM₁

RM₁ RM₁ LM₁ LM₁

Figure 5.19 Right and left upper and lower permanent molars. Mesial, buccal, and occlusal views, two times natural size.

- The hypocone is the smallest (sometimes absent) cusp. It occupies the distolingual crown corner.
- The lingual cusps are occlusally less prominent than the buccal cusps and have heavier wear.
- In occlusal view, the lingual crown surface is more visible than the buccal crown surface.
- The largest of the three roots is often compressed buccolingually and set beneath the protocone and hypocone.
- The two smaller roots are rounder and set buccally (one mesial and one distal), and the mesiobuccal root is usually larger.
- All roots angle distally with respect to the major crown axes.

5.10.8 Lower Molars (Figure 5.19)

The IPCFs (when present) are located on the mesial and distal crown faces.

- The longest crown axis is usually mesiodistal.
- The protoconid is the largest, most heavily worn cusp. It occupies the mesiobuccal crown corner.
- The hypoconulid is the smallest cusp (unless there are additional, smaller cusps labeled C-6, C-7, and so on). It is placed distally and centered on the mesiodistal crown axis.
- The buccal cusps are occlusally less prominent than the lingual cusps and have heavier wear.
- The two major roots are mesiodistally compressed and set under the mesial and distal crown halves.
- All roots angle distally with respect to the major crown axes.

Suggested Further Readings

Because teeth are central to so many studies, an enormous literature on the dentition has developed over several centuries (see, for example, Metress and Conway, 1974, for a partial bibliography through 1974, and Foley and Cruwys, 1986, and Scott and Turner, 1988, for more recent reviews). Anthropologists, paleontologists, forensic analysts, and dental researchers have all contributed to this literature. *Dental Anthropology*, a publication of the Dental Anthropology Association, is a newsletter/journal with articles, reviews, and regular bibliographies.

Avery, J. K. (Ed.) (1987) *Oral Development and Histology.* Baltimore, Maryland: Williams and Wilkins. 380 pp.
    A comprehensive textbook on oral anatomy.

Hillson, S. (1986) *Teeth.* Cambridge: Cambridge University Press. 376 pp.
    An excellent guide to all matters having to do with teeth. Fine illustrations of a variety of mammalian teeth and a full consideration of human teeth from many perspectives.

Hillson, S. (1996) *Dental Anthropology.* Cambridge: Cambridge University Press. 373 pp.

    Everything about teeth, from an anthropological perspective.

Kelley, M. A., and Larsen, C. S. (Eds.) (1991) *Advances in Dental Anthropology.* New York: Wiley-Liss. 389 pp.

    An edited volume covering the spectrum of dental anthropology.

Kieser, J. A. (1990) *Human Adult Odontometrics.* Cambridge: Cambridge University Press. 194 pp.

    A comprehensive guide to what can be done with tooth measurements.

Scott, G. R., and Turner, C. G. (1997) *The Anthropology of Modern Human Teeth: Dental Morphology and Its Variation in Recent Human Populations.* New York: Cambridge University Press. 382 pp.

    A comprehensive guide to the anthropological utility of dental variation in modern humans.

Steele, D. G., and Bramblett, C. A. (1988) *The Anatomy and Biology of the Human Skeleton.* College Station, Texas: Texas A&M University Press. 291 pp.

    A good atlas with written descriptions of the human dentition.

Taylor, R. M. S. (1978) *Variation in Morphology of Teeth: Anthropologic and Forensic Aspects.* Springfield, Illinois: C. C. Thomas. 384 pp.

    An eye-opening guide to variation in human dentition.

Woelfel, J. B., and Scheid, R. C. (1997) *Dental Anatomy:* Its Relevance to Dentistry. Baltimore, Maryland: Williams and Wilkens. 449 pp.

    Everything about teeth, from the dentist's perspective. Excellent illustrations of variation among many individuals for each tooth.

CHAPTER 6

Hyoid and Vertebrae

In early vertebrates, a chain of cartilage-replacement bones formed around the notochord to give it stiffness. Over time, the bones developed processes that wrapped around the dorsal side of the nerve cord. Today we know these bones as vertebrae, elements from which the entire animal subphylum Vertebrata derives its name. In most modern vertebrates the vertebrae have replaced the notochord as the principal means of support for the central part of the body. The sacrum and coccyx are vertebrae, but they function as parts of the bony pelvis. The hyoid bone, an intermediary between the skull and postcranial skeleton, combines skeletal elements of the second and third pharyngeal arches associated with the gills of primitive fish.

6.1
Hyoid (Figures 6.1–6.3)

6.1.1 Anatomy

The hyoid bone is located in the neck and can be palpated immediately above the thyroid cartilage (the protuberance on the neck's anterior surface). It is the only bone in the body that does not articulate with another bone. Instead, it is suspended from the tips of the styloid processes of the temporal bones by the *stylohyoid ligaments*. The hyoid gives attachment to a variety of muscles and ligaments that connect it to the cranium, mandible, tongue, larynx, pharynx, sternum, and shoulder girdle. Its shape is highly variable, and it is often fractured in forensic cases involving strangulation (Pollanen and Ubelaker, 1997). The U-shaped hyoid consists of three major parts that are variably fused.

a. The **body** straddles the midline. It is a thin, posterosuperiorly concave, curved bone that articulates with, or is fused laterally to, the hyoid horns.

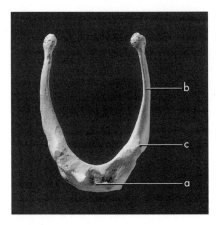

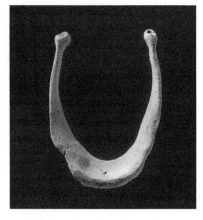

Figure 6.1 **Hyoid, superior.** Posterior is up. Natural size. Key: a, body; b, greater horns; c, lesser horns.

Figure 6.2 **Hyoid, lateral.** Anterior is toward the left, superior is up. Natural size.

Figure 6.3 **Hyoid, inferior.** Posterior is up. Natural size.

b. The **greater horns** are long, thin structures that form the posterior sides of the hyoid bone and project posterolaterally from the body on either side. The tip of each horn is a slightly expanded tubercle that serves as the attachment for the *lateral thyrohyoid ligament.*

c. The **lesser horns** are small, conical eminences on the superior surface of the bone in the area where the body and greater horns join. Their variably ossified apices give attachment to the *stylohyoid ligaments.*

6.1.2 Growth

The hyoid ossifies from six centers, two for the body and one for each of the greater and lesser horns.

6.1.3 Possible Confusion

Hyoids are most frequently mistaken for immature vertebral sections.

- The body is far thinner and more platelike than a vertebral centrum, or even any part of a vertebral arch.
- The horns are longer and thinner than the spinous processes on any vertebral element.

6.1.4 Siding

- The widest part of the greater horn is anterior, and the horn thins as it sweeps posteriorly, superiorly, and laterally to end in the tubercle.
- The superior surface bears the lesser horns.

6.2
General Characteristics of Vertebrae

Because the fused vertebrae of the sacrum and coccyx form part of the bony pelvis, they are described in Chapter 11. We describe here the twenty-four movable vertebrae—the seven cervical, twelve thoracic, and five lumbar vertebrae. In this section we introduce parts common to most vertebrae; in the following sections we concentrate on each class of vertebra.

6.2.1 Anatomy (Figures 6.4–6.31)

The vertebral column is most often composed of thirty-three elements in the adult human. Of these, twenty-four are separate (movable vertebrae), and the others are variably fused within the bony pelvis. It should always be possible to identify isolated, even fragmentary individual vertebrae by

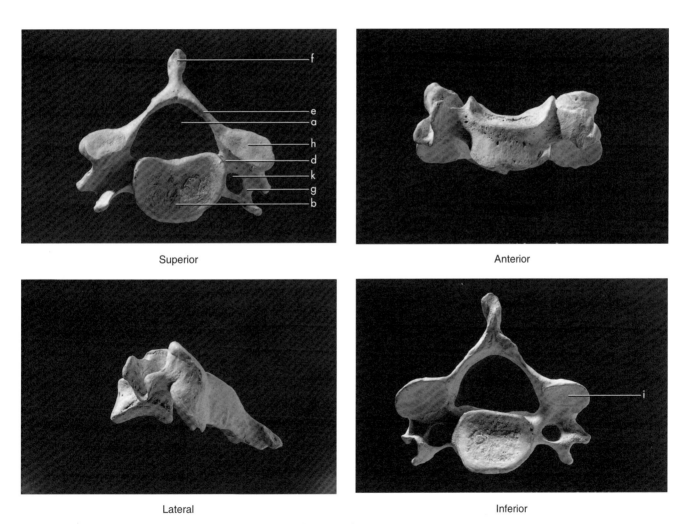

Superior

Anterior

Lateral

Inferior

Figure 6.4 **Fifth cervical vertebra,** a "typical" cervical. Natural size. Key: a, vertebral foramen; b, vertebral body; d, pedicle; e, lamina; f, spinous process; g, transverse process; h, superior articular facet; i, inferior articular facet.

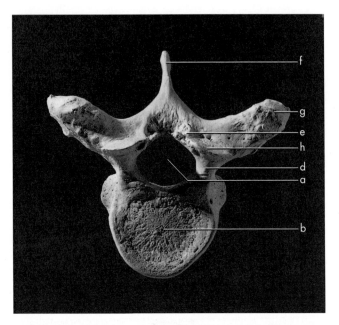

Superior

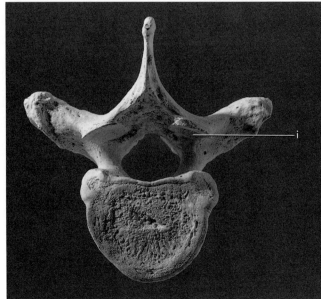

Inferior

Posterior

Anterior

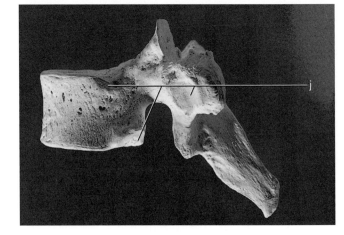

Lateral

Figure 6.5 **Seventh thoracic vertebra,** a "typical" thoracic. Lateral is lit from the lower left to accentuate costal articulations. Natural size. Key: a, vertebral foramen; b, vertebral body; d, pedicle; e, lamina; f, spinous process; g, transverse process; h, superior articular facet; i, inferior articular facet; j, costal foveae.

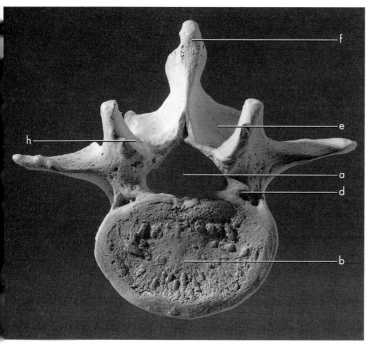

Superior

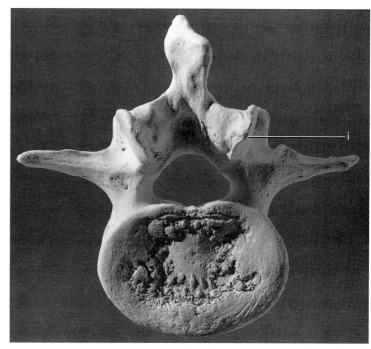

Inferior

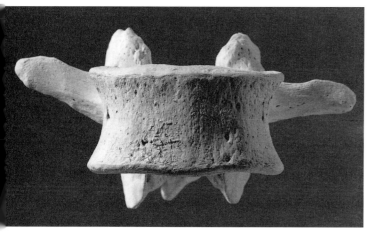

Anterior

Posterior

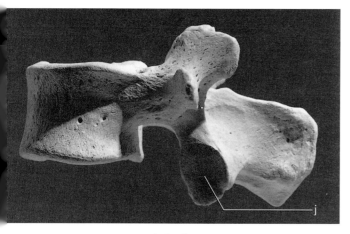

Lateral

Figure 6.6 **Third lumbar vertebra,** a "typical" lumbar. Lateral is lit from the upper right to accentuate the spine and inferior articular facet. Natural size. Key: a, vertebral foramen; b, vertebral body; d, pedicle; e, lamina; f, spinous process; h, superior articular facet; i, inferior articular facets.

type: **cervical, thoracic,** or **lumbar.** Cervical vertebrae are in the neck, thoracic vertebrae in the thorax region, and lumbar vertebrae just superior to the bony pelvis. Within these units, individual vertebrae are designated by letter (L = lumbar, T = thoracic, C = cervical) and identified by number from superior to inferior. For example, the most superior thoracic vertebra is designated T-1.

Successive vertebrae articulate directly with one another across synovial joints. All movable vertebrae have two superior and two inferior articular facets. These four facets control movement between adjacent vertebrae; their shapes are the key to the function and identification of different vertebrae. Individual vertebrae are united within the vertebral column by ligaments and muscles that hold them together in a flexible unit. The three primary functions of each vertebra are to bear body weight, to anchor muscles and ligaments, and to protect the *spinal cord*. When the entire column or large segments of the column are available, it is easy to identify each vertebra by type and number. Successive centra are larger caudally because of successively greater weight-bearing responsibilities. Some individual vertebrae, like the atlas (C-1), are easily diagnosed when isolated because of their unique morphology. Others, particularly the mid-cervicals and mid-thoracics, are far more difficult to identify by number when found disassociated from the other vertebrae.

In life, the *intervertebral disks* are made up of concentric rings of specialized fibrocartilage and lie between adjacent vertebrae. Each disk is composed of a circumference of fibrous tissue and fibrocartilage known as the *anulus fibrosus*. At the center of the disk is a soft substance known as the *nucleus pulposus*. These tissues are surrounded by a *fibrous capsule* that binds together adjacent vertebral bodies and encapsulates the disks. These soft tissue components are critical for movement in the vertebral column. The intervertebral disks make up more than one-fifth of the column's total height in life and contribute to its distinctive longitudinal curvature. Disks are thickest in the lumbar and cervical regions, where the vertebral column is most freely movable.

Individuals can vary in the number of vertebrae they bear in each category. This variation may occur in over 10% of all individuals in a skeletal population. Variation from the usual condition most often involves the addition or subtraction of one vertebral segment. The most frequent deviation from the usual pattern of seven cervical, twelve thoracic, and five lumbar vertebrae is the case of an extra thoracic or lumbar vertebra with an associated shift, as in thirteen thoracic and four lumbar vertebrae. We first describe the basic components of any vertebra before describing particular vertebral categories.

a. The **vertebral foramen** is the hole in each vertebra through which the spinal cord passes. Each vertebra thus forms a segment of the **vertebral canal** passing down the vertebral column.

b. The **vertebral body** is a spool-shaped structure that constitutes the main weight-bearing portion of a vertebra (except for the cases of the atlas and axis). The body is very thin-walled, composed mostly of lightweight, fragile, spongy bone, and it is therefore very susceptible to postmortem damage. It is also a center for blood production. The large foramen at the midline on its posterior surface is for the exit of the *basivertebral vein*.

c. The **vertebral arch** encloses the spinal cord posterior to the vertebral body.

d. The **pedicle** is the short segment of the arch close to the vertebral body, attached more superiorly to the body than inferiorly. Its superior and inferior surfaces are notched for the *spinal nerves* that emerge from the *spinal cord* to innervate corresponding body segments. These nerves pass through the gaps between adjacent articulating vertebrae, the **intervertebral foramina.**

e. Posterior to each pedicle is the **lamina,** the platelike part of the arch that attaches the pedicle to the spinous process.

f. The **spinous process** projects posteriorly on the midline and serves to anchor the *interspinous* and *supraspinous ligaments* and several muscles. These ligaments limit flexion of the vertebral column. Because the spinous processes act as levers for muscle action, the length, size, and slope of individual spines depend on the functional role that various back muscles play.

g. One **transverse process** is found on each side of each vertebra. Like the spinous processes, the transverse processes act as levers for the muscles attached to them. Movements of the axial skeleton are made possible by the muscles acting on these levers, whereas movement is restricted by the ligaments that hold the vertebrae together. Transverse processes in thoracic vertebrae articulate with the ribs.

h. The **superior articular facets** face posterosuperiorly in most cervical vertebrae, posteriorly in the thoracic vertebrae, and medially in the lumbar vertebrae.

i. The **inferior articular facets** face in the opposite directions of the superior facets for each vertebral class.

j. The thoracic vertebrae bear **costal foveae,** articular facets for the ribs, on the sides of the body and on the transverse processes.

6.2.2 Growth (Figure 6.7)

Most vertebrae ossify from three primary and five secondary centers. In immature vertebrae the three primary centers are the centrum and the two halves of the neural arch (the body is composed of this centrum plus a small segment of each neural arch). Secondary centers are found at the tips of each transverse process and the spine. Flattened, ringlike apophyses surround the periphery of the superior and inferior surfaces of each body. The superior and inferior surfaces of immature vertebral bodies have a "billowed" appearance, with a thin bony ring instead of a round plate of bone (plates are found in many other mammals) that fuses to the body upon skeletal maturation.

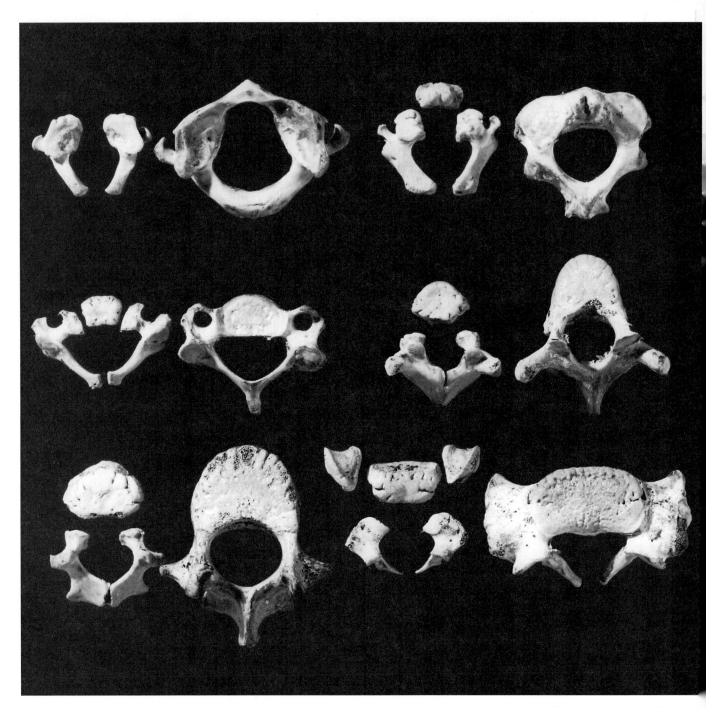

Figure 6.7 **Vertebral growth.** The pairs of immature vertebrae in superior view contrast a one-year-old with a six-year-old in each case. Top row left, atlas. Top row right, axis. Middle row left, cervical vertebrae number five. Middle row right, thoracic vertebrae number five. Bottom row left, lumbar vertebrae number two. Bottom row right, sacral vertebrae number one. Natural size.

6.3
Cervical Vertebrae (*n* = 7) (Figures 6.4 and 6.8)

Cervical vertebrae that share the characteristic cervical pattern are C-3 through C-6. We first describe this pattern and then assess the three remaining cervical vertebrae.

6.3.1 Identification

- Vertebral body. Cervical bodies are interlocking, with saddle-shaped superior and inferior surfaces. Cervical bodies are smaller than those of thoracic or lumbar vertebrae.
- Arch. At the root of the cervical arch and to either side of the cervical vertebral body, there are **transverse foramina** (Figure 6.4k) medial to the transverse processes. These foramina house the vertebral arteries, which pass upward to the posterior part of the brain. Cervical arches delimit triangular vertebral foramina, which are large and wide relative to the size of the vertebral body.
- Spinous process. Cervical spinous processes project fairly horizontally behind the vertebral body (mostly posteriorly, partly inferiorly). They are usually bifurcated (bifid) and often asymmetrical at their posterior tips, shorter than thoracic spinous processes, and not as massive as lumbar spinous processes.
- Transverse processes. Cervical transverse processes are very small, single or paired tubercles that project laterally beyond the transverse foramina.
- Superior and inferior articular facets. Cervical articular facets are cup-shaped or planar. The superior and inferior facets are parallel; both pass from anterosuperior to posteroinferior. The inferior facet is situated more posterior to the vertebral body than the superior facet.

6.3.2 Special Cervical Vertebrae (Figures 6.9–6.14)

- The **atlas** (C-1) lies between the cranium and the axis. Superior articular facets of the atlas are concave and elongate and receive the condyles of the occipital bone. The atlas vertebra has no vertebral body, lacks a spinous process, and has no articular disks superior or inferior to it. On the posterior surface of the anterior bony rim of this vertebra (the anterior edge of the vertebral foramen) there is an oval articulation for the dens (see below) of the axis.
- The **axis** (C-2) also lacks a typical vertebral body but has a projecting process that forms a pivot for the atlas. This projection is called the **dens** or **odontoid process** (Figure 6.12l). Thus, when the head is nodded up and down, movement is mostly at the joint between the occipital condyle and C-1. When the head turns from side to side, the atlas rotates about the dens of the axis.
- The **seventh cervical vertebra** (C-7) is transitional between a typical cervical and a typical thoracic vertebra. Its vertebral body is the largest of the cervicals and has a flat inferior surface. Its spine most closely

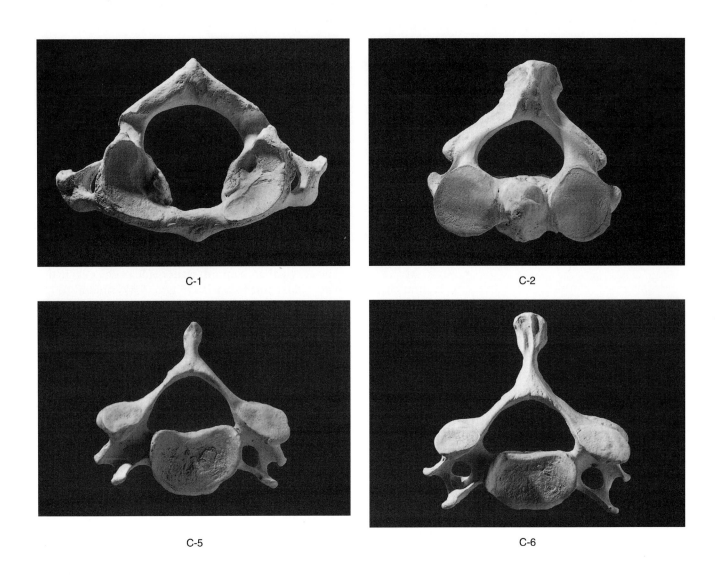

C-1

C-2

C-5

C-6

resembles a thoracic spine and is usually the uppermost vertebral spine palpable in the midline of the back in a living individual.

6.3.3 Siding

- The long axis of the spinous process is directed posteroinferiorly.
- The superior articular facet faces posterosuperiorly, and the inferior one faces the opposite direction.
- The raised articular sides of the vertebral body are superior.

6.4
Thoracic Vertebrae (*n* = 12) (Figures 6.5 and 6.15)

Thoracic vertebrae that share the characteristic thoracic pattern are T-2 through T-9. We first describe this pattern, then assess the four remaining thoracic vertebrae. Each thoracic vertebra articulates with a pair of ribs.

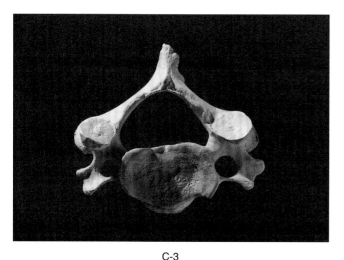

C-3

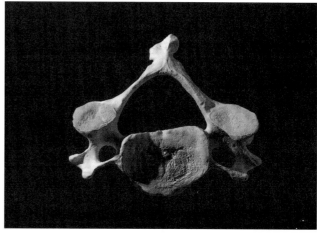

C-4

C-7

Figure 6.8 **Cervical vertebrae, superior.**
Posterior is up. Natural size.

6.4.1 Identification

- Vertebral body. Upper thoracic bodies are roughly triangular in superior outline, whereas lower thoracic vertebral bodies are more circular. Thoracics are intermediate in size between cervical and lumbar vertebrae. Thoracic vertebrae bear articular surfaces on their lateral sides which support the ribs. Each typical rib has two articulations with vertebrae, one on the side of adjacent thoracic bodies, and one at the end of the thoracic transverse process. Vertebrae T-2 through T-9 bear superior and inferior half facets, called **demifacets,** on the superior and inferior edges of the lateral sides of their bodies. Thus, the vertebral body of any typical thoracic vertebra articulates with four rib heads, one superolaterally and one inferolaterally on each side. When rib articulation is not shared between adjacent bodies, "whole" costal articulations, or **facets,** result.

- Arch. The vertebral canals are small relative to the bodies and are more circular in outline than in cervical vertebrae. Thoracic vertebrae lack transverse foramina.

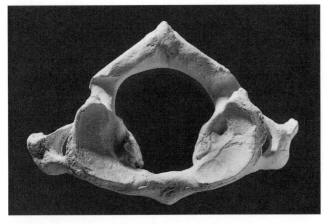

Figure 6.9 **Atlas vertebra, superior.** Natural size.

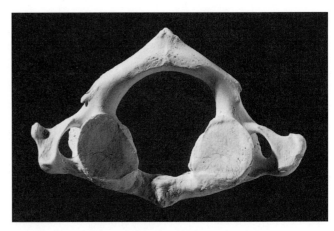

Figure 6.10 **Atlas vertebra, inferior.** Natural size.

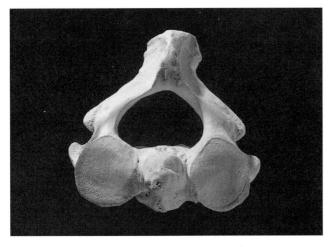

Figure 6.11 **Axis vertebra, superior.** Natural size.

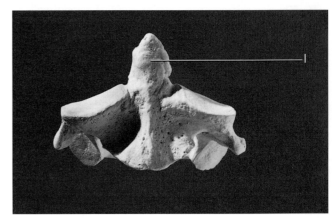

Figure 6.12 **Axis vertebra, anterior.** Natural size. Key: l, dens or odontoid process.

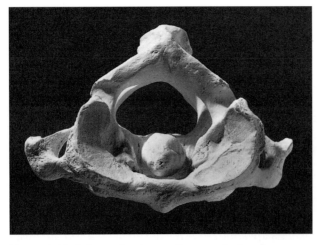

Figure 6.13 **Atlas and axis vertebrae, articulated, superior.** Natural size.

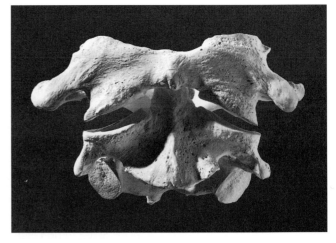

Figure 6.14 **Atlas and axis vertebrae, articulated, anterior.** Natural size.

- Spinous process. Thoracic spinous processes are long, straight, and narrow compared to the short, bifid cervical spinous processes and the hatchet-shaped lumbar spinous processes. The angle between the body and the spinous process becomes more acute and is directed more inferiorly in vertebrae midway between T-2 and T-8.
- Transverse processes. Thoracic transverse processes are prominent lateral projections from the arch. They bear articular **facets** (**foveae**) on their anterolateral corners for articulation with the tubercles of the ribs.
- Superior and inferior articular facets. Thoracic superior and inferior articular facets are very flat (planar) and set vertically, facing directly anterior (the inferior facet) or posterior (the superior facet).

6.4.2 Special Thoracic Vertebrae (Figures 6.16–6.19)

- **The first thoracic vertebra** (T-1) has a whole costal facet superiorly and a half costal facet inferiorly. It retains the most cervical-like characteristics of its spine and body.
- **The tenth thoracic vertebra** (T-10) usually has a complete, superiorly placed costal facet on each side of the vertebral body and costal articulations on the transverse processes.
- **The eleventh thoracic vertebra** (T-11) has an intact, superiorly placed costal facet on each side of the vertebral body but no costal articulation on the transverse processes.
- **The twelfth thoracic vertebra** (T-12) resembles T-11, but the inferior articular surfaces assume the lumbar pattern.

6.4.3 Siding

- The long axis of the spinous process is posterior and is angled sharply inferiorly, particularly in the mid-thoracics.
- The superior articular facet faces posteriorly, and the inferior one faces anteriorly.
- The costal articulation on the transverse processes faces anterolaterally.
- In lateral aspect, the inferior articular facets are separated from the rear half of the vertebral body by a considerable gap.
- The inferior dimensions of the body are greater than its superior dimensions.

6.5
Lumbar Vertebrae (*n* = 5) (Figures 6.6, 6.20, and 6.21)

It is comparatively easy to identify individual positions of isolated lumbar vertebrae. Like the thoracic and cervical vertebrae, the lumbars progressively increase in size from superior to inferior (L-1 to L-5). The lumbars

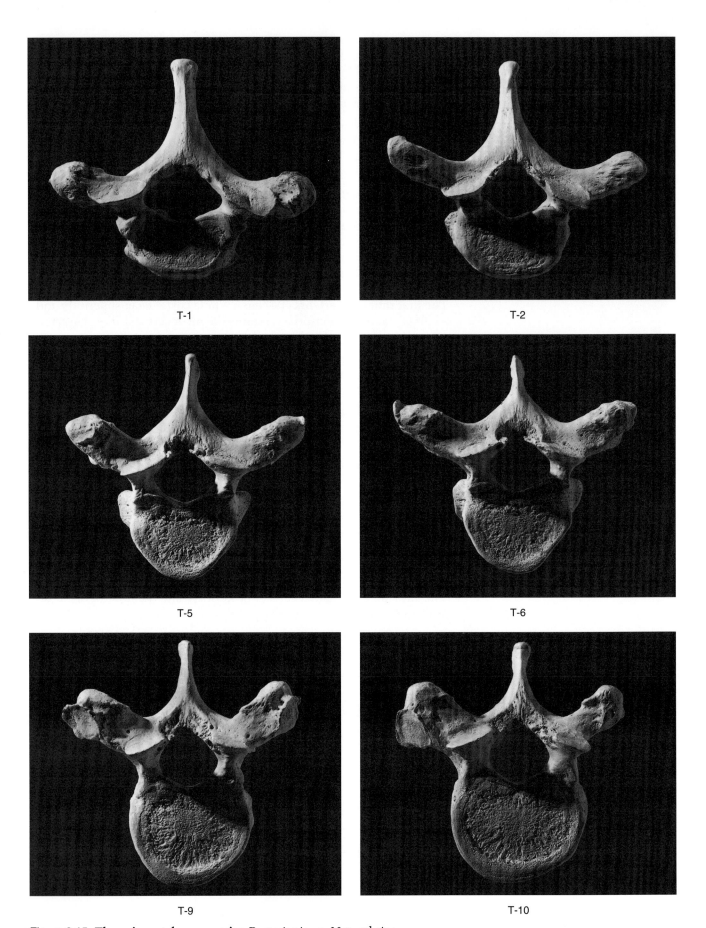

T-1

T-2

T-5

T-6

T-9

T-10

Figure 6.15 **Thoracic vertebrae, superior.** Posterior is up. Natural size.

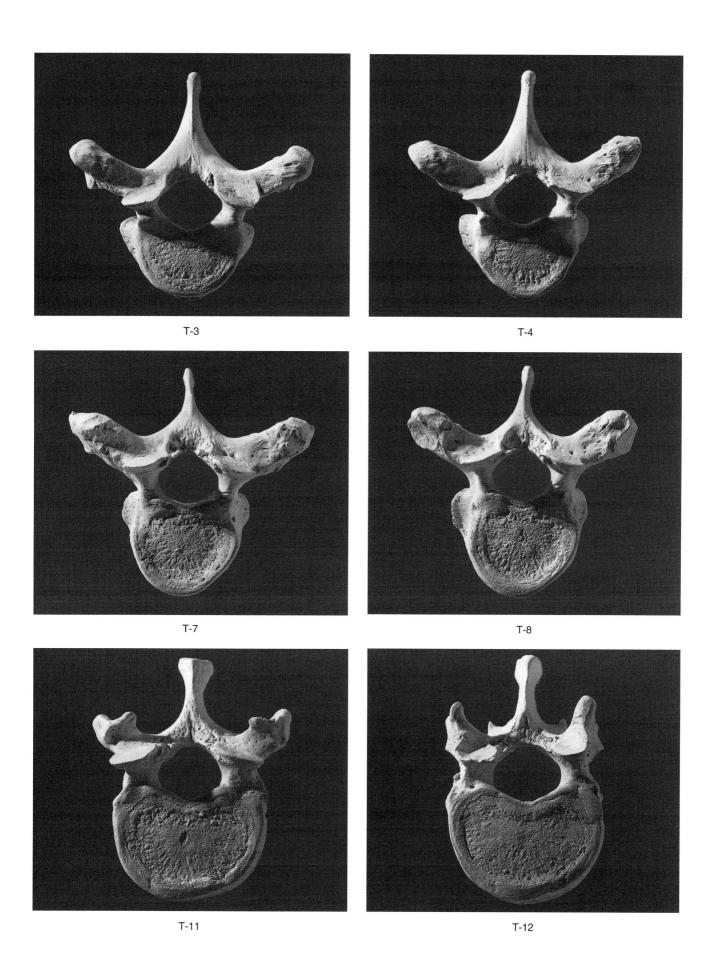

T-3

T-4

T-7

T-8

T-11

T-12

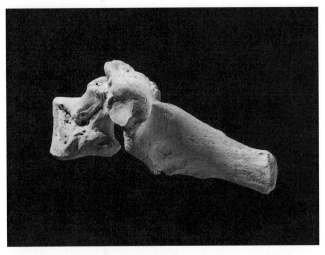

Figure 6.16 **First thoracic vertebra, lateral.** Natural size.

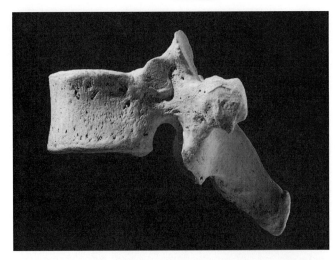

Figure 6.17 **Tenth thoracic vertebra, lateral.** Natural size.

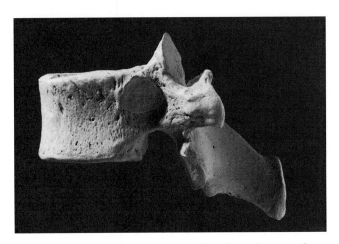

Figure 6.18 **Eleventh thoracic vertebra, lateral.** Natural size.

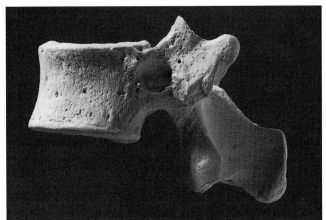

Figure 6.19 **Twelfth thoracic vertebra, lateral.** Natural size.

are the largest of all free vertebrae. Note that the individual illustrated in this text has a sacralized L-5 with a pair of extra sacral articular surfaces.

6.5.1 Identification

- Vertebral body. Lumbar bodies lack costal pits and transverse foramina.
- Arch. Lumbar arches are very small relative to the size of the bodies.
- Spinous process. Lumbar spinous processes are hatchet-shaped, large, blunt, and more horizontally oriented than other vertebral spinous processes (they are perpendicular to the coronal plane).
- Transverse processes. Lumbar transverse processes are relatively smaller and thinner than thoracic transverse processes and lack any articular surfaces. They are rudimentary or absent on L-1 and increase in size and projection inferiorly.

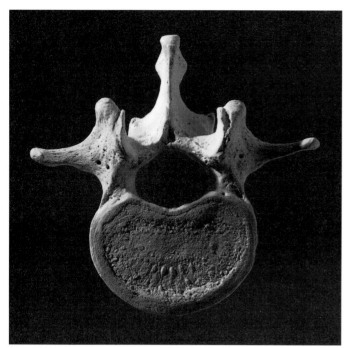

L-1

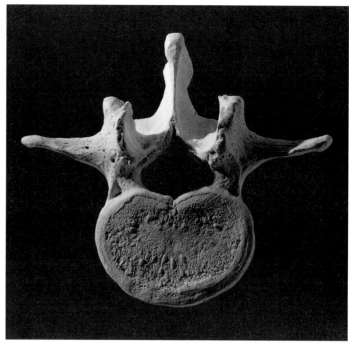

L-2

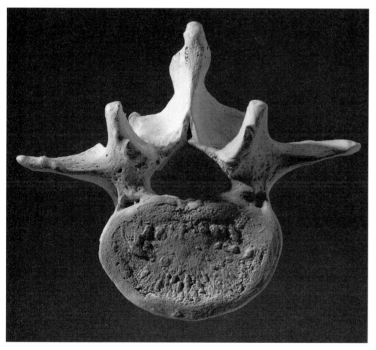

L-3

Figure 6.20 **Lumbar vertebrae, superior.** Posterior is up. Note that L-5 is "sacralized" and articulates five times with the sacrum. Natural size.

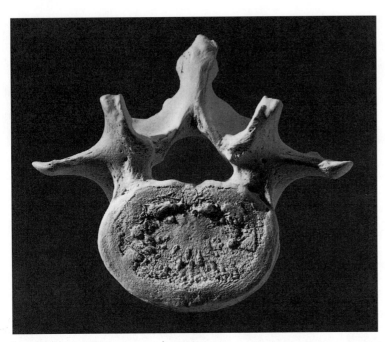

L-4

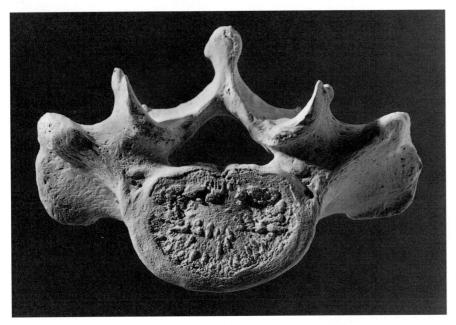

L-5

- Superior and inferior articular facets. The superior and inferior lumbar articular facets are not parallel. Instead, the superior articular facets are concave and face posteromedially. The inferior articular facets are convex and face anterolaterally.

6.5.2 Identifying Lumbar Position

The L-1 is the smallest of the series. When viewing the posterior surface of a lumbar vertebra (see Figure 6.21), imagine a quadrangle connecting

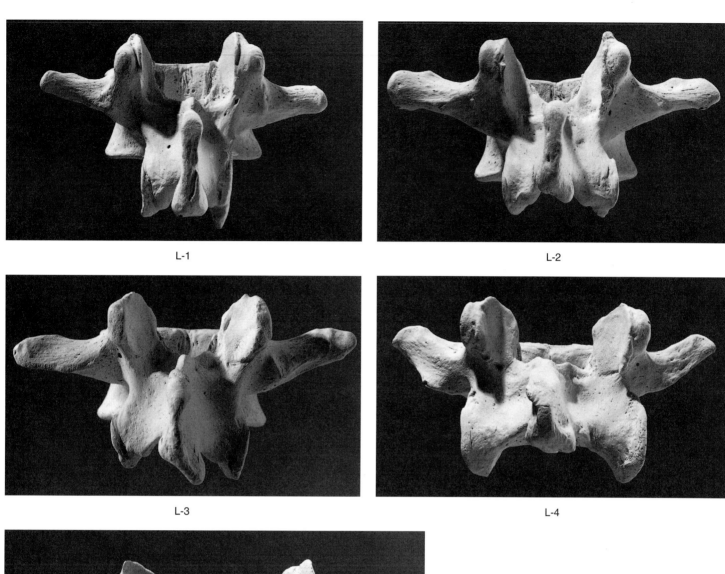

L-1

L-2

L-3

L-4

L-5

Figure 6.21 **Lumbar vertebrae, posterior.** Viewing the bones in this aspect allows their individual identification. Natural size.

the centers of four superior and inferior articular facets. In L-1 and L-2, this outline is a vertically elongate rectangle. In L-3 and L-4 the outline resembles a square. In L-5 the outline is a horizontally elongate rectangle.

6.5.3 Siding

- The long axis of each transverse process passes from the vertebral body and lamina superolaterally.

- The superior articular facets are concave and face posteromedially, whereas the inferior facets are convex and face anterolaterally.

- Note the large gap between the inferior articular facets and the posterior surface of the vertebral body in lateral aspect. The arch originates from the superior half of the vertebral body.

6.6
Functional Aspects of the Vertebrae

Compared to many other animals, humans have a specialized head, large limbs, and a relatively short vertebral column. Different portions of the column perform independent functions, and the shapes of individual vertebrae are therefore different. The vertebral column can be divided into five regions: cervical, thoracic, lumbar, sacral, and coccygeal. The thoracic segment is concave anteriorly. The sacral and coccygeal segments are described in Chapter 11 on the pelvic girdle. Both the cervical and lumbar segments of the vertebral column are concave dorsally, permitting habitually erect posture. Most vertebrae share basic parts: arch, vertebral body, articular processes, transverse processes, and spinous processes. This correspondence of parts in sequential bones is called **serial homology.**

The different shapes of the different parts of each vertebra correspond to the functions they perform in different parts of the vertebral column. The most flexible part of the column is the neck, where cervical bodies are small, intervertebral disks are thick, and vertebral foramina are large, permitting much flexibility of movement. Thoracic disks are much thinner, and the superior and inferior articular facets of these vertebrae are parallel. The lumbar region is second to the cervical in mobility, with thick disks, and articular facets that are cup-shaped. Both thoracic and lumbar vertebrae allow anteroposterior bending of the column, but this movement is more restricted in the thoracic region. Medial and lateral bending of the column is also restricted in the thoracic region, but thoracic vertebrae allow for medial and lateral rotation (twisting of the column), which is limited in the lumbar region.

CHAPTER 7

Thorax: Sternum and Ribs

In primitive, air-breathing fishes, breathing was accomplished by swallowing movements in which air was gulped into the lungs. Early reptiles improved on this system when they evolved a means to respire via the musculoskeletal mechanics of the thoracic skeleton. This was made possible as the thoracic ribs extended ventrally from the vertebral column to reach the sternum. The sternum formed a kind of ventral bony column that fused into a bony bar and anchored the distal ends of the ribs. Further soft tissue specializations led to more sophisticated breathing functions.

The skeleton of the human **thorax,** or chest, is like a basket or cage composed of cartilage and bone. It is attached dorsally to the vertebral column. This structure encloses and protects the principal organs of circulation and respiration, the heart and lungs, and is the base to which the upper limbs are attached. The major bones forming the thorax are the sternum and the twelve ribs on each side. The upper seven ribs on each side connect, via cartilage, directly with the sternum and are sometimes called "true," or "sternal," ribs. Ribs 8–10 attach to the sternum indirectly, also via cartilage, and are sometimes called "false," or "asternal," ribs. The last two ribs ("floating" ribs) have short cartilaginous ends that lie free in the sides of the body wall.

7.1
Sternum (Figures 7.1–7.3)

7.1.1 Anatomy

The sternum, or breastbone, functions at its upper end to connect the shoulder girdle (clavicle and scapula) to the thorax. In addition, it anchors the anterior ends of paired ribs 1–7 via cartilage. The bone is composed of three main parts in adulthood but develops from six segments. The segment joints may all fuse in adulthood, but their location is indicated by the costal notches along each side of the sternum.

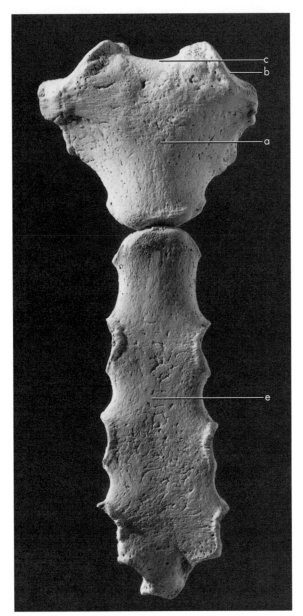

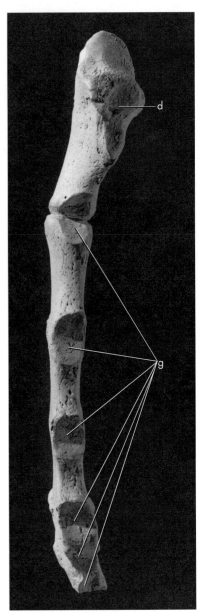

Figure 7.1 **Sternum, anterior.** The xiphoid process on this sternum had not ossified and is not shown. Superior is up. Natural size. Key: a, manubrium; b, clavicular notch; c, jugular (suprasternal) notch; e, corpus sterni.

Figure 7.2 **Sternum, lateral.** Superior is up. Natural size. Key: d, costal notches; g, costal notches.

a. The **manubrium** is the most massive, thickest, and squarest of three main sternal elements. It is the superiormost element of the sternum and is the widest part of this bone.

b. The **clavicular notches** occupy the superior corners of the sternum. It is here that the manubrium articulates with the right and left clavicles.

c. The **jugular (suprasternal) notch** is the midline notch on the superior border of the manubrium.

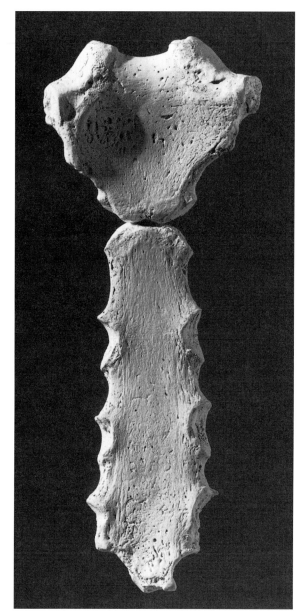

Figure 7.3 **Sternum, posterior.** Superior is up. Natural size.

d. The **costal notches** occupy both sides of the manubrium inferior to the clavicular notches. These notches represent articulations with the costal cartilages of the first ribs. The manubrium shares articulation for the second ribs with the corpus sterni.

e. The **corpus sterni** is the central part, the body, or "blade," of the sternum. It is formed during ontogeny from the fusion of sternal segments (sternebrae) 2–5. The corpus sterni may fuse, partially fuse, or remain unfused with the manubrium in adulthood.

f. The **sternal angle** is the angle formed between the fused manubrium and corpus sterni.

g. The **costal notches** along either side of the corpus sterni are for articulation with the costal cartilages of ribs 2–7.

h. **Lines of fusion** are often apparent. These lines pass horizontally through the right and left costal notches for ribs 3–5.

i. In 5–10% of adult corpora sternorum a midline foramen, the **sternal foramen,** perforates the sternal body.

j. The **xiphoid process** is the variably ossified inferior tip of the sternum. It often fuses with the corpus sterni in older adults. It shares the seventh costal notch with the body. This process can be partially ossified and may ossify into bizarre asymmetrical shapes with odd perforations. In short, the xiphoid is a highly variable element. The xiphoid process of the individual chosen to illustrate this text, for example, was not ossified at the time of death.

7.1.2 Growth

The four superior centers of ossification (manubrium and corpus sterni segments 2–4) appear in fetal life. Fusion between sternal segments is often irregular, but fusion between the lower two centers in the corpus sterni (segments 4 and 5) occurs soon after puberty. Fusion occurs between segments 2, 3, and 4 by early adulthood. As mentioned above, the manubrium and corpus sterni sometimes fail to unite even in adulthood.

7.1.3 Possible Confusion

Fragments of the sternum might be mistaken for fragments of pelvis or immature vertebrae.

- The costal and clavicular notches, and the lines of fusion between sternal segments, should be sufficient to ensure correct identification.
- The sternum is a bone with very low density. Its cortex is paper-thin, perforated by numerous microforamina. This makes even fragmentary sterna easy to sort from other elements, such as the pelvis.
- Infant sternebrae are often confused with vertebral centra. Infant centra have rougher, less mature (billowed, granular) cortical surfaces.

7.1.4 Siding

- Fragments of manubrium and corpus sterni can be sided by noting that the anterior surface of this bone is rougher and more convex than the smooth, gently concave posterior surface.
- The lines of fusion pass horizontally, and the mediolaterally widest point on the corpus sterni is at the third segment inferior to the manubrium.

7.2.1 Anatomy

There are usually twelve ribs on each side of the thorax, for a total of twenty-four in the adult male and female human body. The number of ribs is variable; there may be eleven or thirteen ribs on a side, with supernumerary ribs in either the cervical or lumbar segment (Black and Schever, 1997). The upper seven ribs (numbers 1–7) articulate directly with the sides of the sternum via cartilage. Ribs 8, 9, and 10 are interconnected medially by common cartilages that attach to the sternum. The last two ribs, 11 and 12, have free-floating ventral ends. All ribs articulate via their proximal ends with thoracic vertebrae. The ribs usually increase in length from rib 1 to rib 7, and decrease from rib 7 to rib 12. Recent concern with the use of the sternal rib end to estimate adult age at death (Chapter 17) has led several investigators to refine methods of siding and sequencing human ribs (Mann, 1993; Hoppa and Saunders, 1998). These apply to ribs known to be from a single individual.

a. The **head** of a rib is the swollen proximal part of the rib. It bears two articular surfaces (demifacets) for contact with the bodies of successive thoracic vertebrae. The first rib and ribs 10–12 are unifaceted.

b. The **neck** of a rib is the short segment between the head and the rib's articulation with the transverse process of the thoracic vertebra.

c. The **tubercle** is located on the posteroinferior corner of each rib. It articulates with the transverse process of the thoracic vertebra, presenting a medial articular facet for articulation with the transverse process of the thoracic vertebra, and a nonarticular part for ligamentous attachment.

d. The **angle** is the sharp curve in the bone lateral to the tubercle. It is marked by a prominent line on the external surface of the shaft immediately distal (lateral) to the tubercle. This line marks the attachment of the deep muscles of the back. It also marks the shift from the caudally facing external rib surface to the more cranially oriented external surface. The tubercle to angle distance increases from rib 4 to rib 11.

e. The **shaft** of a rib is the curved, tapering segment between the tubercle and the rib's distal (ventral, anterior) end. Shafts of ribs 3–6 are thicker and rounded in section compared to those of ribs 7–12.

f. The **costal groove** of a rib is the groove along the medial side of the inferior edge of the rib shaft. In life this groove houses an *intercostal artery, vein,* and *nerve.* It is most prominent on ribs 5–7.

g. The **sternal end** of a rib is the anterior (ventral) end of the shaft. This end is a roughened, porous, cupped oval surface for the attachment of cartilage. Its surface changes substantially with increasing age. The sternal ends of ribs 11 and 12 taper to a point.

h. The **cranial (upper) edge** of most ribs is blunt, smooth, and convex.

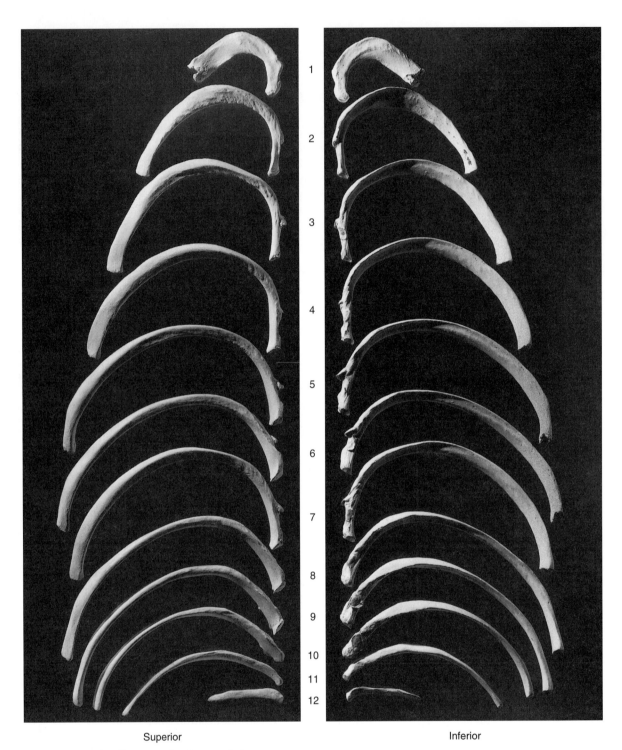

1
2
3
4
5
6
7
8
9
10
11
12

Superior

Inferior

Figure 7.4 **Right ribs.** One-third natural size.

i. The **caudal (lower) edge** of most ribs is sharp, with a costal groove on the medial surface. This groove gives the surface of the rib that faces the body cavity a concave appearance.

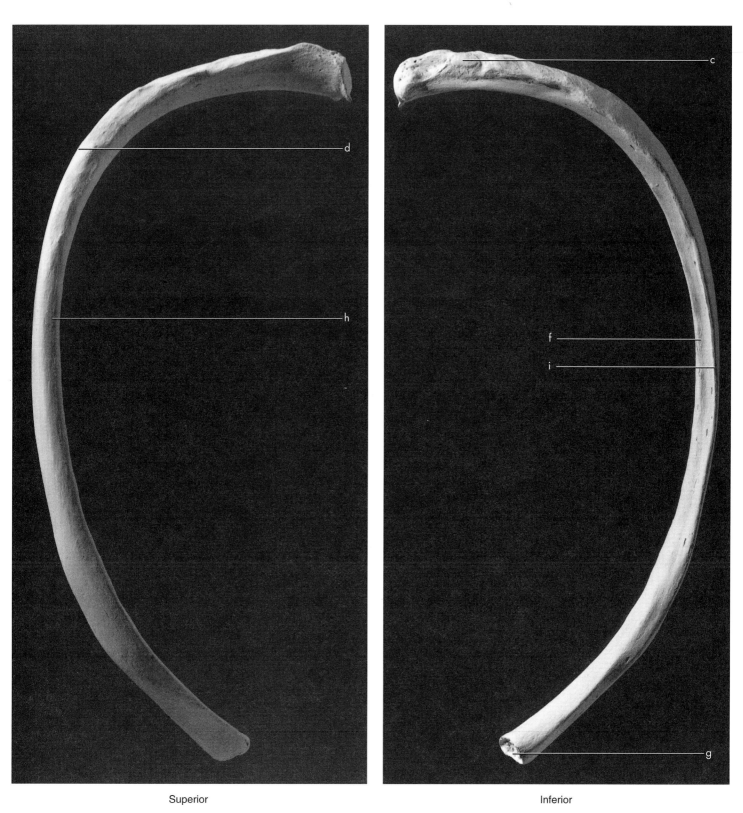

Superior   Inferior

Figure 7.5 **Right eighth rib,** a "typical" rib. Natural size. Key: c, tubercle; d, angle;
f, costal groove; g, sternal end; h, cranial (upper) edge; i, caudal (lower) edge.

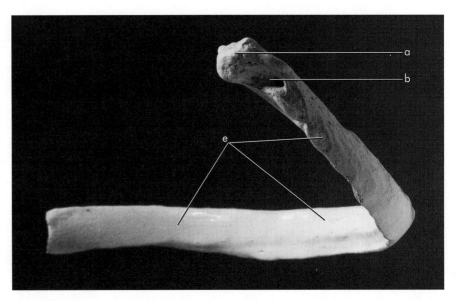

Figure 7.6 **Right eighth rib, posteroinferior.** Natural size. Key: a, head; b, neck; e, shaft.

7.2.2 Special Ribs (Figure 7.7)

- The **first rib** is the most unusual and therefore most easily diagnosed rib. It is a broad, superoinferiorly flattened, short, tightly curved bone with only one articular facet on its small rounded head. Its superior (cranial) surface is roughened by muscle attachments. This surface also bears two shallow grooves, the anterior one for the *subclavian vein* (Figure 7.7a) and the other for the *subclavian artery* and *inferior trunk of the brachial plexus* (medial) (Figure 7.7b). The ridge between these grooves is prolonged ventrally into the **scalene tubercle** for the attachment of the *anterior scalene muscle* (Figure 7.7c). There is no true inferior costal groove.

- The **second rib** is intermediate between the unusual first rib and the more typical ribs 3–9. It has a large tuberosity for the *serratus anterior muscle* near the external midshaft position (Figure 7.7d).

- The **tenth rib** is like ribs 3–9 but usually has only a single articular facet on the head.

- The **eleventh rib** has a single articular facet on the head and lacks a tubercle. Its sternal end is narrow and often pointed. Its shaft has a slight angle and a shallow costal groove.

- The **twelfth rib** is shorter than the eleventh and may be shorter than the first. This rib is similar to the eleventh in morphology and also lacks the angle and the costal groove.

7.2.3 Growth

Ribs, except for the eleventh and twelfth, ossify from four centers. Epiphyses for the head and for the articular and nonarticular parts of the tubercle appear in the teens. They fuse to the rib body in early adulthood.

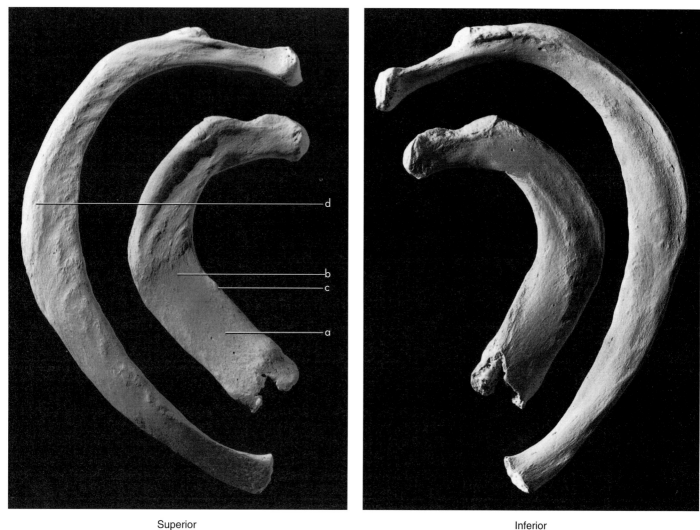

Superior   Inferior

Figure 7.7 **Right first and second ribs.** Natural size. Key: a, groove for subclavian vein; b, groove for subclavian artery and plexus; c, scalene tubercle; d, tuberosity for serratus anterior muscle.

7.2.4 Possible Confusion

- A fragmentary first rib might be mistaken for an inferior ramus of the os coxae. The cortex of the rib, however, is not as thick, the surface is more irregular, and the cross section is much thinner than in the os coxae.

- Proximal ends of other ribs, when broken into short segments, could be mistaken for metatarsal or metacarpal shafts. However, the cross section of a rib is more irregular, with one sharp edge. Usually the tubercle, head, or part of the costal groove is enough to diagnose a broken specimen as a rib.

- The first rib neck and head are flatter and smaller, respectively, than a transverse process from a thoracic vertebra.

- The head of a rib can be confused with a broken infant ulna, but attention to bone maturity can aid in diagnosis of these parts.

7.2.5 Siding

- For the first rib, the head and neck point inferiorly, and the superior surface bears grooves when in correct anatomical position.
- For all other ribs, the heads point toward the midline, and the tubercles are inferior.
- The cranial edge is thicker and blunter than the grooved, sharp inferior edge.
- The inner surface of the twelfth rib faces superiorly.

7.3
Functional Aspects of the Thoracic Skeleton

This introduction to the vertebrae, ribs, and sternum treats these elements individually. It is also important to note the functional interconnections between the elements and the dynamic, coordinated role that they play in the living human. The lungs, which function to transmit gases in and out of the bloodstream, are protected by the ribs, sternum, and vertebrae. However, these skeletal elements do more than just protect the lungs, heart, and great vessels—they provide attachment for respiratory muscles and for muscles that move the forelimbs. Inhalation, bringing oxygen-rich air into the lungs, requires action of the thoracic musculoskeletal system. Expiration can be more passive, except in heavy breathing.

The inhalation of several liters of air is accomplished by increasing the volume of the thorax. During inhalation, the diaphragm is depressed and the ribs, particularly ribs 2–6, rotate about an axis through their heads as the sternal ends and the sternum itself are lifted superiorly and slightly anteriorly. In addition, the lower ribs move into more horizontal positions, widening the transverse diameter of the thorax. These actions increase the volume of the thorax, and atmospheric pressure then forces air into the elastic lungs contained within.

There are several muscles involved in inhalation. The ribs are elevated by the *scaleni muscles*, the *diaphragm* (attached to the xiphoid process, the lower four ribs, and the bodies and arches of lumbar vertebrae 1–2), and the *external intercostal muscles* (which attach between adjacent ribs). When forceful exhalation is called for, the *internal intercostal muscles*, which are oriented at nearly right angles to the external ones, depress the ribs. This is assisted by contraction of *abdominal muscles* to decrease the volume of the thorax and thus expel air.

CHAPTER 8

Shoulder Girdle: Clavicle and Scapula

The departure of fishes from their aquatic habitat brought profound changes in locomotion. The forelimb girdle (shoulder girdle), attached to the rear of the head, detached and moved tailward, leaving a flexible neck. As this girdle moved back, some of the head's dermal armor and gill muscles remained attached to it, and our own shoulder girdle is still attached to the skull by derivatives of some of these primitive gill muscles. The remaining dermal bone element constitutes part of our clavicle. The human shoulder girdle provides support and articulation for the humerus and anchors a variety of muscles. The clavicle's function as a strut for the shoulder is made obvious when fracture of this bone is accompanied by anteromedial collapse of the shoulder. The shoulder girdle embraces the thorax posteriorly, laterally, and anteriorly, providing a platform for movements of the forelimb.

8.1
Clavicle (Figures 8.1–8.3)

8.1.1 Anatomy

The clavicle is a tubular, somewhat S-shaped bone. Its medial end (sternal extremity) articulates, via a synovial joint, with the clavicular notch of the manubrium. Its lateral end (acromial extremity) articulates with the scapula. The clavicle is oval to circular in cross section. The medial end is rounded and flared like a trumpet, and the lateral end is flattened superoinferiorly. This element is easily palpated along its length in a living person.

a. The **costal impression (tuberosity)** lies on the medial end (sternal extremity) of the clavicle's inferior surface. It is a broad, rough surface that anchors the *costoclavicular ligament,* which strengthens the sternoclavicular joint.

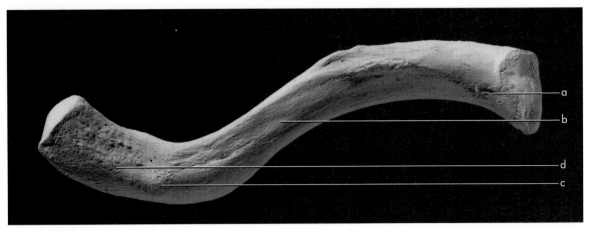

Figure 8.1 **Right clavicle, inferior.** Anterior is up, lateral is toward the left. Natural size. Key: a, costal impression (tuberosity); b, subclavian sulcus (groove); c, conoid tubercle; d, trapezoid line (oblique ridge).

b. The **subclavian sulcus (groove)** runs along the posteroinferior quadrant of the midshaft, providing a roof over the great vessels of the neck, and insertion for the *subclavius muscle* between the clavicle and the rib-cage. In fracture, the subclavius protects these vessels by preventing motion in the free end of the jagged fractured bone.

c. The **conoid tubercle** is found on the lateral end (acromial extremity) of the clavicle, located posteriorly. It is the attachment point for the *conoid ligament,* which attaches to the scapula's coracoid process and reinforces the joint between these two bones.

d. The **trapezoid line (oblique ridge)** leads laterally from the conoid tubercle. It is the attachment site for the *trapezoid ligament,* which functions like the *conoid ligament.*

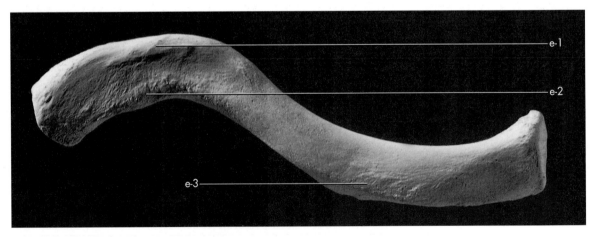

Figure 8.2 **Right clavicle, superior.** Anterior is down, lateral is toward the left. Natural size. Key: e, superior surface with attachment sites for trapezius muscle (e-1), deltoideus muscle (e-2), and pectoralis muscle (e-3).

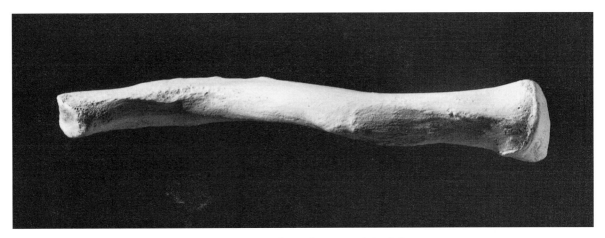

Figure 8.3 **Right clavicle, anterior.** Superior is up, lateral is toward the left. Natural size.

e. The **superior surface** of the clavicle bears less relief than the inferior surface. The rugosities on the posterolateral surface of the bone mark the attachment site for the *trapezius muscle* (e-1), and the rugosity on the anterolateral clavicle surface is for the *deltoideus muscle* (e-2). The rugosity on the anteromedial surface anchors part of the *pectoralis major muscle* (e-3).

8.1.2 Growth (Figure 8.4)

The clavicle ossifies in membrane and secondary cartilage from two primary centers in the shaft, and a secondary center at the sternal end. The bone is of developmental interest because it is the first to ossify *in utero* at week 5. It is the last to fuse, on the sternal end, at 20–25 years of age.

8.1.3 Possible Confusion

The lateral end of the clavicle is most often mistaken for the acromion process of the scapula in fragmentary specimens.

• The scapula's acromion continues medially to become the scapular spine, whereas the clavicle's lateral end becomes increasingly cylindrical as the shaft stretches medially.

• In comparisons of the tips of each bone, the clavicle's facet is lateral and the acromion's facet anteromedial.

8.1.4 Siding

The following criteria should be sufficient to correctly side fragments of clavicle:

• The medial end is round; the lateral end is flattened.

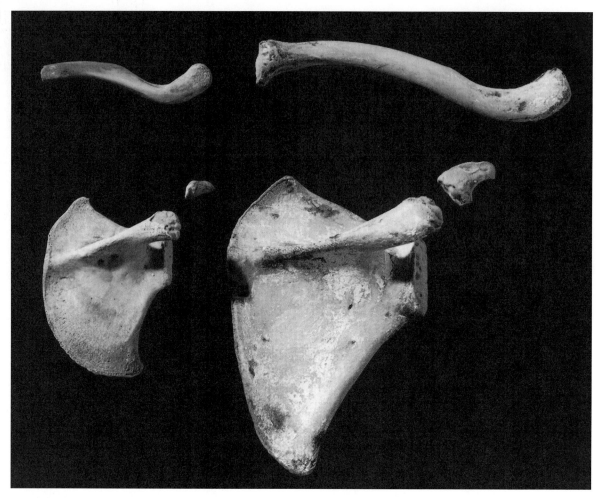

Figure 8.4 Clavicular and scapular growth. Above, superior view of a one-year-old and a six-year-old clavicle. Below, dorsal view of a one-year-old and a six-year-old scapula. Natural size.

- The bone bows anteriorly from the medial end, curves posteriorly at midshaft, and then sweeps anteriorly again as it reaches the lateral, flattened end. Thus, the apex of the first of two curves giving the clavicle its S shape is medial, and the apex of the second curve is lateral.
- Most irregularities and roughenings are on the inferior surface.
- The facet for the first costal cartilage is on the inferior edge of the medial clavicle, and the costal tuberosity is inferior.

8.2
Scapula (Figures 8.5–8.8)

8.2.1 Anatomy

The scapula is a large, flat, triangular bone with two basic surfaces, the posterior (dorsal) and costal (anterior, or ventral). There are three borders

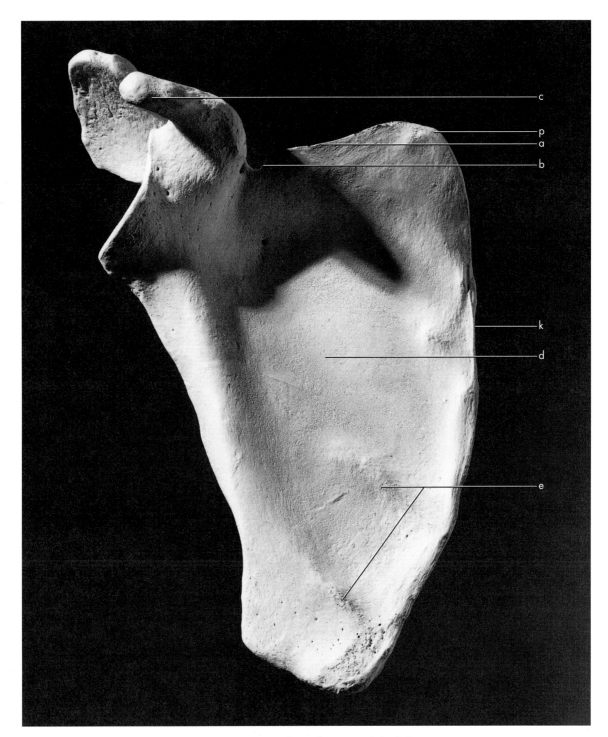

Figure 8.5 **Right scapula, anterior.** Superior is up, lateral is toward the left. Natural size. Key: a, superior (cranial) border; b, scapular notch or foramen; c, coracoid process; d, subscapular fossa; e, oblique ridges; k, medial (vertebral) border; p, superior angle.

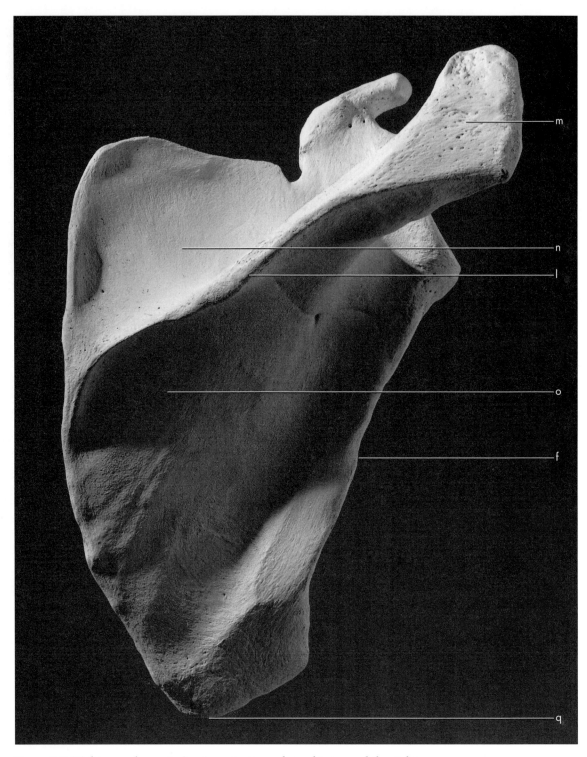

Figure 8.6 **Right scapula, posterior.** Superior is up, lateral is toward the right. Natural size. Key: f, lateral (axillary) border; l, scapular spine; m, acromion (acromion process); n, supraspinous fossa; o, infraspinous fossa; q, inferior angle.

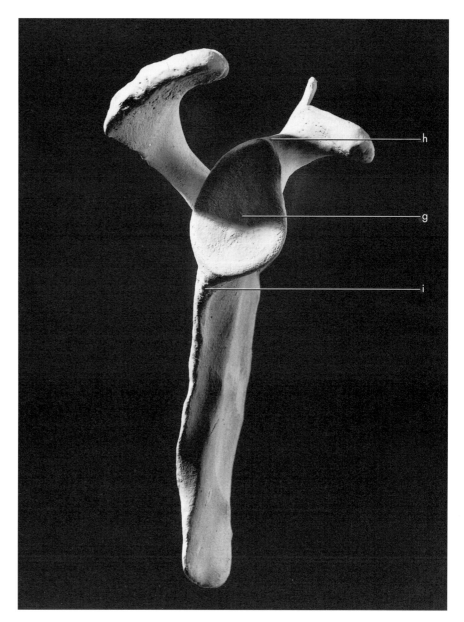

Figure 8.7 **Right scapula, lateral.** Superior is up, anterior is toward the right. Natural size. Key: g, glenoid cavity (glenoid fossa); h, supraglenoid tubercle; i, infraglenoid tubercle.

that meet in three angles. The scapula articulates with the clavicle and the humerus.

a. The **superior (cranial) border** is shortest and most irregular.

b. The **scapular notch** (or **foramen**) is a variable feature on the superior border. This semicircular notch is formed partly by the base of the coracoid process. It transmits the *suprascapular nerve* and may become a foramen if the ligament across its cranial edge ossifies.

c. The **coracoid process** juts anteriorly and superolaterally from the superior border of the scapula. This fingerlike, blunt, rugose projection

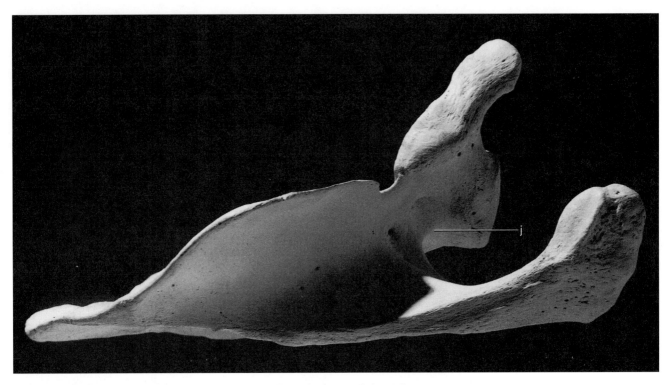

Figure 8.8 **Right scapula, superior.** Anterior is up, lateral is toward the right. Natural size. Key: j, scapular neck.

anchors a variety of muscles, ligaments, and fascial sheets important in the function of the shoulder joint.

d. The **subscapular fossa** is the shallow concavity that dominates the anterior (costal) surface of the scapula.

e. The **oblique ridges** that cross the subscapular fossa from superolateral to inferomedial are formed by *intramuscular tendons* of the *subscapularis muscle*, a major muscle that functions in medial rotation and adduction of the humerus and assists in other movements of the arm at the shoulder.

f. The **lateral (axillary) border** is the thickest (anteroposteriorly) border. It is usually slightly concave.

g. The **glenoid cavity (glenoid fossa)** is a shallow, vertically elongate structure that receives the head of the humerus. The shallow concavity of this joint allows great mobility of the humerus (circumduction comes easily), but the shoulder joint is consequently more prone to dislocation than the hip joint.

h. The **supraglenoid tubercle** sits adjacent to the superior edge of the glenoid cavity, at the base of the coracoid. This anchors the *long head of the biceps brachii muscle*, a flexor of the arm and forearm.

i. The **infraglenoid tubercle** sits just adjacent to the inferior edge of the glenoid cavity. It gives origin to the *long head of the triceps brachii* *muscle,* an extensor of the forearm and extensor and adductor of the arm at the shoulder.

j. The **scapular neck** is the slightly constricted region at the base of the glenoid fossa.

k. The **medial (vertebral) border** is straightest, longest, and thinnest.

l. The **scapular spine** dominates the posterior surface of the scapula. It passes mediolaterally across this surface, merging medially with the vertebral border and projecting laterally into the acromion process.

m. The **acromion (acromion process)** is the lateral projection of the scapular spine. Its cranial surface is very rough, providing attachment for a portion of the *deltoideus muscle,* a major arm abductor whose origins also extend along the inferior edge of the scapular spine. The upper fibers of the *trapezius muscle,* which act as scapular rotators, also insert here. The anteromedial corner of the acromion bears a small articular surface for the distal end of the clavicle.

n. The **supraspinous fossa** is the large, mediolaterally elongate hollowing superior to the base of the spine. It is the site of origin of the *supraspinatus muscle,* a major abductor of the arm.

o. The **infraspinous fossa** is the hollowing inferior to the scapular spine. This extensive, weakly concave area is the site of origin of the *infraspinatus muscle,* a lateral rotator of the arm. Intramuscular tendons of this muscle attach to the ridges on the surface of the fossa.

p. The **superior angle** of the scapula is where the superior and medial (vertebral) borders intersect. The *levator scapulae muscle* attaches to the dorsal surface of the scapula in this region.

q. The **inferior angle** of the scapula is where the vertebral (medial) and axillary (lateral) borders intersect. Rugosities on the costal and dorsal surfaces in this area mark the insertions of the *serratus anterior muscles* and origin of the *teres major muscle* from the medial and lateral borders, respectively.

8.2.2 Growth (Figure 8.4)

The scapula ossifies from seven or more centers. One of these is for the body, two each for the coracoid and acromion, and one each for the vertebral border and inferior angle. Ossification of the borders is variable, with elongated plates appearing and fusing during adolescence.

8.2.3 Possible Confusion

• When fragmentary, the scapula might be mistaken for the pelvis. In all of its flat parts, however, the scapula is thinner than the pelvis. Indeed, the scapular blade is mostly a single, thin layer of bone instead of spongy bone sandwiched between cortices as in the pelvis.

- A broken fragment of glenoid could be mistaken for the hip joint. The glenoid is much shallower and smaller than the acetabulum.
- The coracoid could be mistaken for the transverse process of a thoracic vertebra, but the coracoid is nonarticular.
- The lateral part of the acromion is sometimes mistaken for a fragment of lateral clavicle. However, because the acromion is continuous with the thin, platelike spine, rather than a cylindrical shaft, this misattribution can be avoided. The inferior acromion surface is also smooth and concave; the distal clavicle is not.
- Tiny fragments of scapular blade or infant scapulas could be mistaken for wings of the sphenoid, but the thin bone of the scapula is bounded by broken surfaces, whereas broken sphenoid pieces normally have free or sutural edges.

8.2.4 Siding

When intact, the glenoid is lateral, the spine posterior. When fragmentary, use the following criteria:

- For an isolated glenoid, the glenoid is teardrop-shaped, with the blunt end inferior. When looking directly into the correctly oriented glenoid fossa, note that the anterior edge of the fossa has a broad notch in it. The area of the supraglenoid tubercle at the superior edge of the glenoid is displaced anteriorly. The posterior border of the glenoid is waisted, the edge raised and roughened. The anterior border is not as raised; it gently slopes into the rest of the scapula.
- For an isolated acromion, the inferior surface of the acromion is concave and smoother than the superior. The clavicular facet is placed anteriorly and medial to the tip.
- For an isolated vertebral border, the anterior surface is concave and the posterior is convex. The oblique ridges run from superolateral to inferomedial (parallel to the scapular spine).
- For an isolated inferior angle, the anterior surface is concave, the posterior convex. The thickest border is lateral (axillary).
- For an isolated axillary border, the broad sulcus inferior to the glenoid parallels the border and is displaced anteriorly. The border itself thins inferiorly. The bony thickening is greatest (forming a "bar") on the anterior surface. Thickness of the cortex increases as the glenoid is approached along this border.
- For an isolated coracoid, the smooth surface is inferior, the rough superior. The anterior border is longer. The hollow on the inferior surface faces the glenoid area (posteroinferiorly).
- For an isolated spine, the spine thins medially (vertebrally) and thickens toward the acromion. The inferior border has a tubercle that points inferiorly. Adjacent to the spine, the infraspinous fossa is most deeply excavated medially. The supraspinous fossa is the most deeply excavated laterally. A variably present foramen (or foramina) perforates the scapula at the superolateral base of the spine, at the depth of the supraspinous fossa.

8.3
Functional Aspects of the Shoulder Girdle

Because the glenohumeral joint between scapula and humerus is free to move as the scapula moves, and because the shoulder girdle has only one bony connection to the thorax, the upper limb of a human is far more mobile than the lower limb. In contrast to the shoulder, the hip joint is fixed in relation to the vertebrae. In the glenohumeral joint, there is a great disparity between the large articular surface of the humeral head and the smaller glenoid surface. This provides further mobility to the arm because the ball-shaped humeral head can rotate in any direction in the glenoid fossa. The capsule of the glenohumeral joint is ligamentous and muscular (and relatively weak), and the shoulder is therefore an easily dislocated joint.

Actions of the shoulder girdle are accomplished, for the most part, by muscles inserting on the scapula. The scapula moves, sliding and rotating on the back, in response to muscle contractions that change the orientation and position of the glenoid. One of the major scapular rotators is the *trapezius muscle,* which originates from the nuchal line of the occipital and the spines of cervical and thoracic vertebrae. The trapezius muscle inserts on the scapular spine, acromion process, and superoposterior lateral clavicle. The contraction of its various parts can therefore elevate, suspend, stabilize, and rotate the scapula. The *serratus anterior* works with the trapezius, inserting along the medial edge of the scapula's costal surface. Contraction of the lower fibers of this muscle can therefore rotate the scapula, with the opposite rotation produced by the *rhomboid major muscle.*

In all, sixteen muscles affect movements of the shoulder. The scapula is a sort of mobile foundation for muscles that move the arm. The scapula itself can be moved so that the glenoid faces different directions. Muscles anchored on this mobile platform in turn move the arm via the shoulder joint. A few large, superficial muscles cross both the shoulder and elbow joints and can effect movement there.

CHAPTER 9

Arm: Humerus, Radius, and Ulna

The first vertebrates lacked jaws. These animals, similar to modern lampreys and hagfish, also lacked paired fins. Jaws and fins evolved 400 million years ago, allowing fish to more effectively locomote and feed. Jawed fish have paired fins set on flat plates of bone that are attached to the muscles of their body walls. The paired fins, flexible fans of small bones, are used primarily as aids in stabilizing and steering. The limbs of terrestrial animals evolved from this structural arrangement as fins were transformed into rod-bearing segments. Although the limbs of land vertebrates appear very different from fish fins, the two homologous structures are actually highly comparable.

Each vertebrate limb has a base and three segments. The bases, the limb girdles, are the old basal fin plates of fish, which evolved to take on the function of transferring the weight of the body to the terrestrial tetrapod's limbs. The proximal vertebrate limb segments constitute the upper arm and thigh. The intermediate limb segments, the forearm and foreleg, each bear two bones in humans, the radius and ulna in the upper limb, and their serial homologs, the tibia and fibula, in the leg. In this chapter we consider the three bones of the upper two forelimb segments, the humerus, radius, and ulna.

9.1
Humerus (Figures 9.1–9.6)

9.1.1 Anatomy

The upper arm bone, or humerus, is the largest bone of the upper limb. It comprises a proximal end with a round articular head, a shaft, and an irregular distal end. The humerus articulates proximally with the glenoid fossa of the scapula and distally with both the radius and the ulna.

  a. The **head** is a hemisphere on the proximal end of the humerus that faces medially and articulates with the glenoid fossa of the scapula.

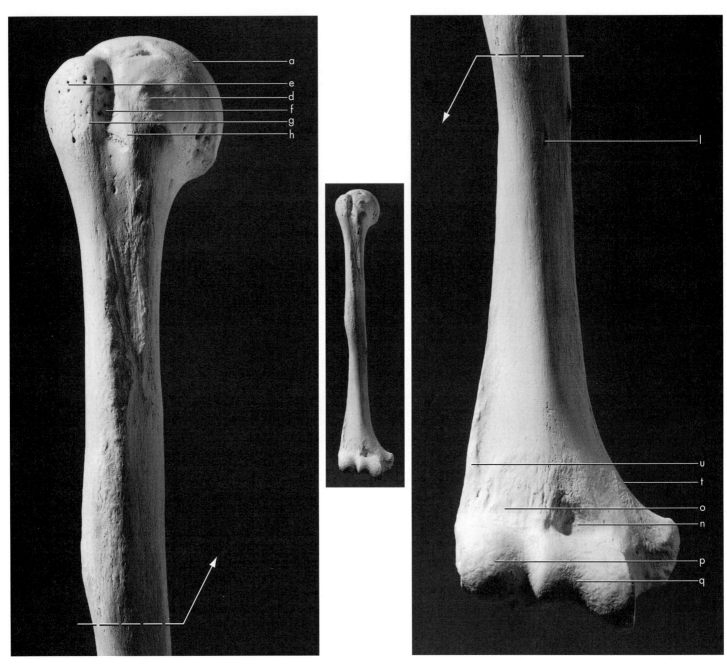

Figure 9.1 **Right humerus, anterior:** *left,* proximal end; *right,* distal end. Natural size. Key: a, head; d, lesser tubercle; e, greater tubercle; f, intertubercular sulcus (bicipital groove); g, crest of the greater tubercle; h, crest of the lesser tubercle; l, nutrient foramen; n, coronoid fossa; o, radial fossa; p, capitulum; q, trochlea; t, medial supracondylar crest (ridge); u, lateral supracondylar crest (ridge).

b. The **anatomical neck** is the groove that circles around the articular surface of the head for the attachment of the *joint capsule.*

c. The **surgical neck** is the short segment inferior to the head which links head and shaft.

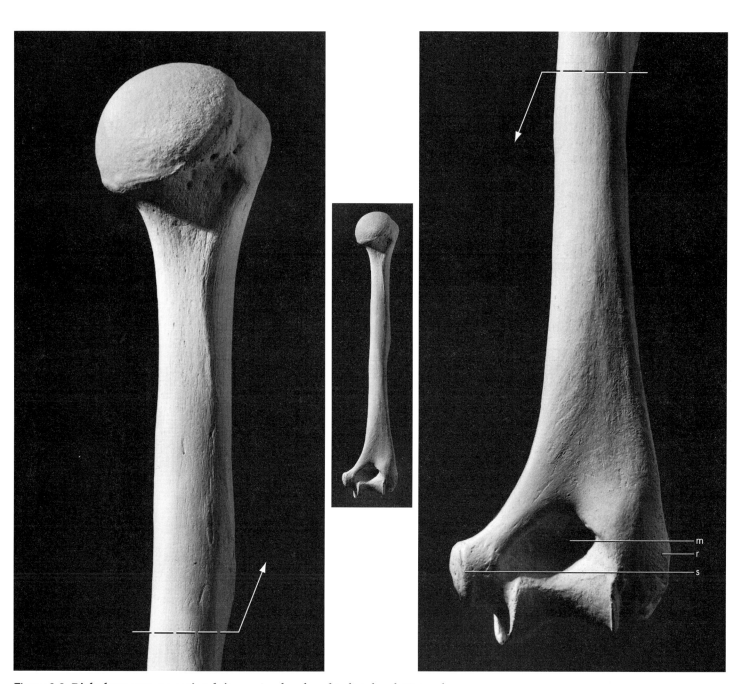

Figure 9.2 **Right humerus, posterior:** *left,* proximal end; *right,* distal end. Natural size. Key: m, olecranon fossa; r, lateral epicondyle; s, medial epicondyle.

d. The **lesser tubercle** is a small, blunt eminence anterolateral to the head on the proximal humeral end. The lesser tubercle marks the insertion of the *subscapularis muscle,* which originates on the costal surface of the scapula and medially rotates the humerus.

e. The **greater tubercle** is larger, more posterior, and projects more laterally than the lesser tubercle. The greater tubercle bears facets for the insertion of the *supraspinatus, infraspinatus,* and *teres minor muscles.* These muscles, together with the *subscapularis muscle,* consti-

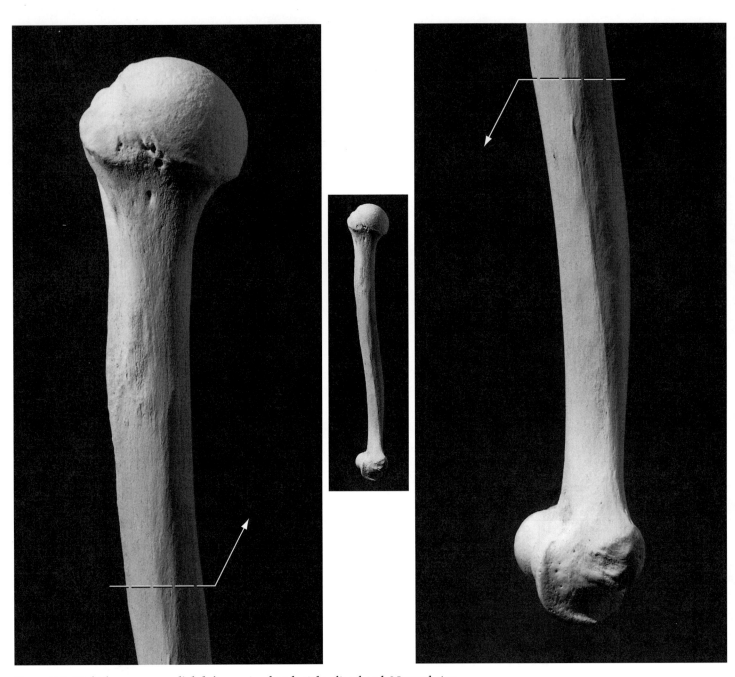

Figure 9.3 **Right humerus, medial:** *left,* proximal end; *right,* distal end. Natural size.

tute the *rotator cuff muscles.* In addition to medial and lateral rotation, these muscles also aid in adduction and abduction of the arm.

f. The **intertubercular sulcus (bicipital groove)** extends longitudinally down the proximal shaft. It begins between the two tubercles and houses the *tendon of the long head of the biceps brachii muscle.* In life, the *transverse humeral ligament* connects the two tubercles to bridge the groove and form a canal.

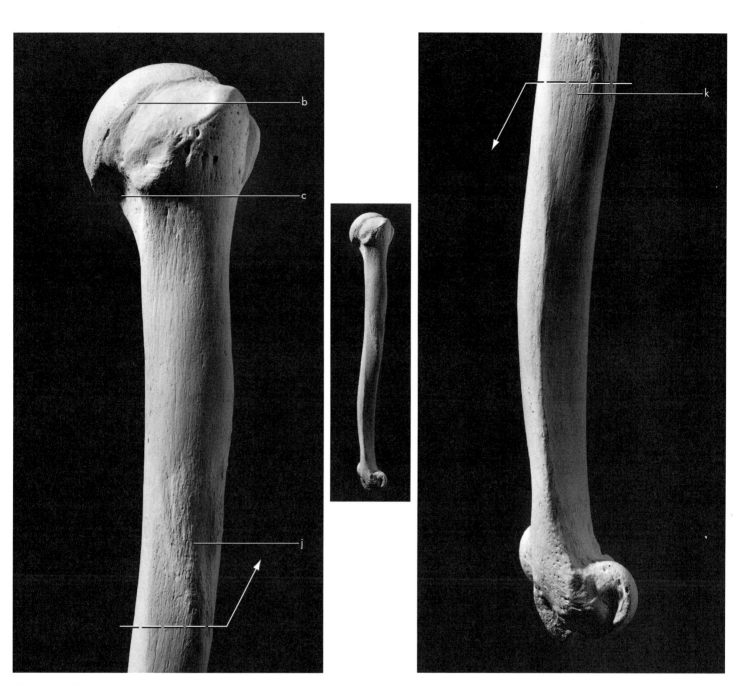

Figure 9.4 **Right humerus, lateral:** *left,* proximal end; *right,* distal end. Natural size. Key: b, anatomical neck; c, surgical neck; j, deltoid tuberosity; k, radial sulcus (spiral groove).

g. The **crest of the greater tubercle** forms the lateral lip of the intertubercular groove. It is the insertion site for the *pectoralis major muscle,* a muscle that originates on the anteromedial clavicle, the sternum, and the cartilage of the true ribs. This muscle acts to flex, adduct, and medially rotate the arm.

h. The **crest of the lesser tubercle** forms the medial lip of the intertubercular groove. It is the insertion site for the *teres major muscle,* a medial rotator of the arm.

Figure 9.5 **Right humerus, proximal.** Anterior is up, lateral is toward the right. Natural size.

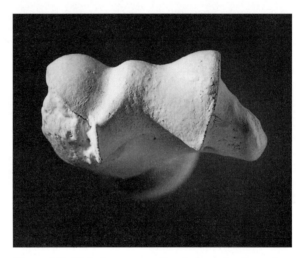

Figure 9.6 **Right humerus, distal.** Anterior is up, lateral is toward the left. Natural size.

i. The humeral **shaft** is variably triangular, ranging from more cylindrical in its proximal section to an anteroposteriorly compressed, rounded triangle distally.

j. The **deltoid tuberosity** is on the lateral surface of the shaft. It is the insertion site of the *deltoideus muscle,* a major abductor (among other functions) of the arm which originates from the anterior border and superior surface of the clavicle, the lateral margin and superior surface of the acromion, and the scapular spine. The deltoid tuberosity is recognized by its roughened surface. It spirals around the shaft from posterosuperior to anteroinferior.

k. The **radial sulcus** (**spiral groove**) is found on the posterior surface of the shaft. It is a shallow, oblique groove for the *radial nerve* and deep vessels that pass parallel and immediately posteroinferior to the deltoid tuberosity. Its inferior boundary is continuous distally with the lateral border of the shaft.

l. The **nutrient foramen** is located anteromedially and exits the shaft from distal to proximal. A good way to remember the direction of entry of nutrient foramina into all of the long bones is to imagine tightly flexing your own arms (at the elbows) and legs (at the knees) in front of you. In this position you can look into the bones via the foramina. In long bones, these foramina carry the *nutrient arteries.*

m. The **olecranon fossa** is the largest of three hollows on the distal humeral end. It is posterior, accommodating the olecranon process of the ulna during forearm extension. The deepest area of this fossa may be perforated, forming a foramen, or **septal aperture.**

n. The **coronoid fossa** is the larger, medially placed hollow on the anterior surface of the distal humerus. It receives the coronoid process of the ulna during maximum flexion of the forearm.

o. The **radial fossa** is the smaller, laterally placed hollow on the anterior surface of the distal humerus. It receives the head of the radius during maximum flexion of the forearm.

p. The **capitulum** is the rounded eminence that forms the lateral portion of the distal humeral surface. It articulates with the head of the radius.

q. The **trochlea** is the notch- or spool-shaped medial portion of the distal humeral surface. It articulates with the ulna.

r. The **lateral epicondyle** is the small, nonarticular lateral bulge of bone superolateral to the capitulum. It serves as a site of attachment to the *radial collateral ligament* of the elbow, and it is the origin of a common tendon of origin of supinator and extensor muscles in the forearm.

s. The **medial epicondyle** is the nonarticular, medial projection of bone superomedial to the trochlea. It is more prominent than the lateral epicondyle. It provides attachment to the *ulnar collateral ligament,* to many flexor muscles in the forearm, and to the *pronator teres muscle.*

t. The **medial supracondylar crest** (**ridge**) is superior to the medial epicondyle and forms the sharp medial border of the distal humerus.

u. The **lateral supracondylar crest** (**ridge**) is superior to the lateral epicondyle and forms the sharp lateral border of the distal humerus.

9.1.2 Growth (Figure 9.7)

The humerus ossifies from eight centers: the shaft, the head and both tubercles (a composite of several early centers including the head itself and each tubercle; a "conjoint"), the capitulum and trochlea (the capitulum fuses to the trochlea before either fuses to the shaft), and each epicondyle.

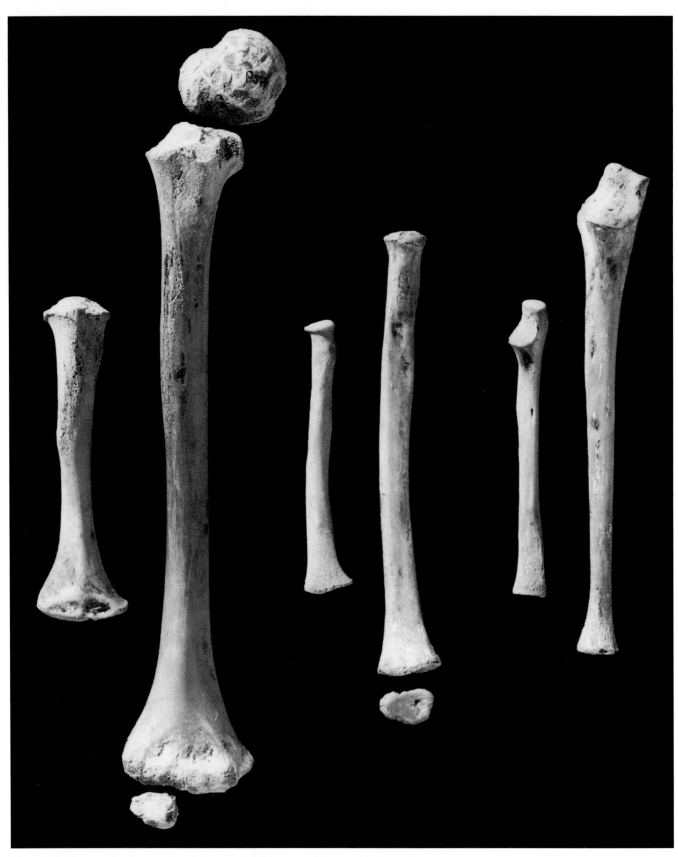

Figure 9.7 Humeral, ulnar, and radial growth. The pairs of immature humeri (left), radii (center), and ulnae (right), shown here in anterior view, are from a one-year-old and a six-year-old. Natural size.

9.1.3 Possible Confusion

- The humeral head cannot be mistaken for a femoral head because the former is only half of a sphere, whereas the latter is substantially more than half.
- The femoral head has a distinct depression, or **fovea capitis,** which the humerus lacks.
- The humeral shaft is larger and more circular in section than the radial, ulnar, or fibular shafts.
- The humeral shaft is smaller and more irregular in section than the femoral shaft.
- The humeral shaft is smaller and less triangular than the tibial shaft.

9.1.4 Siding

- For an intact bone, the head faces medially, the capitulum is lateral, and the olecranon fossa is posterior.
- For an isolated proximal end, the head is medial, and the lesser tubercle and intertubercular groove are anterior.
- For an isolated distal end, the olecranon fossa is posterior, the medial epicondyle is larger, and the capitulum is lateral and oriented anteriorly. If the articular end is missing, the coronoid fossa is larger and more medial than the radial fossa.
- For an isolated shaft fragment, the deltoid tuberosity is lateral, passing from posterosuperior to anteroinferior, and the nutrient foramen exits the bone toward its proximal end. A small thin ridge runs along the entire medial edge of the shaft, and the nutrient foramen is found on this edge. The lateral lip of the intertubercular groove is stronger and longer.

9.2
Radius (Figures 9.8–9.12)

9.2.1 Anatomy

The radius is the shortest of the three arm bones. It is named for its action, a turning movement about the capitulum of the humerus, which allows the bone to rotate relative to the more fixed ulna. The radius articulates proximally with the humerus at the capitulum and medially with the ulna on both proximal and distal ends. Distally, the radius articulates with two carpal bones of the wrist.

a. The **head** is a round articular structure on the proximal end of the radius. It articulates, via its cupped proximal surface (articular fovea), with the humeral capitulum, whereas the edge of the radial head (circumferential articulation) articulates with the radial notch of the ulna.

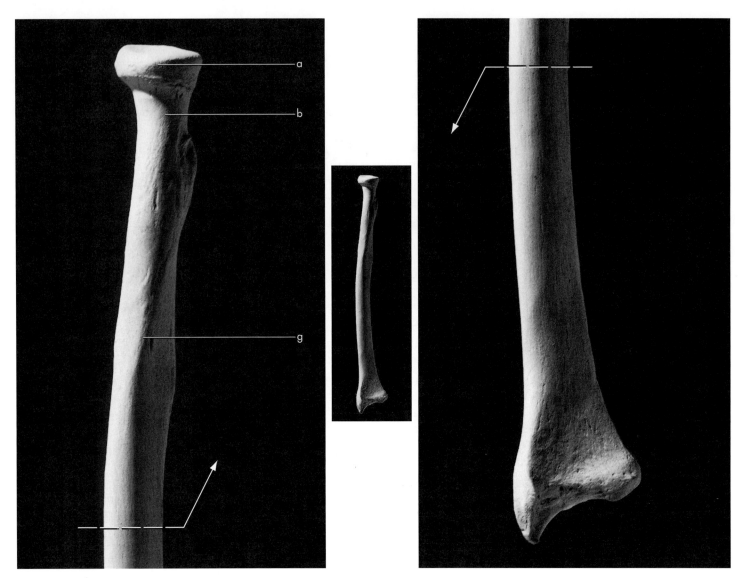

Figure 9.8 **Right radius, anterior:** *left,* proximal end; *right,* distal end. Natural size. Key: a, head; b, neck; g, oblique line.

b. The **neck** is the slender segment of the radius between the head and the radial tuberosity.

c. The **radial tuberosity** (**bicipital tuberosity**) is a blunt, rugose, variably shaped structure on the anteromedial side of the proximal radius. It marks the insertion of the *biceps brachii muscle,* a flexor of the forearm, and a large biceps *bursa* that underlies this muscle.

d. The **shaft** of the radius is the long thin section below the radial tuberosity and proximal to the expanded distal end.

e. The **nutrient foramen** exits the bone toward its distal end and is located on the anterior surface of the proximal half of the radius.

f. The **interosseus crest** is the sharp medial edge of the radius shaft. It serves as the attachment site for a fibrous membrane, the *interosseus membrane,* which divides the forearm into two compartments, ante-

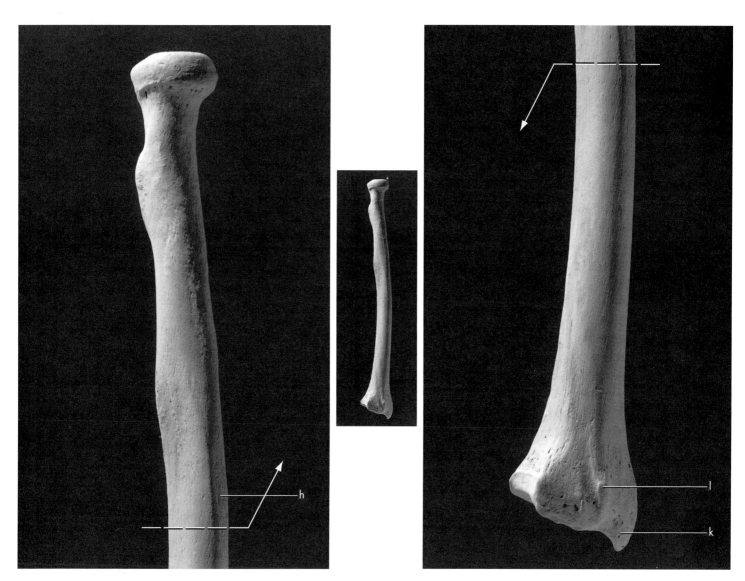

Figure 9.9 **Right radius, posterior:** *left,* proximal end; *right,* distal end. Natural size. Key: h, pronator teres insertion; k, styloid process; l, dorsal tubercle (Lister's tubercle).

rior and posterior. These house the flexor and extensor groups of muscles that act across the wrist.

g. The **oblique line** is on the anterior surface of the shaft. It spirals inferolaterally from its origin at the base of the radial tuberosity. This line gives origin to extrinsic muscles of the hand.

h. The **pronator teres insertion** is a midshaft roughening on the lateral surface.

i. The **ulnar notch** is a concave articular hollow on the medial distal corner of the radius. It articulates with the distal end of the ulna.

j. The **distal radial articular surface** articulates with carpal bones—the lunate on the medial side, and the scaphoid on the lateral side.

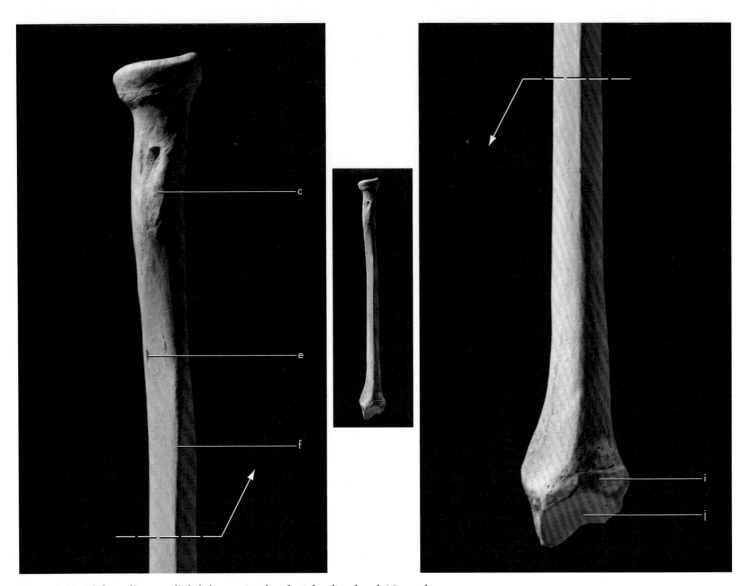

Figure 9.10 **Right radius, medial:** *left,* proximal end; *right,* distal end. Natural size. Key: c, radial tuberosity (bicipital tuberosity); e, nutrient foramen; f, interosseus crest; i, ulnar notch; j, distal radial articular surface.

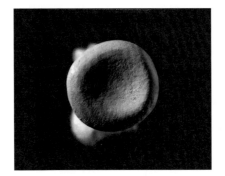

Figure 9.11 **Right radius, proximal.** Lateral is up, anterior is toward the left. Natural size.

Figure 9.12 **Right radius, distal.** Dorsal is up, lateral is toward the right. Natural size.

k. The **styloid process** is a sharp projection on the lateral side of the distal radius.

l. The **dorsal tubercle (Lister's tubercle)** is a large tuberosity on the posterior surface of the distal radius. The grooves between this and other tuberosities on the dorsum of the distal radius house the tendons for extrinsic extensor muscles of the hand.

9.2.2 Growth

The radius ossifies from three centers: the shaft, the head, and the distal end.

9.2.3 Possible Confusion

Radial shaft segments might be mistaken for the ulna or fibula, and distal ends might be mistaken for the clavicular notch of the manubrium, but the following features help to sort fragments of radius.

- The ulnar shaft tapers continuously (the circumference decreases) from proximal to distal, whereas the radius does not.

- The ulnar shaft has a sharp interosseus crest, but the two other corners are not so evenly rounded as they are in the radius, which has a teardrop shape in cross section and a smoother, more uniform crest.

- Most of the ulnar shaft is triangular. The ulnar shaft only becomes round in cross section at its distal end. In contrast, the radial shaft is circular proximally and a rounded triangle at midshaft. It is a broad, anteroposteriorly compressed oval with thin cortex in a more distal cross section.

- The fibula, also a long slender bone with crests, is much more irregular in cross section and much longer than the radius.

- The distal radial articulation is made up of two discernible portions, whereas the clavicular notch of the manubrium is singular.

9.2.4 Siding

- For an intact radius, the ulnar notch is medial, the radial tuberosity and interosseus crest are medial, the dorsal tubercles are posterior, and the styloid process is lateral.

- For an isolated proximal end, the tuberosity faces anteromedially. The medial portion of the proximal ulnar articular surface has the greatest proximodistal dimension.

- For an isolated segment of shaft, the interosseus crest is medial, and the oblique line is anterior. The nutrient foramen exits the bone distally and is situated anteriorly on the shaft. The posterolateral surface has the greatest rugosity at about midshaft.

- For an isolated distal end, the anterior surface is smooth and flat, the posterior surface has extensor grooves, the ulnar notch is medial, and the styloid process is lateral. The styloid process is smooth on its anterior surface.

9.3
Ulna (Figures 9.13–9.18)

9.3.1 Anatomy

The ulna is the longest, thinnest bone of the forearm. It articulates proximally with the trochlea of the humerus and head of the radius. Distally it articulates with the ulnar notch of the radius and with an articular disk that separates it from the carpal bones and provides freer rotation of the hand and radius around the ulna than is seen in many other mammals.

a. The **olecranon (olecranon process)** of the ulna is the most proximal part of the bone. It is a massive, blunt process. The *triceps brachii muscle*, the primary extensor of the forearm, inserts on the tuberosity of the process.

b. The **trochlear notch (semilunar notch)** of the ulna articulates with the trochlear articular surface of the distal humerus. In contrast to the more mobile radius, rotary motion is very restricted at the ulnar part of the elbow joint, sharply limiting the ulna in its ability to rotate around its long axis.

c. The vertical **guiding ridge** separates the trochlear notch into medial and lateral portions.

d. The **coronoid process** is the anterior, beak-shaped projection at the base of the semilunar notch.

e. Immediately inferior to the coronoid process is a roughened depression called the **ulnar (brachial) tuberosity** for the insertion of the *brachialis muscle*, a flexor of the elbow that originates from the anterior surface of the humerus.

f. The **radial notch** is the small articular surface for the radius. It is located along the lateral margin of the coronoid process.

g. The **shaft** is the long segment of bone between the brachial tuberosity and the inflated distal end of the ulna.

h. The **nutrient foramen** exits the bone in a distal direction and is found on the anteromedial ulnar shaft.

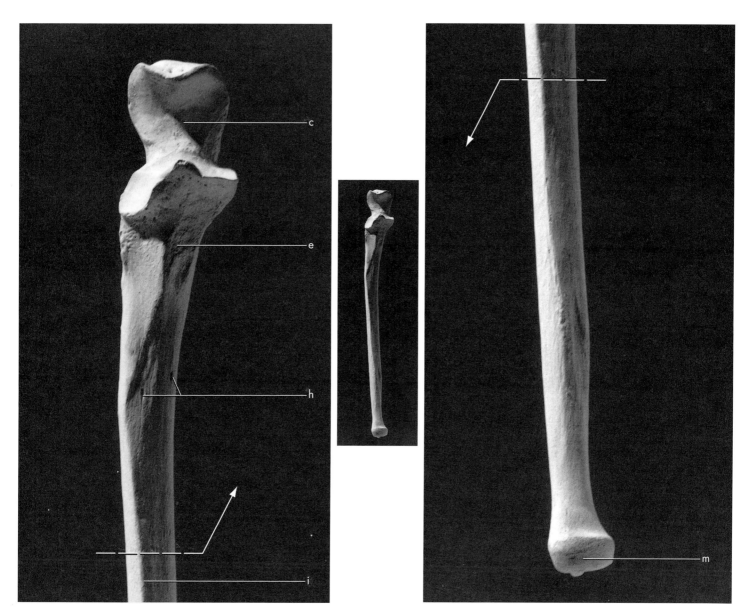

Figure 9.13 **Right ulna, anterior:** *left,* proximal end; *right,* distal end. Natural size. Key: c, guiding ridge; e, ulnar (brachial) tuberosity; h, nutrient foramen (double here); i, interosseus crest; m, radial (circumferential) articulation.

i. The **interosseus crest** lies opposite the radius, on the lateral surface of the ulnar shaft.

j. The **pronator ridge** is a short, variably expressed ridge on the distal quarter of the shaft. It is located anteromedially and is the site of origin for the *pronator quadratus muscle.*

k. The **styloid process** is the sharp, distalmost projection of the ulna. It is set on the posteromedial corner of the bone. Its end gives attachment to the *ulnar collateral ligament* of the wrist. It is separated from the remainder of the head by a deep groove or pit, the fovea.

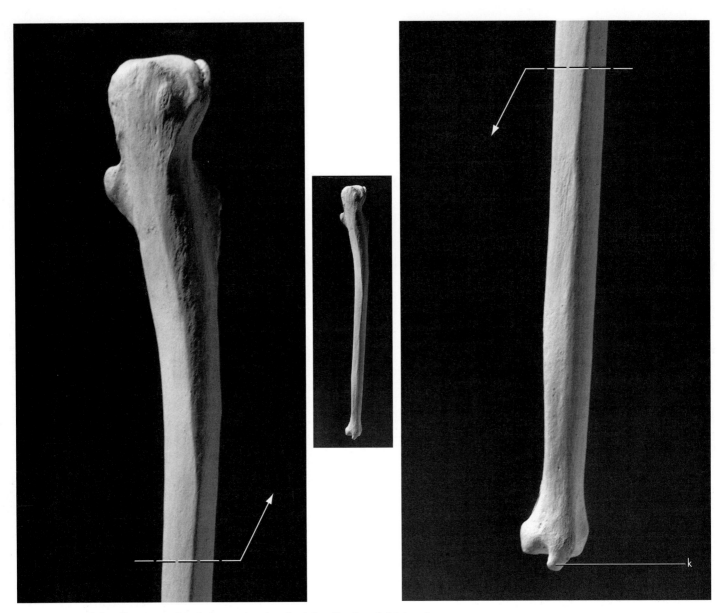

Figure 9.14 **Right ulna, posterior:** *left,* proximal end; *right,* distal end. Natural size. Key: k, styloid process.

l. The **extensor carpi ulnaris groove** is adjacent to the styloid process, located proximolaterally to it. It houses the *tendon of the extensor carpi ulnaris muscle,* a dorsiflexor and adductor of the hand at the wrist.

m. The **radial (circumferential) articulation** is the distal, lateral, round articulation that conforms to the ulnar notch of the radius in the same way that the radial head conforms to the radial notch of the proximal ulna.

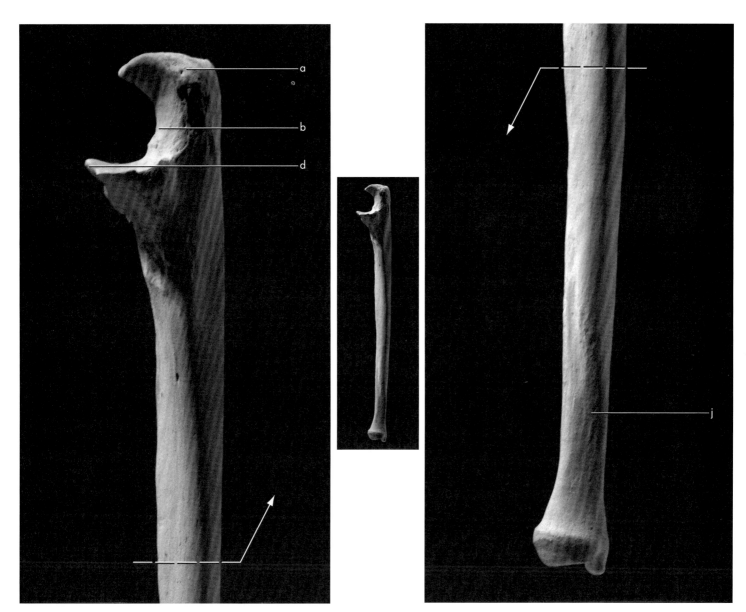

Figure 9.15 **Right ulna, medial:** *left,* proximal end; *right,* distal end. Natural size. Key: a, olecranon (olecranon process); b, trochlear notch (semilunar notch); d, coronoid process; j, pronator ridge.

9.3.2 Growth

The ulna ossifies from three centers: the shaft, the olecranon process, and the distal end.

9.3.3 Possible Confusion

- The ulnar proximal and distal ends are diagnostic, but isolated shafts could be mistaken for radial or fibular shafts. The radial shaft is more triangular, or teardrop-shaped, in section, with two rounded corners

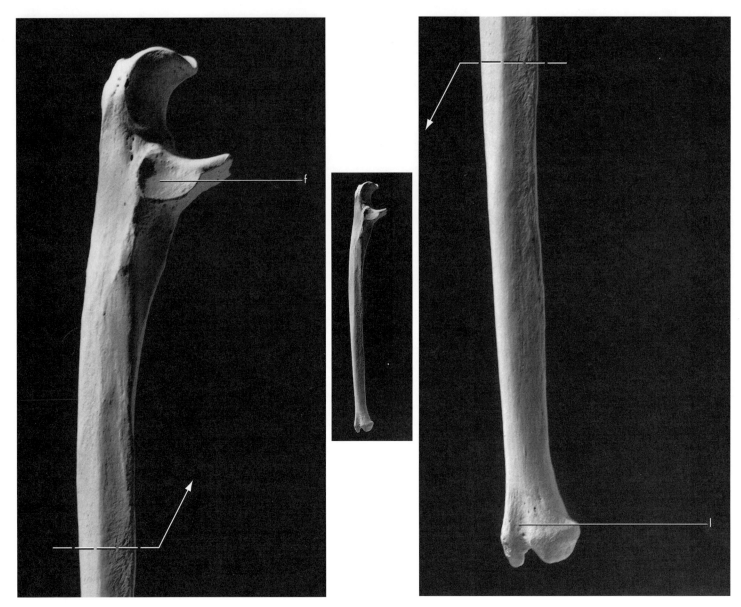

Figure 9.16 **Right ulna, lateral:** *left,* proximal end; *right,* distal end. Natural size.
Key: f, radial notch; l, extensor carpi ulnaris groove.

Figure 9.17 **Right ulna, proximal.** Lateral is up, anterior is toward the left. Natural size.

Figure 9.18 **Right ulna, distal.** Anterior is down, lateral is toward the right. Natural size.

and one sharp corner. Radial shafts do not taper distally as the less regular ulnar shaft does.

- The fibular shaft is much more irregular in cross section, with multiple sharp corners.
- See the radius shaft description for further details.

9.3.4 Siding

- For an intact ulna, the olecranon process is proximal and posterior, the radial notch is lateral, and the interosseus crest is lateral.
- For an isolated proximal end, use the same criteria given above for an intact ulna. Note also that the brachial tuberosity is medially displaced.
- For an isolated shaft segment, the shaft tapers distally, the nutrient foramen exits the bone distally and is located on the anterior shaft surface. The interosseus crest is lateral. The shaft surface anterior to the crest is more hollowed.
- For an isolated distal end, the styloid process is posterior, and the groove for the *extensor carpi ulnaris* is lateral to the process.

9.4
Functional Aspects of the Elbow and Wrist

The elbow joint has a single joint capsule, but its three different bony elements operate differently within the capsule. The humeroulnar joint is a simple hinge, whereas the humeroradial joint is a pivot joint resembling a ball-and-socket joint. The proximal and distal radioulnar joints are mirror images, allowing the radius to spin during pronation and supination. The axis of rotation passes obliquely across the forearm through the proximal radius and distal ulna. The hand articulates with the forearm through the radiocarpal articulation at the wrist.

Actions of flexion and extension at the elbow joint are accomplished by contraction of two major antagonists, the *biceps brachii* (flexor) and *triceps brachii* (extensor). The former has two origins: the long head from the supraglenoid tubercle of the scapula (the tendon passes through the intertubercular groove of the humerus) and the short head from the tip of the scapular coracoid process. The insertion of this muscle on the bicipital tuberosity of the radius makes it a powerful flexor of the forearm at the elbow. It can also supinate the forearm. The *triceps brachii* is also a complex muscle, with three heads of origin: the long head from the infraglenoid tubercle of the scapula and the lateral and short heads from the posterior surface of the humeral shaft. This major extensor of the forearm at the elbow inserts on the olecranon process of the ulna.

In addition to flexion and extension at the elbow, the unique articulation of the elbow allows for pronation and supination of the forearm. The *pronator teres muscle* originates from the medial epicondyle and medial supracondylar ridge of the humerus and inserts on the lateral radial shaft. Supination occurs when the *biceps brachii* works on the already pronated forearm. In addition, the *supinator muscle* originates on the lateral epicondyle of the humerus and the lateral surface of the proximal ulna. It crosses to the anterior oblique line of the proximal radius, and its contraction thus causes supination.

Many muscles surround the radius and ulna. Most of these function via tendons to cause flexion and extension at the wrist and within the hand. Flexors are found in a compartment anterior to the radius and ulna, whereas extensors are located in a compartment posterior to these bones. These extrinsic hand muscles and their tendons can be easily palpated.

CHAPTER 10

Hand: Carpals, Metacarpals, and Phalanges

The hand is a complex structure that represents the distal tetrapod limb segment. It is the modified end of the ancestral fish fin, a structure based on jointed bony rays. In the generalized reptilian hand, a set of small wrist bones (carpals) forms the foundation for five digits. Each digit is composed of one large proximal segment (a metacarpal) and a chain of additional bones (the phalanges). Digital reduction and modification have occurred in a great variety of mammals, from the wings of bats to the single toes of modern horses. Humans have retained the generalized pattern of five digits. There are a total of twenty-seven bones in each human hand, eight carpal bones arranged in two rows, followed distally by a single row of five metacarpals. Farther distally, there is a single row of five proximal phalanges, a single row of four intermediate phalanges, and a single row of five distal, or terminal, phalanges.

In addition to the twenty-seven major hand bones, there are small bones called **sesamoid** bones that lie within tendons of the hand. These are not usually recovered and are rarely studied by osteologists, who should nevertheless always be alert to their presence as they are of considerable functional significance. In the hand, a pair of sesamoids is usually found at the palmar surface of the first metacarpal head. Figures 10.1–10.3 summarize and illustrate articulations within the hand.

Elements of the hand skeleton are described below in three categories: the **carpals,** the **metacarpals,** and the **hand phalanges.** In the carpal region of the hand, as in the tarsal region of the foot, a variety of different names have been applied to each bone as anatomical nomenclature has changed through the years. For readers curious about this history, O'Rahilly (1989) provides a good summary.

Before analyzing the various elements that make up the hand, it is useful to note the importance of anatomical nomenclature in study of the hand. In dealing with elements of the hand, it is easy to become confused by the terms "anterior," "posterior," "medial," and "lateral" because these terms can only be applied when the specimen is in proper anatomical position. For this reason, it is useful to supplement the directional terms when possible, using the following sets of synonyms:

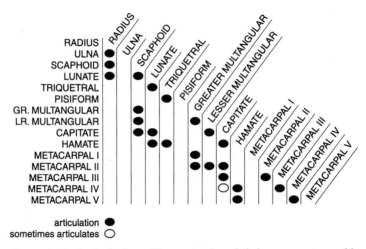

Figure 10.1 Articulation of bones in the adult human wrist and hand.

- anterior = palmar
- posterior = dorsal
- medial = ulnar = little finger side
- lateral = radial = thumb side

The term **ray** is often applied to each finger, or toe, including the digit's phalanges and metacarpal. By convention, the thumb ray, or **pollex**, is identified by the number 1. The index finger is ray 2, the middle finger is ray 3, the ring finger is ray 4, and the little finger is ray 5.

For hand and foot elements, it is often easiest to use memory-based, or positional, "tools" in order to side the bone quickly. The text provides a set of these tools but also employs standard anatomical techniques to side each bone. Many of the positional techniques detailed here are adopted from Bass (1995).

10.1
Carpals

The eight bones of the adult wrist are often described as cubical in shape with six surfaces, but this is a misleading characterization. Each bone has a unique, diagnostic shape. For this reason, identification is straightforward. An introduction to the functional anatomy of the wrist facilitates study of the individual wrist elements. The palmar surface of the carpus, or wrist, bears four major carpal projections. The hook of the hamate and the pisiform underlie the medial edge of the palm at the base of metacarpal 5 (MC 5). The scaphoid tubercle and trapezium crest underlie the lateral edge of the palm, at the base of the thumb metacarpal (MC 1). In life there is a fibrous band stretched transversely between these carpal elevations. This band, the *flexor retinaculum*, creates a **carpal tunnel** through which flexor tendons of the wrist pass.

The carpals are divided into a proximal row incorporating (from radial to ulnar) the **scaphoid,** the **lunate,** the **triquetral,** and the **pisiform.** The

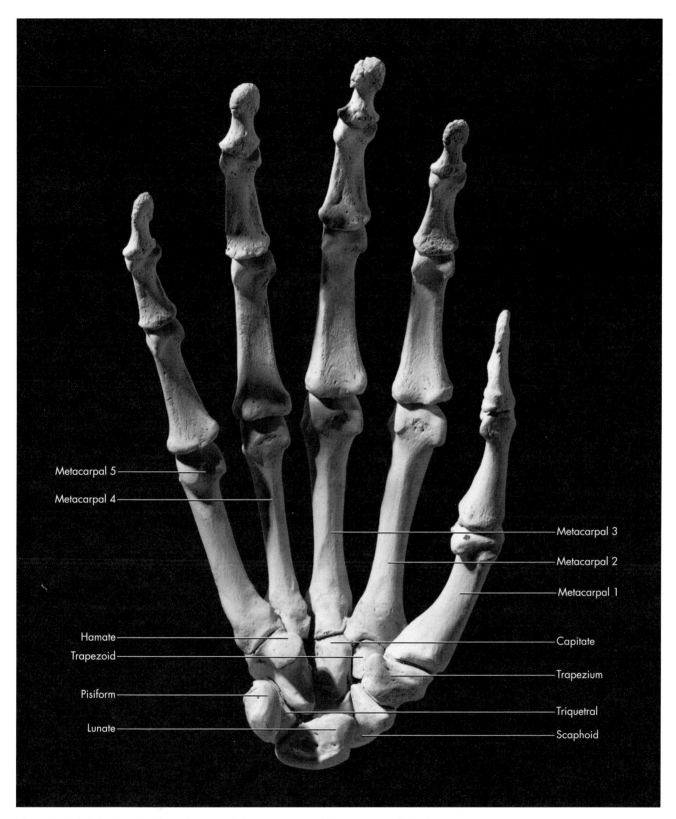

Figure 10.2 **Right hand, palmar** (anterior) (anterior). Small sesamoid bones not included. Natural size.

Labels on figure:
Metacarpal 5
Metacarpal 4
Metacarpal 3
Metacarpal 2
Metacarpal 1
Hamate
Trapezoid
Pisiform
Lunate
Capitate
Trapezium
Triquetral
Scaphoid

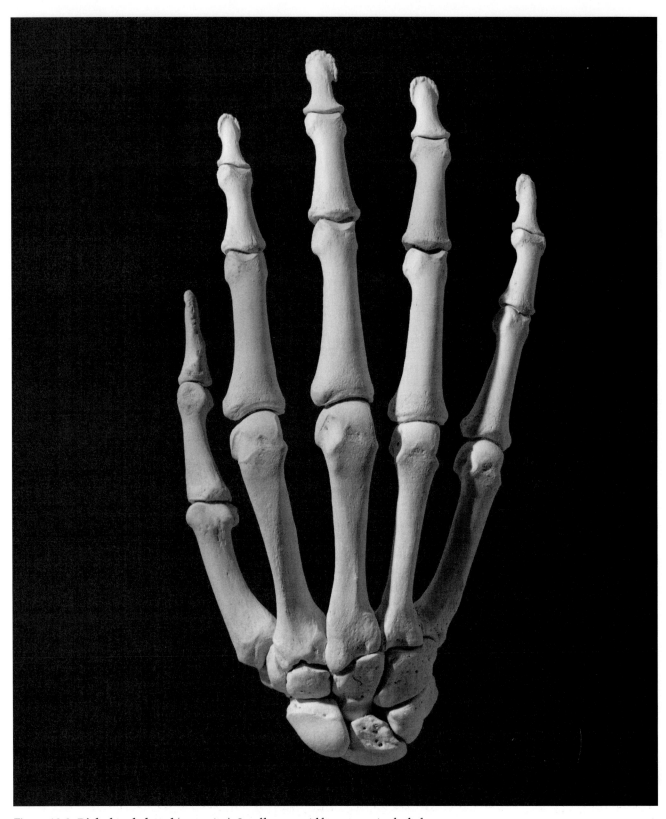

Figure 10.3 **Right hand, dorsal** (posterior). Small sesamoid bones not included.
Natural size.

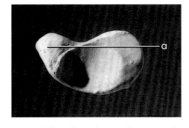 

View from the capitate   View from the radius

Figure 10.4 **Right scaphoid.** Palmar is up. Natural size.

scaphoid and lunate both articulate with the radius. The distal row of carpals, again from radial to ulnar, is composed of the **trapezium (greater multangular)**, the **trapezoid (lesser multangular)**, the **capitate,** and the **hamate.**

10.1.1 Scaphoid (Figure 10.4)

The scaphoid bone (also known as the hand navicular) is shaped like a boat, with a major concave surface for the head of the capitate and a major convex surface that articulates with the distal radius. One of the largest carpal bones, it is the most lateral and proximal carpal, interposed between the radius and the trapezium, at the base of the thumb. The **tubercle** is a blunt, nonarticular projection adjacent to the hollowed capitate facet. It is one of four attachments for the *flexor retinaculum,* a fibrous band at the wrist.

- **Anatomical siding.** The concave articular surface houses the capitate head and is therefore distal. The tubercle is on the palmar surface and is lateral (toward the thumb).
- **Positional siding.** Hold the concave (capitate) facet toward you and the tubercle up. The tubercle leans toward the side from which the bone comes.

10.1.2 Lunate (Figures 10.5–10.6)

The lunate is shaped in the form of a half-moon, with the very concave surface articulating with the capitate, and the large broad articulation opposite this sharing the distal radial articular surface with the scaphoid.

- **Anatomical siding.** The radial facet is proximal, and the capitate facet is distal. The long, narrow, crescent-shaped facet on the flat surface of the bone articulates laterally (on the thumb side) with the scaphoid. The remaining facet, for the triquetral, is displaced dorsally. The largest nonarticular surface is palmar.
- **Positional siding.** Place the flat side on the table and the most concave facet toward you. The remaining facet rises up and toward the side from which the bone comes.

Figure 10.5 **Right lunate.** Dorsal is up. Natural size.

View from the capitate

Figure 10.6 **Right lunate.** Palmar is up. Natural size.

View from the scaphoid   View from the triquetral

10.1.3 Triquetral (Figures 10.7–10.8)

The triquetral is the third bone from the thumb side in the proximal carpal row. It has three main articular surfaces (hence its name). The bone is distinguished from the trapezoid by its pisiform articulation, a single, circular, isolated, and elevated facet.

- **Anatomical siding.** The smallest of the three major articular surfaces is for the pisiform, on the bone's palmar, medial surface. The largest facet is for the hamate and is distal. The lunate facet is lateral to, but continuous with, this facet.

- **Positional siding.** Hold the common edge between the two largest facets toward you and vertical. When the third facet (for the pisiform) is up, this facet points toward the side from which the bone comes.

Palmar view   View from the hamate

Figure 10.7 **Right triquetral.** Lateral is up. Natural size.

Figure 10.8 **Right triquetral.** Palmar is up. Natural size.

10.1.4 Pisiform (Figures 10.9–10.10)

The pisiform bone is pea-shaped, with one side flattened by the triquetral articular facet. The nonarticular body is an attachment point for the *flexor*

View from the triquetral

View from the proximal,
palmar hamate end (lateral)

Figure 10.9 **Right pisiform.**
Distal is toward the left.
Natural size.

Figure 10.10 **Right pisiform.** The
triquetral facet is up and faces
the left (dorsal). Natural size.

retinaculum. The pisiform is the smallest of the carpals. Because it develops within a tendon, it is actually a sesamoid bone. There are other, much smaller sesamoid bones found embedded in flexor tendons, for example, at some metacarpophalangeal and interphalangeal joints.

- **Anatomical siding.** The pisiform's nonarticular body underlies the ulnar corner of the palm's base. The bone's morphological variation makes siding accurate in only about 85–90% of all cases.

- **Positional siding.** Hold the facet toward you and turn the bone until the bulk of the nonarticular surface that is visible in this view is up. The groove and the bulk of this visible surface is displaced toward the side from which the bone comes.

10.1.5 Trapezium (Greater Multangular) (Figure 10.11)

The trapezium is an irregularly sided bone of medium size. It is distinguished by its largest facet, a saddle-shaped articular surface for the base of MC 1 (thumb), and by a long, raised, narrow tubercle, or crest.

- **Anatomical siding.** The bone has an elongate, proximodistally oriented crest, or tubercle, on its palmar surface. This is an attachment point for the *flexor retinaculum.* The groove adjacent to this tubercle is medial

View from the MC-2
base

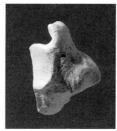

View from the lateral
scaphoid end

View from the
scaphoid-trapezoid
boundary (from medial)

Figure 10.11 **Right trapezium.** Palmar is up. Natural size.

(toward the center of the hand). The thumb articulation is distal and faces laterally.

- **Positional siding.** Place the bone on a flat surface with the tubercle on top and away from you and the concave facets on either side. The groove adjacent to the tubercle is on the side from which the bone comes.

10.1.6 Trapezoid (Lesser Multangular) (Figure 10.12)

The trapezoid is the second, boot-shaped, smallest carpal bone of the distal row. It articulates distally with the base of MC 2.

- **Anatomical siding.** The largest nonarticular surface is dorsal, and its most pointed corner is proximal and lateral (on the thumb side). Just palmar to this corner there is a sharp ridge where the lateral, more convex articular facet for the trapezium meets the proximal, more concave articular facet for the scaphoid.
- **Positional siding.** Place the boot's "sole" on the table, with the narrow, V-shaped space between the articular facets (or the largest articular double facet) toward you. The toe of the boot then points toward the side from which the bone comes.

View from the distal-most trapezium

View from the capitate–scaphoid boundary (from proximal)

Figure 10.12 **Right trapezoid.** Palmar is up. Natural size.

10.1.7 Capitate (Figures 10.13–10.14)

The capitate is a large carpal bone that articulates distally with the bases of MC 3, MC 2, and (sometimes) MC 4. Its distal end is therefore squared off, while the proximal end is rounded. The **head** (Figure 10.13a) of the capitate is the rounded end of the bone that articulates proximally with the hollow formed by the lunate and scaphoid. The **base** (Figure 10.13b) of the capitate is the more squared-off end that articulates distally most directly with the base of MC 3.

- **Anatomical siding.** Proximally the head articulates in the hollow formed by the lunate and scaphoid. The largest, flattest nonarticular surface is dorsal. The more concave surface of the head is for the hamate and is thus medial (on the little finger side).
- **Positional siding.** When the head is up and the base rests on the table, place the long, narrow articulation that runs up one side of the bone

View from the hamate

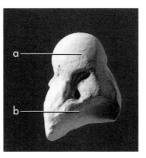
View from the scaphoid and trapezoid (from lateral)

Palmar view

Dorsal view

Figure 10.13 **Right capitate.** Proximal is up. Natural size.

Figure 10.14 **Right capitate.** Palmar is up. Natural size.

View from the MC-3 base

(from the base to the head) toward you. This articulation is on the side from which the bone comes.

10.1.8 Hamate (Figure 10.15)

The hamate is the carpal bone with the **hamulus** (Figure 10.15a) or hook-shaped, nonarticular projection on the palmar surface. This projection is the fourth attachment point for the *flexor retinaculum.*

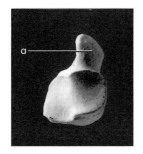
View from the MC-4 and MC-5 bases

View from the capitate

View from the triquetral

View from the lunate

Figure 10.15 **Right hamate.** Palmar is up. Natural size.

- **Anatomical siding.** The bone articulates distally with the bases of MC 4 and MC 5 via a double facet at the base of the hamulus. The hamulus is placed on the distal, palmar surface of the bone and is medial, hooking over the edge of the carpal tunnel in this position.
- **Positional Siding.** Place the flat nonarticular surface down with the hook and two adjacent metacarpal facets away from you. The hook leans toward the side from which the bone comes.

10.1.9 Growth

The carpal bones each ossify from a single center.

10.1.10 Possible Confusion

Because of their small size and compact construction, the identification of fragmentary carpal bones is not usually called for. If the hand has been carefully collected, intact bones are usually available. Each of the carpal bones is distinctive and impossible to confuse with another. Confusion of the adult carpal bones with the tarsal bones is improbable because the former are all smaller than the latter, and the shapes are all distinctive.

Many skeletons, especially those from archeological contexts, have incomplete hands because of postmortem disturbance of the skeleton before excavation by, for instance, rodents. These animals often move smaller skeletal elements during their burrowing activities. During archeological excavation, very small bones like the sesamoids, pisiform, and terminal phalanges are sometimes lost if care and fine screening are not employed in recovery.

10.2
Metacarpals (Figures 10.16–10.20)

The metacarpals are numbered MC 1 (the thumb) through MC 5 (the little finger), according to the five rays of the hand. They are all tubular bones, with round distal articular surfaces (**heads**) and more squarish proximal ends (**bases**). They are identified and sided most effectively according to the morphology of the bases.

The bases of the metacarpals articulate with their neighbors in positions 2–5. All four of the carpals in the distal row articulate with one or more metacarpal bases: the trapezium with MC 1 and MC 2, the trapezoid with MC 2, the capitate with MC 2, MC 3, and MC 4, and the hamate with MC 4 and MC 5.

10.2.1 First Metacarpal (Thumb)

The first metacarpal is the shortest metacarpal, broader and more robust in its shaft than the others. Its single proximal articular surface is saddle-shaped, corresponding to the facet on the trapezium.

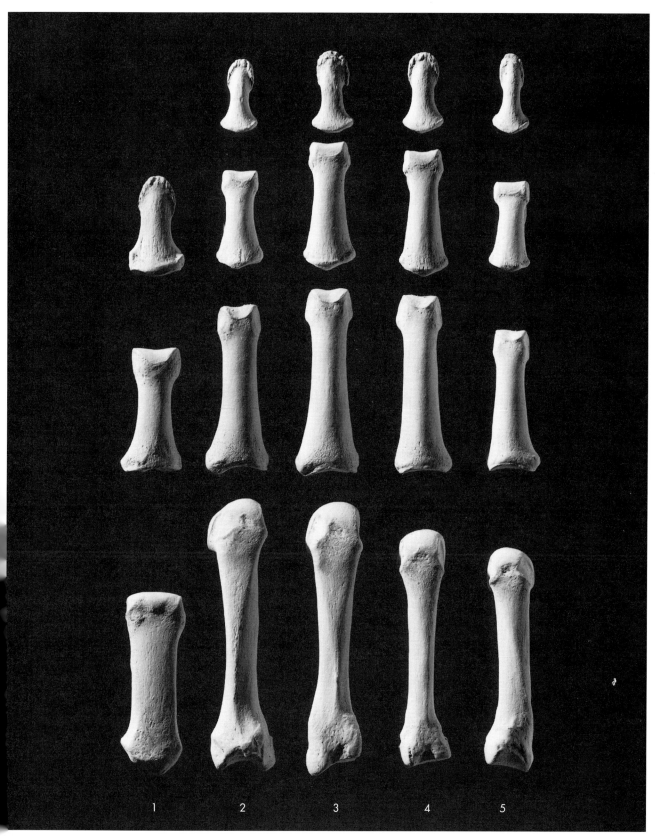

Figure 10.16 **Right hand, dorsal** (posterior), rays 1–5, showing the metacarpals and the proximal, intermediate, and distal hand phalanges. Distal is up, lateral is toward the left. Natural size.

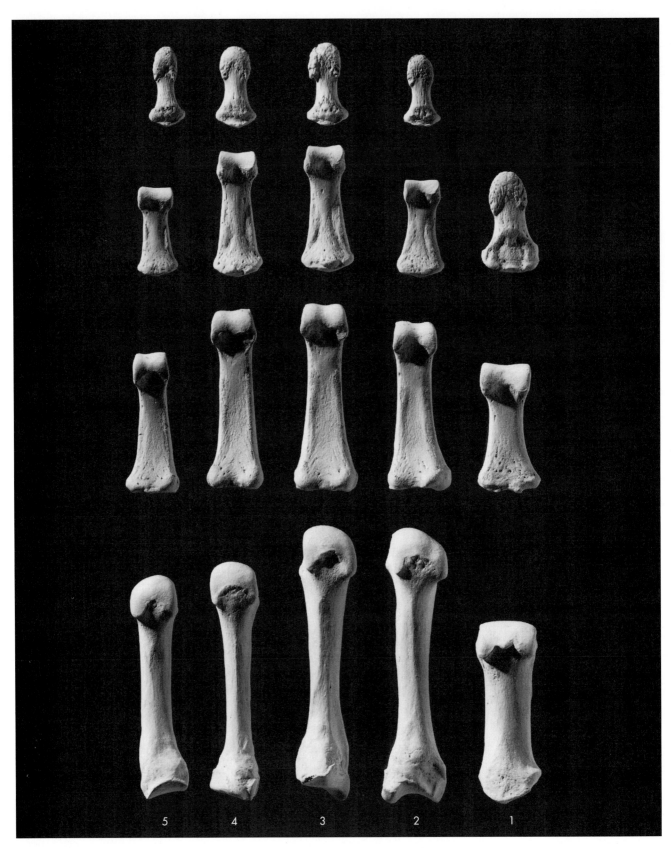

Figure 10.17 **Right hand, palmar** (anterior), rays 1–5, showing the metacarpals and the proximal, intermediate, and distal hand phalanges. Distal is up, lateral is toward the right. Natural size.

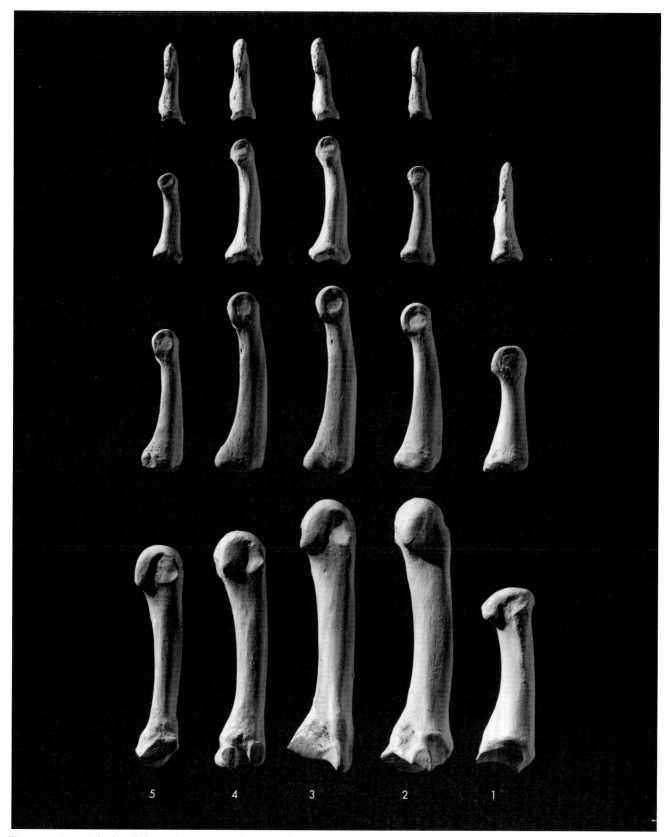

Figure 10.18 **Right hand, lateral,** rays 1–5, showing the metacarpals and the proximal, intermediate, and distal hand phalanges. Natural size.

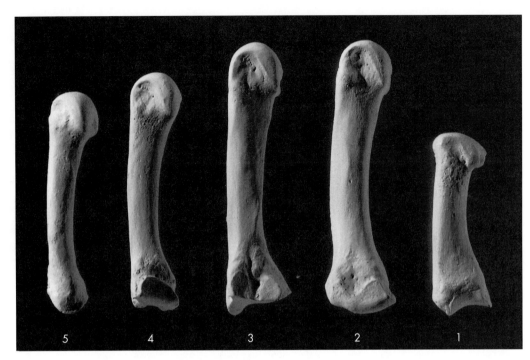

Figure 10.19 **Right-hand metacarpals, medial.** Natural size.

Figure 10.20 **Right-hand metacarpal bases, proximal.** Dorsal is up, lateral is toward the left. Natural size.

- **Siding.** The maximum palmar projection of the bone at the base is always toward the medial side. Therefore, in a proximal view, imagine dividing the saddle-shaped proximal facet into medial and lateral portions. The medial portion of the articular surface is always smaller in area. The lateral palmar surface of the shaft is larger and more excavated than the medial palmar surface. Viewed dorsally, with distal end up, the axis of maximum length is skewed basally toward the side the bone is from.

10.2.2 Second Metacarpal

The second metacarpal is normally the longest metacarpal, at the base of the index finger. The base presents a long, curved, bladelike wedge that articulates with the trapezoid, capitate, trapezium, and MC 3.

- **Siding.** The most proximal part of the base is a broad, bladelike, medially positioned wedge that bears the articulation for MC 3.

10.2.3 Third Metacarpal

The third metacarpal lies at the base of the middle finger. It is the only metacarpal with a sharp projection, the **styloid process,** at its base. It articulates with the capitate and MC 2 and MC 4 at the base.

- **Siding.** The styloid process is on the lateral, or MC 2, side of the bone.

10.2.4 Fourth Metacarpal

The fourth metacarpal lies at the base of the ring finger and is shorter and more gracile than MC 2 or MC 3. It has a fairly square base with three or four articular facets. It articulates (at its base) with the capitate (sometimes), hamate, MC 3, and MC 5.

- **Siding.** The proximal and medial basal facets share a common, right-angle articular edge.

10.2.5 Fifth Metacarpal

At the base of the little finger, the fifth metacarpal is the thinnest and shortest of the MC 2, MC 3, MC 4, and MC 5 series. It bears only two basal facets, one for the hamate and one for MC 4.

- **Siding.** The nonarticular side of the base faces medially, away from MC 4.

10.2.6 Growth (Figure 10.21)

Each metacarpal except MC 1 ossifies from two centers, a primary one for the shaft (body) and base, and a secondary one for the distal extremity (the head). The thumb metacarpal has a separate center for its base (but none for its distal extremity).

10.2.7 Possible Confusion

Metacarpals 2–5 are stouter than metatarsals 2–5. Metacarpal shafts are larger in diameter relative to length and not as straight and slender as metatarsal shafts. Metacarpal heads are more rounded than the mediolaterally compressed metatarsal heads.

10.2.8 Siding

In siding the metacarpals, the bases are always proximal, and the palmar shaft surfaces are always more concave than the dorsal surfaces in lateral

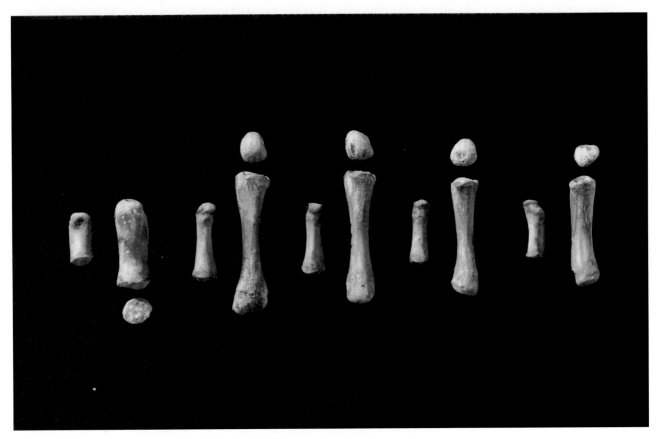

Figure 10.21 Metacarpal growth. The pairs of immature metacarpals are shown here in anterior (palmar) view, with ray 1 on the left and ray 5 on the right. They are from one-year-old and six-year-old individuals. Natural size.

view. Features of the base are used to side metacarpals as outlined in the preceding sections.

10.3
Hand Phalanges (Figures 10.16–10.18)

The phalanges are all shorter than metacarpals, lack rounded heads, and are anteroposteriorly flattened in their shafts. The thumb phalanges are shorter and squatter than the others, and the thumb lacks an intermediate phalanx. The expanded proximal end of each phalanx is the **base.** The distal end is the **head** (proximal or intermediate phalanges) or the **distal tip** (distal phalanges only). The nonarticular tubercles adjacent to the metacarpal heads and the phalangeal joints are attachment points for the *collateral ligaments.*

Dorsal surfaces of the hand phalanges are smooth and rounded. The palmar surfaces, in contrast, are flat and more roughened, especially along either side of the **shaft,** where raised ridges mark attachment sites for the *fibrous flexor sheaths,* tissues that prevent the flexor tendons from "bow stringing" away from the bones as the fingers are flexed.

10.3.1 Proximal Hand Phalanges

Each proximal hand phalanx displays a single, concave proximal (basal) articular facet for the metacarpal head. The proximal thumb phalanx is readily recognizable by its short, stout appearance.

10.3.2 Intermediate Hand Phalanges

Each intermediate hand phalanx displays a double proximal articular facet for the head of the proximal phalanx, and each also has a distal articular facet. The thumb ray bears only two phalanges, lacking a morphologically intermediate phalanx.

10.3.3 Distal Hand Phalanges

Each distal hand phalanx displays a double proximal articular facet for the head of the intermediate phalanx. The terminal end of each has a nonarticular pad, the **distal phalangeal tuberosity.** The thumb phalanx is readily recognizable because of its short, stout appearance. The dorsal surfaces of these phalanges are more rounded, the palmar surfaces more rugose.

10.3.4 Growth

Hand phalanges each ossify from two centers, a primary one for the shaft plus distal end, and a secondary one for the base.

10.3.5 Possible Confusion

Hand phalanges have shafts whose palmar surfaces are flattened, forming a half-circle in cross section. Foot phalanx shafts are circular in cross section.

10.3.6 Siding

For siding hand and foot phalanges it is best to work with whole specimens and comparative materials, particularly *in vivo* radiographs.

10.4
Functional Aspects of the Hand

Humans have effectively abandoned the use of their forelimbs as supports during locomotion. Primates in general, and humans in particular, have evolved **hands,** which provide the ability for these organisms to manipulate their environment in complex ways. As noted in section 10.1, forearm muscles operate the digits of the hand via tendons that pass across the wrist. The metacarpal heads form foundations from which the thumb and fingers work. The thumb bears only two phalanges, but the saddle-shaped, sellar joint at the base of its metacarpal allows this digit great mobility and the ability to oppose the other digits. Joints between the phalanges are hinge joints whose extension is checked by a *palmar ligament,* and whose abduction, adduction, and rotation are checked by *collateral ligaments.*

Most of the force in the grip or extension of the fingers comes from forearm muscles that send tendons across the wrist to insert on the digits. These forearm muscles are called *extrinsic hand muscles.* As Cartmill *et al.* (1987) point out, forearm muscles control much of the hand's movement and, in a functional sense, the forearm is best thought of as an appendage of the hand rather than the other way around. *Intrinsic muscles of the hand* lie within the palm and produce abduction and adduction of the fingers as well as special movements, particularly of the thumb.

CHAPTER 11

Pelvic Girdle: Sacrum, Coccyx, and Os Coxae

The bony structure at the base of the front limbs is the shoulder girdle and the one at the base of the hind limbs is the pelvic girdle. The pelvic girdles of terrestrial vertebrates are connected to the vertebral column and are much larger than their homologs in fish. These adaptations are required for weight bearing and muscle attachment in the terrestrial forms. In early land-dwelling vertebrates, the right and left limb girdles joined dorsally with the sacral vertebrae to form a bony ring around the rear of the trunk.

The adult human bony pelvis is composed of three main elements: the right and left **os coxae** and the **sacrum** and **coccyx.** The sacrum and coccyx are part of the axial skeleton and are actually variably fused vertebrae. The bony pelvis functions to support and protect the abdominal and pelvic organs. In addition, it anchors muscles of the abdomen and leg. Unlike the shoulder girdle, which is a movable platform, the pelvic girdle is firmly fixed to the axial skeleton via its vertebral element, the sacrum.

11.1
Sacrum (Figures 11.1–11.4)

11.1.1 Anatomy

The sacral vertebrae fuse during adolescence into one immobile, wedge-shaped bone, the sacrum. This bone is typically formed from five segments but sometimes is made up of four or six. The sacrum is located at the base of the vertebral column and articulates on the left and right with the two os coxae and inferiorly with the small coccyx.

a. The **promontory** is a point at the superior midline edge of the sacral plateau which articulates with the terminal, or most inferior, lumbar vertebra.

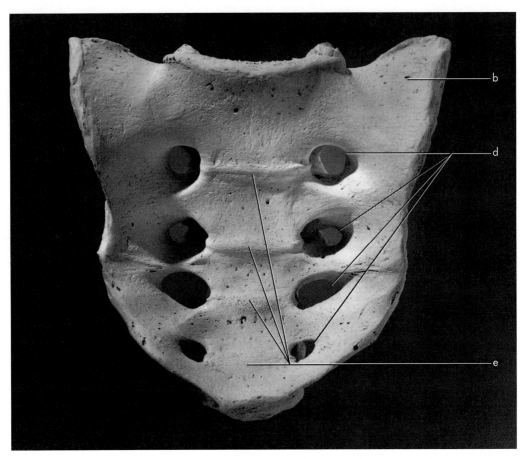

Figure 11.1 **Sacrum, anterior.** Natural size. Key: b, ala; d, anterior (pelvic) sacral foramina; e, transverse lines.

b. The **alae,** or "wings," sweep laterally from the first sacral element on either side. They articulate with the medial, posterior surface of the os coxae.

c. The **sacroiliac joint (auricular surface)** is the articulation between the sacrum and the os coxae. This articular surface is best seen in lateral aspect. The articulation is the most immobile synovial joint in the body. The "ear-shaped" surface articulates with the auricular surface of the os coxae.

d. The **anterior (pelvic) sacral foramina** are perforations in the sacrum's concave anterior surface through which the *sacral nerves* pass.

e. The **transverse lines** are horizontal ridges on the concave anterior sacral surface. They mark the edges of the fused sacral vertebrae.

f. The **superior articular facets** of the sacrum articulate with the inferior articular processes of the most inferior lumbar vertebra.

g. The **dorsal wall** of the sacrum is a rough, irregular, variable plate of bone made up of the ossified laminae and articular processes of the fused sacral vertebrae.

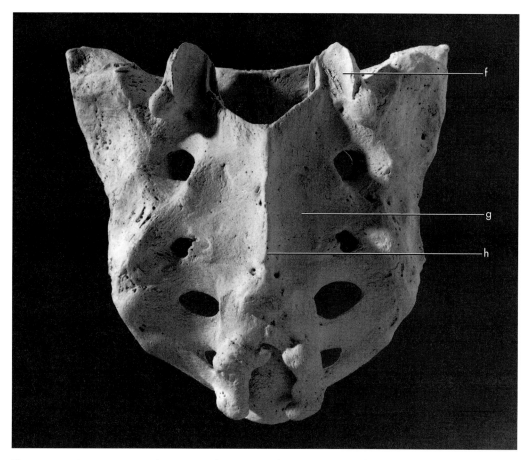

Figure 11.2 **Sacrum, posterior.** Natural size. Key: f, superior articular facet; g, dorsal wall; h, median crest (spine).

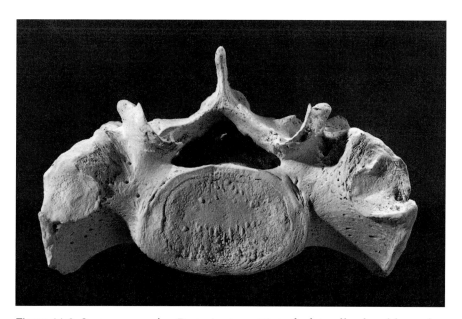

Figure 11.3 **Sacrum, superior.** Posterior is up. Note the laterally placed facets for the sacralized L-5 of this individual. Natural size.

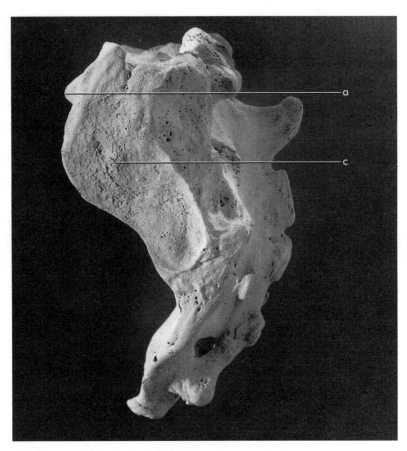

Figure 11.4 **Sacrum, left lateral.** Anterior is toward the left. Natural size. Key: a, promontory; c, sacroiliac joint (auricular surface).

h. The **median crest (spine)** is the highly variable midline projection of the dorsal wall composed of fused spines of the sacral vertebrae.

11.1.2 Growth

The sacrum ossifies from a total of thirty-five centers in individuals with five sacral vertebrae. Each sacral body develops from a primary center with two epiphyseal plates. Corresponding arches ossify from two centers, and an additional pair of costal elements form the anterolateral projections of each sacral vertebra. Two epiphyseal plates form at each sacroiliac articular region, one forming the auricular surface, and one the lateral margin inferior to this. See Figure 6.7.

11.1.3 Possible Confusion

- The promontory region might be mistaken for a lumbar vertebra when fragmented. Lumbars, however, lack attached alae.

- The auricular surface might be mistaken for an os coxae fragment when broken. The sacral auricular, however, has virtually no adjacent outer bone surface surrounding it as the os coxae does.

11.1.4 Siding

When fragmentary, parts of the sacrum can be sided as follows:
- The anterior sacral surface is smooth and concave, with transverse lines.
- The size of sacral vertebrae diminishes inferiorly.
- The auricular (sacroiliac) surface is lateral, and the apex of this V-shaped feature is anterior.

11.2
Coccyx (Figures 11.5–11.6)

The coccyx, the vestigial tail, is highly variable in shape, with three to five (most often four) variably fused segments. The rudimentary vertebrae of the coccyx show articular and transverse processes superiorly, but they lack pedicles, laminae, and spinous processes. The sacral articulation is via the superior coccygeal body as well as a relatively large pair of tubercles called the **cornua** (Figure 11.6a). The latter are rudimentary articular processes that contact the sacrum. The coccyx may fuse with the sacrum late in life.

As with the sacrum, the individual vertebral elements of the coccyx decrease in size inferiorly, and horizontal lines of fusion can be seen between adjacent coccygeal vertebrae. The coccyx serves to anchor pelvic muscles and ligaments.

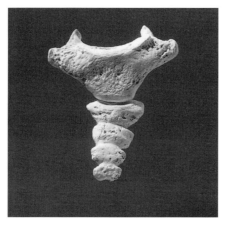

Figure 11.5 **Coccyx, anterior.**
Superior is up. Natural size.

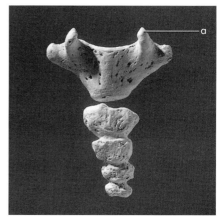

Figure 11.6 **Coccyx, posterior.**
Superior is up. Natural size.

11.3
Os Coxae (Figures 11.7–11.10)

11.3.1 Anatomy

Unlike many bones that gain their names because of perceived similarities to common objects, the os coxae resembles no common object and thus has earned the informal name **innominate**—the "bone with no name." The os coxae differs in males and females, its anatomy representing a compromise between the demands of locomotion and birthing. The os coxae is a part of the bony pelvis and is formed ontogenetically from three different parts, the **ilium, ischium,** and **pubis,** which fuse in early adolescence. Anatomical orientation of the os coxae is accomplished by placing the hip socket laterally and the ilium superiorly; this allows the plane of the **pubic symphysis** (the only place where right and left os coxae nearly meet) to define the sagittal plane.

The features identified here occur on both the surfaces and the edges of the os coxae. Many of the features are visible from different views of the bone.

a. The **ilium** is the thin, bladelike section superior to the hip socket.

b. The **ischium** is the massive, blunt, posteroinferior part of the bone that one sits on.

c. The **pubis** is the anteroinferior part of the bone that approaches the opposite os coxa at the midline.

d. The **acetabulum** is the laterally facing, round hollow that forms the socket of the hip, articulating with the head of the femur.

e. The **acetabular fossa,** or **notch,** is the nonarticular surface within the acetabulum. It is the attachment point for the *ligamentum teres,* a ligament that limits femoral mobility and accompanies the vessel that supplies blood to the femoral head.

f. The **lunate surface** is the surface within the acetabulum where the femoral head actually articulates.

g. The **iliac pillar (acetabulo-cristal buttress)** is the bony thickening, or buttress, located vertically above the acetabulum on the lateral iliac surface. This pillar extends to the superior margin of the ilium.

h. The **iliac (cristal) tubercle** is the thickening at the superior terminus of the iliac pillar.

i. The **iliac crest** is the superior border of the ilium. It is S-shaped when viewed superiorly. Many of the *abdominal muscles* originate on the crest.

j. The **gluteal lines** are rough, irregular lines that demarcate the attachment of the *gluteal muscles* on the lateral surface of the ilium. They vary from prominent to imperceptible between individuals and across their paths. The **inferior gluteal line** (j-1) is a horizontal line just superior to the acetabulum. The **anterior gluteal line** (j-2) is a line that curves posteroinferiorly through the fossa posterior to the

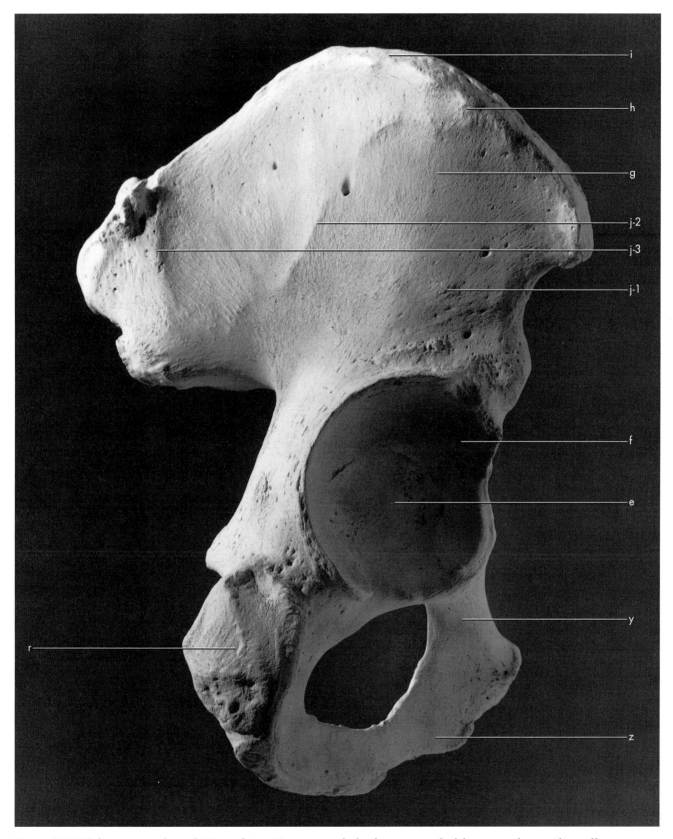

Figure 11.7 **Right os coxae, lateral.** Natural size. Key: e, acetabular fossa or notch; f, lunate surface; g, iliac pillar (acetabulo-cristal buttress); h, iliac (cristal) tubercle; i, iliac crest; j-1, inferior gluteal line; j-2, anterior gluteal line; j-3, posterior gluteal line; r, ischial tuberosity; y, iliopubic (superior pubic) ramus; z, ischiopubic (inferior pubic) ramus.

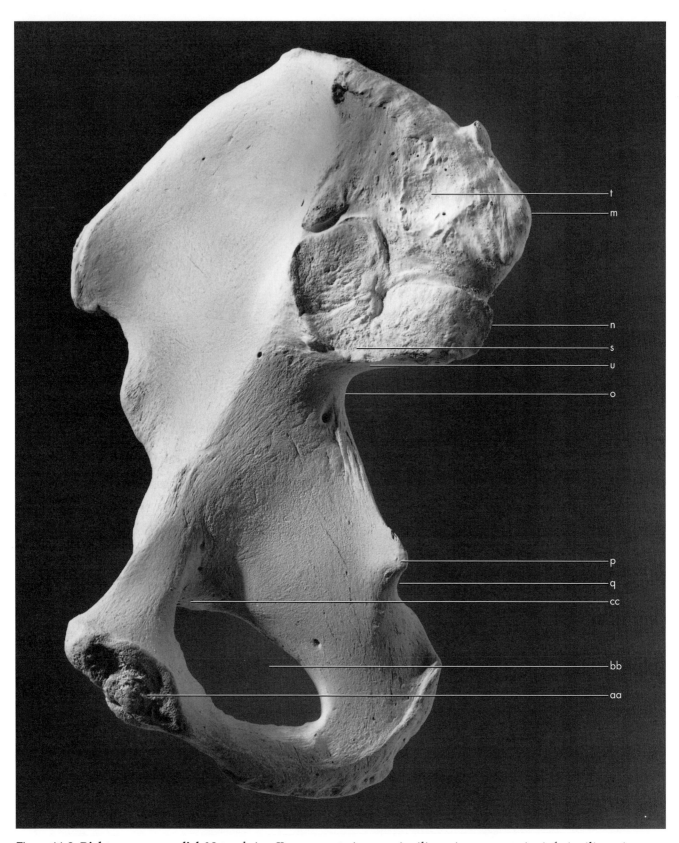

Figure 11.8 **Right os coxae, medial.** Natural size. Key: m, posterior superior iliac spine; n, posterior inferior iliac spine; o, greater sciatic notch; p, ischial spine; q, lesser sciatic notch; s, auricular surface; t, iliac tuberosity; u, preauricular sulcus; aa, pubic symphysis; bb, obturator foramen; cc, obturator groove.

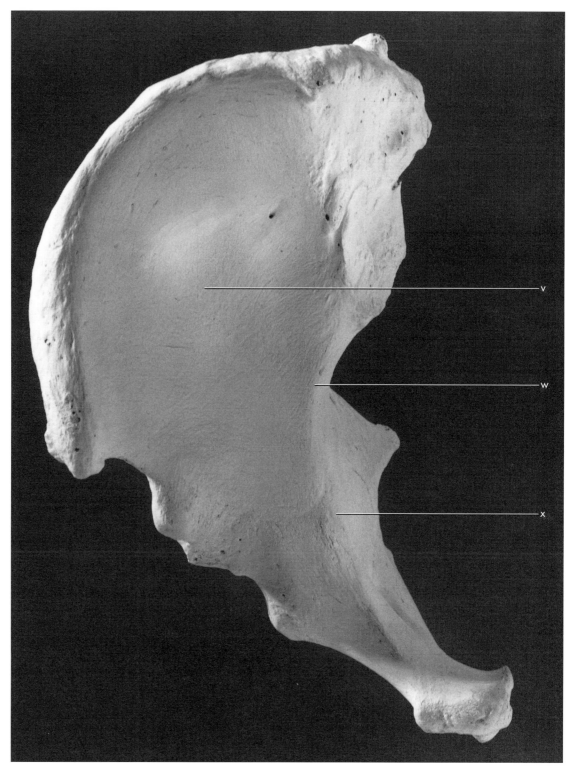

Figure 11.9 **Right os coxae, superior.** Anterior is down. Natural size. Key: v, iliac fossa; w, arcuate line; x, iliopubic (iliopectineal) eminence.

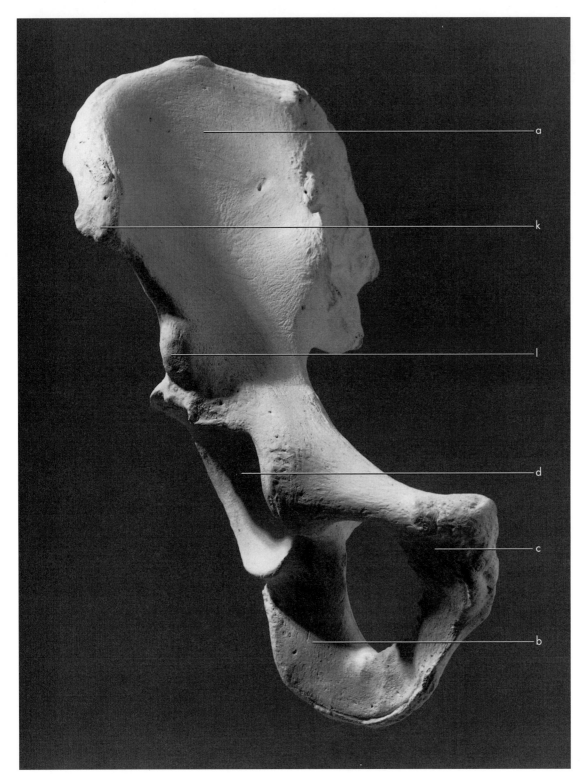

Figure 11.10 **Right os coxae, anterior.** Natural size. Key: a, ilium; b, ischium; c, pubis; d, acetabulum; k, anterior superior iliac spine; l, anterior inferior iliac spine.

iliac pillar. The **posterior gluteal line** (j-3) is more vertically placed, near the posterior edge of the ilium. The *gluteus minimus muscle* originates between the inferior and anterior lines, and the *gluteus medius muscle* arises between the anterior and posterior lines. The *gluteus maximus muscle* originates posterior to the posterior gluteal line. The first two gluteal muscles, *minimus* and *medius*, are abductors and medial rotators of the femur at the hip, and the *gluteus maximus* is a lateral rotator and extensor of the femur at the hip.

k. The **anterior superior iliac spine** is located, according to its name, at the anterior end of the iliac crest. It anchors the *sartorius muscle* and the *inguinal ligament.*

l. The **anterior inferior iliac spine** is a blunt projection on the anterior border of the os coxae, just superior to the level of the acetabulum. It is the origin for the straight head of the *rectus femoris muscle,* a strong leg flexor at the hip. Its lower extent serves as the attachment site for the *iliofemoral ligament.*

m. The **posterior superior iliac spine** is the posterior terminus of the iliac crest. It is an attachment for part of the *gluteus maximus muscle,* an extensor and lateral rotator of the leg at the hip.

n. The **posterior inferior iliac spine** is a sharp projection just posteroinferior to the sacral articular surface. It partially anchors the *sacrotuberus ligament,* which acts to connect the sacrum and os coxae.

o. The **greater sciatic notch** is the wide notch just inferior to the posterior inferior iliac spine. The *piriformis muscle,* a lateral rotator of the thigh at the hip, and the nerves leaving the pelvis for the lower limb pass through this notch. Bone cortex in the os coxae is thickest in this area.

p. The **ischial spine** for attachment of the *sacrospinous ligament* is located just inferior to the greater sciatic notch.

q. The **lesser sciatic notch** is the notch between the ischial spine superiorly and the rest of the ischium inferiorly. The *obturator internus muscle,* a lateral rotator and sometimes abductor of the femur at the hip, passes through this notch.

r. The **ischial tuberosity** is the blunt, rough, massive posteroinferior corner of the os coxae. It anchors the extensor muscles of the thigh at the hip, including the *semitendinosus, semimembranosus, biceps femoris* (*long head*), and *quadratus femoris.*

s. The **auricular surface** is the ear-shaped sacral articulation on the medial surface of the ilium.

t. The **iliac tuberosity** is the roughened surface just superior to the auricular surface. It is the attachment site for *sacroiliac ligaments.*

u. The **preauricular sulcus** is a variable groove along the anteroinferior edge of the auricular surface.

v. The **iliac fossa** is the smooth hollow on the medial surface of the iliac blade.

w. The **arcuate line** is an elevation that sweeps anteroinferiorly across the medial os coxae from the apex of the auricular surface toward the pubis.

x. The **iliopubic (iliopectineal) eminence** marks the point of union of the ilium and pubis at the anterior extension of the arcuate line.

y. The **iliopubic (superior pubic) ramus** connects the pubis to the ilium at the acetabulum.

z. The **ischiopubic (inferior pubic) ramus** is the thin, flat strip of bone connecting the pubis to the ischium.

aa. The **pubic symphysis** is the near midline surface of the pubis where the two os coxae most closely approach. In life there is a fibrocartilaginous disk between the symphyseal faces.

bb. The **obturator foramen** is the large foramen enclosed by the two pubic rami and the ischium. In life it is mostly closed by a membrane across it.

cc. The **obturator groove** is the wide groove on the medial surface of the iliopubic ramus, at the superior and lateral corner of the obturator foramen. The *obturator vessels* and *nerve* pass through this groove.

11.3.2 Growth (Figure 11.11)

There are three primary and five secondary centers of ossification in each os coxae. The ilium, ischium, and pubis form the primary centers, fusing around the acetabulum. The ilium has one secondary center at the anterior inferior spine and one across the iliac crest. The pubis has one center at the symphysis (the "ventral rampart"), and the ischium has one at the tuberosity that extends along the inferior pubic ramus. The eighth center ("os cotyledon") is located in the depth of the acetabulum.

11.3.3 Possible Confusion

- Fragmentary iliac blades might be mistaken for cranial or scapular fragments. The cranial bones are, however, of more uniform thickness. They have cortices of about equal thickness around the diploë.
- Scapular blades are thinner than iliac blades and display subscapular ridges.
- Fragmentary auricular areas could be mistaken for sacra, but in the latter bone there are attached alae, and the adjacent surfaces have no evidence of sacroiliac roughening or sciatic notches.

11.3.4 Siding

When intact, the os coxae is easily sided because the pubis is anterior, the iliac crest is superior, and the acetabulum is lateral. When fragmentary, various parts of the os coxae can be sided as follows:

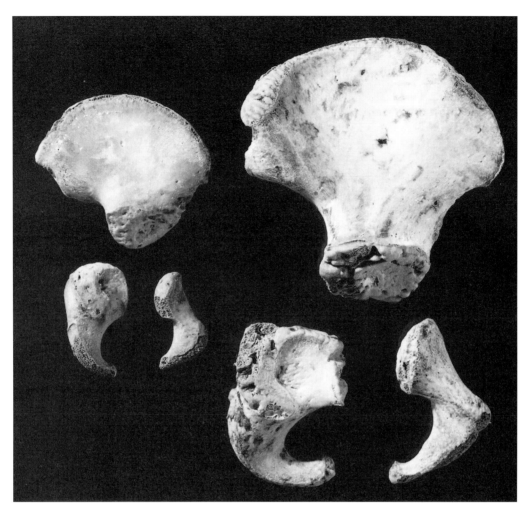

Figure 11.11 **Os coxae growth.** The three elements of the os coxae, shown here in lateral view, are from a one-year-old and a six-year-old. Natural size.

- For isolated pubic regions, the ventral surface is rough, the dorsal surface is smooth and convex, the symphysis faces the midline, and the superior pubic ramus is more robust than the inferior pubic ramus.
- For isolated ischial regions, the thicker ramus faces the acetabulum. The thinner ramus is therefore anteroinferior. The surface of the ischial tuberosity faces posterolaterally.
- For isolated iliac blades, the iliac pillar is lateral and anteriorly displaced. The auricular surface and related structures are posterior and medial.
- For isolated iliac crests, the iliac tubercle is anterior and lateral, and the surface anterior to it is more concave than the surface posterior to it. The crest sweeps posteromedially from this point until it reaches the level of the anterior edge of the auricular surface and turns laterally.
- For isolated acetabula, the acetabular notch is inferior and faces slightly anterior. The inferior end of the "c" made by the lunate surface is broader and more blunt than the superior end. The ischial ramus is

posterior, and the superior pubic ramus is anterior. The ilium is superior.

- For isolated auricular surfaces, the auricular is posterior on the ilium and faces medially. Its apex points anteriorly, and the roughened surface for the sacroiliac ligaments is superior. The greater and lesser sciatic notches are posteroinferior.

11.4
Functional Aspects of the Pelvic Girdle

The human pelvis is the distinctive foundation for a unique locomotor mode among primates, habitual bipedality. When this locomotor mode was adopted over 4 million years ago, most muscle groups attached to the pelvis altered their function. The bony architecture of the pelvis shows the effects of these mechanical changes (Lovejoy, 1988).

The three axes and six possible directions of rotation at the hip joint are the same as those in the shoulder joint—abduction and adduction, medial and lateral rotation, and flexion and extension. Like the *deltoideus muscle* of the shoulder, the major hip abductors *gluteus medius* and *gluteus minimus* form a hood across the top of the joint. Pulling between the iliac blade and the femur's greater trochanter, these muscles perform the key role of stabilizing the pelvis and superincumbent trunk during walking. The largest muscle in the human body, *gluteus maximus*, is defined as a hip extensor, and in humans this muscle's primary role is to keep the trunk from pitching forward during running. The forward swing of the leg during walking is produced by the *iliopsoas muscle*, and the leg is decelerated by the *hamstring muscles*. These and other muscles that control the movements of the hip joint take their origin from various surfaces and projections of the os coxae.

CHAPTER 12

Leg: Femur, Patella, Tibia, and Fibula

The evolution of the leg mirrors that of the arm as described in Chapter 9. The single thigh bone, the **femur,** is the serial homolog of the upper arm bone, the humerus. Likewise, the lower bones of the leg, the **tibia** and **fibula,** are serial homologs of the radius and ulna. The largest sesamoid bone in the body, the **patella,** lies at the knee joint. As we find when we introduce the functional anatomy of the human leg, the bipedal locomotor mode practiced by hominids has resulted in major specializations of the leg bones.

12.1
Femur (Figures 12.1–12.6)

12.1.1 Anatomy

The femur is the longest, heaviest, and strongest bone in the body. It supports all of the body's weight during standing, walking, and running. Because of its strength and density, it is frequently recovered in forensic, archeological, and paleontological contexts. The femur is a valuable bone because of the information it can provide on the stature of an individual (see Chapter 16).

The femur articulates with the acetabulum of the os coxae. Distally, it articulates with the patella and proximal tibia. The leg's actions at the hip include medial and lateral rotation, abduction, adduction, flexion, and extension. At the knee, motion is far more restricted, confined mostly to flexion and extension. Although the main knee action is that of a sliding hinge, this joint is one of the most complex in the body.

   a. The **head** is the rounded proximal part of the bone that fits into the acetabulum. It constitutes more of a sphere than the hemispherical humeral head.

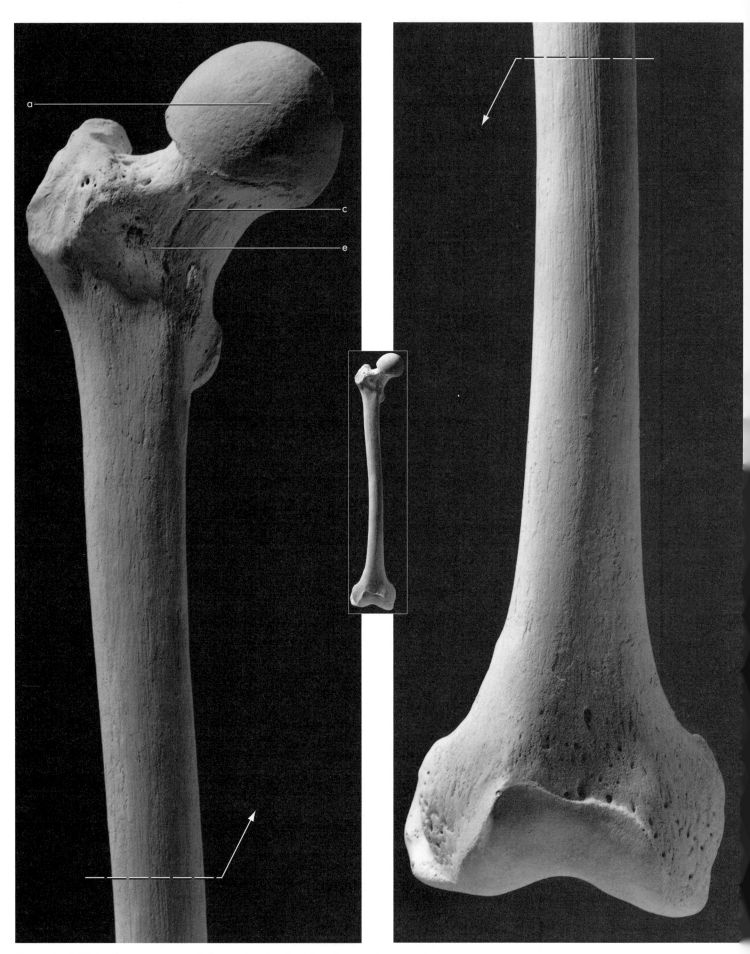

Figure 12.1 **Right femur, anterior:** *left,* proximal end; *right,* distal end. Natural size. Key: a, head; c, neck; e, intertrochanteric line.

b. The **fovea capitis** is the small, nonarticular depression near the center of the femur's head. It receives the *ligamentum teres* from the acetabular notch of the os coxae.

c. The **neck** of the femur connects the head with the shaft.

d. The **greater trochanter** is the large, blunt, nonarticular prominence on the lateral, proximal part of the femur. It is the insertion site for the *gluteus minimus* (anterior aspect of the trochanter) and *gluteus medius muscles* (posterior aspect), both major abductors and medial rotators of the thigh at the hip. Their origins are on the broad, flaring iliac blade of the os coxae. These muscles are crucial in stabilizing the trunk when one leg is lifted from the ground during bipedal locomotion.

e. The **intertrochanteric line** is a variable, fairly vertical, roughened line that passes between the lesser and greater trochanters on the anterior surface of the base of the neck of the femur. Superiorly, this line anchors the *iliofemoral ligament,* which is the largest ligament in the human frame. It acts to strengthen the joint capsule of the hip.

f. The **trochanteric fossa** is the pit excavated into the posteromedial wall of the greater trochanter. This pit is for the insertion of the tendon of *obturator externus,* a muscle that originates around and across the membrane that stretches across the obturator foramen of the os coxae. This muscle acts to laterally rotate the thigh at the hip. Just above its insertion, the medial tip of the greater trochanter receives several hip muscles: the *superior* and *inferior gemellus,* the *obturator internus,* and the *piriformis.* The latter two are important abductors, and all of these muscles can laterally rotate the femur.

g. The **obturator externus groove** is a shallow depression aligned laterally and superiorly across the posterior surface of the femoral neck. In hominids, erect posture brings the tendon of insertion for the *obturator externus muscle* into contact with the posterior surface of the femoral neck, creating the groove.

h. The **lesser trochanter** is the blunt, prominent tubercle on the posterior femoral surface just inferior to the point where the neck joins the shaft. This is the insertion point of the *iliopsoas tendon* and the common insertion of the *iliacus muscle,* originating in the iliac fossa, and the *psoas major muscle,* originating from the lumbar vertebrae and their disks. These muscles are major flexors of the thigh at the hip.

i. The **intertrochanteric crest** is the elevated line on the posterior surface of the proximal femur between the greater and lesser trochanters. It passes from superolateral to inferomedial. At its midpoint is a small tubercle (the **quadrate tubercle**), which is the site of insertion of the *quadratus femoris muscle,* a lateral rotator of the femur.

j. The **gluteal line,** or **tuberosity,** is a long, wide, roughened, posterolaterally placed feature that extends from the base of the greater trochanter to the lip of the linea aspera (see 12.1.1n). It can be a depression or assume the form of a true tuberosity. If the latter is present, it is often referred to as the **third trochanter.** It is the insertion for part of the *gluteus maximus muscle,* an extensor and lateral

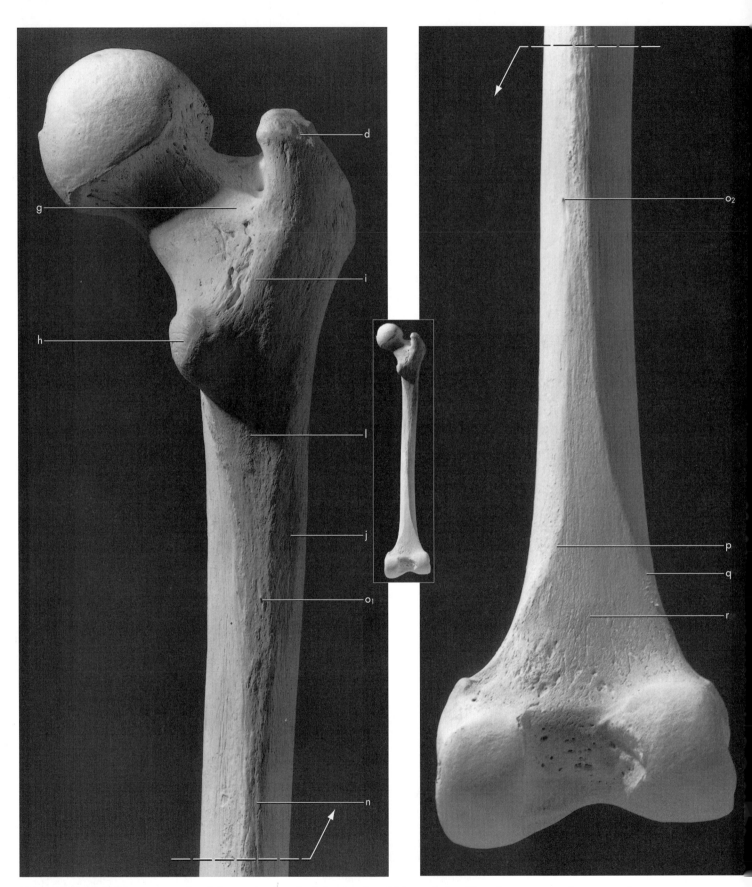

Figure 12.2 **Right femur, posterior:** *left,* proximal end; *right,* distal end. Natural size. Key: d, greater trochanter; g, obturator externus groove; h, lesser trochanter; i, intertrochanteric crest; j, gluteal line or tuberosity; l, pectineal line; n, linea aspera; o, nutrient foramen; p, medial supracondylar line (ridge); q, lateral supracondylar line (ridge); r, popliteal surface.

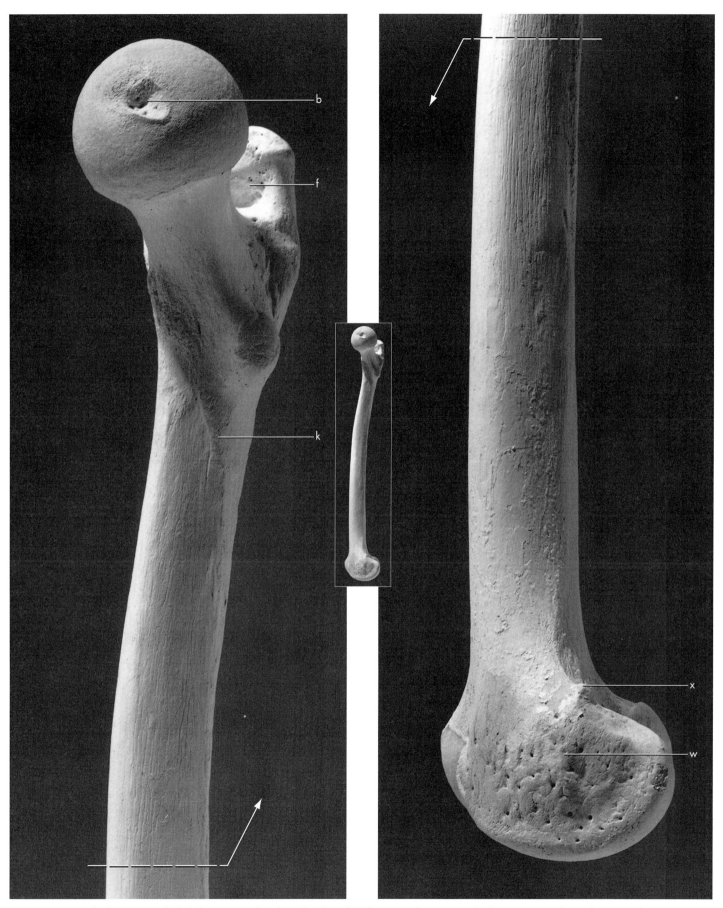

Figure 12.3 **Right femur, medial:** *left,* proximal end; *right,* distal end. Natural size. Key: b, fovea capitis; f, trochanteric fossa; k, spiral line; w, medial epicondyle; x, adductor tubercle.

rotator of the thigh at the hip that originates on the posterior half of the os coxae, sacrum, and coccyx.

k. The **spiral line,** spiraling inferior to the lesser trochanter, connects the inferior end of the intertrochanteric line with the medial lip of the linea aspera. It is the origin of the *vastus medialis muscle,* a part of the *quadriceps femoris muscle,* a knee extensor that inserts on the anterior tibia.

l. The **pectineal line** is a short, curved line that passes inferolaterally from the base of the lesser trochanter, between the spiral line and gluteal tuberosity. It is the insertion of the *pectineus muscle,* which originates from the pubic part of the os coxae and acts to adduct, laterally rotate, and flex the thigh at the hip.

m. The **femoral shaft** is the long section between the expanded proximal and distal ends of the bone.

n. The **linea aspera** is the long, wide, roughened, and elevated ridge that runs along the posterior shaft surface. It merges into the spiral, pectineal, and gluteal lines proximally and divides into the supracondyloid ridges distally. The linea aspera is a primary origin site for the *vastus muscles* and the primary insertion site of the adductors (*longis, brevis,* and *magnus*) of the hip.

o. The **nutrient foramen** is located about midshaft level on the posterior surface of the bone, adjacent to or on the linea aspera. This foramen exits the bone distally.

p. The **medial supracondylar line (ridge)** is the inferior, medial extension of the linea aspera, marking the distal, medial corner of the shaft. It is fainter than the lateral supracondylar ridge.

q. The **lateral supracondylar line (ridge)** is the inferior (distal), lateral extension of the linea aspera. It is more pronounced than the medial supracondylar ridge.

r. The **popliteal surface** is the wide, flat, triangular area of the posterior, distal femur. It is bounded by the condyles inferiorly, and by the supracondylar lines medially and laterally.

s. The **lateral condyle** is the large protruding knob on the lateral side of the distal femur.

t. The **lateral epicondyle** is the convexity on the lateral side of the lateral condyle. It is an attachment point for the *lateral collateral ligament* of the knee. Its upper surface bears a facet that is an attachment point for one head of the *gastrocnemius muscle,* a flexor of the knee and plantarflexor of the foot at the ankle.

u. The **popliteal groove,** a smooth hollow on the posterolateral side of the lateral condyle, is a groove for the *tendon of the popliteus muscle.* This muscle inserts on the posterior tibial surface and is a medial rotator of the tibia at the knee.

v. The **medial condyle** is the large knob on the medial side of the distal femur. Its medial surface bulges away from the axis of the shaft. The medial condyle extends more distally than the lateral condyle.

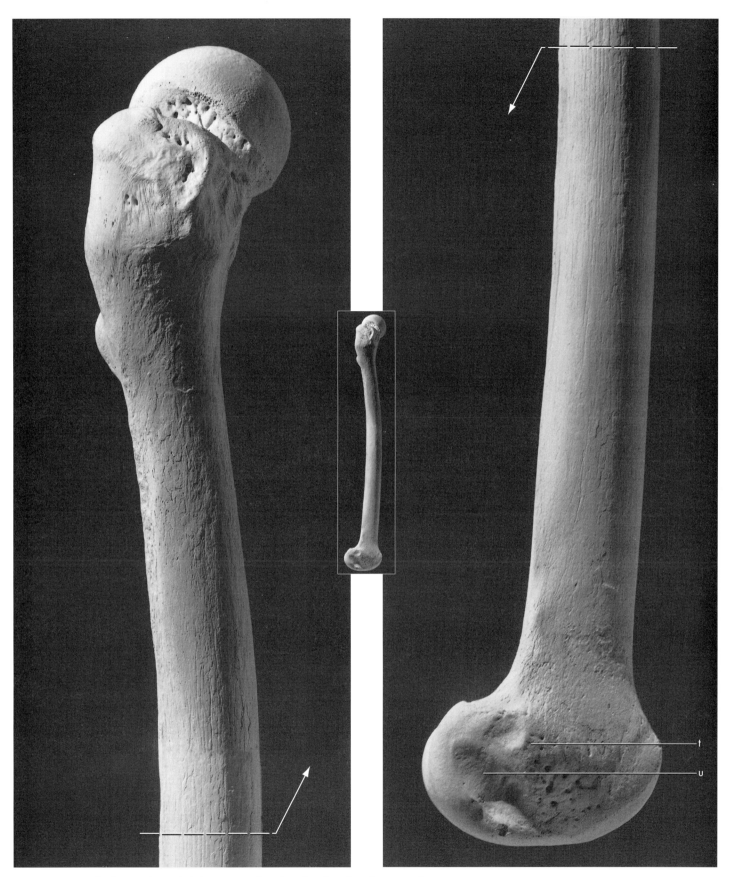

Figure 12.4 **Right femur, lateral:** *left,* proximal end; *right,* distal end. Natural size. Key: t, lateral epicondyle; u, popliteal groove.

Figure 12.5 **Right femur, proximal.** Posterior is up, lateral is toward the left. Natural size.

w. The **medial epicondyle** is the convexity on the medial side of the medial condyle. It is a point of attachment for the *medial collateral ligament* of the knee.

x. The **adductor tubercle** is a variable, raised tubercle on the medial supracondylar ridge just superior to the medial epicondyle. It is an

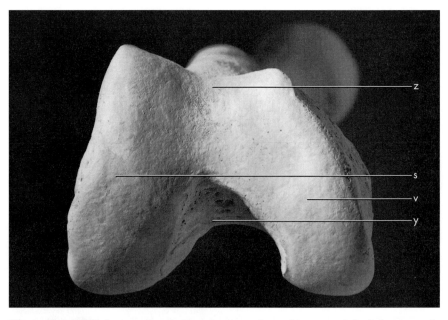

Figure 12.6 **Right femur, distal.** Anterior is up, lateral is toward the left. Natural size. Key: s, lateral condyle; v, medial condyle; y, intercondylar fossa or notch; z, patellar surface.

attachment point for the *adductor magnus,* a muscle originating on the lower edge of the obturator foramen and ischial tuberosity. This muscle adducts the thigh at the hip.

y. The **intercondylar fossa,** or **notch,** is the nonarticular, excavated surface between the distal and posterior articular surfaces of the condyles. Within the fossa are two facets that are the femoral attachment sites for the *anterior* and *posterior cruciate ligaments,* a pair of crossed ligaments linking the femur and tibia. These ligaments strengthen the knee joint.

z. The **patellar surface** is a notched, articular area on the anterior surface of the distal femur, over which the patella glides during flexion of the knee. The lateral surface of this notch is elevated, projecting more anteriorly than the medial boundary of the notch. This helps to prevent lateral dislocation of the patella during full extension of the knee.

12.1.2 Growth (Figure 12.7)

The femur ossifies from five centers: one for the shaft, one for the head, one for the distal end, and one for each trochanter.

12.1.3 Possible Confusion

Neither intact femora nor femoral fragments are easily confused with other bones.

*   The femoral head has a fovea and is a more complete sphere than the humeral head.
*   The femoral shaft is larger, has a thicker cortex, and is rounder in cross section than any other shaft. It has only one sharp corner, the linea aspera.

12.1.4 Siding

*   For intact femora or proximal ends, the head is proximal and faces medially. The lesser trochanter and linea aspera are posterior.
*   For isolated femoral heads, the fovea is medial and displaced posteriorly and inferiorly. The posteroinferior head-neck junction is more deeply excavated than the anterosuperior junction.
*   For proximal femoral shafts, the nutrient foramen opens distally, and the linea aspera is posterior and thins inferiorly. The gluteal tuberosity is superior and faces posterolaterally.
*   For femoral midshafts, the nutrient foramen opens distally, the bone widens distally, and the lateral posterior surface is usually more concave than the medial posterior surface.
*   For distal femoral shafts, the shaft widens distally and the lateral supracondylar ridge is more prominent than the medial. The medial condyle extends more distally than the lateral.

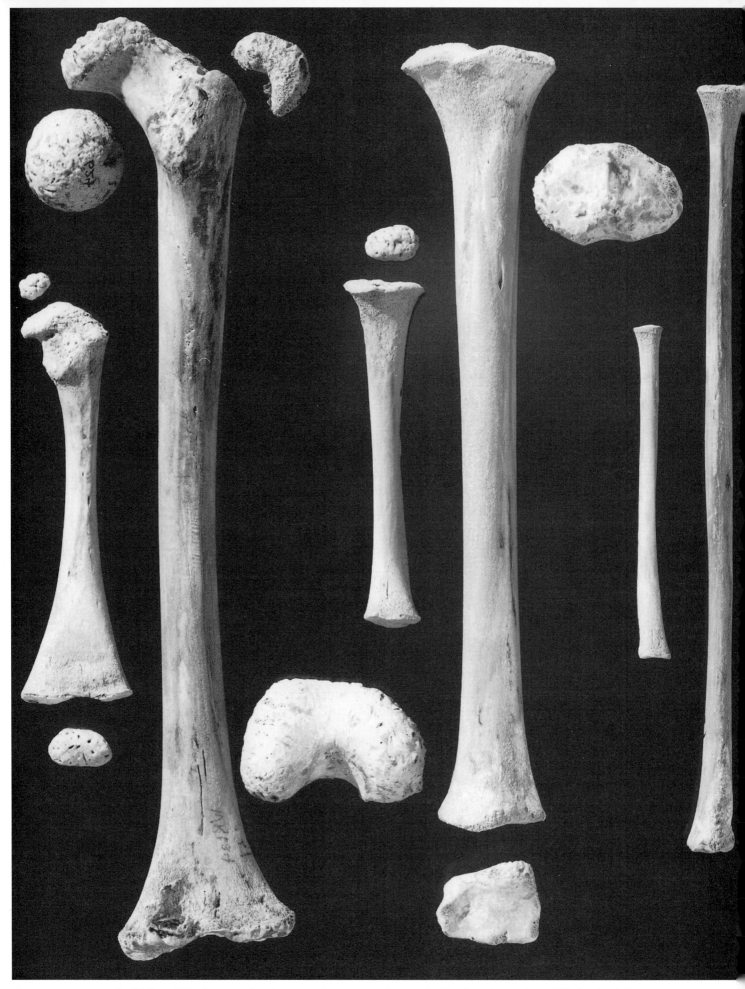

Figure 12.7 **Femoral, tibial, and fibular growth.** The pairs of immature femora (left), tibiae (center), and fibulae (right), shown here in anterior view, are from a one-year-old and a six-year-old. Natural size.

- For femoral distal ends, the intercondylar notch is posterior and distal, and the lateral border of the patellar notch is more elevated. The lateral condyle bears the popliteal groove, and the medial condyle bulges away from the line of the shaft. Relative to the shaft axis, the lateral condyle extends more posteriorly than the medial. The medial condyle extends more distally than the lateral because in anatomical position the femur angles beneath the body.

12.2
Patella (Figures 12.8–12.9)

12.2.1 Anatomy

The patella, the largest sesamoid bone in the body, articulates only with the patellar surface of the distal femur (patellar notch). The patella rides in the *tendon of the quadriceps femoris*—the largest muscle of the thigh and the primary extensor of the knee. The patella functions to protect the knee joint, to lengthen the lever arm of the quadriceps, and to increase the area of contact between the patellar tendon and the femur.

a. The **apex** (nonarticular point) of the patella is the nonarticular point on the bone. It points distally.

b. The **lateral articular facet** for the distal femur faces posteriorly and is the largest part of the large articular surface of the patella.

c. The **medial articular facet** for the distal femur faces posteriorly and is smaller than the lateral articular facet.

12.2.2 Growth

The patella ossifies from a single center.

Figure 12.8 **Right patella, anterior.** Superior is up; lateral is toward the left. Natural size.

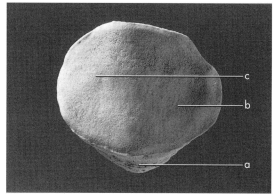

Figure 12.9 **Right patella, posterior.** Superior is up; lateral is toward the right. Natural size. Key: a, apex; b, lateral articular facet; c, medial articular facet.

12.2.3 Possible Confusion

This bone might be mistaken for an os coxae fragment, but only in a very fragmentary state. The acetabulum of the os coxae is strongly hollowed, as opposed to the much flatter articular surface of the patella.

12.2.4 Siding

- **Anatomical siding.** The patella is triangular in shape. Its thin, pointed apex is distal, and the thicker, blunter end is proximal. The lateral articular facet, which articulates with the lateral condyle of the femur, is the larger of the two facets.
- **Positional siding.** Place the apex away from you and the articular surface on the table. The bone falls toward the side from which it comes.

12.3
Tibia (Figures 12.10–12.15)

12.3.1 Anatomy

The tibia is the major weight-bearing bone of the lower leg. It articulates proximally with the distal femur, twice laterally with the fibula (once proximally and once distally), and distally with the talus.

a. The **tibial plateau** is the proximal tibial surface on which the femur rests. It is divided into two articular sections, one for each femoral condyle. In life there are fibrocartilage rings around the periphery of these articular facets, the *medial* and *lateral menisci.*

b. The **medial condyle** is the medial part of the tibial plateau. Its femoral articulation is oval, with the long axis oriented anteroposteriorly. Its lateral edge is straight.

c. The **lateral condyle** is the lateral part of the tibial plateau. Its femoral articulation is smaller and rounder than the medial articulation.

d. The **intercondylar eminence** is the raised area on the proximal tibial surface between the articular facets.

e. The **medial intercondylar tubercle** forms the medial part of the intercondylar eminence.

f. The **lateral intercondylar tubercle** forms the lateral part of the intercondylar eminence. The anterior and posterior *cruciate ligaments* and the anterior and posterior extremities of the *menisci* insert into the nonarticular areas between the condyles, which are just anterior and posterior to the medial and lateral intercondylar tubercles, respectively.

g. The **superior fibular articular facet** is located on the posteroinferior edge of the lateral condyle.

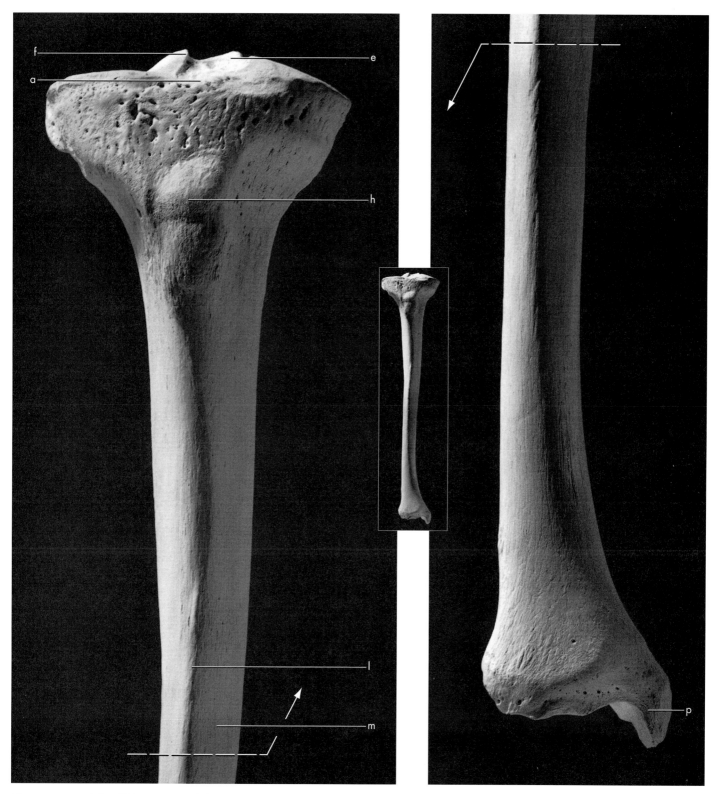

Figure 12.10 **Right tibia, anterior:** *left,* proximal end; *right,* distal end. Natural size. Key: a, tibial plateau; e, medial inter-condylar tubercle; f, lateral intercondylar tubercle; h, tibial tuberosity; l, anterior surface (anterior crest); m, medial surface; p, medial malleolus.

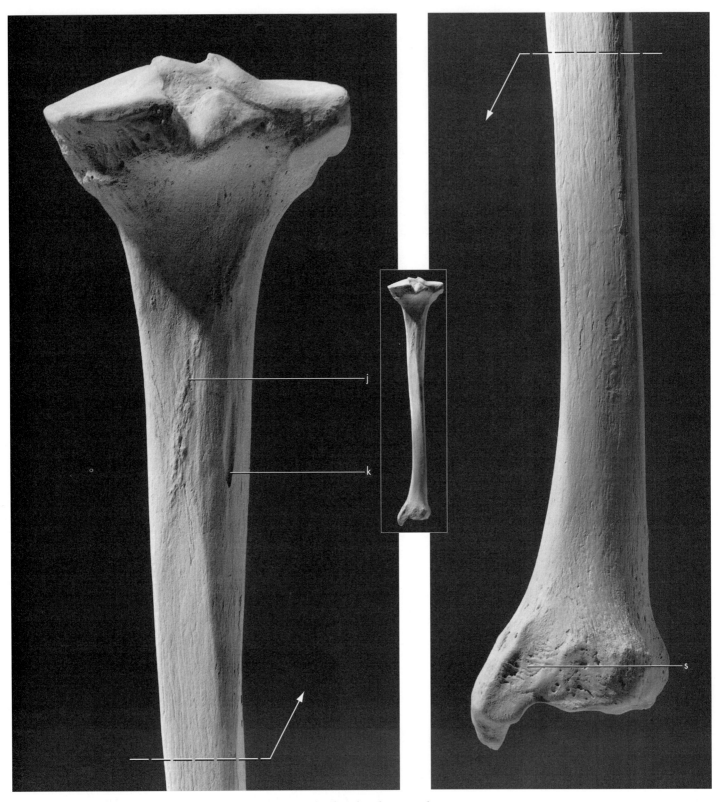

Figure 12.11 **Right tibia, posterior:** *left,* proximal end; *right,* distal end. Natural size. Key: j, soleal (popliteal) line; k, nutrient foramen; s, malleolar groove.

h. The **tibial tuberosity** is the rugose area on the anterior surface of the proximal tibia. Its superior part is smoothest and widest. The patellar ligament of the *quadriceps femoris muscle,* a major lower leg extensor at the knee, inserts here.

i. The tibial **shaft** is the fairly straight segment of the tibia between the expanded proximal and distal ends.

j. The **soleal (popliteal) line** crosses the proximal half of the posterior tibial surface from superolateral to inferomedial. The line demarcates the inferior boundary of the *popliteus muscle* insertion. This muscle is a flexor and medial rotator of the tibia and originates from the popliteal groove on the lateral femoral condyle. The line itself gives rise to the *popliteus fascia* and *soleus muscle,* a plantarflexor of the foot at the ankle.

k. The **nutrient foramen** is just inferolateral to the popliteal line. It is a large foramen that exits the bone proximally.

l. The **anterior surface (anterior crest)** of the shaft forms the anterior edge of the "shin."

m. The **medial surface** of the shaft forms the medial edge of the "shin" of the lower leg. This subcutaneous border is the widest tibial shaft surface.

n. The **interosseus surface** of the shaft is lateral, opposite the fibula. It is the most concave of the three tibial surfaces.

o. The **interosseus crest** is the lateral crest of the shaft which faces the fibula. It is the attachment area for the *interosseus membrane,* a sheet of tissue that functions to bind the tibia and fibula together and to compartmentalize lower leg muscles into anterior and posterior groups, just as its serial homolog does in the forearm.

p. The **medial malleolus** is the projection on the medial side of the distal tibia that forms the subcutaneous medial knob at the ankle. Its lateral surface is articular, for the talar body.

q. The **fibular notch** is the distolateral corner of the tibia. It is a triangular nonarticular area for the thick, short interosseus *tibiofibular ligament.* This ligament binds the distal tibia and fibula together as a unit at this syndesmosis. The proximal ankle, or **talocrural,** joint is formed by the tightly bound distal tibia and fibula, which articulate with the superior, medial, and lateral talar surfaces.

r. The **inferior fibular articular surface** is a thin articular surface for the fibula which faces laterally at the base of the fibular notch.

s. The **malleolar groove** on the posterior aspect of the medial malleolus transmits the tendons of the *tibialis posterior* and *flexor digitorum longus muscles,* plantar flexors.

12.3.2 Growth (Figure 12.7)

The tibia ossifies from three centers: one for the shaft and one for each end of the bone. Separate centers for the tibial tuberosity sometimes occur.

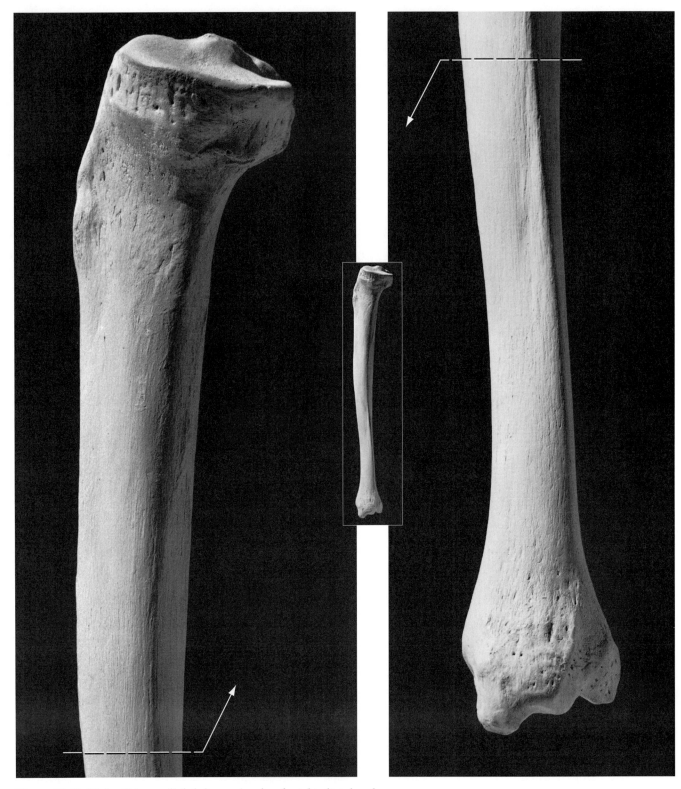

Figure 12.12 **Right tibia, medial:** *left,* proximal end; *right,* distal end.
Natural size.

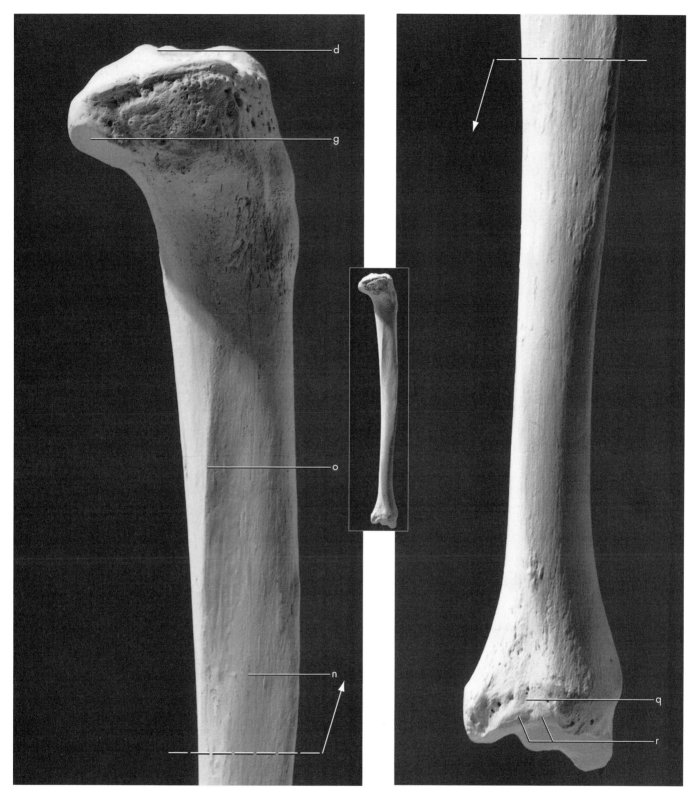

Figure 12.13 **Right tibia, lateral:** *left,* proximal end; *right,* distal end. Natural size. Key: d, intercondylar eminence; g, superior fibular articular facet; n, interosseus surface; o, interosseus crest; q, fibular notch; r, inferior fibular articular surface.

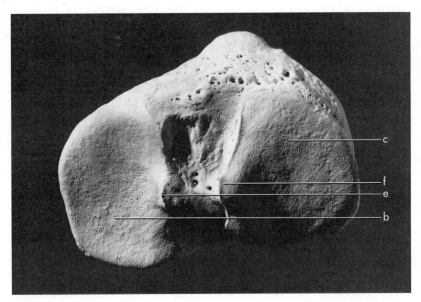

Figure 12.14 **Right tibia, proximal.** Anterior is up, lateral is toward the right. Natural size. Key: b, medial condyle; c, lateral condyle; e, medial intercondylar tubercle; f, lateral intercondylar tubercle.

Figure 12.15 **Right tibia, distal.** Anterior is up, lateral is toward the left. Natural size.

12.3.3 Possible Confusion

- The triangular tibial cross section differentiates fragments of this bone from the femur or the much smaller humerus. The tibial shaft is much larger than the radial or ulnar shafts.

- Proximal and distal ends of the tibia are diagnostic, and the only possibility of confusion arises in mistaking a segment of the proximal articular surface for the body of a vertebra. The tibia's articular surface is much denser and smoother than the articular surface of a vertebral body.

12.3.4 Siding

- For an intact tibia, the tibial tuberosity is proximal and anterior. The medial malleolus is on the distal end and is medial.

- For the proximal tibia, the tibial tuberosity is anterolateral, the fibular articulation is placed posterolaterally, and the lateral femoral articular surface is smaller, rounder, and set laterally. The intercondylar eminence is set posteriorly, and the axis of the nonarticular strip on the plateau runs from anterolateral to posteromedial. This strip is wider anteriorly than posteriorly. The intercondylar eminence has a more concave medial border and a more evenly sloping lateral border.

- For fragments of shaft, the entire shaft tapers distally, and the interosseus crest is lateral and posterior. The medial surface is the widest and faces anteriorly. The nutrient foramen is posterior and exits proximally. The cortex is thickest at midshaft.

- For the distal end, the malleolus is medial and its distalmost projection is anterior. The grooves for the plantar flexor tendons are posterior.

The fibular notch is lateral and the interosseus crest runs toward its anterior surface. The margin of the articular surface for the superior talus is grooved on the anterior surface but not the posterior surface.

12.4
Fibula (Figures 12.16–12.21)

12.4.1 Anatomy

The fibula is a long, thin bone that lies lateral to the tibia, articulating twice with it and once with the talus. Although this bone plays only an indirect role in the knee joint, serving to anchor ligaments, it plays a key role in forming the lateral border of the ankle joint. The fibula bears very little weight, not even touching the femur at its superior end.

a. The **head** is the swollen proximal end of the fibula, more massive and less mediolaterally flattened than the distal end. It is the attachment point for the *biceps femoris muscle* (a flexor and lateral rotator at the knee) and the *lateral collateral ligament* of the knee joint.

b. The **styloid process** is the most proximal projection of the bone, forming the posterior part of the head.

c. The **proximal fibular articular surface** is a round, flat, medially oriented surface that corresponds to a similar surface on the lateral proximal tibia.

d. The fibular **shaft** is the long, thin, fairly straight segment of the bone between the expanded proximal and distal ends.

e. The **interosseus crest** is an elevated crest that runs down the medial, slightly anteriorly facing surface of the shaft. Attached to it is the *interosseus membrane*, a fibrous sheet that binds the fibula and tibia and divides the lower leg musculature into anterior and posterior compartments.

f. The **nutrient foramen** opens proximally on the posteromedial surface, at about midshaft level. The cross section of the fibular shaft can be extremely variable in this region.

g. The **lateral malleolus** is the inferiormost (distalmost) projection of the fibula. Its lateral, nonarticular surface is subcutaneous, forming the lateral knob of the ankle.

h. The **malleolar (distal) articular surface** is a flat, medially facing, triangular surface whose apex faces inferiorly. The surface articulates with the lateral surface of the talus.

i. The **malleolar fossa** is located just posterior to the distal articular surface. It is the attachment site of the *transverse tibiofibular* and *posterior talofibular ligaments*, which strengthen the ankle joint.

j. The **peroneal groove,** for tendons of the *peroneus longus* and *peroneus brevis muscles*, marks the posterior surface of the distal fibula. These muscles plantarflex and evert the foot at the ankle, originating in the

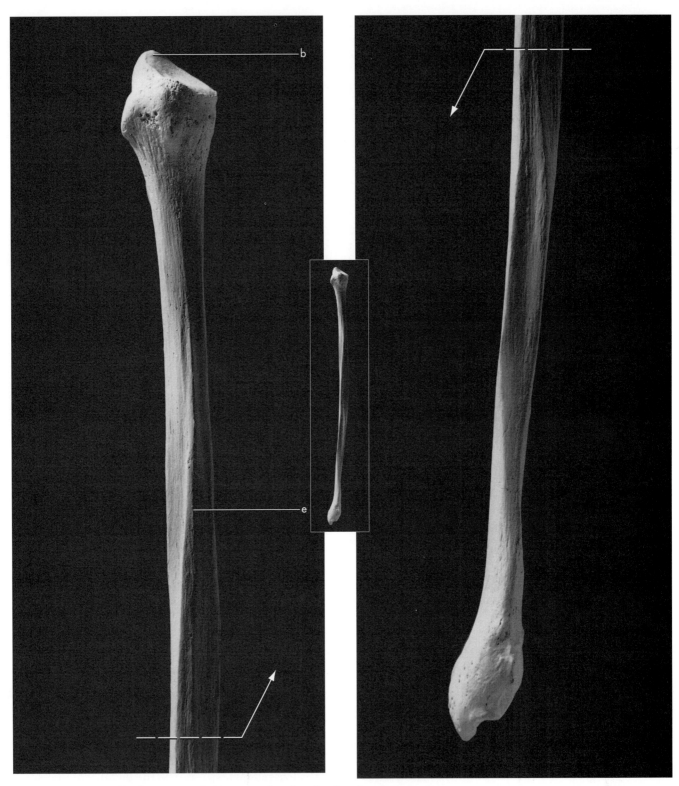

Figure 12.16 **Right fibula, anterior:** *left,* proximal end; *right,* distal end. Natural size. Key: b, styloid process; e, interosseus crest.

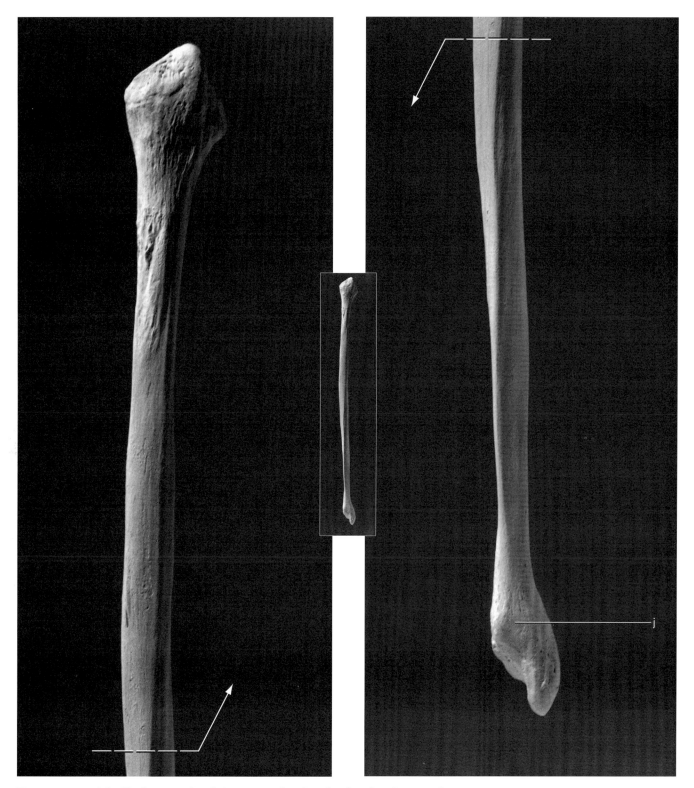

Figure 12.17 **Right fibula, posterior:** *left,* proximal end; *right,* distal end. Natural size. Key: j, peroneal groove.

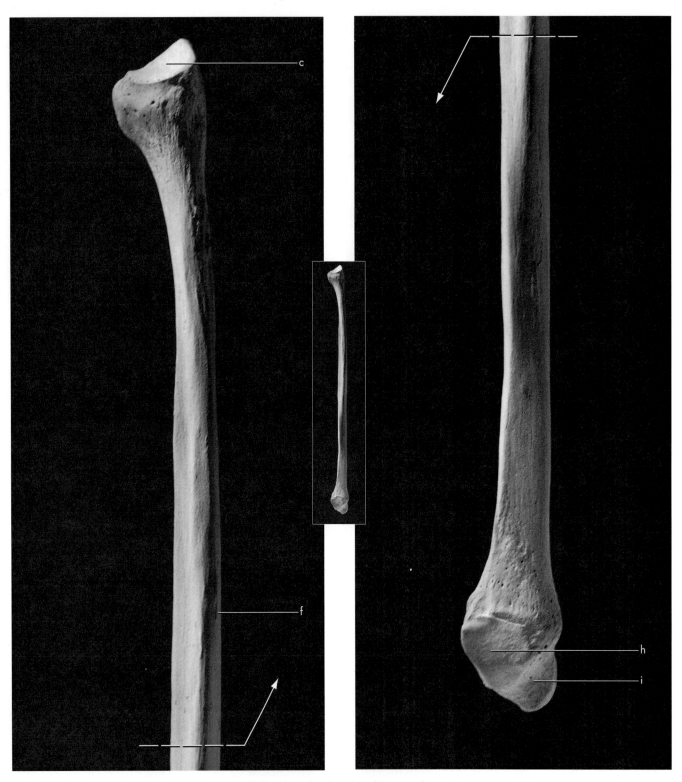

Figure 12.18 **Right fibula, medial:** *left,* proximal end; *right,* distal end. Natural size. Key: c, proximal fibular articular surface; f, nutrient foramen; h, malleolar (distal) articular surface; i, malleolar fossa.

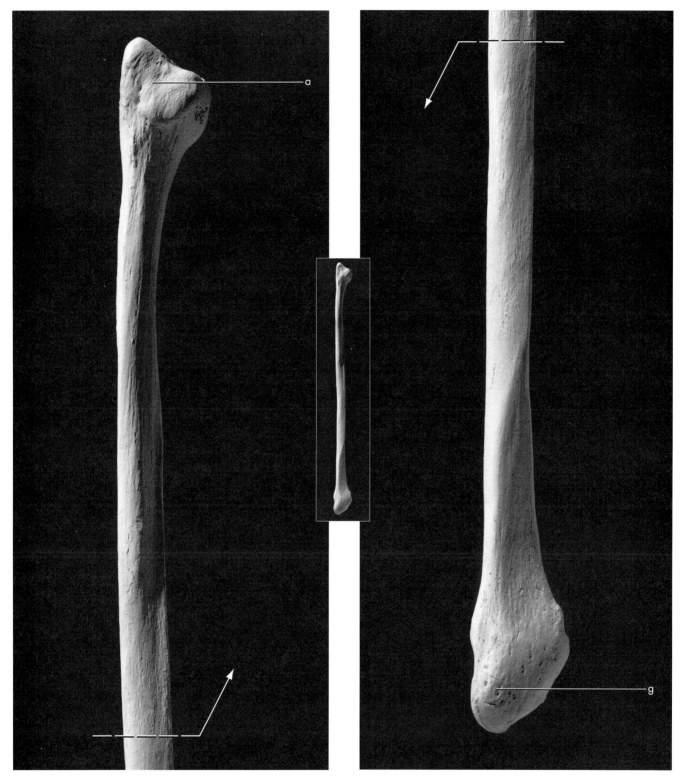

Figure 12.19 **Right fibula, lateral:** *left,* proximal end; *right,* distal end. Natural size. Key: a, head; g, lateral malleolus.

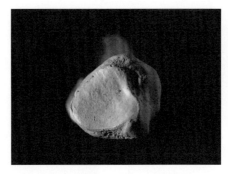

Figure 12.20 **Right fibula, proximal.** Anterior is up; lateral is toward the right. Natural size.

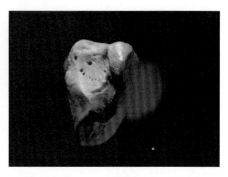

Figure 12.21 **Right fibula, distal.** Posterior is up; lateral is toward the right. Natural size.

leg and inserting at the bases of the first (*longus*) and fifth (*brevis*) metatarsals.

12.4.2 Growth (Figure 12.7)

The fibula, like the tibia, ossifies from three centers: one for the shaft and one for each end.

12.4.3 Possible Confusion

The proximal and distal fibular ends are distinctive and are rarely confused with other bones.

- The distal end is flattened along the plane of the articular facet, whereas the proximal end of the fifth metatarsal, for which a distal fibular end might be mistaken, has two facets and is flattened perpendicular to the planes of each.
- Fibular shafts are thin, straight, and usually quadrilateral (sometimes triangular), with sharp crests and corners. They are thus more irregular in cross section than either radial or ulnar shafts.

12.4.4 Siding

- For an intact fibula, the articular surfaces for the tibia are medial, the head is proximal, the flattened end is distal, and the malleolar fossa is posterior.
- For the proximal end, the styloid process is lateral, proximal, and displaced posteriorly. The articular surface faces medially and is also displaced posteriorly. The neck is roughest laterally.
- For the shaft, try to use intact specimens for comparison. The nutrient foramen opens proximally. The sharpest crest on the triangular proximal end is the interosseus crest. This expands downshaft to a rough, flattened surface.

- For the distal end, the malleolar fossa is always posterior, and the articular facet is always medial, with its apex pointed inferiorly.

12.5
Functional Aspects of the Knee and Ankle

The knee joint is the most complex joint in the human body, whereas the ankle joint is considerably more simple. The main actions at both the knee and ankle joints are flexion and extension. Knee extension is accomplished primarily by the *quadriceps muscles.* This group of muscles originates from the os coxae and proximal femur and inserts on the tuberosity of the tibia via the tendon in which the patella is embedded. Flexion at the knee is primarily accomplished by the *hamstring muscles.* These also originate from the os coxae and proximal femur but insert on the proximal tibia and fibula, just distal to the knee joint.

In walking and running, propulsion of the body is created by contracting muscles of the lower limb. Muscles in the anterior and posterior compartments of the lower leg (separated by the interosseus membrane) act to move the skeleton of the foot, much as forearm muscles move the hand. *Plantarflexors* occupy the posterior compartment, and *dorsiflexors* are found anterolateral to the tibia. *Plantarflexor muscles,* originating on the posterior side of the lower leg and attaching to the calcaneus via the *Achilles tendon,* contract at the same time that the *quadriceps,* the primary extensor of the lower leg, contracts. The coordinated action of these two muscle groups straightens the leg and plantarflexes the foot at the ankle, producing a strong ground reaction force and propelling the body forward. A comparison of the bony anatomy of the human pelvic girdle and leg with that of our closest living relatives, the African apes, reveals profound evolutionary changes related to the acquisition of bipedality more than 4 million years ago.

CHAPTER 13

Foot: Tarsals, Metatarsals, and Phalanges

The bones of the foot are obvious serial homologs of the hand bones. The deep evolutionary history of the foot elements is similar to that described for those of the hand in Chapter 10. There are a total of twenty-six bones in each human foot, one less than in each hand. Of the seven tarsal bones, two occupy a proximal row, four occupy a distal row, and one is centered between rows. The tarsals are followed distally by a single row of five metatarsals. Farther distally, there is a single row of five proximal phalanges, a single row of four intermediate phalanges, and a single row of five distal, or terminal, phalanges.

In addition to these twenty-six, as with the hand, there are small **sesamoid** bones that lie within tendons of the foot. In the foot, a pair of

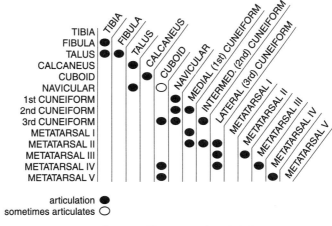

Figure 13.1 Articulation of bones in the adult human ankle and foot.

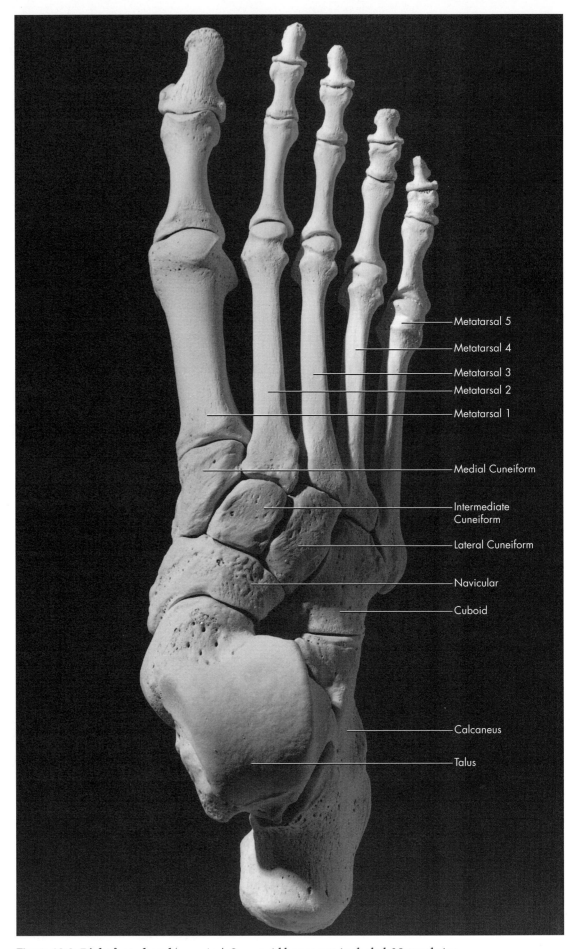

Metatarsal 5

Metatarsal 4

Metatarsal 3

Metatarsal 2

Metatarsal 1

Medial Cuneiform

Intermediate
Cuneiform

Lateral Cuneiform

Navicular

Cuboid

Calcaneus

Talus

Figure 13.2 **Right foot, dorsal** (superior). Sesamoid bones not included. Natural size.

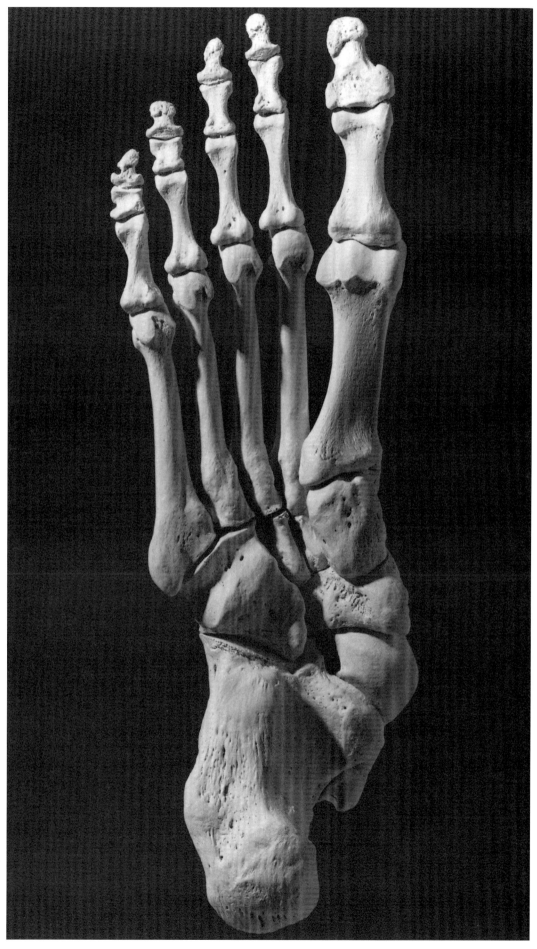

Figure 13.3 **Right foot, plantar** (inferior). Sesamoid bones not included. Natural size.

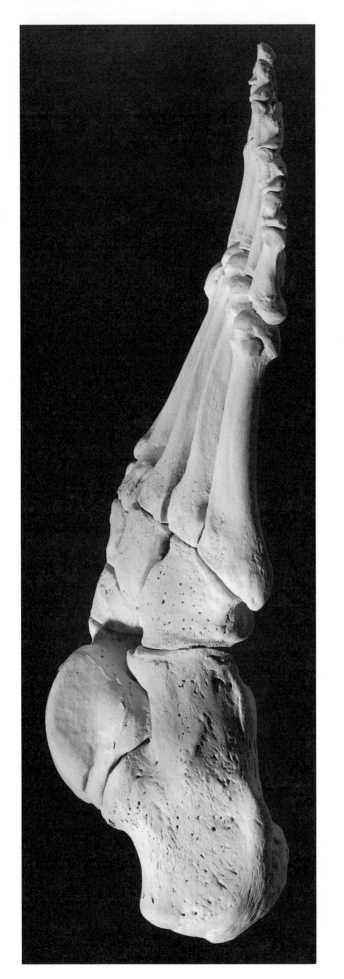

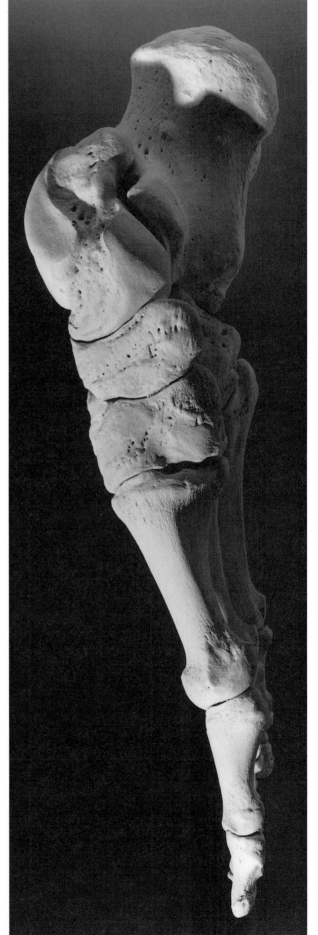

Figure 13.4 **Right foot:** *top*, **lateral**; *bottom*, **medial**. Sesamoid bones not included. Natural size.

sesamoids is usually found below the head of the first metatarsal. Figures 13.1–13.4 summarize and illustrate articulations within the foot. Elements of the foot skeleton can be divided into three segments, the **tarsals, metatarsals,** and **phalanges.**

It is convenient to apply specific directional terms to the foot. **Plantar** refers to the sole of the foot, its inferior surface in standard anatomical position. **Dorsal** is the opposite, superior surface. **Proximal** (posterior) is toward the tibia, **distal** (anterior) is toward the toe tips, and the distalmost phalanges are referred to as "terminal." In addition, the big toe is sometimes called the **hallux,** and its ray is identified as ray 1. Other rays are numbered as in the hand.

The human foot has dramatically changed during its evolution from a grasping organ to a structure adapted to bipedal locomotion. Most of the foot's mobility, flexibility, and grasping abilities have been lost in humans as the foot had adapted to shock absorption and propulsion.

13.1
Tarsals

The seven tarsals combine with the five metatarsals to form the longitudinal and transverse arches of the foot. The **talus** articulates superiorly with the distal tibia and fibula at the ankle joint. The **calcaneus** forms the heel of the foot, supports the talus, and articulates anteriorly with the **cuboid,** the third largest tarsal bone. The metatarsals articulate proximally with the cuboid and three **cuneiforms.** The seventh tarsal, the **navicular,** is interposed between the head of the talus and these cuneiforms.

13.1.1 Talus (Figures 13.5–13.6)

The talus is called the **astragalus** in other animals. It is the second largest of the tarsals and is placed between the tibia and fibula superiorly and the calcaneus inferiorly. No muscles attach to this bone. It rests atop the calcaneus and articulates distally with the navicular. It forms the lower member of the **talocrural joint.** Talar variation is illustrated in Figure 2.1.

a. The **head** is the rounded, convex, distal articular surface of the talus. It fits into the hollow of the navicular.

b. The **body** is the squarish bulk of the bone posterior to the talar head.

c. The **trochlea** is the saddle-shaped articular surface of the body. Its sides are the lateral and medial **malleolar surfaces,** which articulate with the fibula and tibia, respectively.

d. The **neck** connects the head of the talus to the body. Occasionally there are small articular facets on the neck, formed by contact with the anterior surface of the distal tibia during strong dorsiflexion of the foot at the ankle joint. These facets are called **squatting facets.**

e. The **groove for flexor hallucis longus** is the short, nearly vertical groove on the posterior surface of the talar body. It is so named

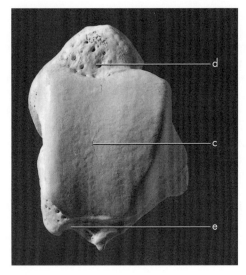
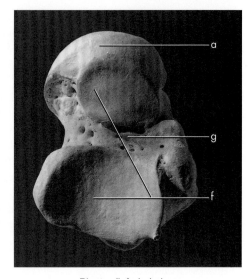

Dorsal (superior) view

Plantar (inferior) view

Figure 13.5 **Right talus.** Distal is up. Natural size. Key: a, head; c, trochlea; d, neck; e, groove for flexor hallucis longus; f, calcaneal, or subtalar articular, surfaces; g, sulcus tali.

because it transmits the tendon of this muscle, a calf muscle that plantarflexes the foot and hallux.

f. The **calcaneal,** or **subtalar articular, surfaces** on the inferior aspect of the talus are usually three in number and variable in shape.

g. The **sulcus tali** is a deep groove between the posterior and middle calcaneal articular surfaces.

- **Anatomical siding.** The saddle-shaped articular surface for the distal tibia is superior, and the talar head is anterior. The larger malleolar surface (for the fibula) is lateral.

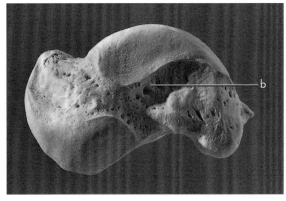

Medial view

Lateral view

Figure 13.6 **Right talus.** Dorsal is up. Natural size. Key: b, body.

 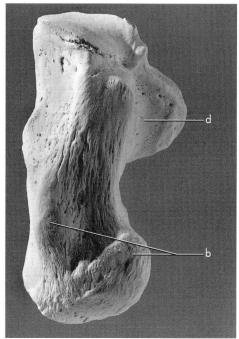

Dorsal (superior) view   Plantar (inferior) view

Figure 13.7 **Right calcaneus.** Distal is up. Natural size. Key: a, calcaneal tuber;
b, lateral and medial processes; d, sustentacular sulcus (groove).

- **Positional siding.** The head is medial when viewed from above and
  aligns with the hallux.

13.1.2 Calcaneus (Figures 13.7–13.9)

The calcaneus, or "heel bone," is the largest of the tarsal bones, and the
largest bone of the foot. It is located inferior to the talus and articulates
anteriorly (distally) with the cuboid.

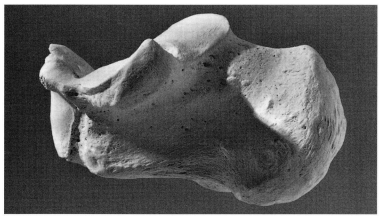

Medial view   Posterior view

Figure 13.8 **Right calcaneus.** Dorsal is up. Natural size.

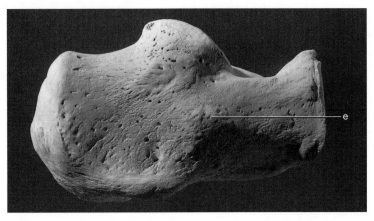

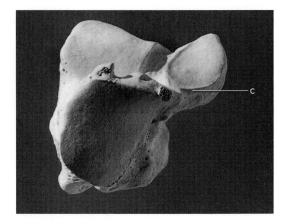

Lateral view   Anterior view

Figure 13.9 **Right calcaneus.** Dorsal is up. Natural size. Key: c, sustentaculum tali; e, peroneal tubercle.

a. The **calcaneal tuber** is the large, blunt, nonarticular, posterior process of the heel. It is the insertion point of the *calcaneal,* or *Achilles, tendon.* Contraction of the *gastrocnemius* and *soleus* calf muscles causes plantarflexion of the foot, with the calcaneal tuber serving as a lever arm and the talar body as the fulcrum.

b. The **lateral** and **medial processes,** on the plantar portion of the calcaneal tuberosity, serve to anchor several *intrinsic muscles* of the foot. The lateral process is much smaller than the medial.

c. The **sustentaculum tali** is the shelf on the medial side of the calcaneus. It supports the talar head.

d. The **sustentacular sulcus (groove)**, just inferior to the sustentaculum, is a pronounced groove. The tendon of the *flexor hallucis longus muscle*, a plantar flexor of the big toe, travels through this groove. It is continuous posterosuperiorly with the groove on the posterior extremity of the talus.

e. The **peroneal tubercle** is a rounded projection low on the lateral surface of the calcaneal body. It is closely associated with tendons of the *peroneus longus* and *brevis muscles*. These muscles plantarflex and evert the foot, inserting on the base of the first and fifth metatarsals, respectively.

- **Possible confusion.** It is not possible to confuse an intact calcaneus with another bone. A broken calcaneal body might sometimes be mistaken for a section of femoral greater trochanter, but the only articular surface on the proximal femur is the spherical head.

- **Anatomical siding.** The tuberosity is posterior and the inferior surface is nonarticular. The sustentaculum tali projects medially, inferior to the talar head.

- **Positional siding.** With the heel toward you and the articular surfaces up, the shelf projects opposite the side from which the bone comes.

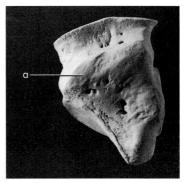

Figure 13.10 **Right cuboid.** Distal is up. Natural size. Key: a, cuboid tuberosity.

13.1.3 Cuboid (Figures 13.10–13.11)

The cuboid bone sits on the lateral side of the foot, sandwiched between the calcaneus and the fourth and fifth metatarsals, articulating with the navicular and third cuneiform. It is recognized by its large size and projecting, pointed, proximal articular surface. It is the most cuboidal, or cube-shaped, of the tarsal bones. The **cuboid tuberosity** (Figure 13.10a) is a large tuberosity on the inferolateral surface of the bone. The tendon for the *peroneus longus muscle* enters the foot via the groove superior to this tuberosity.

- **Anatomical siding.** The wide, flat nonarticular surface is superolateral, and the pointed calcaneal facet is proximal. The tuberosity is inferolateral. There is an articulation on the medial, but not the lateral, surface.

- **Positional siding.** Look directly at the flat nonarticular surface. With the calcaneal facet toward you, the tuberosity is on the side from which the bone comes.

13.1.4 Navicular (Figures 13.12–13.13)

The navicular is named for the strongly concave proximal surface that articulates with the head of the talus. On the distal surface the navicular

| View from the lateral cuneiform | View from the calcaneus | View from the MT-4 and MT-5 bases |

Figure 13.11 **Right cuboid.** Dorsal is up. Natural size.

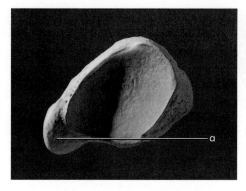
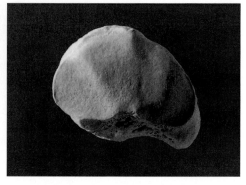

View from the talus (proximal view)   View from the cuneiforms (distal view)

Figure 13.12 **Right foot navicular.** Dorsal is up. Natural size. Key: a, tubercle.

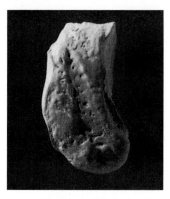

Superomedial view

Figure 13.13 **Right foot navicular.** Distal is toward the left. Natural size.

has a large facet divided by two ridges. These set off the articular planes of the three cuneiforms. In addition, the navicular often articulates with a corner of the cuboid. The **tubercle** (Figure 13.12a) of the navicular is a large, blunt projection on the medial side of the bone. This tubercle is the main insertion of the *tibialis posterior muscle,* a plantar flexor of the foot and toes.

- **Possible confusion.** Although generally similar in gross shape, this bone is much larger than the scaphoid and has a flat side with three facets.

- **Anatomical siding.** The concave talar facet is proximal. The large, flat nonarticular surface is dorsal, and the tubercle is medial.

- **Positional siding.** Hold the bone by the tubercle, with the concave articular surface facing you and the flat nonarticular side up. The tubercle is opposite the side from which the bone comes.

13.1.5 Medial (First) Cuneiform (Figure 13.14)

The first cuneiform is the largest of three cuneiforms. It sits between the navicular and the base of the first metatarsal, articulating with these as

Medial view   View from the intermediate cuneiform   View from the foot navicular

Figure 13.14 **Right medial (first) cuneiform.** Dorsal is up. Natural size.

well as with the second cuneiform and the base of metatarsal 2 (MT 2). It is less wedge-shaped than the other cuneiforms and is distinguished by the kidney-shaped facet for the base of the first metatarsal.

- **Anatomical siding.** The longest, kidney-shaped articular surface is distal, its long axis vertical. The broad, rough, nonarticular surface is medial, and the facet for the intermediate cuneiform is superior. The proximal facet for the navicular is concave.

- **Positional siding.** Place the large, kidney-shaped articular facet away from you and orient its long axis vertically, with the smaller, more concave navicular facet toward you. With the bone resting on its blunter end, the only other facet is on top and faces toward the side from which the bone comes.

13.1.6 Intermediate (Second) Cuneiform (Figure 13.15)

The intermediate cuneiform is the smallest of three cuneiforms. It is located between the navicular and second metatarsal. In addition, it articulates on either side with the first and third cuneiforms.

- **Anatomical siding.** The bone's nonarticular dorsal surface is broadest, and the bone wedges inferiorly, participating in the transverse arch of the foot. The proximal articular surface is usually the most concave

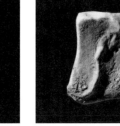

View from the medial cuneiform   View from the lateral cuneiform   View from the foot navicular

Figure 13.15 **Right intermediate (second) cuneiform.** Dorsal is up. Natural size.

View from the intermediate cuneiform   View from the cuboid   View from the foot navicular

Figure 13.16 **Right lateral (third) cuneiform.** Dorsal is up. Natural size.

(for the navicular). The lateral edge of this facet (proximal, lateral corner of the bone) is concave in profile. The lateral facet is double.

- **Positional siding.** Place the flat, nonarticular surface up, and the concave facet toward you. The outline of the superior (dorsal) surface is a square whose most projecting corner points opposite the side from which the bone comes.

13.1.7 Lateral (Third) Cuneiform (Figure 13.16)

The lateral cuneiform is intermediate in size between the other cuneiforms. It is located in the center of the foot, articulating distally with the second, third, and fourth metatarsals. Medially it contacts the intermediate cuneiform, laterally the cuboid, and proximally the navicular.

- **Anatomical siding.** The dorsal surface is a rectangular, nonarticular platform, and the bone wedges inferiorly to this. The proximal (navicular) articulation is wider but smaller than the elongate, wedge-shaped distal MT-3 facet. The border between the navicular and cuboid facets projects in a V shape as the base of the third cuneiform wedges between these two bones.

- **Positional siding.** Place the flat, nonarticular surface up (wedge down), with the small end-facet toward you. The Africa-shaped facet is away from you. The longest boundary of the upper surface is toward the side from which the bone comes.

13.1.8 Growth

The tarsals each ossify from a single center, with the exception of the calcaneus, which has an epiphysis at its posterior end.

13.1.9 Possible Confusion

As with the carpals, most of the tarsals are compactly and robustly constructed. Identification of fragmentary tarsals is therefore not usually required. The exception to this is the calcaneus, which is less dense and often broken in its nonarticular areas. Since most of the tarsals are larger than carpals, they are more often recovered from archeological contexts.

Side identification of tarsals, as with carpals, is facilitated by positional techniques. Some of the techniques presented here are adopted from Bass (1995).

13.2
Metatarsals (Figures 13.17–13.21)

The metatarsals, like the metacarpals, are numbered from MT 1 (the hallux, or big toe) through MT 5, according to the five rays of the foot. They are all tubular bones with round distal articular surfaces (**heads**) and more squarish proximal ends (**bases**). As with the metacarpals, metatarsals are identified and sided most effectively according to the morphology of their bases.

The plantar metatarsal **shaft** (body) surfaces are always more concave in lateral view than the dorsal shaft surfaces. The metatarsal bases articulate with adjacent metatarsals in positions 2–5. All of the tarsals in the distal row articulate with at least one metatarsal base.

13.2.1 First Metatarsal

The first metatarsal is the shortest but most massive metatarsal. It articulates at its base with the medial cuneiform. The **sesamoid grooves** at the base of the head correspond to sesamoid bones (Figure 13.23) in the tendons of *flexor hallucis brevis*, a short plantar flexor of the big toe.

- **Siding.** The basal facet has a convex medial profile and a straight lateral profile.

13.2.2 Second Metatarsal

The second metatarsal is the longest and narrowest metatarsal. It has two lateral facets at the base, for the lateral cuneiform and MT 3. It articulates proximally with the intermediate cuneiform and medially with the medial cuneiform. Its base has slightly more of a "styloid" appearance than the base of MT 3.

- **Siding.** The most proximal point on the base is lateral to the main shaft axis.

13.2.3 Third Metatarsal

The third metatarsal is very similar to MT 2, but its base has two medial basal facets that are smaller than MT-2 lateral facets. The lateral basal facet is single and large. The base is squarer than the MT-2 base, and there is a large, bulging tubercle distal to the lateral articular facet. The base articulates with MT 2 and MT 4 and with the lateral cuneiform.

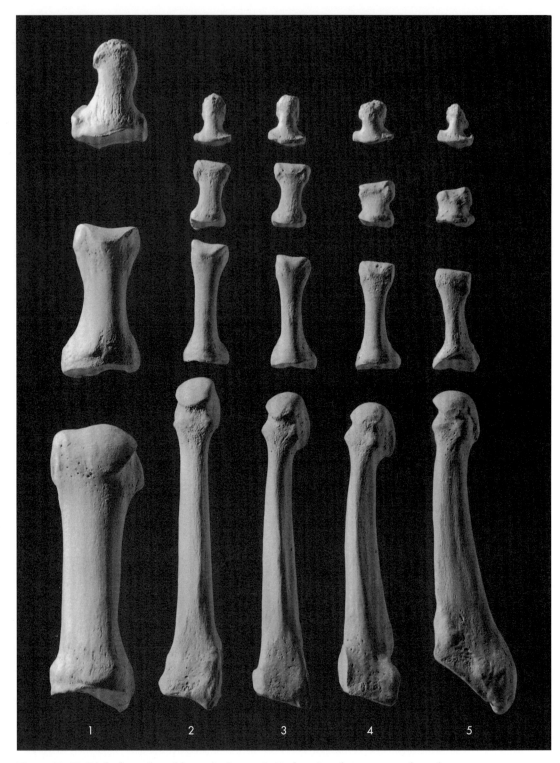

Figure 13.17 **Right foot, dorsal** (superior), rays 1–5, showing the metatarsals and the proximal, intermediate, and distal foot phalanges. Distal is up; lateral is toward the right. Natural size.

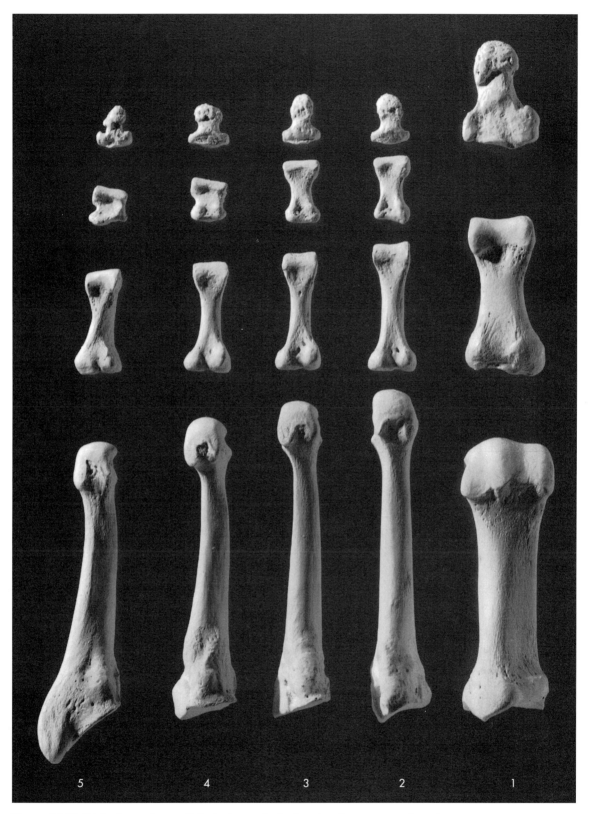

Figure 13.18 **Right foot, plantar** (inferior), rays 1–5, showing the metatarsals and the proximal, intermediate, and distal foot phalanges. Distal is up; lateral is toward the left. Natural size.

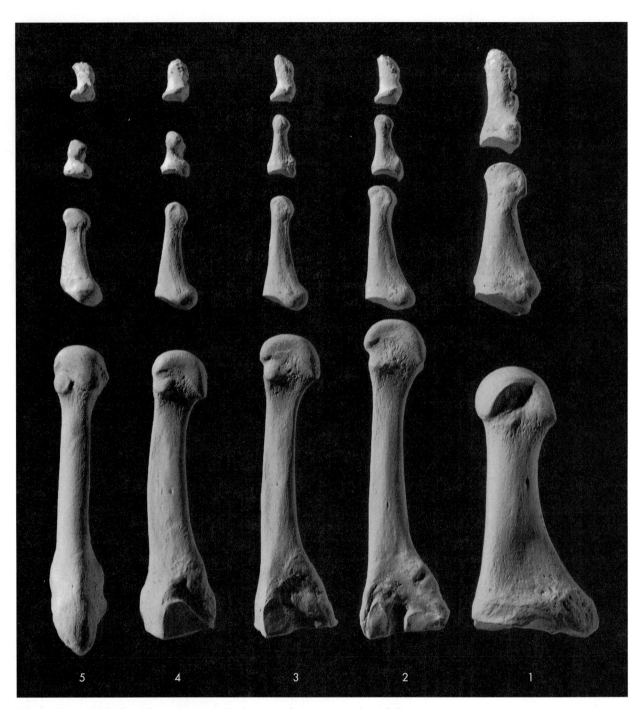

Figure 13.19 **Right foot, lateral,** rays 1–5, showing the metatarsals and the proximal, intermediate, and distal foot phalanges. Natural size.

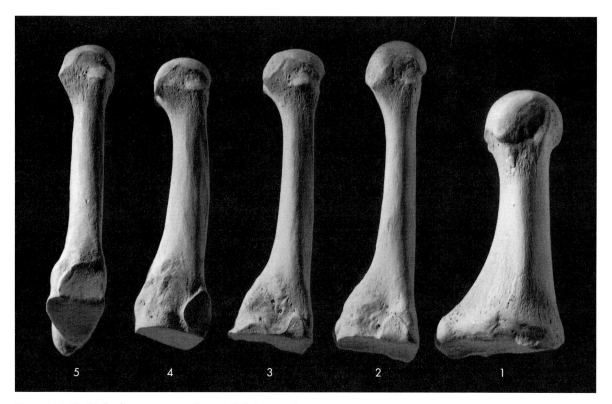

Figure 13.20 **Right foot metatarsals, medial.** Natural size.

- **Siding.** The most proximal point on the base is lateral to the main shaft axis.

13.2.4 Fourth Metatarsal

The fourth metatarsal is shorter than MT 2 or MT 3. It has single medial and lateral basal facets for articulation with MT 3 and MT 5. The proximal facet for the cuboid is fairly oval.

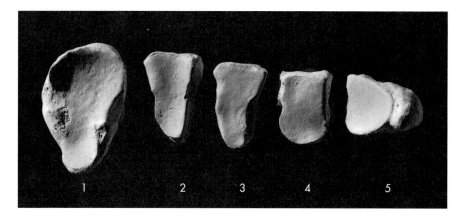

Figure 13.21 **Right foot metatarsal bases, proximal.** Dorsal is up. Natural size.

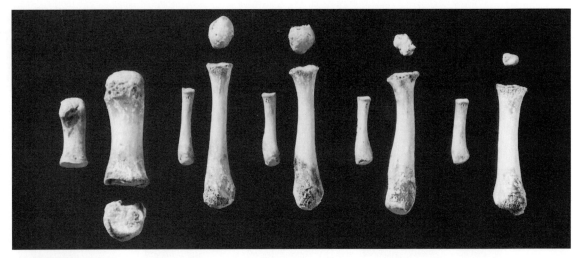

Figure 13.22 **Metatarsal growth.** The pairs of immature metatarsals are shown here in plantar view, with ray 1 on the left and ray 5 on the right. They are from one-year-old and six-year-old individuals. Natural size.

- **Siding.** The most proximal point on the base is lateral to the main shaft axis.

13.2.5 Fifth Metatarsal

The fifth metatarsal bears a **styloid process,** a distinctive, blunt, nonarticular basal projection. It is lateral, opposite the MT-4 facet, and projects proximally. It is the insertion point for the tendon of the *peroneus brevis muscle,* a calf muscle that acts to plantarflex and evert the foot at the ankle. The proximal basal articulation is for the cuboid.

- **Siding.** The styloid process is lateral.

13.2.6 Growth (Figure 13.22)

Metatarsals each ossify from two centers. For MT 2, MT 3, MT 4, and MT 5, there is one center for both the base and shaft and a second for the distal end. For the MT 1, there is a center for the body and one for the proximal end.

Figure 13.23 **Sesamoids of the right foot:** *left,* superior (dorsal) view; *right,* lateral view.

13.2.7 Possible Confusion

Metatarsals and metacarpals are similar in gross size and shape but are easily distinguished from each other. Metatarsal 2, MT 3, MT 4, and MT 5 are longer than MC 2, MC 3, MC 4, and MC 5, with straighter and narrower shafts. Metatarsal heads are more mediolaterally compressed and smaller relative to their bases than the metacarpal heads.

13.2.8 Siding

Metatarsal bases are always proximal, and the most proximal part of the base is always lateral on MT 2, MT 3, MT 4, and MT 5.

13.3
Foot Phalanges (Figures 13.17–13.19)

The foot phalanges share their basic features, heads, bases, and shafts, with the hand phalanges (see section 10.3).

13.3.1 Proximal Foot Phalanges

Each proximal foot phalanx displays a single, concave proximal facet for the metatarsal head, and a spool-shaped, or trochlear, surface distally. The hallucial proximal phalanx is larger and stouter than the others.

13.3.2 Intermediate Foot Phalanges

Each intermediate foot phalanx displays a double proximal articular facet for the head of the proximal phalanx. Each also has a trochlea-shaped distal articular facet. These are "stunted," squat versions of their analogs in the hand.

13.3.3 Distal Foot Phalanges

Each distal foot phalanx displays a double proximal articular facet for the head of the intermediate phalanx, but the terminal tip of the bone is a nonarticular pad, the **distal phalangeal tubercle.** These are very small and stubby compared to distal hand phalanges.

13.3.4 Growth

Foot phalanges each ossify from two centers, one for the shaft plus distal end and one for the base.

13.3.5 Possible Confusion

Hand and foot phalanges are easily distinguished, even when isolated.

- The foot phalanges are all much shorter than their analogs in the hand.
- The proximal foot phalanges are the only ones that might be mistaken for hand phalanges.
- The hand phalanx shafts are dorsoventrally compressed, forming a D shape in cross section, whereas foot phalanx shafts are more circular in cross section.
- Foot phalanges display more constriction at midshaft than do hand phalanges.
- The hallux has only a proximal and a morphologically distal phalanx, and these squat, massive bones are very distinctive.

13.3.6 Siding

For siding hand and foot phalanges it is best to work with whole specimens and comparative materials. The head is distal, and the base proximal. The dorsal phalangeal shaft surfaces are smooth and straight, whereas the plantar surfaces are more irregular and curved.

13.4
Functional Aspects of the Foot

The rigid, transversely and longitudinally arched form of the human foot is a radical departure from the grasping appendage that characterizes the order Primates. This anatomy evolved to meet the peculiar demands of habitual, striding, bipedal locomotion. Cartmill *et al.* (1987) note that anatomy of the human foot is best appreciated by considering it as a transformed hand in which the thumb is tied to the second digit, the metacarpals elongated to form a longer lever, the phalanges shortened, and the serial homolog of the triquetral (the foot's calcaneus) enlarged into a massive lever arm. Foot phalanges have fingerlike movements, but these are comparatively restricted. Most foot movement occurs among the ankle bones. *Extrinsic foot muscles* in the foreleg, like those of the forearm, move the foot and toes. These are compartmentalized, with plantar flexors posteriorly and dorsiflexors anterolaterally. These muscles operate mostly via tendons across the ankle. As in the hand, there are also *intrinsic foot muscles*.

CHAPTER 14

Recovery, Preparation, and Curation of Skeletal Remains

In this chapter we introduce procedures useful in recovering skeletal material. The chapter is organized in an approximately chronological manner. We first review aspects of discovery and techniques of retrieval and excavation. Then we discuss transport of skeletal material, primary laboratory preparation, and restoration. Analysis of remains is covered in Chapter 15.

In every aspect of osteological analysis, whether in the field or in the laboratory, common sense is critical. If there is any one overriding rule, it is to think before you act. There is no single formula, recipe, or procedure to apply in every field situation. There are simply too many different discovery contexts and too much preservational variation involving osteological remains. Cemetery excavation is different from isolated skeleton excavation, fossil bones are different from modern ones, forensic cases are different from archeological projects, and waterlogged burial conditions are different from mummified remains in the driest deserts. There are, however, some general principles that apply in most instances where skeletal remains are concerned.

14.1 Search

Osteological remains may be discovered in forensic, archeological, or paleontological contexts either as a result of intentional professional or amateur survey or as a result of an accident. Intentional search methodologies vary widely, from paleontological expeditions to murder or disappearance investigations. Sometimes the search is large scale, aimed at recovery of scattered hominid remains in large fossil fields as at Maka in Ethiopia (see Chapter 27). Other times, the search can be very localized, either as in the charred rubble of the Branch Davidian Compound in Texas (Owsley *et al.*, 1995) or on a larger scale as in the search for military MIAs

277

within decades-old craters created by high-speed military impact (Hoshower, 1998). Sometimes remains from archeological or historical cemetery contexts may be brought to the attention of law enforcement agents or medical examiners by vandalism or natural causes (Berryman *et al.*, 1991). Other times, forensic osteologists are involved in the search for clandestine graves using methods ranging from aerial photographs to trained scent-detection dogs (France *et al.*, 1992). Of course, not all searches result in discovery, and many a paleontologist returns empty handed. Search methods will vary widely in osteology depending on the unique context of each case, but once discovery is made, a series of steps are set in motion.

14.2
Discovery

Skeletal remains are often found by accident. For example, hikers and construction crews often find osteological material. When they do, they usually report it to local law-enforcement authorities. Since there are many more dog, horse, cow, and goat bones than human bones on the surface of most places on earth, these are often mistaken for human bones by laypersons and amateurs. A general rule for the practicing human osteologist is to assume that law-enforcement personnel who often first encounter such remains (including, in some cases, coroners) are not qualified to render accurate opinions on isolated, fragmentary skeletal remains. For example, pet rabbits that died in trailer home fires have been identified by official coroners as year-old human infants. Forensic human osteologists usually encounter situations in which morphological identification is easily accomplished (but see Ubelaker *et al.*, 1991). On the other hand, a coordinated effort between physical anthropologists and law-enforcement specialists at a crime scene is absolutely essential for the acquisition of all available clues (Wolf, 1986; Maples and Browning, 1994; Dirkmaat and Adovasio, 1997).

On being introduced to skeletal material, the human osteologist is faced with three critical questions:

- Is the material human?
- How many individuals are represented?
- Of what antiquity is the material?

Experience is the most valuable commodity in answering the first two questions. When in any doubt, consult comparative skeletal material or the illustrations in this book.

The third question is usually more difficult to answer accurately, particularly if contextual information is not available. To give an accurate answer it is necessary to engage in a little detective work. The condition of bones themselves does not tell very much, because the physical condition of the bones is largely controlled by the physical environment in which they were deposited. Bone weathering and deterioration are accelerated by direct sunlight, high heat, fluctuating temperature and humidity, biotic influences, and soil acidity. When these variables are held to a minimum, bone deterioration can progress very slowly. To assess the **an-**

tiquity of skeletal remains (not individual age at death), it is necessary to give primary consideration to the context. Has the skeleton been recently disturbed? What kinds of artifacts appear with the remains? False teeth, dental fillings, coins, beads, pottery, coffin nails, and other evidence of material culture (if real association can be established with the skeletal remains) may be critical in determining the antiquity of the bones. Wear of the teeth can sometimes provide clues to the origin of the skeletal material; in many places, the teeth of recently dead individuals usually show far less wear than that seen under aboriginal conditions. Because contextual information is so critical to the accurate determination of the antiquity and origin of skeletal remains, the osteologist should always make every effort to visit the discovery site and make a firsthand assessment and record of the depositional history and associations of osteological material.

14.3
Excavation and Retrieval

Proper evaluation of any skeletal remains normally requires collection of the bones and subsequent laboratory analysis. Upon discovery, the osteologist's natural inclination (particularly when the remains were the object of a search and especially if they are hominid fossils) is to get the bones out—to collect them. This immediate lifting of the specimen from its context is the worst strategy under most circumstances, particularly when the bones are found in an archeological or paleontological context where they have been for a long time. They can no longer "walk away" on their own. Before disturbing the context, move away from the site and carefully develop a strategy for recording and recovery. If celebration of the discovery is called for, hold it off-site. On the site, it is advisable to show patience and restraint, while thoughtfully devising an appropriate strategy to extract the remains. Writing impressions down in the form of field notes is necessary and helps in this planning.

After the thrill and excitement of discovery have abated, it is time for serious, objective assessment of the situation. The following questions should be carefully considered:

- What are the political and legal constraints under which recovery must proceed? Goldstein (1995) provides an illuminating case study in this regard. Osteologists should confer with project management to ensure that all applicable laws are followed and all concerned parties informed.

- In what condition is the bone?

- What has happened to the bone as it has been exposed? How has natural or human-induced erosion uncovered and scattered the bone across the landscape?

- What contextual information is available?

- What options are there for recovering the bone? Consider the available time, labor, and equipment. As Hoshower (1998) notes, flexible excavation strategy is a key component of successful recovery.

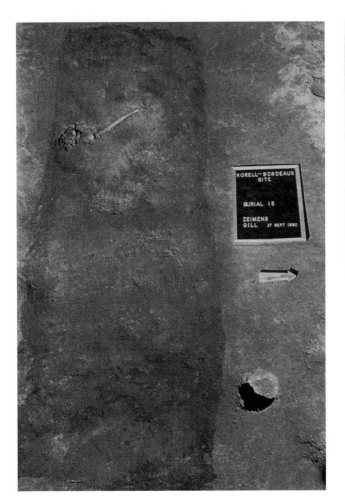

Figure 14.1 A pioneer burial from near the historic Bordeaux Trading Post, southeastern Wyoming. The grave, which seemed to be associated with the trading post or the nearby Oregon Trail, was found in 1980 by a field archeology crew under the direction of George Gill. *Left,* oblique view of the grave, outlined by the perimeter of the darker in-filling soil. The right ulna is exposed near the upper end of the grave filling. Artifacts found with the skeleton include a wedding ring and adjacent black ring of mourning and three coins. Seen in this photograph are the remains of boots and a wide-brimmed black hat over the face. A displaced fragment of right radius is pedestaled next to the right patella. *Right,* The grave after excavation (from Gill, Fisher, and Ziemans, 1984).

Figures 14.1 through 14.12 illustrate some aspects of skeletal recovery in archeological situations. Ubelaker (1989) is a good source of additional illustrations, and Chapters 25–27 present case studies that involve the recovery of skeletal parts. Chapter 21 reviews special considerations involving biomolecular sampling and precautions.

The following general steps should be taken in the recovery of skeletal material:

- If skeletal parts have been scattered by erosion, mark each with a pin flag and assess the distribution to predict where more pieces might be found.

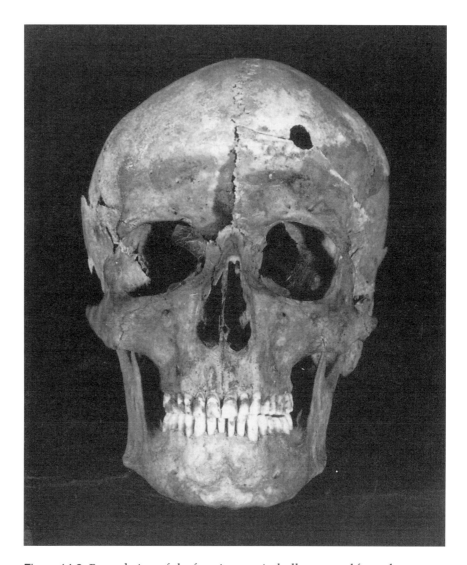

Figure 14.2 Frontal view of the frontiersman's skull excavated from the grave shown in Figure 14.1. Note the cranial gunshot wound caused by a .44- or .45-caliber weapon. A second perimortem gunshot wound to the hip and at least three healing rib fractures were noted by the investigators. This individual, a male who probably lost his wife prior to his own death (indicated by the black ring next to the wedding band), most likely was shot in 1869 or 1870 (from Gill, Fisher, and Ziemans, 1984).

- The overriding concern in all work subsequent to the discovery of the specimen, in the field or the laboratory, is to allow no further damage to occur. Excavation and extraction damage to osteological remains is common but unnecessary. Steps should be taken to consolidate fragile bone ***in situ*** (in place) with preservatives, if practical.

- Lose as little information as possible, especially concerning **context.** The remains, whether in a forensic or an archeological context, are one of a kind. They are nonrenewable resources. There is only one such bone, individual, burial, or cemetery, and this means that there is only one chance to extract the remains completely and correctly. Actions

Table 14.1
Equipment and Supplies for Osteological Fieldwork

TEAM EQUIPMENT
  Compass
  Tape measures
  Hammers (carpenter's and geological)
  Nails or stakes
  Twine (yellow or white)
  Line levels
  Shovels
  Buckets
  Picks
  Trowels
  Digging probes (dental and wood)
  Brushes (various sizes and stiffnesses)
  Screens
  Axes and saws
  Metal detector (forensic cases)
  Step ladder
  Camera tripod
  Cameras (video, polaroid, 35-mm)
  Camera flash attachment and cable release
  Photo scale and directional arrow
  Photo information board
  Photo gray card
  Pliers, wrenches, cutters, files
  Ropes
  Tarpaulins
  Stereo photo viewer
  Aerial photographs
  Field umbrellas
  Travel and excavation permits
  Tape recorder
  Portable tables & chairs
  Bush clippers
  Global positioning instrument
  Portable computer

TEAM SUPPLIES
  Film
  Glue
  Tape (masking, gaffer's, transparent)
  Preservative
  Solvent for preservative
  Packing material
  Containers (boxes, bags, vials)
  Aluminum foil
  Tissue paper
  Newspaper
  Notebooks
  Writing utensils
  Labeling pens
  Labeling ink
  Batteries (flashlight and camera)
  Wire
  Plaster

  Gauze
  Water
  Acetate sheets
  Nail polish
  Syringes
  Video tapes
  Audio tapes

PERSONAL EQUIPMENT
  Digging instruments
  Pocket knife
  Scissors
  Writing utensils
  Radio
  Hand lens
  Flashlight, extra bulbs
  Notebook
  Calculator and batteries
  Camera, film, batteries, flash, lenses
  Caliper
  Measuring tape
  Photo background, cable release, scale, strap
  Small photo mirrors
  Gloves
  Travel documents (passport, tickets, health
    certificate, insurance forms, permission letters)
  Hat
  Tent
  Sunshower
  Canteen
  Plate, cup, utensils
  Towel
  Sleeping bag
  Needle and thread
  Small preparation kit
  Extra glasses or lenses
  Sunglasses

PERSONAL SUPPLIES
  Money
  Lens cleaning fluid and papers
  Film
  Institutional letterhead and envelopes
  Gaffer's tape
  Sunscreen
  Medicines, water purifier
  Plastic bags
  Permanent ink marking pens
  Rubber bands
  Paper clips
  Toiletries
  Soap and shampoo
  Toothbrush and paste
  Chapstick
  Insect repellent

Figure 14.3 Prehistoric archeological site in northern California, before and after excavation. *Above,* the outlines and darker in-fillings of circular grave pits can be seen on the floor of the excavation trench. The scales are in inches. *Below,* flexed burials were revealed within the slightly superimposed grave pits of Burial 2. Here, the two individuals in this burial are shown, both flexed. The same large stone appears in the lower left corners of the two photographs.

Figure 14.4 Flexed burial of an adult accompanied by grave goods. Shell beads adhere to the cranium, stones are in the mouth, and other beads and pendants are seen around the postcranial skeleton. Prehistoric, northern California. The scale is in inches.

taken during recovery have consequences that long outlive any investigator, rendering the osteologist's responsibility a weighty one.

- Obtain proper equipment for recovery. Table 14.1 is a supply and equipment checklist that osteologists may find useful to consult before heading into the field. Provided that the bones and their context are not immediately jeopardized, recovery should be delayed until the proper tools are available. Use your judgment on how precarious the situation is.

- Before disturbing the scene make written and photographic records of everything important. Never rely on memory. It is often useful to establish a datum point on the site to control a grid laid out with stakes and cord across the surface. All recovered objects can be related to the grid. Wolf's (1986) advice on approaching a crime scene with scattered material as if you were clearing a minefield is appropriate in this regard. A Polaroid camera is valuable, and 35-mm black-and-white film is inexpensive and easily processed. A scale and directional arrow should be included in all drawings and photographs. Subsequent stages of the excavation should be photographed from as many angles as appropriate. Remove all tools, obscuring roots, and lumps of earth before

Figure 14.5 Prehistoric cremation exposed by an archeological excavation. Associated with the cremation are obsidian artifacts. The cremated bone fragments are seen in the area above the LAS. 7 indicator. Prehistoric, northern California. The scale is in inches.

photographing. If lighting is a problem, particularly in an excavation, use a white sheet or aluminum foil on cardboard, or a flash attachment, to illuminate the specimen. The key to successful photography is the control of light (see Chapter 15). If a specimen is of critical im-

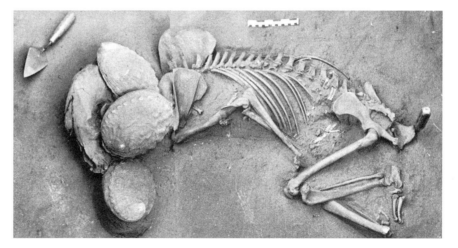

Figure 14.6 Ceremonial bear burial. Three large abalone shells cover the bear's skull. Because the bones of large mammals were usually exploited for their nutritive value, large mammal bones from archeological refuse deposits are usually very fragmentary; this case is an obvious exception. Prehistoric, northern California.

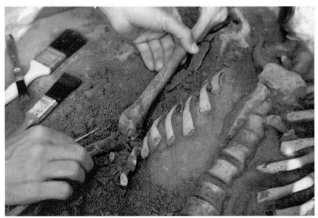

Figures 14.7 and 14.8 An extended burial from the prehistoric Libben site in Ohio. The excavator (*left*), having cleared the central part of this burial, carefully exposes the hand phalanges with a dental probe. After full recording, the archeologist (*right*) carefully removes the ulna and humerus. The ribs have been exposed for photographic recording and drawing, but they were not endangered by undercutting during excavation.

portance, it is good to make a video record of it for teaching, lecturing, curatorial, and forensic purposes.

- Begin preservation measures if necessary (for hardeners and other preservatives, see section 14.7).

- Collect all bone exposed on the surface, even fragments that do not seem to be hominid. Remove your shoes if necessary and get down on your belly for a close look. Make sure the light is adequate before doing this. It is to your advantage that rains have washed the bone surface clean; as you disturb the soil it becomes more difficult to recognize small bone fragments. Move slowly and carefully, not trampling remains or artifacts underfoot.

- Screen earth from the abdominal region of all skeletons to recover dietary or fetal skeletal remains. Screen all of the loose surface earth left over from each skeleton. A 1.0-mm mesh size (window screen) will recover most important fragments. Water wash excavated material through a screen to make the small fragments more visible and easy to recover.

- For burials, or other articulated *in situ* material, expose the bones one at a time. In an archeological context it is important to recognize that there is a very large culturally determined and ethnographically observed range of variation in human mortuary practice. For most applications, however, there are some general kinds of burials to which the osteologist should be alert: A **primary interment** is a burial in which all

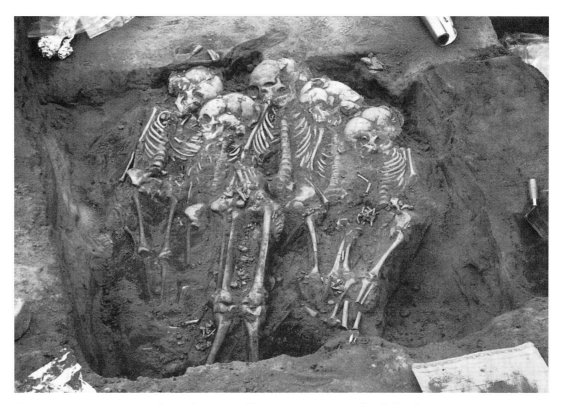

Figure 14.9 Mass grave at the prehistoric Libben site containing the skeletal remains of four children and an adult. The legs and feet of a later extended adult burial protrude into the excavation from the bottom of the photograph. Such superimposition of different burial events is commonly encountered in aboriginal cemeteries, usually because graves were not marked with permanent surface monuments.

Figure 14.10 Extended burial of an immature individual from the prehistoric Libben site. Great care must be taken with such young individuals to recover the many unfused bones.

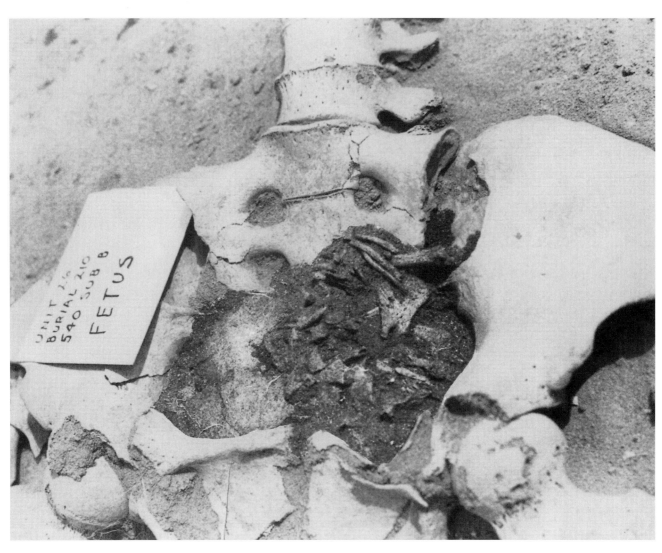

Figure 14.11 Bones of a fetus were found within the pelvic cavity of this female skeleton from the prehistoric Libben site. The humerus, ribs, and scapula of the fetus are just anterior to the sacroiliac articulation.

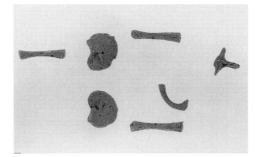

Figure 14.12 This fetal skeletal material from the prehistoric Libben site represents one of the smallest burials ever recovered. Good preservation combined with great concern for detail during excavation made the skeletal collection from this archeological site one of the best available for the study of prehistoric demography (see Chapter 20). Natural size.

of the bones are in an anatomically "natural" arrangement. Such burials are sometimes classified according to whether the extremities are extended or flexed. There are no neat categorizations here, and one photograph is worth many words of description. A **secondary interment** is a burial in which the bones of a skeleton are not in "natural" anatomical relationship but have been gathered together some time after complete or partial disarticulation of the skeleton and then buried. Sometimes this burial comprises a bundle of bones. A **multiple interment** is a burial in which more than one individual is present. These burials include ossuaries, burial urns containing more than one individual, and a variety of other possibilities. **Cremation** is a mortuary practice involving the intentional burning of the body. Cremations can often be informative—the less efficient the fire, the more informative the specimen. Micozzi (1991) reviews mortuary practices worldwide.

- In exposing burials, use appropriate tools carefully. Dental picks are sharp and efficient, but they can easily damage the bone. Wooden or bamboo tools may sometimes be suitable, and a range of brushes of various sizes and stiffness is indispensable. Work from the rib cage outward where possible. Do not use the trowel in a sweeping motion unless you are doing exploratory work. Try to leave bones supported as you clean, clearing the foot and hand bones last. Watch for soil color and texture changes, rodent and root disturbance, mat impressions, rotted vegetation, wood, insect remains, charcoal, and associated artifacts such as lip plugs or beads. Be alert to all soft tissue that might remain, including hair, skin, fingerprints, and ligaments. Write or tape-record your notes; memory will not suffice. Record angles of flexion, the orientation of the body and head, the depth of the bones from the surface or the datum, and any other contextual details. Take soil samples where appropriate. Remember that all details of context should be retained in an archeological or forensic excavation. In archeological situations, context often yields the greatest amount of behavioral information. Excavation of a site destroys it, and contextual data left unrecorded are lost forever.

- Samples for biochemical analyses and histology (see Chapter 21) should be taken at the earliest time, with clean tools and gloves, in sterile containers, to avoid contamination.

- The actual removal of the bones themselves is one of the last steps of recovery, after exposure, photography, and drawing. To aid in their hardening, let the exposed bones dry completely in an area shaded from the sun. Free each bone gently—do not use force. Earth (matrix) may be left on the thin parts (scapula, pelvis) to avoid damage in transport. Do not attempt fine cleaning in the field; this should be performed in situations where light, tools, water, comfort, advice, comparative material, and time are in greater supply—in other words, in the laboratory. Take each bone out individually, and label as you go along, particularly the ribs and vertebrae. Keep right and left hands, feet, and rib bones in separate containers. Keep unfused epiphyses with their associated bones. Be observant as you remove the elements, watching for fetal bones, sesamoids, kidney and gall bladder stones, and small artifacts. Save everything, even if you think that it might not be human. It is easier to do accurate identification in the laboratory. Do not disregard immature skeletal parts or disturbed burials; such disregard will skew the cemetery representations and ultimately have an adverse effect on any demographic reconstruction.

- After removal, if the bone is still wet, let it dry completely in the shade. Never mix the specimens during washing or packing at the site. Screen all earth that remains at the spot of the burial, watching for beads, teeth, and other items that you may have missed as you removed the larger bones.

- Decisions about washing of material at the recovery site are best left to the judgment of the investigator. As Brothwell (1981) notes, there is generally no problem in deciding about bone strength; bones are not generally deceptive in this regard. It is usually evident during recovery whether they are likely to disintegrate when handled. If water is available, and the bone is well preserved, it may be convenient to wash parts of each skeleton in the field camp. It is often best to remove the earth from the cranial cavity and orbits while the soil is still damp. Do not try to reconstruct bones in the field—just keep them together.

- Decisions about the application of preservatives (section 14.7) must be made during exposure, and this application will often be necessary before removal is possible.

- Aluminum foil is an inexpensive but very effective material that can be used to stabilize, protect, and keep bones and their parts in place during lifting and transport. Press the foil firmly around all the bone and matrix irregularities.

- Some investigators use burial record forms. These can be useful, provided that caution is used in determining sex, age, and other features often prompted by such forms. Be sure to mark these determinations of identity as preliminary when they are made in the field.

14.4
Transport

When exposure, photography, and drawing are complete, the skeletal remains can be removed from the archaeological context. In the forensic context the decision to remove is made by the chief of the crime scene. This begins a "chain of custody" in which the osteologist must participate (Melbye and Jimenez, 1997). If the bone is well preserved, the elements can be lifted individually. Once bones have been extracted and field-cleaned, they are ready for packing and transport to the laboratory. Available vehicles and packing materials for transport vary greatly. Transport thus depends on what is available. However, under all conditions, one primary rule should always be observed: Do not let any more damage occur.

Occasionally it is necessary to remove the burial or its parts as a unit, in a supporting block of matrix, for study or display. To remove an entire burial, employ the paleontological technique of **jacketing.** To do this, first isolate the specimen on a pedestal of earth. Cover it with several layers of wet tissue paper where the bone is exposed. Next, use burlap bandages soaked in plaster to form a cast or jacket around the specimen. Reinforce this jacket as necessary with strengthening rods of metal or wood. After the plaster has hardened, undercut the specimen and lift it out. This is an expensive operation in terms of material, time, and personnel. It requires experience and should be used only when necessary.

Once the burial has been removed, washed, and dried, the hands, feet, and ribs should be bagged by side and the vertebrae by type. Labeling these elements at the time of removal greatly facilitates later sorting in the laboratory. Fine cleaning and gluing should be left for the laboratory. Label all bags and boxes used for collection with waterproof ink. Paper bags are prone to deterioration if used for long-term storage, but they breathe moisture and are therefore better than plastic bags for post-excavation sorting and transport of skeletal material that retains residual moisture. In transport, keep the bones away from water. It is most important at this point to label and maintain organization as the bones are extracted from their context and moved to the laboratory.

When packing bones, pack tightly with lots of padding (plastic bags or newspaper) to avoid movement in the container. Pack heavier, denser bones at the bottom of containers and the more fragile bones such as scapula, pelvis, and crania at the top. Be sure that all bones stay in their assigned containers during the jostling that inevitably accompanies transport. Be particularly careful about the cranium, whose face is fragile. The cranium and mandible should each be packed separately, and care should be taken to ensure that teeth that might dislodge from their sockets during transport remain with the jaw that lost them.

14.5
Sorting

The osteologist is often faced with the challenge of sorting out individuals from a collection in which more than one skeleton is represented. Of primary importance in this sorting are age, size, and sex differences as well as bilateral nonmetric traits. Matching articular or interproximal facets often provides clues about association. Preservational factors such as bone color, weathering, or integrity are of secondary importance, but sometimes of use in sorting individuals.

The **minimum number of individuals (MNI)** in any assemblage of bones is the minimum number of individuals necessary to account for all of the elements in the assemblage. A very simple example might be an assemblage consisting of two specimens, a fragment of left humeral head and a fragment of left distal humerus. These two specimens, even though the intervening shaft is missing, *could* represent the same individual (unless they are of patently different individual ages). Even though there might be two individuals involved the MNI value would, in this case, be 1.

This basic logic is used to determine the MNI value for any assemblage of human remains, following these procedures: First, remove all nonhuman elements. Next, separate the bones according to element and side. Within each right-side element category, count the minimum number of individuals, not pieces, represented. Consider all possible joins between fragments and assess the age of each fragment. Perform the same minimum number count for left-side bones within each element category, then check for individuals represented by left-side bones that either do match, or possibly belong, to those from the right. These do not increase the count. Left-side bones that do not match corresponding right-side ones in age or morphology are added to the minimum number count. After this is done for all paired and unpaired elements, the greatest minimum number of individuals determined for any element constitutes the minimum number of individuals in the assemblage.

Consider, for example, an assemblage consisting of two right maxillae with full adult dentitions, three left femora of adults, one right femur of an infant, two sacra, four adult right calcanei, and three permanent right upper central incisors. The minimum number of individuals (MNI) in this assemblage is six—the infant plus five adults (determined by two right central incisors in maxillae, and three isolated upper right central incisors). The maximum number of individuals is determined by counting all nonjoining, nonmatching elements—in this case, a total of fifteen. For another example, see Figure 14.13.

14.6
Preparation

Techniques used in skeletal preparation (cleaning) vary according to the condition of the bone and the context of its discovery. Forensic osteology requires special treatment (see below). Continue to use common sense. If the bone is well preserved, it can be washed in lukewarm water (without detergents or any other additives) using brushes, wooden probes, and spray bottles. Never wash more than one skeleton at a time. Use a screen in all washing, field and laboratory, to keep small bones from being lost. As the washing water becomes muddy, small fragments may detach and become lost in the sediment at the bottom of the basin or disappear down the drain. Clean the basin and screen frequently, making sure that both are checked between processing each burial. Depending on humidity, the washed bones dry in 24 to 48 hours on wire racks in the shade. In the lab, you can speed this up by using a fan to blow air across racks of drying bones. Never use a heat source due to the danger of bone surface exfoliation.

For fossils, more sophisticated preparation techniques are often called for. Fossils are sometimes encased in a very hard matrix that may be even harder than the bone itself. Sometimes the matrix can be softened with acetone, paint thinner, or even water. Very important fossil specimens should be molded (see section 14.9) before cleaning to make a record of prepreparation status. When cleaning, matrix samples should be kept so that future investigators might be able to establish **provenance**, the stratigraphic and spatial position of the specimen.

Some commonly used paleontological preparation tools and techniques include the following:

- Hammer and chisel. This technique has a long history, and many fossils have the scars to prove it. Speed is an asset, but the shock imposed on the specimen and the lack of fine control are negative points.

- Dental drill. This fast but dangerous technique has been used for many years. Positive points include good cutting power, more control than a hammer and chisel, and lack of shock to the specimen. On the other hand, extreme care must be taken to keep the grinder's surface from drilling into the surface of the object being cleaned.

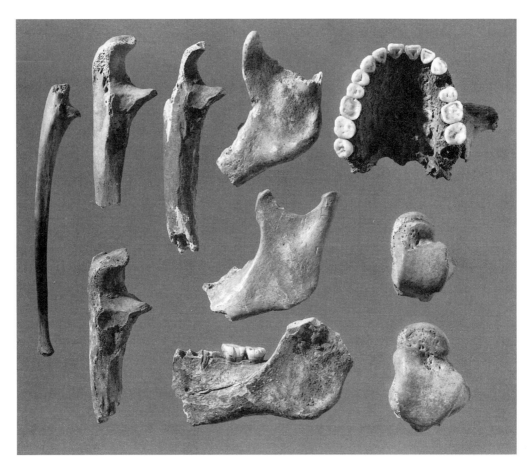

Figure 14.13 To determine the minimum number of individuals for this sample of ten specimens (shown one-half natural size), first note that there are no nonhuman pieces. Sort the pieces by skeletal element and side. There are two right tali, three left mandibles, one maxilla, and four right ulnae. One of the ulnae is immature. The MNI is thus equal to at least four people. Since no pieces join or are antimeres (opposite sides of the same individual), it is possible that each piece represents a different individual, so the maximum number of individuals indicated for this sample is ten.

- Dental pick (or needle held in vise) under binocular microscope. This is often the most effective way to clean a fossil; it gives the preparator much control and limits potential damage to tiny areas. On the other hand, this work requires an enormous amount of time and patience.

- Acid treatment. Dilute acetic, formic, or hydrochloric acids can be used to dissolve matrix holding some fossils. Excellent detail may be obtained by this technique, but it calls for patient, extended monitoring to keep the acid from attacking the specimen itself, etching its surface, or weakening its structural integrity. Take all standard laboratory precautions when using acids.

- Air abrasion. This preparation technique uses a tool that shoots a stream of particles at the matrix like a miniature sandblaster. This provides speed and control without shock. On the other hand, the abrasive particles may obscure detail and frost the bone and tooth surfaces. In

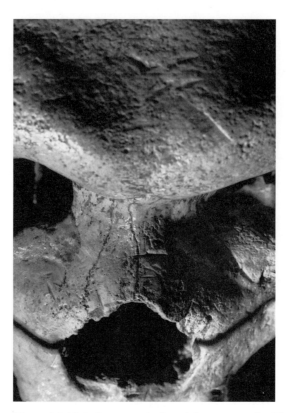

Figure 14.14 Laboratory-induced damage on the Pleistocene fossil *Homo* cranium from Petralona, Greece. Removal of matrix by a high-speed grinding wheel has produced damage on the original surfaces of the left nasal bone, and chiseling marks are seen above the left orbit. Natural size.

addition, it is difficult to control the abrasive stream as it ramifies through cracks below the surface of a specimen.

- Air scribe/electric scribe. This vibrating steel point is driven by air or electricity and acts as a miniature jackhammer. It is very useful in some cases, demonstrating a kind of microchiseling action that exfoliates matrix from bone. Care must be taken to avoid excessive vibration, and the sharp instrument tip should touch only matrix, never the fossil itself.

The most essential ingredients in successful fossil preparation are patience and experience, but help and advice from skilled preparators with both is crucial. Note that all of the techniques discussed carry with them hazards for the fossils. It is better to be safe than sorry in preparing important fossils. Preparation trauma readily observable on already cleaned fossils shows that many past workers were not cautious enough (Figures 14.14–14.15).

Other specialized preparation techniques in osteology involve forensic material with adhering flesh. Here, most of the flesh can be carefully removed with cutting tools. The specimen can then be boiled with or without chemicals such as enzyme-based detergents. For discussion of a variety of defleshing techniques useful in building comparative mammalian skeletal collections, consult Hildebrand (1968) and Mori (1970). Be sure to consult a soft tissue pathologist before removing soft tissues—

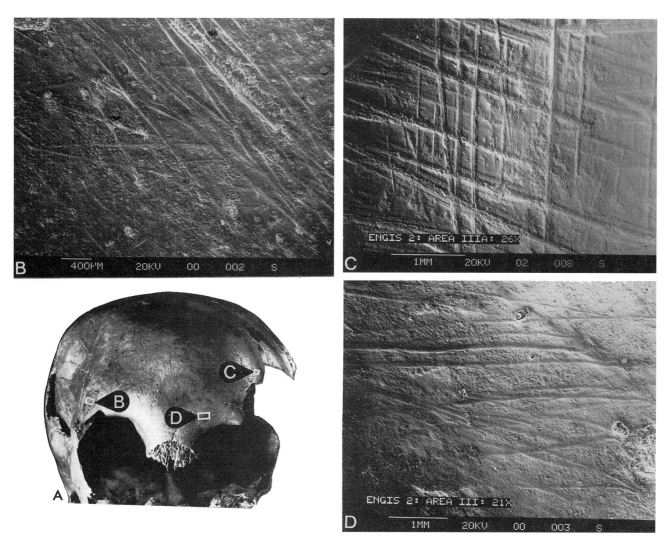

Figure 14.15 Frontal view of an immature Neanderthal cranium from Engis, Belgium (A). The superficial scratches along the midline were made by a diagraph instrument needle, and those along the broken edge of the left frontal were made by sandpaper used to smooth the previously plaster-reconstructed area behind. These marks were interpreted as evidence of Neanderthal mortuary practice (Russell and LeMort, 1986). This interpretation was shown to be mistaken by White and Toth (1989). Scanning electron micrographs (B–D) show the sandpaper striations. Note the "doubling back" at the end of each sandpaper grain's path as the end of the sanding stroke was reached. These figures illustrate the utility of the scanning electron microscope in investigating bone modification at magnification.

never destroy evidence of any kind. Even the insect larvae contained in these tissues may provide important clues in a forensic case.

14.7 Preservation

The strengthening of bones can be accomplished in several ways during and after recovery. Various hardeners, or preservatives, are available.

These are usually either water- or acetone-soluble; they include Glyptal (a product that, when undiluted, has the consistency of thick glue), polyvinylacrylate (PVA, or Vinac, a plastic dissolved in acetone), Paraloid B72, Bedacryl (which can be purchased in hard chunks which are then dissolved in an organic solvent), Butvar (a powder mixed in solvent), and many others. Cyanoacrylate glues of various viscosities are irreversible but may be necessary to apply before removal of remains, particularly heavily burned or ashed tooth crowns in forensic contexts (Mincer *et al.*, 1990). Consult with the conservator of the institution that will house the osteological material about the kind of preservative preferred.

The key to using any preservative is correct dilution. The most frequent failure in preservative use is the failure to dilute the solution enough, which results in poor penetration and the formation of a hard outer skin on the specimen but a lack of internal hardening. Impregnation with preservative having the consistency of water is recommended, usually about a 5–10% solution. It is usually best to dip whole specimens into the solution and then let them dry on a wire screen. In the case of more fragile remains that will not stand up to immersion, drip the solution onto the specimen. Use organic solvents (thinners) with extreme caution. Many of these chemicals are dangerous; avoid breathing their fumes and remember that they are often extremely dangerous because of their flammability.

14.8
Restoration

Restoration involves putting pieces of broken bones back together. A detailed knowledge of osteology greatly simplifies this procedure; the ability to identify the side and position of fragments allows the quick identification of joins. The restoration of fragmentary bones is often described as more difficult than doing a jigsaw puzzle. This is an exaggeration. A broken skeleton has one more dimension and far more information than a picture puzzle of a polar bear in a snowstorm. Restoration is often quick and easy for the competent osteologist (see Figure 14.16 for an illustration of restoration in progress). The following are valuable guidelines for restoration:

- Use a glue that may be dissolved later. This ensures future workers the ability to correct any unintentional mistakes of restoration.

- Do not hurry. As in preparation, patience and experience are essential in good skeletal restoration.

- Restore the face and vault separately before joining them. Use the mandibular condyles as a guide to restoring the correct cranial breadth when this is in doubt.

- Be sure the bones to be glued together are dry unless a water-soluble glue is being used.

- Do not glue until you are positive of a good join. Check under the microscope if necessary.

Figure 14.16 In the laboratory, osteological specimens are fully labeled and restored. Here the osteologist checks for joins. When joins are found, the bones are glued together and temporarily supported in a sandbox while the glue sets. Comparative material should be accessible during these operations. The washing and drying of additional specimens proceeds in the background.

- Make sure the joining surfaces are clean of debris. Adhering grit, preservative, and flakes of bone can result in misalignment of broken pieces.

- Use color, texture, and, most importantly, anatomy to match the pieces.

- Never glue teeth into the sockets until you are absolutely positive that they belong there. Interproximal contact facets are an invaluable guide to correct tooth placement.

- Do not glue yourself into a corner by leaving unfilled holes. Instead, use masking tape to make temporary joins. Do not leave the masking tape on for more than a couple of months and be careful that the bone surface can release the tape without exfoliating. When satisfied that the joins are correct, progressively remove the tape and glue the broken surfaces together.

- Reconstruct only where necessary. Use soft plaster or a 50:50 mixture of paraffin wax and dry plaster heated to a liquid in a saucepan on a hotplate (not an open flame). Do not ignite the paraffin. This restoration material is easy to work with and remove. In contrast, modeling clays (plasticine) tend to be more greasy and should generally not be used except as temporary props. After the restoration is complete, be sure to demarcate the reconstructed from the real surfaces. Reconstruction (as opposed to restoration) is rarely justifiable for an original because it is subjective and obscures valuable cross-sectional information.

- Use a sandbox and gravity to position pieces while rebuilding them. Anchor one piece in the sand and balance the other piece on top of it, perhaps temporarily supporting the glue join with masking tape. Be sure to let the glue completely harden before removing the piece from the sandbox.

- Where contacts are limited and weak, brace the parts by using struts made of wooden or glass rods.

- Do not use glues, preservatives, or reconstruction material that will inhibit molding rubbers that may be used on the specimen at a later time. Check any such substances for compatibility with molding rubber before applying them to the specimen.

- For some specimens complete restoration is extremely difficult; some distortion is the result of warping rather than fracture. Such warping is impossible to correct.

14.9
Molding and Casting

Casts of skeletal material are used in osteology and paleontology for several purposes. They provide a good three-dimensional archival record of the object under study (Mann and Monge, 1987; Smith and Latimer, 1989). In addition, they are useful in communicating findings with colleagues and for slicing into cross sections for comparative purposes. Specialized techniques and materials used in molding and casting teeth are outlined in Hillson (1992, 1996). Several molding methods may be used for bones, depending on the needs of the investigator. Some of these are listed here:

- Dental impression compound. Powders mixed with water form material that dentists use to make impressions of teeth. Alginate, Geltrate, and many other brands are good for making quick, one-sided molds and are widely available, inexpensive, and easy to use. They are not recommended for holding very fine detail or for the production of more than one or two casts. The molds deteriorate within days even when kept moist.

- Latex rubber molds. This material gives higher resolution and the ability to make two-part molds. The material is more expensive than dental impression compounds but lasts much longer.

- Silastic (silicon) rubber molds. For mold detail and longevity, silastic rubber is the material of choice for molding. This material, however, is the most costly and time consuming to use.

- Scanning electron microscope replicas. A wide variety of other quick-setting dental impression rubbers are used for making epoxy replicas of small portions of bone surfaces for work in the scanning chamber.

The easiest casting material to use is plaster, which comes in many varieties. Some casting materials include the following:

- Common plaster of paris is inexpensive but soft and easily scratched.
- High-strength gypsum cements and dental plasters are harder, especially when mixed with hardeners instead of water. Detail on casts in these materials can be excellent, and shrinkage is minimal.
- Many plastics and epoxys are also available, but shrinkage of these materials combined with their tendency to shorten mold lives sometimes rules against them.

There are various techniques for coloring casts of all materials to bring out detail. All involve a vehicle such as alcohol for dissolving and spreading artists' pigments. Artists' fixatives and spray lacquer add a long-lasting, appealing, and protective finish to the final product.

14.10
Curation

During the initial processing of skeletal material, it is advisable to label all bones individually with a prefix designating the site and a number representing the skeletal individual. For example, BOD-VP-1/1 is a specimen number for the first vertebrate paleontological specimen from the first locality collected in the Bodo area of Ethiopia in 1981. It is crucial that this labeling be legible, with numerals that anyone can read. Be very careful not to confuse 9s with 2s or 1s with 7s. Specimen numbers should be written in permanent, waterproof ink and protected under nail polish. For softer bones, it may be necessary to let a drop of preservative dry and harden the bone surface before putting the label on the bone. This treatment prevents the ink from diffusing into an illegible blob. Specimen numbers are essential; they represent the critical links between the bones and information on their original context. Mixing of labeled material is very bad practice in the laboratory, but mixing of unlabeled material is often irreversible, and therefore unforgivable. Labels should be put on bones early in the curatorial process.

There are two main objectives of curation. The first, as in other steps outlined above, is to prevent loss of information. Information loss can come in the form of actual physical destruction of the bones and teeth, in the mixing of unlabeled elements in the collection, or in the loss or destruction of the records (including computer records) for the skeletal material. Almost all simple breakage of bones can be repaired with glue. The objective of curation, however, is to prevent breakage in the first place by properly handling and storing bony material. Untrained or unqualified persons should not handle osteological material without supervision. Metric or photographic analysis of material should not be allowed to damage the specimens. Bones are fairly tolerant of a range of storage conditions, but their containers (boxes, trays, bags, padding) should be composed of a nondeteriorating, acid-free material. Bones should be stored in areas in which humidity, incident light, and extreme temperatures are kept to a minimum. Steps should be taken to see that insects and rodents are kept away from stored skeletal material and records. To prevent accidental loss of records from flooding, fire, and theft, it is advisable to make a copy of all skeletal records (whether computer disk or hard copy) and to store this copy in a separate location.

The second major role of curation is the provision of research access to the collection. To provide this access, it is necessary to impose and maintain a high degree of organization in the skeletal collection. A researcher should be able to move quickly and efficiently between the bony remains and their records. Computer databases are important, not only for collection organization, but also as a means of enhancing research access.

Suggested Further Readings

Some additional published sources describing skeletal recovery in archeological and forensic contexts are provided here. Note that there is no written substitute for experience; the inexperienced osteologist charged with retrieving skeletal remains should always enlist the aid of more experienced colleagues, particularly archeologists.

Brothwell, D. R. (1981) *Digging Up Bones* (3rd Edition). Ithaca, New York: Cornell University Press. 208 pp.
> Chapter 1 discusses excavation of skeletal material in archeological context.

Feldmann R. M., Chapman, R. E. and Hannibal, J. T. (Eds.) (1989) *Paleotechniques*. Knoxville, Tennessee: Paleontological Society Publication Number 4. 358 pp.
> A wide variety of technique papers about preparation, replication and illustration of fossils.

France, D. L., Griffin, T. J., Swanburg, J. G., Lindemann, J. W., Davenport, G. C., Tramell, V., Armbrust, C. T., Kondratieff, B., Nelson, A., Castellano, K., and Hopkins, D. (1992) A multidisciplinary approach to the detection of clandestine graves. *Journal of Forensic Sciences* 37:1445–1458.
> Methods, results, recommendations on searching for clandestine graves. With a good literature review.

Haglund, W. D., and Sorg, M. H. (Eds.) (1997) *Forensic Taphonomy*. Boca Ratton, Florida: CRC Press. 636 pp.
> An excellent edited volume with reviews and case studies covering a wide array of topics.

Joukowsky, M. (1980) *A Complete Manual of Field Archaeology: Tools and Techniques of Field Work for Archaeologists*. Englewood Cliffs, New Jersey: Prentice-Hall. 630 pp.
> A good guide to field and laboratory techniques involving recovery of archeological remains.

Killiam, E. W. (1990) *The Detection of Human Remains*. Springfield, Illinois: C. C. Thomas. 255 pp.

The most comprehensive guide available; even has a chapter on "parapsychological methods."

Krogman, W. M., and İşcan, M. Y. (1986) *The Human Skeleton in Forensic Medicine* (2nd Edition). Springfield, Illinois: C. C. Thomas. 551 pp.
Chapter 2 provides a good summary of crime scene investigation procedures.

Leiggi, P., and May, P. J. (Eds.) (1994) *Vertebrate Paleontological Techniques.* Cambridge: Cambridge University Press. 421 pp.
A guide to a variety of field and laboratory techniques used by vertebrate paleontologists.

Smith, J., and Latimer, B. (1989) A method for making three-dimensional reproductions of bones and fossils. *Kirtlandia* (Cleveland Museum of Natural History) 44:3–16.
A good introduction to molding and casting techniques used with modern and fossil osteological material.

Ubelaker, D. H. (1989) *Human Skeletal Remains: Excavation, Analysis, Interpretation* (2nd Edition). Washington, D.C.: Taraxacum. 172 pp.
This text provides excellent illustrations and descriptions on the subject of archeological excavation of skeletal material.

Wolf, D. J. (1986) Forensic anthropology scene investigations. In: K. J. Reichs (Ed.) *Forensic Osteology: Advances in the Identification of Human Remains.* pp. 3–23. Springfield, Illinois: C. C. Thomas.
A guide on how to approach skeletal remains in a forensic context.

CHAPTER 15

Analysis and Reporting of Skeletal Remains

Analysis begins in the field and extends to the laboratory. The procedures discussed in Chapter 14 are primarily related to recovery and preparation of skeletal remains. There, we traced remains from their point of discovery into the laboratory. Now we introduce analytical procedures commonly applied to human osteological material. Osteological analysis involves both observation and measurement, and we review some techniques for each. We also discuss laboratory and field photography. Finally, because a primary goal of the osteologist's work is usually documentation and communication, we consider the elements of effective reporting of the results of osteological analysis.

15.1 Analysis

15.1.1 Setting

Sound procedures, an appropriate setting, and careful use of the proper equipment are essential in osteological analysis, not only to ensure accurate results but also to safeguard the skeletal remains. Of primary importance is that work be conducted over a padded surface. Care should be taken to prevent specimens from contacting hard surfaces or rolling onto the floor. Even well-preserved bones that seem sturdy may be fragile when compared to the instruments used to measure them. The osteologist must therefore be careful not to crush, pierce, scratch, or otherwise damage the specimen with the instruments. Poking or prodding the skeletal material with the fingers can also easily damage the bone, especially the more fragile parts of the cranium.

The study of bones is best done in a well-lit laboratory. Lighting is critical. Overhead fluorescent lights are poor for osteological work because they tend to fill the room with diffuse light. Observation of osteological detail depends on control of incident light on the specimen, and for this reason a swing-arm, incandescent light source is recommended for osteological analysis. A unidirectional light source makes it possible for the researcher to highlight subtle bony features or modifications by angling the light to enhance the visibility of surface detail.

It must be noted that there may be several biological and chemical hazards involved with recovery and analysis of skeletal remains, particularly in forensic contexts (Galloway and Snodgrass, 1998).

15.1.2 Metric Analysis

Recent advances in computer scanning technology (see sections 15.3 and 15.4) have made it possible for osteologists to acquire metrics digitally, and display them via the computer screen, remote to the specimen(s) under study. At first glance, this technological capability brings with it the prospect of fast, easy, and accurate data acquisition and distribution. Rare or valuable specimens can now be digitized by laser or CT scans and "held" in computer memory (Zollikofer *et al.*, 1998). So alluring are the siren songs of these new capabilities that some have suggested that they obviate the need to curate original specimens (see also Chapter 16). This is a mistaken and extremely dangerous philosophy in osteology.

History holds lessons here. When it became possible to "capture" morphology of fossil specimens via silicon rubber and accurate plastics, replicas of the fossil originals were distributed throughout the world. These casts are immensely valuable as teaching and preliminary research tools, but experience has shown that they are no substitute for the originals. Distortion, matrix cover, and internal morphology are all lost in a cast. Workers unfamiliar with originals can be badly mislead by such features, features only visible with reference to the original. Misinterpretations based on inaccurate observations and measurements on photographs and casts are an embarrassing part of the published literature of human paleontology. See Clarke and Howell (1972) for an analysis of problems inherent in observations and measurements from photographs and casts.

With either two- or three-dimensional digital images scanned from original osteological specimens, these problems are compounded dramatically as the investigator moves another step away from the original. Color, texture, internal anatomy, matrix cover, preservative cover, preparation damage, erosions, and distortions of all kinds may be faithfully recorded by such imaging of the original, but for the osteologist seated at a computer monitor on the other side of the planet, these features are often not digitally distinguishable from actual bony anatomy. Humans routinely make mistakes, and these technologies seem poised to increase dramatically the number and impact of these mistakes in human osteology. For obvious reasons, osteologists should always refer to the original specimen, even when acquiring metric data from digital sources.

The popular image of an osteologist at work is one of a person in a white laboratory coat, measuring instrument in hand, manipulating some bone (usually a cranium). As indicated in Chapter 14, the role of the professional osteologist is far broader than this. In the formative days of osteology, quantification of bony anatomy was the focus of most work. Elaborate sets of measuring points were defined, and vast quantities of metric data were compiled. Just because a certain measurement has been reported, however, does not mean that it is either useful or even reproducible (Stirland, 1994). Conversely, new metrics are routinely developed to quantify morphological observations. The traditional days in which measurement was done for measurement's sake are thankfully over now, but metric analysis does continue to play a primary role in osteology. Howells (1969a) provides an interesting background for the selection of skeletal metric points.

Because osteological work is part of the scientific enterprise, it is necessary to communicate results to other researchers in an unambiguous and precise manner. One of the most convenient and effective ways to communicate osteological observations is to quantify them—to express them in numbers. Thus, to inform colleagues and others about a particular tooth, it is a simple matter to measure and count characteristics of that tooth. All scientific measurements, including those in osteology, should be taken in the metric system. This system expresses linear osteological measurements in millimeters, centimeters, and meters. Some observations may be quantified even though they are difficult or inappropriate to record as actual measurements. For example, traits such as the Carabelli's cusp on upper molars may be recorded as **absent** or **present.** Variables recorded in this way are called **discrete variables,** as opposed to **continuous variables** such as linear measurements.

Many measuring tools have been invented and developed for osteological analysis. Figure 15.1 illustrates some of them. The most frequently used is the **sliding caliper,** which has a pair of jaws whose variable gape is measured via a dial or scale on the caliper shaft. A **spreading caliper** is usually used for work on cranial anatomy. **Osteometric boards** are useful in measuring lengths and angles of postcranial elements. Metal cubes for holding crania are also available, and instruments called **diagraphs** are used to trace certain profiles of crania held in these devices. Most of these precision instruments are made of steel and are expensive. They are also sharp. Care should be exercised to see that the instruments are not damaged during use. More important, since bone is softer than steel, these instruments can scratch or perforate bone surfaces (see Figure 14.15), and care should be taken to see that such damage does not occur during analysis. Many of these instruments have now been linked to microcomputers, allowing the metric data accumulated on large samples of osteological material to be entered and manipulated electronically.

Metric analysis in osteology requires more than simply measuring a given element. It is critically important to provide precise definitions of each measurement. Furthermore, the degree to which measurements can be reproduced is important in metric analysis. The example in Table 15.1 shows how to calculate and report the technique-based measurement error (intraobserver error). Heathcote (1981) and Buikstra and Ubelaker (1994) provide further details on measurement error in human osteology.

Metric data in osteology are usually compiled as a result of measuring arcs, chords, or volumes. Indices are made by combining these values. For example, the cranial index is the product of the maximum cranial breadth (*bi-euryon*) and 100, divided by the maximum cranial length (*glabella* to *opisthocranion*). Indices are convenient because they express shape as a single variable. For example, a short or "broad" skull has a higher index

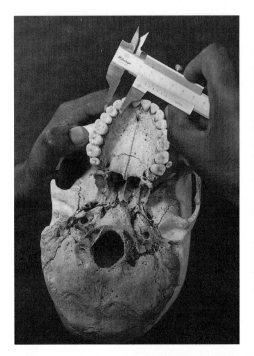

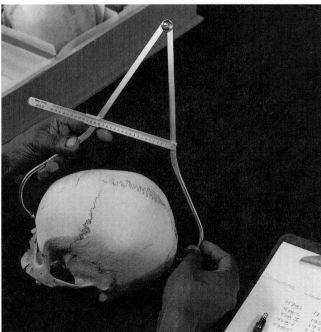

Figure 15.1 Tools for osteometric measurement: *Upper left,* the basic sliding vernier caliper is used to measure the buccolingual breadth of the right maxillary third premolar. Note the rare nonmetric anomaly seen in this dentition, bilaterally present fourth molars. *Upper right,* measurement of cranial length with the spreading caliper. *Left,* measurement of femur length with the osteometric board.

than a long, narrow skull. Bass (1995) provides a concise guide to standard measurements and indices most widely used in human osteology. Cranial metrics and indices are summarized in Table 15.2.

Expression of skeletal element shape, or **morphology,** can be examined by **univariate** (single measurement) or **bivariate** (two measurements, as in an index) means. The use of multiple measurements to express skeletal shape has been called **multivariate morphometrics.** The successes and failures of multivariate analysis in human osteology and paleontology are examined by Lovejoy (1978), Howells (1973), Frayer (1985), Reyment *et al.*

Table 15.1
Procedure for Estimating Measurement Error

A. An osteologist measures a tooth for
the first time and gets a reading of 10.8 mm .... **Measurement A = 10.8 mm**

B. Time is allowed to pass
(best if a few days) so that knowledge
of the original measurement does not
influence the remeasurement. The
same investigator then remeasures the
same tooth at 11.0 mm .... **Measurement B = 11.0 mm**

C. After another hiatus, the same
investigator measures the tooth at 11.5 mm .... **Measurement C = 11.5 mm**

D. There are now three measurements.
To calculate the mean (average) of these,
add the three measurements .... **Measurements A + B + C = 33.3 mm**

. . . and divide the total by the
number of measurements (3). The
result is the mean, or average .... **33.3 ÷ 3 = 11.1 mm**

E. To determine the degree to which
the original measures deviate from this
average, figure the difference between the
mean and each measurement. Add
these differences .... **A: 11.1 - 10.8 = 0.3**
B: 11.1 - 11.0 = 0.1
C: 11.5 - 11.1 = 0.4
0.8

. . . then divide the summed differences
by the number of measurements taken (3) .... **0.8 ÷ 3 = 0.267**

The result reflects the degree to
which the actual measurements differ
from the mean. Convert to a percentage
of the average value .... **0.267 ÷ 11.1 = 0.024**
= 2.4%

The average measurement error from the 11.1 mm mean is about 2%. By reporting this value, the osteologist informs the audience about how repeatable any given measurement is.

(1984), and Corruccini (1978, 1987). Richtsmeier *et al.* (1992) provide a good review of anthropological morphometrics. Rapid advances in laser-scanning, digitization, and computing are bringing powerful tools into human osteology.

Table 15.2
Cranial Metrics and Indices

CRANIAL METRICS[a]

Cranium

Maximum cranial length	(g) to (op)
Maximum cranial breadth	(eu) to (eu)
Maximum cranial height	(ba) to (b)
Maximum frontal breadth	(ft) to (ft)

Face

Total facial height	(gn) to (n)
Upper facial height	(ids) to (n)
Bizygomatic breadth	(zy) to (zy)
Nasal aperture height	(ns) to (n)
Nasal aperture breadth	(al) to (al)
Orbital height	(or to rim, perpendicular to orbital breadth)
Orbital breadth	(mf) to (ek)

Palate

Palate length (internal)	(ol) to (sta)
Palate breadth (internal)	(enm) to (enm)

Mandible

Bicondylar breadth	(cdl) to (cdl)
Bigonial breadth	(go) to (go)
Symphyseal height	(gn) to (id)

CRANIAL INDICES

Cranium

Cranial index

$$\frac{\text{cranial breadth} \times 100}{\text{cranial length}}$$

Cranial module

$$\frac{\text{cranial (length + breadth + height)}}{3}$$

Cranial length/height index

$$\frac{\text{cranial height} \times 100}{\text{cranial length}}$$

Cranial breadth/height index

$$\frac{\text{cranial height} \times 100}{\text{cranial breadth}}$$

Face

Total facial index

$$\frac{\text{total facial height} \times 100}{\text{bizygomatic breadth}}$$

Upper facial index

$$\frac{\text{upper facial height} \times 100}{\text{bizygomatic breadth}}$$

Nasal aperture index

$$\frac{\text{nasal aperture breadth} \times 100}{\text{nasal aperture height}}$$

Orbital index

$$\frac{\text{orbital height} \times 100}{\text{orbital breadth}}$$

Palatal index

$$\frac{\text{palate breadth} \times 100}{\text{palate length}}$$

[a]All abbreviations for osteometric points follow the definitions in Chapter 4, pp. 58–64. Buikstra and Ubelaker (1994) recommend that 34 cranial and 44 postcranial standard measurements be taken on intact skeletons.

15.2
Photography

The osteologist employs archival photography to document (record) skeletal material and its context. This medium is also used to communicate information in published form (as illustrations accompanying text) and in presentations (as slides or prints). Maximizing its usefulness in osteology requires an understanding of the medium that goes beyond the abilities of the casual photographer. For an in-depth treatment of science photography, see Morton (1984) and the first edition of this book (White and Folkens, 1991).

Appendix 1 describes the special techniques used to capture the 1:1 images in Chapters 4–13 of this book.

15.2.1 Equipment

For the situations routinely encountered by osteologists, a single lens reflex camera with interchangeable lenses is required.

A second format that has distinct advantages in fieldwork is "instant" (or Polaroid) photography, which provides an immediate record. These prints can be written on and inserted as records in field notebooks. These are useful when reconstructing scenes in the laboratory, often revealing associations not immediately obvious in the field. They also provide insurance against the loss, damage, or failure of the photograph taken by conventional means. Many professionals make instant photography an integral part of their field notes.

A third format gaining wide acceptance is the digital camera, in both still and video form. The new electronic instruments can become useless on a hot day in the desert, with the nearest replacement lithium battery hundreds of kilometers away. Some low-quality electronic cameras also stop functioning in extremely cold climates. Every photographer should always carry spare batteries.

Camera maintenance is extremely important in the field. Dry and windy conditions require diligent protection from blowing sand. Fungus growing on the lens elements frustrates many photographers in tropical environments. An air-tight case for stowing all of the equipment is indicated for both of these conditions. While exposed, a camera must be kept dry and clean. Leaving a camera in the hot sun may cause the lubricants in the lens to seep onto the optical elements, thus blurring the image.

The ideal lens in osteology for recording specimens has a focal length about twice that of "normal." The optical qualities of such lenses provide an extremely flat field of focus and sharp images without the linear distortion found in wide-angle lenses (15–24 mm) or the depth compression of long-focus lenses (180–600 mm). A 105-mm lens with macro focusing capabilities (35-mm format) is ideal. Macro focusing lenses (also called "close-up lenses") have a longer focusing range, which allows the focal point of the lens to extend farther from the film plane, in effect enlarging small subjects. Some macro lenses have a 1:1 capability. With such a lens, for example, a tooth can be recorded life size on the film, thus maximizing detail. Wide-angle lenses from 20 to 35 mm (in 35-mm format; 50-mm in 6 × 6) are useful for getting wide views of sites and excavations in progress, but because of inherent linear distortion they are only marginally useful in recording specimens for later study.

15.2.2 Film and Exposure

Films range from high-contrast, very slow films (ISO 6) to fast, but grainy, surveillance films (ISO 3200+). In general, lower ISO films (formerly ASA or DIN rated) have finer grain, higher contrast, higher resolution, and require longer exposures. Higher speed ISO films tend to exhibit graininess but require shorter exposures (useful in low-light conditions or for rapidly moving subjects).

Transparencies (slides) have a wider contrast range than prints and are recommended for publication in color and for projected presentations. Print film is better if the anticipated application is for mounted prints. In general, slides are used more widely than prints. Color films must be matched to the color temperature of the light source in order to obtain the desired results. Most films are balanced to daylight, which includes electronic flash. Film labeled as Tungsten requires a 3200 K tungsten light source. Photolamps produce a 3400 K light. Incandescent lights produce a distinct orange cast that becomes more prominent as the wattage of the light source decreases. Fluorescent light creates an undesired green cast that requires a compensating filter to remove it.

Despite what the signs at the airport security checkpoints say, x-rays can damage any film. Exposures to airport x-rays are cumulative and affect faster ISO rated films the most. One international trip may expose a film supply a dozen times or more. Industrial strength lead foil pouches are readily available to protect against this hazard. Hand-carrying the film is another means of protection and can also avoid loss. Security officers will usually hand-check film from carry-on luggage if asked, but not always.

Film also requires a certain degree of care. All color films experience a color shift as they age. Films labeled as Professional are manufactured and released from the factory so that the optimum color rendering is achieved if the film is exposed and developed within about two weeks. Regular film has a longer latitude of time before a color shift is noticeable. The expiration date on the film box is a good indicator. However, if film is purchased relatively fresh from the factory and refrigerated, the aging process is halted. Also, a film subjected to extremely hot conditions shifts rapidly. All exposed film should be developed as quickly as possible to avoid any degradation of the latent image.

If the photographer can anticipate the need for duplicates of given situations or subjects, it is more practical to shoot several originals rather than making duplicates later. A complete roll of 36 exposures with processing is less expensive than 6 duplicates, and the quality is immensely superior.

Miscalculation of exposure is a common mistake. The photographer must first determine the amount of light, or exposure value, and then select the best combination of shutter speed and lens aperture (expressed as the f/stop number) that maximizes the desired effect.

A light meter is often used to determine exposure. Most new 35-mm cameras have an internal light meter, which is most useful when the photographer knows the fundamentals behind the meter. By convention, all light meters are calibrated to a value of 18% gray. Light meters work best when the scene contains a relatively equal balance between shadow (dark) and highlight (light) areas. This is not usually the case in osteological photography. The photographer can mitigate exposure problems in several ways. One of the best is to meter off of an 18% gray card, available in major camera stores. These cards can be cut down to a convenient pocket size and easily placed in front of the lens (or hand-held meter) to measure the light reflected to the camera. The card should be evenly illuminated and devoid of shadows caused by the camera or the photographer. Another way to determine the exposure value is the "sunny 16" rule. Simply, "sunny 16" means that on a clear sunny day with an aperture of f/16, the shutter speed will equal the film's ISO rating.

The choice of aperture and shutter speed depends on the purpose of the photograph and the prevailing conditions. Although f/16 at 1/30 sec has the same exposure value as f/2.8 at 1/1000 sec, the optical effects are quite different. Faster shutter speeds (1/250 sec and faster) stop action and minimize the effects of hand-held camera movement. As a rule, a photographer should hand-hold a camera at a shutter speed no lower than the focal length of the lens—55-mm lens at 1/60 sec, for example. Slower shutter speeds require a tripod or other means of camera support. Small apertures (f/22) create a greater depth of field—that is, a deeper area of the picture is in focus. Larger apertures (f/1.4) give a very shallow depth of field. If the photograph must show a large area in focus, as in an exposed bone bed, the photographer should choose a small aperture with a slow shutter speed.

15.2.3 Lighting and Setup

Photography has been described as painting with light. Good lighting is fundamental to good photography. The photographer must carefully control the light that falls on a specimen to the extent possible. In any situation the photographer can add or remove light and adjust its character from soft to harsh. This is achieved with a variety of tools including electronic flash units, reflectors (mirror, metal foil, white card), and diffusers (cheese cloth, translucent plastics, bouncing light off of flat surfaces).

Most photography in the field utilizes the available natural light, but the quality of this light changes with weather and time of day. Photographs taken just after dawn or before sundown show stronger shadows than those taken at midday. This is often desirable to show more detail. Cloudy conditions diffuse natural light and soften shadows, reducing contrast and the perception of detail. Problems of high contrast and dark shadows on bright sunny days can be minimized with the use of reflectors in addition to a fill flash. The basic reflector is a white card or cloth positioned to reflect sunlight into the shadow. The effect can be increased by wrapping a sheet of crinkled aluminum foil around a piece of cardboard to reflect a harsher light.

Artificial light is generally required in the laboratory. The source can be an electronic flash or photolamps that emit a constant luminance. It is easier to control the image and meter under photolamps (quartz halogen or tungsten), but they generate a great amount of heat.

A gray card is highly recommended for light metering to avoid underexposure, a common mistake when photographing light-colored documents or white bones against dark backgrounds. A quick and inexpensive approach to static photo lighting uses a hand-held camera and direct sunlight and produces serviceable results, but be careful about shadows and reflections.

Setting up bones to be photographed can be simple. Place bones on a black velvet background to improve color accuracy and eliminate shadows cast by the specimen. Properly exposed, this setup is ideal for slide presentations. Specimens can be stabilized in a variety of ways. Rubber washers and table leg end caps make good props. Plasticine and other oil-based clays contaminate many specimens, as well as the background material, and should be avoided. Some soft plastic substances are nonstaining and come in very tacky forms that are additionally useful for propping up specimens.

A measuring scale always adds important information. A metric scale is preferred and should be easy to read. Rock hammers, coins, and lens caps are poor substitutes for a proper metric scale. Place the scale in the plane of the specimen and within the visual frame at the edge. Devices that adjust the height and angle of the scale are easy to construct. A pile of books placed out of frame provides an easy solution, with the scale placed between the pages at the correct height (focal plane).

Proper orientation and lighting separates the professional-looking photograph from the amateurish. The long-standing convention in science illustration has the light coming from the upper left-hand corner relative to the viewer. Skull orientation should follow the Frankfurt Horizontal.

15.3
Radiography

Analysis of bones in the living individual is usually accomplished by passing x-rays through the body and exposing a film placed behind the body part. The bone tissue blocks some x-rays, resulting in a negative image on the film called a **radiograph.** Because bones, including internal parts of bones, block some of the rays, the radiograph can be a valuable aid in diagnosing bone condition in medicine.

Radiography is also a valuable tool for the osteologist. Because there is no risk to a dry bone specimen, various exposures and orientations may be made to show the internal architecture of a bone or the developmental status of an unerupted dentition. Osteologists often use fully enclosed x-ray devices within shielded, benchtop cases with external controls (Faxitron or other). The specimen should be oriented so that the x-ray beam passes through the center of the area of interest, and so that this area is perpendicular to the beam and parallel with the film plane. The specimen and film should be as far as practical from the x-ray source and as close to one another as possible. Computer-assisted enhancement of radiographs may be useful after processing (Odwak and Schulting, 1996). The recent development of computed tomography (CT) scanning has added a potent, nondestructive bone-sectioning tool to the osteologist's kit. CT scanners are usually only found in hospital radiography departments, and their use requires collaboration with personnel therein. Spoor *et al.* (1993) discuss applications and problems involved in the derivation of osteometric data from CT scans. With the advent of industrial CT scanning, accuracies of less than a tenth of a millimeter are routine. When possible, however, it is always better to take measurements directly from the specimen because measuring a laser or CT scan greatly increases the chance that matrix, clothing, distortion, or erosion will be overlooked and inaccurate measurements generated. To use any of these techniques in investigating human skeletal material, consult a specialist in radiography. See Ortner and Putschar (1981) and Hillson (1996) for a discussion of radiography in osteological and dental analyses.

15.4
Microscopy

Fine details on a bone's or tooth's surface may be best investigated with a binocular dissecting microscope. Intense, unidirectional light sources can be used to emphasize microscopic detail. For discriminating between various kinds of surface alteration on bones (rootmarks, cutmarks, pathology), the binocular microscope is a most valuable aid. To photograph microscopic structure or trauma on a bone's surface, the scanning electron microscope (SEM) can provide excellent images with great depth of field. For large specimens that do not fit in the microscope's vacuum chamber, or for specimens housed in institutions without SEM facilities, it may be necessary to replicate the object's surface by molding it with dental impression rubbers and pouring an epoxy positive to be used in the analysis. For work on bone surfaces at high magnification, the SEM is the tool of choice (and, unfortunately, expense). Standard histological microscopic techniques are used to study the microscopic structure of bone below the surface or the internal structure of teeth (Hillson, 1996). Schultz (1997) provides two reviews of how microscopy can be applied in human osteological studies.

15.5
Computing

The computer revolution has impacted all areas of science, including human osteology. Satellites communicate with hand-held computers (Global Positioning System) to determine latitude and longitude. Laptop computers connected to electronic distance-measuring devices allow laser-precision in plotting specimens in the field. Desktop computers receive, process, and output our thoughts, our data, and our images. Laser scanners connected to computers allow the external form of objects to be imaged in three dimensions, imported to the computer, and digitally manipulated in many ways (Figure 15.2). Medical computers allow us to peer deep within osseus structures to see formerly hidden evidence of ancient pathology. The exploding global communications network makes it possible to exchange ideas and data rapidly across international frontiers and between field and laboratory. As a result of all these developments, it is impossible to think of working in human osteology without the aid and working knowledge of computer technologies.

Figure 15.2 A three-dimensional laser-scanning unit at the University of California, Berkeley. The machine accurately measures three-dimensional surface topography and enters these data to a computer where measurements, restorations, and other manipulations are possible.

In addition to the basic word processing skills and programs necessary for scholarly communication, all osteologists should learn the basics of database and spreadsheet programs that allow for the rapid and easy manipulation of large osteological data sets. Computers are tools that help the imagination, but they do not substitute for it. Unfortunately, human osteologists have not been immune to the false hope that computers and technology can replace real specimens and real expertise (see Chapter 16). The current hi-tech craze has created some interesting exercises. For example, bones have been imaged by CT and laser scans, input to desktop and mainframe systems, manipulated therein (with attendant beautiful colors and bones floating and rotating in space), and then copied by sculpting the digitized bony form into plastic. Some such exercises are useful to surgeons in customizing prosthetic devices and to investigators assessing fragile remains in matrix (Lynnerup *et al.,* 1997). However, some of this hi-tech wizardry applied to archeological and paleontological specimens leaves the observer to conclude, "That was really awesome, but so what?" Computers can do incredible things, but they are tools that help us investigate, organize, and document. They do not substitute for our own imagination and critical facilities when assessing osteological remains.

15.6
Reporting

After the usually unpublished initial field reports to granting agencies and various governmental regulatory agencies, published reporting of hominid osteological remains from paleontological contexts often occurs in three stages, announcement in a prominent international journal such as *Science* or *Nature*, followed by anatomical description in a more specialized journal such as the *American Journal of Physical Anthropology*, and finally, years later, full monographic treatment. Basic metric, preservational, and contextual data are reported, along with interpretations, in all three publication venues. Chapter 27 illustrates this by means of a case study.

In forensic human osteology, the reporting of skeletal remains usually follows a different series of steps. Here, because of the rigorous procedures adopted by law enforcement agencies and medical examiners (Di Maio and Di Maio, 1989), the osteologist's most important report becomes part of the legal record instead of moving straight toward publication. The format for forensic reporting in the United States and beyond is established by the National Forensic Anthropology Data Bank at the University of Tennessee, Knoxville (Moore-Jansen and Jantz, 1989). The inventory encompasses the entire skeleton, and bones are scored for presence/absence and condition, with notes on things like pathology. Sundick (1984) provides a good review of the osteologist's participation in forensic cases.

In a forensic setting, the pressure for accurate and immediate reporting is sometimes very intense. Osteologists may be forced to conduct their examinations in suboptimal conditions—at morgues, in criminal laboratories, and even in refrigerated trucks or warehouses at the disaster scene. Pressure may come from the sensitivity of the case, from relatives wishing to conduct funerary rites, or from law enforcement agents requiring quick answers to pursue their investigations or hold their suspects. Under these conditions of inadequate facilities (including inadequate comparative materials) and intense pressure, osteologists are more prone to make mistakes. Suffice it to say that there is no tolerance for such mistakes in a forensic context, whatever the conditions. The forensic osteologist should always state what is defensible in a court of law, keep speculation to a minimum, and work closely with others on the multidisciplinary investigation team.

In archeological osteology, the collaborating archeologist and osteologist usually work out a reporting procedure in advance of the excavations and determine what information should be made available in reports or publications. The recent publication of the volume *Standards for Data Collection from Human Skeletal Remains* is a milestone in the standardization of data collection for osteological remains from archeological contexts. This 1994 volume, realized only under the pressure of federal legislation forcing imminent destruction of osteological collections (see Chapter 16), contains a series of chapters and appendices (inventory forms for adult and immature remains) that provide a framework for the observation and recording of osteological attributes. It is an invaluable resource for the osteologist practicing in an archeological context.

The following points are offered as a general guide to reporting on human osteological material. Most osteological reports, particularly in forensic settings, cover the points outlined here.

- **Introduction.** The osteologist should note when and how first contact was made regarding the case. The nature of the materials received or observed should be noted here. Any steps taken by the osteologist to preserve or otherwise alter the material should be outlined.

- **Bones present.** This is simply a listing of what bony remains were analyzed, sometimes with MNI (minimum number of individuals) determinations and their explanations included.

- **Context and condition of the remains.** This is important in forensic work. Note should be made of the context in which the bones were found. Remember that all of the remains that you are given for analysis constitute evidence, often crucial and always irreplaceable. In particular, any cultural or biological remains associated with the bones should be noted. Soft tissue adhering to the bones should be described. Before removal of any soft tissue remains, check with a forensic pathologist about sampling of this material. Any soft tissue present should be radiographed extensively before removal to check for objects within (bullets, clothing, etc.). Never dispose of any associated material without consulting the officials involved in the investigation.

- **Pathology.** Assessment should be limited to the hard tissue. Note any evidence of bony pathology, and leave the soft tissue to other experts. Note healed fractures and other osteological manifestations of disease.

- **Anomalies.** Report anything unusual about the skeletal remains, such as six fingers, and other nonmetric traits. These facts may help in identification. In assessing radiographs, note any features that might help correlate with premortem films and thereby establish identity.

- **Trauma.** Report any sign of osteological trauma, ranging from healed fractures to excavation-related fractures. Try to determine how recent the fractures are by noting evidence of healing, color differences, or rootmark etching on broken surfaces. Express an opinion on whether the bone was fresh when broken (perimortem fracture) or dry (nonvital). Distinguish between pre- and postdepositional trauma when possible (Maples, 1986).

- **Age, sex, race, stature, and weight.** For these, be as specific as possible, but do not give estimates whose precision is not warranted. Give the appropriate limits of confidence in all determinations. Tell what methods were used to make the estimates and why these methods were used.

- **Time and cause of death.** Osteologists are almost never able to make these estimates with certainty. Whereas experienced investigators may speculate on time of death by using odor, grease, tissue, or bone weathering, these attributes all vary according to temperature, humidity, and cover. And how can the osteologist examining a gunshot through the head know that the victim was poisoned before being shot? For these reasons, the osteologist must work closely with a professional forensic pathologist and strictly avoid speculations about death based on bony evidence in isolation. By studying healed lesions the osteologist can sometimes say whether a person lived beyond skeletal trauma, but unhealed lesions often do not, by themselves, indicate the cause of death. The osteologist's legal contribution is usually limited to identification, sometimes including individuation. The skeleton itself gives little evidence relevant to questions about the time and cause of death.

- **Individuation** is sometimes referred to as "personal identification" (Rogers, 1986). This is the determination of the personal identity of the remains. The best hope for individuation, without soft tissue indicators such as fingerprints, is in dental records. For skeletal material lacking dental evidence for individuation, it is often possible to match pathological lesions or premortem photographs or radiographs with postmortem images of the bone. Positive identifications may be based on old fractures or discrete trabecular or sinus patterns (Webster *et al.*, 1986). See Caldwell (1986) for a discussion of techniques used in facial reproduction from a dry skull. Individuation is often important in legal and insurance matters.

- **Metrics and nonmetrics.** Report standard dental, cranial, and postcranial measurements, as well as observations on nonmetric traits.

- **Summary.** Simply summarize the most significant conclusions reached for the sections above.

Osteological findings of general interest are usually reported in a scientific publication that makes the data available to the scientific community as a permanent record. In describing the results of osteological analysis, communication must be unambiguous. The osteologist should specify exactly what materials were analyzed, what procedures were used in the analysis, and what results were achieved. Most scientific publications have basic sections of "Introduction," "Materials and Methods," "Results," "Conclusions," and "Bibliography." Scientific papers on human osteology are frequently found in book or monograph form as well as in journals such as *American Journal of Physical Anthropology* and *Journal of Human Evolution*.

Suggested Further Readings

Buikstra, J. E., and Ubelaker, D. H. (1994) *Standards for Data Collection from Human Skeletal Remains.* Fayetteville, Arkansas: Arkansas Archeological Survey Report No. 44. 206 pp.
> The essential standards volume in North America.

Hillson, S. (1996) *Dental Anthropology.* Cambridge: Cambridge University Press. 373 pp.
> Appendix A provides a good guide to field and laboratory methods used to extract, dissect, replicate, image, section, and preserve dental remains.

Krogman, W. M. and İşcan, M. Y. (1986) *The Human Skeleton in Forensic Medicine* (2nd Edition). Springfield, Illinois: C. C. Thomas. 551 pp.
> An introduction to radiographic analysis is provided in Chapter 12. Chapter 10 is about the restoration of physiognomy.

Maples, W. R. (1986) Trauma analysis by the forensic anthropologist. In: K. J. Reichs (Ed.) *Forensic Osteology: Advances in the Identification of Human Remains.* Springfield, Illinois: C. C. Thomas. pp. 218–228.
> Review of the limitations under which the forensic osteologist operates.

Mead, E. M. and Meeks, S. (1989) Photography of archaeological and paleontological bone specimens. In: R. Bonnichsen and M. H. Sorg, (Eds.)

Bone Modification. Orono, Maine: Center for the Study of the First Americans. pp. 267–281.

> This paper has a special orientation to the photography of bones and is therefore of value to the osteologist.

Morton, R. A. (Ed.) (1984) *Photography for the Scientist* (2nd Edition). London: Academic Press. 542 pp.

> A complete guide to the subject, with many advanced techniques.

Reyment, R. A., Blackith, R. E., and Campbell, N. A. (1984) *Multivariate Morphometrics* (2nd Edition). London: Academic Press. 233 pp.

> A comprehensive introduction to morphometric analysis.

Stewart, T. D. (1979) *Essentials of Forensic Anthropology.* Springfield, Illinois: C. C. Thomas. 300 pp.

> The classic book on the subject, written by America's most experienced forensic osteologist.

Thomas, D. H. (1986) *Refiguring Anthropology: First Principles of Probability and Statistics.* Prospect Heights, Illinois: Waveland Press. 532 pp.

> An engaging, anthropologically oriented introduction to statistics.

CHAPTER **16**

Ethics in Osteology

Ethics is the study of standards of conduct and moral judgment. Ethics can be seen as a system of values that specify a code of conduct. There are multiple value systems, each with definitions of what is considered right and wrong. These definitions are cultural, so what might be ethical for one scientist, for example, might be unethical for a religious leader. Different ethical systems collide across the spectrum of osteological endeavors—from the forensic, to the archeological, to the paleontological.

Professionals who study human skeletal remains are frequently called on to make ethical judgments. In this chapter we examine some of the ethical standards peculiar to osteology and universally practiced by osteologists. For some osteological issues there are no easy answers and no prescribed codes of conduct to guide the practitioner (Figure 16.1).

Human osteologists are routinely called on to practice in the glare of the media spotlight, and within legal, political, social, and economic arenas where science may be misconstrued and misrepresented. Whether the issue is the number of perished individuals in the Branch Davidian compound at Waco (Owsley *et al.*, 1995) or the affinities of an individual who died thousands of years ago at Kennewick (Morell, 1998), the human osteologist is obliged to be guided by facts and the scientific approach rather than by speculation, superstition, economics, preconception, or political expediency.

16.1
Ethics in Forensic Osteology

Human skeletal remains often figure prominently in legal matters. Osteologists are routinely asked to identify skeletal remains—to determine whether or not they are human and, if they are, to determine the age, sex, identity, and antiquity of the remains. Information is provided on how these determinations may be made for bony remains in Chapter 17. Sometimes the osteologist is asked by law-enforcement representatives to make identifications, a report (see Chapter 15) is filed, and the matter ends

Figure 16.1 **A human fetal skull.** Is it right or wrong for this unborn individual's remains to be curated in a museum? Different people give different answers to this question depending on what ethical perspective they bring to the issue.

there. On occasion, however, the osteologist becomes more deeply enmeshed in the legal system.

In many countries law is practiced in an adversarial system in which prosecuting and defense attorneys and their teams square off in a court of law. There are often serious questions of criminality or inheritance involved in cases involving identification of the deceased. The stakes may be high. Osteologists may be retained by either side to offer their expert testimony in such legal matters. In fact, the American Academy of Forensic Sciences has formally incorporated physical anthropology as one of its disciplines.

16.1.1 Boundaries of Evidence

In forensic work it is important for the osteologist to keep two things clearly in mind at all times. First, any analytical conclusion drawn in an osteological report must be defensible. In other words, the osteologist must prepare for a challenge by employing the most sound and up-to-date analytical methods available. Second, the osteologist should always avoid stepping beyond the boundaries set by the osteological evidence itself. In other words, he or she is an expert in osteology and not necessarily an expert in criminalistics, pathology, toxicology, engineering, or detective work.

The forensic osteologist must always report and testify within the bounds of the bony evidence and according to the principles of the scientific discipline that he or she represents. The osteologist should explicitly draw the attention of all concerned parties to limitations of the evidence itself and to uncertainties associated with the identifications that have been performed.

One example suffices to show the tragic toll that can be taken by a failure to observe these basic rules. The Vietnam War and associated conflict in Southeast Asia resulted in the deaths of hundreds of thousands of people, among whom were American military personnel. The Americans who never returned, and whose bodies were not accounted for, were listed as MIA—missing in action. Thirteen men aboard an American C-130 gunship shot down over Laos in 1972 were counted among the MIAs. Over ten years later an excavation at the crash site recovered 50,000 pieces of bone; the largest bone was 13 cm long, and most fragments had a maximum dimension of around 1 cm. After analysis by the U.S. Army, it was announced that positive identifications had been made and that all thirteen men had been accounted for by these bone fragments. The skeletal remains were then forwarded to the families for burial. Relatives of the crew members pressed the issue of identification, and an independent investigation of these bones was made. It became clear that the analysis, although done by professional osteologists, had made conclusions about age, sex, race, and individuation which went far beyond the evidence (Getlin, 1986).

Science, fortunately, is self-correcting in cases like this one in which the evidence can be examined by several investigators and faulty analysis thereby exposed. In the meantime, however, the lives of many people can be deeply affected by the conclusions of the osteologist, who clearly has a responsibility to respect the limits of the hard evidence.

16.1.2 The Expert Witness

Witnesses in court proceedings are sworn to tell the truth, the whole truth, and nothing but the truth. The expert witness in forensic osteology must also, of course, adhere to these rules. See Kogan (1978), Stewart (1979), and Feder (1991) for further details on expert testimony. Testimony by scientist expert witnesses is on the increase in courts of law throughout the world, although different judicial systems handle experts in different ways. For example, in Germany the expert witness is called by the court, whereas in the United States experts are often retained by lawyers on either side of a case. Testimony by scientists may have dramatic impact on the outcome of a judicial proceeding, particularly when the testimony is seen as ethical and articulate and comes from a person expert in his or her discipline. Hollien (1990) observes that several major problems surround scientists serving as expert witnesses. Among the most important are the lack of training of most scientists for the courtroom setting, the great variance in the qualifications of "experts" admitted into the courtroom, and the pressures upon the scientist, both overt and subtle, to adopt an advocacy position for the side paying the bills.

Many of these problems are insignificant to individual osteological expert witnesses who are ethical, well-trained, experienced, and stay within the boundaries imposed by the evidence. Unfortunately, history has shown that this is too often not the case. Courts have different standards for admitting scientific evidence. State and U.S. courts traditionally followed the "Frye rule," named after a landmark 1923 decision. This held that expert testimony must be based on a well-recognized scientific principle or discovery that is "sufficiently established to have gained general acceptance in the particular field in which it belongs." The general acceptance test is only one of several questions that can help the court in evaluating expert testimony. The 1993 Daubert U.S. Supreme Court decision suggests that other factors to consider are whether a method or technique can be (and has been) tested, whether it has been subjected to peer review and publication, and what the error rates are.

The late American physical anthropologist and "expert witness" Dr. Louise Robbins (Hansen, 1993) may serve as an example. Robbins was a self-appointed expert in footprint identification who testified at numerous trials about her abilities to individuate people based on impressions left by their shoes, socks, or bare feet. She claimed that her techniques allowed her to tell whether a person made a particular print by examining any other shoes belonging to that individual. She stated that footprints were better indicators for identifying people than fingerprints. Other expert witnesses testified on her behalf, and against her, during a forensic career that spanned a decade and saw many convictions. Only years later, after her death, were her techniques and conclusions exposed. Physical anthropologist Owen Lovejoy of Kent State University noted, "She may well have believed what she was saying, but the scientific basis of her conclusions was completely fraudulent" (p. 66 in Hansen, 1993).

To her own eyes, and to the eyes of the attorneys who retained her, Dr. Robbins was acting in a professional and ethical manner. In the eyes of other scientists, she was unethical. In the end, her conclusions were shown to be unreliable, but years of litigation were involved, and years of incarceration resulted from convictions aided by her interpretations.

Forensic experts often disagree, and not always because one of them is unethical or untrained. Nordby (1992, p. 1116) asks, "How can we understand the grounds for genuine disagreement between two honest, qualified forensic experts?" He argues that we must distinguish between seeing and observing. We all see, but we observe different things based on the contexts of our knowledge, beliefs, values, and goals. Nordby argues convincingly that it is the role of the expert witness to refine the context of observation based on expert understanding, always examining hidden observational expectations that may influence supplied interpretations. The expert must always be self-critical, and always ready to defend what may turn out to be the only supportable conclusion—the conclusion of "I don't know." The expert supplies good reasons to support that opinion. Nordby (p. 1124) concludes:

Both knowing and not knowing are informed positions reached by careful application of scientifically defensible methods. When the results of those methods do not rationally allow us to prefer one conclusion over an alternative, we must settle for knowing why we do not know.

16.2
Ethics in Archeological Osteology

There is a stark contrast between the widely accepted and easily deline-ated ethical and legal guidelines for the osteologist working in a forensic setting and the ambiguities of ethics and law that haunt the osteologist working in an archeological setting. Evidence, logic, reason, and the sci-entific method are all held in high esteem in the forensic realm. When claims of entitlement enter the legal system, for example, there is rarely an attempt to "balance" scientific and spiritual evidence—the former takes precedence. This has traditionally been the case in the realm of scientific archeology. Now, however, the situation involving bones from archeological contexts throughout the world has become complex, fluid, ambiguous, politicized, and confusing due to the promulgation of laws which aim to redress what are seen as religious injustices undertaken in the name of science. It is necessary to examine the causes for this situa-tion and to consider some of its implications for osteological research involving human remains.

16.2.1 Ethics in Collision

Research in human osteology necessarily involves the study of hard tis-sues that represent people. A variety of living people have objected to osteological excavation and research. Anatomical study of the deceased has been controversial from its very inception. Early anatomists were forced to retrieve and dissect their cadavers in secrecy. Today, as every first-year medical student knows, remains of the dead are vital resources for teaching the living. So it is in osteology.

Death has a high emotional value, as indicated by the fact that both modern and prehistoric humans have developed a wide range of customs and rituals for dealing with it, customs and rituals which change through time. Archeologists and physical anthropologists have learned and con-tinue to learn about past human mortuary practices by excavating skeletal remains. As outlined elsewhere in this book, careful analysis of the bones themselves has led to insights into the diet, living conditions, population structure, genetic relationships, health, and evolution of hominids in both the recent and remote past. The emotive value of death has combined with the information content of human bones to form a combustible mix in the modern world. The debate over human skeletal remains poses eth-ical dilemmas for practicing osteologists and archeologists.

In North America and Australia large populations of indigenous, abo-riginal people met European explorers several hundred years ago. Subse-quent to this contact, both continents saw the decline and sometimes extinction of native peoples and their cultural heritage as European colo-nization proceeded. Native Australian and American survivors of these invasions suffered and continue to suffer great injustices. A callous disre-gard for surviving native people has sometimes been demonstrated by developers, museologists, governments, and anthropologists in the recov-ery and disposition of aboriginal skeletal remains. A double standard has sometimes been applied in the disturbance and subsequent treatment and

Figure 16.2 The Crow Creek massacre bone bed.

disposition of European versus indigenous skeletal remains (McGuire, 1989; Hubert, 1989).

In Israel, ultraorthodox Jews have sought to restrict archaeological research. In New York, African-Americans have insisted that skeletal remains of slaves only be studied by African-American osteologists. Archeological excavation and analysis of skeletal remains have been seriously curtailed and even stopped at the insistence of some native North American and Australian groups, who have used the issue as a forum from which to express their grievances. The question of excavation and post-excavation handling and disposition of osteological remains has rapidly gathered considerable symbolic importance to many people. In some regions research into prehistory has suffered setbacks, with excavations being halted while cultural and skeletal material is reburied. There is no better way to illustrate and explore the ethical issues of excavation, analysis, and reburial of human skeletal remains than to present a case history (for additional views see Meighan, 1992; Klesert and Powell, 1993; Goldstein and Kintigh, 1990; Jones and Harris, 1998; Webb, 1987; and Zimmerman, 1987a and b, 1989, 1997).

Just over six hundred years ago, long before Columbus explored the "New World," a fortified village site at Crow Creek, on what is now the Sioux Indian Reservation in South Dakota, was inhabited by nearly one thousand Native Americans. This prehistoric, probably Arikara, village was attacked and nearly five hundred of the inhabitants killed. Although there is no historic record of this event, human skeletal remains documenting the massacre were discovered eroding out of the site in 1978 (Figure 16.2) (Willey and Emerson, 1993). An agreement was reached between the Sioux Tribe (on whose land the site was located), the U.S. Army

Corps of Engineers, and the project archeologists. Project member L. Zimmerman of the University of South Dakota saw the situation as follows:

Many Native American peoples, through some sort of Pan-Indian re-definition of sacredness, now consider all human skeletal material to be sacred. Arguments raised by archeologists and physical anthropologists that the people represented by certain remains are only distantly related to peoples who show concern—or even not related to them at all—make no difference under the present Native American views of sacredness. As anthropologists, we should be able to appreciate the right of a people to hold such beliefs, and, recognizing that cultures and their belief systems change, to respect the Native American definition of sacredness. . . . we must be concerned with the rights and political and religious viewpoints of the living people, as well as show concern for the people whose remains we are excavating and analyzing. (Zimmerman, 1981:25–26)

Accordingly, analysis of the huge skeletal sample excavated at Crow Creek was limited by the amount of time available before it was placed in gold-painted, concrete coffins and reburied. The alternative to study and reburial was no excavation at all. The project osteologist, P. Willey of the University of Tennessee, summarized the situation as follows:

Only five months were permitted for the analysis, and I and the other physical anthropologists did as complete a job as we could. As the analysis proceeded, however, research questions came up which we could not pursue answers to within the period of time designated for the study. Continuing analyses of the information obtained from our study of the bones have pointed us in even more productive directions, yet the bones were returned to the Sioux Tribal Council as required by contract. . . . Crow Creek is a unique site. The massacre offers an extremely rare opportunity to study the sample of a population at one point in time. Few other collections exist that are so potentially revealing as the one from Crow Creek concerning the diseases a prehistoric people lived with. . . . Crow Creek is also a crucial site for studies of prehistoric stress and biological affinities with other populations.

Indians living today stand to benefit from our conclusions. Additional study of the remains might aid Native Americans further. Examples of the applications of our analyses include determinations of biological relationships which support Indian land claims and understandings of prehistoric disease which could alleviate suffering among present-day Native Americans. If our studies of disease could result in understandings which saved just one child's life, then surely retaining skeletons for complete study is warranted. When we all stand to benefit, the interests of one ethnic group should not be permitted to stand in the way. We must be able to do the most complete and comprehensive study of all human skeletons of all human groups. (Willey, 1981:26)

The issue of excavation, analysis, and reburial stands out clearly at Crow Creek, but it goes far beyond this one occurrence. Some Native Americans have claimed that all archeological research is racist and in violation of the sacred nature of prehistoric sites. Some archeologists have responded by reburying excavated remains, or not excavating at all, as a means of lessening tensions or securing contracts. Several archeologists and physical anthropologists have gone so far as to sign binding legal agreements to rebury any bone material (including nonhuman bone) on the spot without analysis, to bury all photographs and negatives made during the research, and to rebury any artifacts found in the vicinity of the skeletal remains. Skeletal samples have been taken from museum collections and reburied, a phenomenon some physical anthropologists have likened to the destruction of single-copy manuscripts.

The American Committee for Preservation of Archaeological Collections considers the maintenance of archeological collections, including skeletal remains, to be part of the professional and ethical duty of scholars in archeology. This group urges archeologists to abstain from participation in any field project, contract, or other program in which collections obtained from research will be given up for destruction. They recommend that scholarly organizations in archeology treat knowing acts of destruction of archeological materials, or complicity in such acts, as grounds for expulsion from the profession of archeology.

Many physical anthropologists have actively opposed efforts to rebury skeletal material, probably because these scientists are most aware of the potential information in these remains—and are most sensitive to how this information is lost through reburial. Jane Buikstra, a physical anthropologist, has addressed several misconceptions that have characterized the issue (Buikstra, 1981a, 1983). White archeologists have been accused of "never digging up their own ancestors," in effect, practicing a kind of archeological apartheid. As Buikstra notes, the fact that a mostly white set of professional archeologists study a set of mostly aboriginal bones in North America and Australia is not a manifestation of racism. From the excavation of remains of white frontiersmen in Wyoming (Gill *et al.*, 1984; see Chapter 14) to analysis of remains from seventeenth- and eighteenth-century graves of European whalers on Spitsbergen (Maat, 1981, 1987), archeologists and physical anthropologists routinely work on skeletal remains associated with recent Western culture. In fact, the largest and best-studied skeletal series in use by physical anthropologists is composed of mostly white and black individuals from medical school dissections at Case Western Reserve University. Many of these skeletons come from known, named, specific individuals whose religion is often recorded. Furthermore, some of the skeletal sexing and aging techniques most widely used by osteologists were developed, in part, from analysis of the remains of white Americans killed in the Korean War (see Chapter 17). The misconception that continued curation of remains does not help living people has repeatedly been addressed. For example, Ubelaker's (1990) success in positively identifying Native American murder victims from the Pine Ridge Reservation was a direct result of his use of comparative collections of Native Americans.

16.2.2 NAGPRA and Its Effects

In North America, some Native Americans argued during the 1970s and 1980s that because some contemporary Indian people have descendant relationships to some skeletal remains housed in museum collections, the disposition of these remains should be controlled exclusively by modern Native Americans. Numerous state laws addressed the issue, and on November 16, 1990, President Bush approved Public Law 101-601, the Native American Graves Protection and Repatriation Act (NAGPRA), an Act of Congress that directed all museums and laboratories within the United States (which receive federal funding) to inventory all human remains and associated funerary items, to determine which among them can be ancestrally linked to existing federally recognized tribes, to consult with those affected tribes, and to follow the wishes of the tribes regarding those collections. Contrary to popular assumption, NAGPRA does not require reburial but rather gives control over the final disposition of remains to the most appropriate, federally recognized Native American or Hawaiian claimant. Groups are free to choose long-term curation, and the law attempted to set up a process of consultation whereupon this option might be considered, among others.

The NAGPRA legislation came at a time when large numbers of American anthropologists were questioning the rationalist and empiricist roots of their discipline. As Zimmerman (1994, p. 65) notes, "Part of the rift between archeologists and Native Americans stems from a fundamentally different conception of the past. . . . To Native Americans, the idea that discovery is the only way to know the past is absurd." Many anthropologists are sympathetic to the more "spiritually oriented," culturally relativist, anti-scientific view. Indeed, some anthropologists are sympathetic to the view that there is no objective reality at all. The NAGPRA statute is based on something known as "cultural affiliation." The law instructs institutions and potential claimants for "repatriation" to assess "cultural affiliation" by a "preponderance of the evidence," evidence which includes everything from biology to geography to folklore and oral tradition. How are these weighed? Remains may only be returned after such "cultural affiliation" is established. Thus, the NAGPRA legislation has opened institutional skeletal collections up to potential claims by individuals who feel spiritual connections to the remains in question.

The NAGPRA law is administered under the National Park Service of the Department of the Interior. Note that this law does not require institutions to return all skeletal remains to Native Americans, just those which are "culturally affiliated." Eight years after the law was passed, the administering regulations regarding the disposition of "culturally unidentifiable," mostly prehistoric remains, has not been promulgated. Nor has the constitutionality or other aspects of the law been tested in court.

The intent of Congress was to redress documented injustices and provide a means by which aggrieved parties could obtain information about, and custodianship of, ancestral skeletal remains. The law is intentionally vague on how ancestral/descendant status is to be ascertained. Unfortunately, the implementation of the legislation has created bureaucracies that are now determined to extend their own existence by broadening the scope of the law. Even professional osteologists have jumped on this bandwagon, joyously proclaiming that "NAGPRA is forever!" (Rose *et al.*, 1996). It was not the intent of the legislation to create bureaucratic positions at the federal, state, and local levels. Neither was it the intent of the law to pump funds into museums for out-of-work archeologists to conduct further research with collections or to create positions for osteologists. However, these have been among the effects of the law's implementation.

The law required that inventories of all remains be reported in 1995 (within five years of the NAGPRA legislation's passage). That intent was subverted by blanket extensions of the reporting deadline to any institution which asked for one—awarded by the very government agency charged with regulating the law! In its slick, expensive, and politically correct *Common Ground* magazine, this very agency (the National Park Service's Departmental Consulting Archaeologist and Archaeology and Ethnography Program) recently published an article on a new Boy Scout merit badge for archeology. In this example of political correctness run amok, one of the "ethical responsibilities" of Boy Scout counselors under this program is described as follows: The counselor "Avoids all osteological research (in the field and in the lab)" (Skinner *et al.*, 1998).

Human osteologists interested in continuing to curate and study human skeletal remains have been marginalized and demonized within their own intellectual settings and denied access to the very collections they once curated. Curators of human osteological collections have literally been locked out of collections by their own anthropologist colleagues and isolated from the NAGPRA consultation process with Native Americans. Meanwhile, untrained, formerly unemployed archeologists pad their pockets with federal and institutional money under the guise of conducting inventories of remains required by NAGPRA and under bogus extensions to the NAGPRA deadline granted by the National Park Service. No wonder this unforseen and unfortunate turn of events has resulted in the widespread abandonment of skeletal biology by museums and other institutions of higher learning and a parallel exodus of researchers and students from human osteology. One unintended result has been the shift to osteological analyses in more recent cemeteries (Grauer, 1995; Saunders and Herring, 1995). And all this has come with virtually no legal challenges. With no one to regulate regulators bent on reburial, only a few institutions have had the courage to challenge the law in court.

Like any law, NAGPRA will be tested in the judicial system as claimants and institutions disagree over the ultimate disposition of remains. Unfortunately, for reasons of economics and politics, many museums and other institutions are following the National Park Service's lead in extending NAGPRA beyond the intent of Congress. One of the first legal challenges to such broad interpretations of NAGPRA has been mounted in the case of a 9300-year-old skeleton found along the Columbia River. The Kennewick skeleton is of such antiquity that a direct relationship with the modern tribe claiming it under NAGPRA, the Umatilla Tribe, is unlikely. The specimen is said to display a host of non-Native American anatomical features. Yet the Umatilla tribe has adopted the view expressed by the tribal chair of the Pyramid Lake Paiute: "It's a fundamental problem with science. Scientists . . . need to accept what we know: that we've always been here. They don't need to look at any skeletons to determine this." (Morell, 1998, p. 192). Umatilla religious leader Armand Minthorn, a Native American Fundamentalist, states "From our oral histories we know that our people have been part of this land since the beginning of time" (Preston, 1997, p. 74). It is clear that this belief is an explicitly creationist viewpoint, one completely unsupported by any available information outside of the religious or spiritual realm. Yet, under federal statute, the regulatory agency and institutions are instructed to take it seriously in their assessments of "cultural affiliation."

16.2.3 The Future

A common misconception about analysis of skeletal remains from archeological contexts has to do with what constitutes adequate analysis. As Buikstra notes, the notion that continued curation of human skeletal collections is unimportant in scientific studies is simply false. She illustrates this by noting the amount of information that would have been lost if collections made before 1952 had been reburied after one year of analysis. Questions concerning demography and disease were not answerable or remained unasked just a few decades ago. Radiographic technology was inadequate to fully analyze bones. To rebury skeletal remains is to assume that no more questions will be asked and that no further developments in analytical techniques or instruments will occur. These are poor assumptions. A recent review by physical anthropologist C. Turner is worth citing as a summary of the foregoing considerations:

Scientific information about past peoples and their lifeways will be lost with reburial of human skeletons. This is because even a single skeletal series has more kinds of information than one worker can reliably extract, and because new techniques for skeletal research are constantly being devised. I explicitly assume that no living culture, religion, interest group, or biological population has any moral or legal right to the exclusive use or regulation of ancient human skeletons since all humans are members of a single species, and ancient skeletons are the remnants of unduplicable evolutionary events which all living and future peoples have the right to know about and understand. (Turner, 1986:1)

For science to be self-correcting, the scientific databases, whether they are composed of one discovery like Piltdown or a large series like that from Crow Creek, must continue to be available to the scientific community. See Chapters 15 and 18 for a discussion of how inter- and intraobserver error in osteology can influence results of any study. To rebury skeletons is to bury whatever future information they may yield and to deny future researchers the possibility of assessing the work of their predecessors. In short, it is to deny future generations the ability to know their past. Meighan puts it this way: "Reburying bones and artifacts is the equivalent of the historian burning documents after he has studied them" (1994, p. 68).

The imminent threat of reburial has led many osteologists to abandon the idea of keeping original specimens for posterity and instead to turn to alternatives. The compilation of the "standards" volume (see section 15.6) represents such a desperation move. Some have rejoiced in the thought that the inventory and repatriation process has "increased the number of skeletons studied from about 30% to nearly 100%" (Rose *et al.*, 1996), but both the numbers and the sentiment are poorly based. Even the "books in a library" analogy breaks down with original osteological specimens. These remains defy accurate and adequate copying. The bones comprise not just external morphology but internal form and chemical composition as well. No cast, no image, no measurement, no description can adequately record the information potential held by an original osteological specimen, and to suggest that this is not the case is to take false comfort in the face of impending destruction. One thing that sets science apart from other areas of human endeavor is the character of self-correction. As new techniques and new observers allow evidence to be examined in new ways, old errors can be corrected and truth better approximated. Reproducibility of observation is an essential ingredient of science. This, of course, is lost when skeletal remains are destroyed by reburial. Long-term curation of skeletal remains has repeatedly been demonstrated to be essential in forensic, archeological, and paleontological investigations. The very understanding of human diversity rests on this continued curation (Tobias, 1991).

Museums are the institutions most often entrusted to act as repositories for human skeletal remains. Yet many museums have deaccessioned skeletal collections in the face of vocal activists demanding reburial, arguably violating the public trust in the process (after all, among the missions of a museum is the preservation of its collections for posterity). Many reburials have been precipitous, undertaken even before legislation was enacted or tested in court. Reburial is often seen on both institutional and personal levels as an expedient, politically safe, cheap, and therefore an easy way for the public and politicians to assuage imagined guilt and for institutional administrators to escape "negative publicity." But the costs to knowledge of politically correct reburials are high, and the effects are permanent.

Zimmerman (1997, p. 105) has argued, "Quite simply, anthropologists must learn to share control over the past." It is an interesting perspective which holds that the past can be controlled at all! Most of us reckon that the past happened, that there is some evidence of it having happened, and that we should do our best to accurately interpret that evidence and thereby understand what happened. Zimmerman's adoption of the cultural relativist position (or ethic) is widely shared among practicing anthropologists and reflects the ethical dilemmas facing anthropologists of all subdisciplines (Meighan, 1992). How can scientists accommodate explicitly religious viewpoints at odds with the evidence they study? Meighan argues that the destruction of skeletal collections by reburial represents a conflict between religion and science, and when scientists "compromise" in this conflict, by definition, they abandon science itself, with all of the rights and duties inherent to this system of knowing.

Skeletons in museum collections represent only the remains of the few individuals who, largely by chance of burial and discovery, have managed to elude the ravages of time and to open windows on the distant past (Figure 16.3). These bones have the potential to inform all people about the past. When reburied, they will join the vast majority of ancestors missing without a trace. Who looks out for the unborn great-great-grandchildren of contemporary Native Americans or Australians? Who shall tell the generations to come that during the late 20th century, some of their relatives decided to deprive them of the only scientific means of knowing the past? The Iroquois had a tradition that would guide wise people through troubling times, if wise people were prepared to listen. They considered the impact of today's decision on the seventh future generation. History convinces us that the reburial of archeological skeletal collections precipitated by activists, politicians, and museums during the latter decades of the 20th century will be condemned by unborn generations of all people, before seven generations have passed.

Given the issues involved, what steps can the practicing osteologist or archeologist take to ensure that osteological science does not fall victim to political or religious agendas? First and foremost, all skeletal remains should be treated with respect and dignity in their excavation, analysis, and curation. Second, potential descendants should be fully informed, as a common courtesy, about any steps involving human remains being contemplated or undertaken. As a general rule, control of osteological material should be invested in parties who can show direct lineal affinity to the remains in question. In some cases this descent is difficult or impossible to determine accurately. When excavation, study, and curation can be shown to violate traditional values and beliefs of living descendants at the tribal or family level, these values and beliefs should be honored. Most often this occurs in historical archeological contexts. Prehistoric remains are usually more ambiguous in their relationship to modern groups, and the older the specimen, the less likely it is that direct, exclusive descent

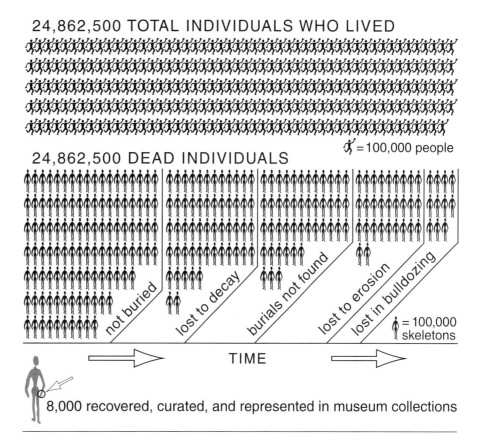

24,862,500 TOTAL INDIVIDUALS WHO LIVED

💃 = 100,000 people

24,862,500 DEAD INDIVIDUALS

not buried

lost to decay

burials not found

lost to erosion

lost in bulldozing

👤 = 100,000 skeletons

TIME

8,000 recovered, curated, and represented in museum collections

Figure 16.3 **The ravages of time.** It is often forgotten that the skeletons comprising modern museum collections represent only tiny fractions of once-living populations. These skeletons, ambassadors from the past, represent those few who have miraculously escaped the ravages of erosion, burial, decay, and modern development to reach us and inform us about prehistory. For example, at the time of this writing, Berkeley's Hearst Museum houses around 8000 mostly partial prehistoric skeletons representing some 9000 years of human occupation of the state of California. This is the world's most important collection of human skeletal remains documenting populations of people who subsisted through hunting and gathering rather than agriculture. These archeologically derived skeletal remains of early Californians have yielded, and continue to yield, an amazing record of human accomplishment, diversification, and adaptation.

This book makes it clear that a great deal of information would be lost if these remains were to be reburied. But what fraction of Native Californians does the Hearst collection really represent? The best estimate for the California Native American population at European contact is 221,000 people (Ubelaker, 1992a). If we make the most conservative estimates on length of occupation (9000 years), average lifespan (40 years; average life expectancy did not reach 40 years in most world populations until the 19th century), and the most simple, linear estimates of population size increase, a total of about 25 million people died in prehistoric California over 9000 years. Therefore, even the largest museum skeletal collection in the state comprises a mere three one-hundredths of one percent (.00032) of the total number, 24,862,500, of Native Californians deceased since 7000 B.C. This tiny, miraculously preserved set of skeletal remains is all that remains to inform us about all the other individuals whose remains were lost to the ravages of time. This precious heritage is what lies in the balance of current legal and ethical debates regarding reburial.

of anyone is demonstrable. Even when it is, however, all parties should be informed of the important information that may become available as a result of analysis of the skeletal remains.

The debate concerning excavation, study, and reburial of human skeletal remains will continue to be heard in the courts. We hope that the ethical concerns of all the various sides to the debate are reflected in the laws that will continue to be written and tested in such settings. Communication, however, is most effective outside the legal system. Professional archeologists and osteologists bear a great responsibility in education. All groups involved in the debate should be engaged in an ongoing dialogue, people talking with each other instead of past each other. These issues will only be defused through public education and through the long-overdue graduate-level education of Native Americans and Australians in physical anthropology.

Any group of people, from the smallest family to the human race, is best equipped to deal with its present and future when it most fully understands and appreciates its heritage. Prehistoric research, including osteological study, is one way that heritage can be fully revealed. The great tragedy in the debate over skeletal remains from archeological contexts is that the issue has sometimes pitted archeologists against descendants and relatives of the people they wish to study. This has occurred at a time when the very archeological resources in North America and Australia which both parties seek to preserve are disappearing at an unprecedented rate at the hands of the developer and the looter (Figure 16.4). In the face of this catastrophic and irreversible decimation of the past's only record, the reburial issue is a costly diversion for all parties. The scientific community and native groups need to redirect their energies in a concerted effort to save and protect the heritage of the past before it disappears.

16.3
Ethics in Human Paleontology

As a general rule, time's eraser is more and more effective the longer it has to operate. As a result, there is a tendency for the fossilized remains of humans and human ancestors to be rare compared to the skeletons of people who have died during the last several thousand years. There are few places on earth where conditions have been conducive to the deep-time preservation of hominid skeletal remains. Such places must have afforded protection from the nearly ubiquitous presence of continual erosion. Hence, sediments accumulated in protected places such as caves or lake basins are the usual discovery sites of fossil hominids.

Ever since fossil hominids were found in the late 1800s, they have received an inordinate amount of scientific attention and public curiosity. In some ways, it was almost as if these fossilized remains of distant relatives became icons. A good deal of nationalistic fervor was devoted to the recovery of fossils in various parts of the world, and it has been argued that Piltdown's status as "the earliest Englishman" played a role in blinding the leading scientists of the day to the obvious fraud that the "specimen" represented. The Taung cranium is a cultural and national icon in South Africa, the Neanderthal skeleton is prominently displayed in

Figure 16.4 Vandalism of osteological remains in an archeological context. Evidence of looting is seen in these two photographs (*above* and *next page*) taken at the site of Nuvakwewtaqa, a large (1000+ room) pueblo in central Arizona occupied between A.D. 1280 and 1425. Vandals have haphazardly discarded the human skeletal remains encountered in their search for grave goods, damaging the bones and forever losing their context. The nonrenewable archeological record is rapidly disappearing due to such plunder. Photos courtesy of Peter Pilles and the Coconino National Forest.

Germany, and Ethiopia is known throughout the world as the home of "Lucy." Today, hominid fossils are viewed as important parts of the cultural heritage of many developing countries.

In many parts of the world where the most important hominid fossils were found, the academic and scientific infrastructures remain little developed. As a result, fossils have often been exported to the nations that had "colonized" these lands, to be studied by foreign experts, and often displayed and curated indefinitely in foreign museums and universities.

Figure 16.4 (*Continued*)

The demise of colonial rule across Africa and Asia ended this situation, but attention to the development of facilities and personnel to support ongoing human paleontological research in these areas was slow in coming. Today, particularly in Africa, a first generation of scholars is taking important steps toward developing the necessary platform from which to conduct world-class research into human paleontology. There are continued instances of exploitative relationships and hit-and-run fossil hunting by foreigners from the developed world, but there are also several models of international collaboration between local and foreign scholars, in both Africa and Asia.

It is essential for osteologists interested in conducting laboratory and field research in foreign countries to make early and open contact with the governmental administrators and local scholars in any country in which they intend to work. Research must go hand in hand with development in these situations, ensuring meaningful, uninterrupted progress and productive science.

Suggested Further Readings

Grauer, A. L. (1995) *Bodies of Evidence: Reconstructing History Through Skeletal Analysis.* New York: Wiley-Liss. 247 pp.
> An edited volume concerned with the analysis of skeletal remains from historic cemeteries.

Jones, D. G., and Harris, R. J. (1998) Archeological human remains. *Current Anthropology* 39:253–264.
> A global view of the issues surrounding reburial, by anatomists.

Krogman, W. M., and İşcan, M. Y. (1986) *The Human Skeleton in Forensic Medicine* (2nd Edition). Springfield, Illinois: C. C. Thomas. 551 pp.
> A comprehensive look at forensic applications in osteology.

Layton, R. (Ed.) (1989) *Conflict in the Archaeology of Living Traditions.* London: Unwin Hyman. 243 pp.
> An edited volume with a wide range but a skewed collection of contributions that address ethics as it relates to archeology, particularly the reburial issue.

Stewart, T. D. (1979) *Essentials of Forensic Anthropology.* Springfield, Illinois: C. C. Thomas. 300 pp.
> Advice to the osteologist operating in the forensic arena.

Ubelaker, D. H., and Grant, L. G. (1989) Human skeletal remains: Preservation or reburial. *Yearbook of Physical Anthropology* 32:260–287.
> A comprehensive review of the issue from the perspective of human osteologists at the Smithsonian. Scientific, legal, political, and ethical issues are discussed.

Vitelli, K. D. (Ed.) (1996) *Archaeological Ethics.* Walnut Creek, California: Altamira Press. 272 pp.
> A compilation of articles from *Archaeology* magazine, covering looting, reburial, and professional behavior, with an appendix made up of statements on ethics from professional organizations.

CHAPTER 17

Assessment of Age, Sex, Stature, Ancestry, and Identity

When osteological remains are recovered in forensic and archeological situations, the osteologist is often called on to make more than taxonomic identifications. Human skeletal remains often reach the osteologist without any documentation about their individual age, sex, stature, or racial affinity. The bulk of the literature on human osteology is composed of hundreds of books and articles describing the development of methods to allow accurate and precise identification of individual traits in skeletal remains. This research continues today, even after a century of intensive study. All of this research and publication has been driven by the need for basic biological information about skeletal material from forensic and archeological contexts. In bioarcheology, individual attributes of a skeletal individual become the fundamental components of work in the investigation of mortuary practices (Chapter 14), paleopathology (Chapter 18), and paleodemography (Chapter 20). In forensic osteology, these individual biological attributes are important in narrowing the field of investigation to certain subsets of people and in establishing the individual identity of the remains themselves. This chapter is an introduction to the progress that has been made in assessing the age, sex, stature, ancestry ("racial," or geographic, affinity), and individuation of human skeletal remains.

Our focus on the human skeletal elements in Chapters 4–13 was aimed at recognition, providing a guide to diagnostic aspects of human bones. Size and shape characteristics usually allow for the unambiguous sorting of human from nonhuman bone, even in very fragmentary material. Although the determination of sex from skeletal remains appears to be an analogous either/or decision, only a few skeletal characters allow the osteologist to make this choice. Furthermore, the other characteristics discussed in this chapter—individual age, stature, and ancestry—do not lend themselves to such easy and simple divisions as human/nonhuman or male/female. Rather, they grade continuously from prenatal to elderly, from short to tall, and from one geographic group to another. For this reason, it is often best to think of our assessments of these characteristics as estimations rather than determinations. Chapter 21 reviews recent work in molecular osteology that also bears on these issues.

17.1
Accuracy and Precision of Estimation

Accuracy is the degree to which an estimate conforms to reality. **Precision** is the degree of refinement with which an estimate is made. We might be accurate in aging a mandible as subadult, for example, but we would still be imprecise. How accurately and precisely can the osteologist estimate the age, sex, stature, and affinity of human skeletal remains? There is no simple, standard answer to such a question. Any identification of a biological quality such as age, sex, stature, or ancestry is, in effect, a probability statement. The likelihood that a given identification is accurate depends on a number of different factors, which are worth general consideration before we turn to the analytical methods themselves.

- Accuracy and precision of identification depend on **age categories.** Because the growing human undergoes a progressive development of the bones and teeth, younger individuals can in general be aged more precisely than older individuals. For example, the ends of the limb bones form and fuse at known ages. Tooth formation and eruption is well documented although somewhat variable. After these growth processes end at maturity, there is little continuing skeletal change to monitor. Subsequent changes in the adult skeleton are often degenerative and task- or health-specific and therefore not so well correlated with elapsed time. Although precise skeletal aging becomes more difficult with adults, establishing the sex of an individual becomes easier. This is because sexual characteristics of the skeleton often develop only when sexual maturity is attained. Most of the criteria established for deducing ancestry are only useful in comparisons between adults. Krogman and İşcan (1986) provide a more detailed overview of these concerns in their text on forensic osteology.

- Accuracy and precision of identification depend on **available skeletal elements.** Different elements have different developmental stages. Some, such as tooth eruption, correspond more closely to chronological age than do others, such as cranial suture closure. Some skeletal elements, such as the pubis of the os coxae, display sexually diagnostic characters, and others do not. Some elements, such as the femur, show high correlations with stature, whereas other elements do not. Some bones of the cranium are useful in discriminating between modern human groups, and others are not.

- Accuracy and precision of identification depend on **sample composition.** Accuracy of identification diminishes when the osteologist is forced to identify isolated individuals by means of age and sex standards derived from other populations. The most accurate and precise sexing and aging are obtained when it is possible to arrange many skeletal specimens in a series (to **seriate**) and to compare within a single biological population. For estimating sex, age, or racial affinity, it is always a great advantage to work with populations of skeletons rather than with isolated finds. This is sometimes the case in archeological settings, but forensic settings rarely provide the opportunity to work with large unknown samples.

- Accuracy and precision of identification depend on **analytical methods.** Different methods yield determinations of sex and age which have different reliability. For example, sexing a pubis with the Phenice technique (see section 17.4.2) is highly reliable, whereas use of the sciatic notch is far less reliable.

- Accuracy and precision of identification depend on the **applicability of the analytical method to the unknown individual or sample.** Most standards used for sexing and aging skeletal remains have been established on the basis of European and American skeletal series. These standards have not been shown to apply equally to human populations in other parts of the world or from prehistoric contexts (see Mensforth and Lovejoy, 1985; Ubelaker, 1987; King *et al.*, 1998). Not only is there variation within single populations in the rate of skeletal maturation, but there is also significant variation between populations (Lampl and Johnson, 1996). This factor is significant due to the limited number of populations on which the currently used methods have been based (see section 17.2).

- Accuracy and precision of identification depend on **research context.** The degree of accuracy needed in a particular analysis depends on the questions being asked and the problems being investigated. If the problem involves merely sorting subadult mandibles from adult mandibles, accuracy should be 100%. On the other hand, if the investigation necessitates separation of 35-year-old from 36-year-old mandibles, no known method will be accurate.

17.2
From Known to Unknown

The very issue of aging and sexing skeletal remains and of determining their stature and affinity implies that these biological qualities are *unknown* for the specimens under analysis. To solve for such unknowns, the osteologist must proceed by comparing the unknown skeletal elements with a standard series of skeletal individuals whose age and sex are *known.* Where do such series exist? Not in many places—certainly not in archeological cemeteries lacking written records. Age, sex, stature, and affinity of skeletons from such contexts must therefore be treated as unknown qualities. For living individuals, variables are more often known. For this reason, radiographic study of modern human development has proven important in establishing aging standards for use by osteologists. Unfortunately, many surface features of bones in the living human are not visible by radiography.

In attempting to solve for unknown biological qualities of skeletal material, osteologists have made intensive use of five major skeletal collections in which there are more or less adequate records of age, sex, stature, and affinity. The Hamann-Todd collection was accumulated between 1912 and 1938 at Case Western Reserve University's Department of Anatomy. Here, 3592 human individuals from low socio-economic status were collected from area hospitals. According to a recent study of the Todd material and its records, only about 16% of the individuals in this collection have sufficiently reliable ages at death to be used in skeletal aging studies (Lovejoy and colleagues, 1985). A second sample that has seen wide use in estimating age for bony remains is a set of skeletons of American military personnel killed in the Korean War (McKern and Stewart, 1957). Unfortunately, this set consisted primarily of male individuals with a limited age distribution. A third source is the Terry collection, some 1600 U.S. white and black adult skeletons from Washington University's Anatomy Department (1920–1965), now housed at the National Museum of Natural History, Smithsonian Institution, Washington, D.C. The fourth collection that has begun to produce data helpful in analysis of skeletal remains is composed of skeletal parts gathered from autopsied individuals in Los Angeles County (Suchey et al., 1986). The use of coroner-based samples to test and develop aging methods is increasing (Pfau and Sciulli, 1994). The fifth collection is the Cobb collection at Howard University, composed of about 600 skeletons of blacks from the Washington, D.C. area.

Standards and methods developed from these collections all suffer some limitations. Except for some infants, most of the skeletons are from adults over 25 years of age. Racial categories are, for the most part, "black" and "white," which are legal and social terms based on local custom rather than biological ancestry (see Chapter 20). Admixture is unaccounted for. These mostly dissection room populations are often of below average socioeconomic status. Many times the ages of death recorded in the collection records are only estimates made by the coroner.

To summarize, the accuracy and precision of an osteologist's attributions of an age, sex, or stature to a skeleton for which these variables are unknown always depend on standards derived from a series of skeletons originally accompanied by independent records of these biological attributes. There are significant problems involved with controlling biological attributes of archeologically derived skeletal material with these and other collections of modern human bones. Most important among them is the fact that none of the series contains people who lived under aboriginal, nonwestern subsistence conditions. The effects of different lifestyles on individuals that make up skeletal series can be dramatic. For example, the rate and degree of tooth wear is higher in aboriginal populations, as is the amount of muscular stress and osteological reaction to that stress.

17.3
Estimation of Age

Individual age determination in skeletal remains involves estimating the individual's age at the time of death rather than the amount of time that has elapsed since death. Chapter 19 is a guide to the latter determination; this section deals with the former. Ubelaker (1989, p. 63) succinctly encapsulates the procedures and problems inherent in aging skeletal remains:

Estimation of age at death involves observing morphological features in the skeletal remains, comparing the information with changes recorded for recent populations of known age, and then estimating any sources of variability likely to exist between the prehistoric and the recent population furnishing the documented data. This third step is seldom recognized or discussed in osteological studies, but it represents a significant element.

On the other hand, many of the attributes that are used to determine age do not seem to be environmentally plastic. The degree to which age standards derived from modern osteological collections may be applied to prehistoric populations is a matter of continuing debate, but available studies indicate that individual variation often swamps populational differences.

Over the course of a lifetime, elements of the skeleton undergo sequential chronological change. In infancy these changes mostly involve the appearance of various skeletal elements. During childhood and adolescence, bones and teeth continue to appear, and epiphyses form and fuse. Even after age 20, bones continue to fuse, metamorphose, and degenerate. This progression forms the foundation for studies of skeletal aging. However, it is important to note that even normal development of the infant is discontinuous and saltatory (Lampl *et al.*, 1992) and that there is substantial variation among different individuals in the rate and timing of developmental changes.

Sex identification in skeletal remains is dichotomous, but determination of an individual's age at death is more complex because it involves dividing the continuum of growth. Individuals of the same chronological age can show different degrees of development. This is true for anatomy of the skeleton as well as for behavior. Thus, even when osteological standards based on known samples are perfect, there is always a degree of imprecision in aging skeletal remains. What is the magnitude of this imprecision?

As noted above, whether dealing with cranial bones, teeth, or postcrania, an already established "system" is used for osteological aging; criteria for aging are identified based on a population whose individuals have known ages. It is possible, for example, to use radiographs of people in living populations to establish that human permanent molars erupt at about 6, 12, and 18 years of age. This control series may then be used to age each individual in an unknown skeletal series, under the assumption that dental eruption followed the same periodicity in both groups.

One drawback in such an approach to the unknown skeletal series is that age assignments are made on an individual basis, without reference to other individuals in the unknown series. Such assignments place individuals into an **age class,** for instance, of 12–18 years. This aging is not as precise as assigning an absolute age, for example, of 15 years. In other words, dividing a continuum leads to imprecision. As Lovejoy and colleagues (1997) note, there are two major sources of error in any estimate of age-at-death. These are the inherent variation within the biological process of aging itself and the investigator's skill in estimating the biological age of the unknown specimen.

An approach that should always be taken with a large sample of unknown individuals to help overcome these imprecisions is **seriation.** Prior to estimating each unknown individual's age, all of the individuals represented by the skeletal element under analysis in the unknown series are arranged in a sequence of increasing age. This approach has many benefits (Lovejoy and colleagues, 1985): seriation may be done quickly, with little fatigue; there is no observer error due to time-shift effects (for example, having to stop in the middle of the analysis because the work day ends); there is constant monitoring of results, with ability to correct observer errors; there is no loss of accuracy as a result of pooling individuals into age categories. Once the sample under analysis is seriated, at least the individuals have been aged *relative to each other.*

Seven age classes commonly used to segregate human osteological remains are as follows: **fetal** (before birth), **infant** (0–3 years), **child**

(3–12 years), **adolescent** (12–20 years), **young adult** (20–35 years), **middle adult** (35–50 years), and **old adult** (50+ years) (Buikstra and Ubelaker, 1994).

17.3.1 Estimating Subadult Age from the Dentition

Eruption and wear of the teeth have been used extensively in aging the human skeleton. Tooth development is more closely associated with chronological age than is development of most other skeletal parts, and it seems to be under tighter genetic control. Because of the regular formation and eruption times for teeth, and because these elements are the remains most commonly found in forensic, archeological, and paleontological contexts, dental development is the most widely used technique for aging subadult remains. Smith (1991) provides a review of the various techniques available. It is important to note that "regular" does not mean "constant." For example, some infants erupt their teeth earlier in their lives, and different individuals erupt their teeth in different orders (Smith and Garn, 1987). It has been shown that some American blacks and whites differ in both rate and sexual dimorphism of tooth mineralization (Harris and McKee, 1990).

Tooth formation begins in the embryo a mere 14–16 weeks after conception (Hillson, 1996). There are four distinct periods of emergence of the human dentition. First, most deciduous teeth emerge during the second year of life. The two permanent incisors and the first permanent molar usually emerge between 6 and 8 years. Most permanent canines, premolars, and second molars emerge between 10 and 12 years. Finally, the third molar emerges around 18 years. Of course, there is interindividual variation in all tooth development and eruption. Age may be determined from developing teeth in several ways. One means is by comparing the unknown individual with a chart or atlas showing the mean stage of development of the entire dentition (Figures 17.1 and 17.2). Another is through comparison of the stages of formation with each individual tooth (Table 17.1). Liversidge (1994) discusses the pros and cons of these methods and recommends the atlas approach for both accuracy and ease of use. Hillson (1996) provides an excellent review of all methods of aging the skeleton through use of dental development. The third molar is the most variable tooth in formation and eruption. Mincer *et al.* (1993) provide data on this tooth's formation and its use in age estimation.

Ubelaker (1989) provides a graphic summary of data on dental development in Native Americans, which we reproduce in Figure 17.1. Note the possible ranges associated with each stage in the diagram. Figure 17.2 shows the source of these errors. Sex-based variation in development and eruption of teeth is most apparent at the canine position, and this tooth should be afforded less attention when aging erupting dentitions. When assessing the age of a subadult individual based on dentition, note all aspects of development, including the completeness of all crowns and roots (**formation**) and the place of each tooth relative to the alveolar margin (**eruption**). When using published standards, be sure to discriminate between emergence through the alveolar margin (bone) or through the gum (soft tissue). Also note that dental development is sensitive to sex and population differences (İşcan, 1988). For more details on dental development and eruption, see Trodden (1982) and Smith (1991). Table 17.1 is a

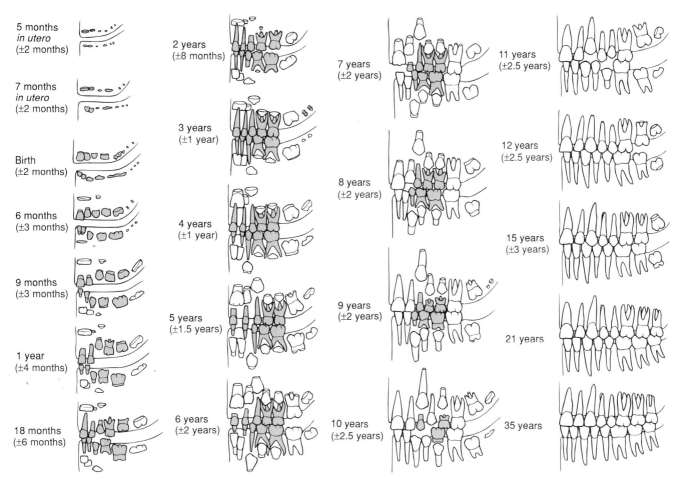

Figure 17.1 Dental development in Native Americans (adapted from Ubelaker, 1989; note that data on the deciduous teeth come from non-Native Americans).

compilation of useful information in age assessment based on the developing dentition.

17.3.2 Estimating Adult Age from the Dentition

Once a permanent tooth erupts, it begins to wear. Rate and patterns of wear are governed by tooth developmental sequences, tooth morphology, tooth size, internal crown structure, tooth angulation, nondietary tooth use, the biomechanics of chewing, and diet (McKee and Molnar, 1988; Walker *et al.*, 1991). If the rate of wear within a population is fairly homogeneous, it follows that the extent of wear is a function of age. This fact can be used in assigning dental ages to adult specimens. Where this has been tested on modern populations, the correlations between known age and tooth wear have been shown to be good (Lovejoy *et al.*, 1985; Richards and Miller, 1991). However, the osteologist should always be on the lookout for cases of accelerated wear due to pathology or use of the teeth as tools (Milner and Larsen, 1991).

The first step in assessing age by dentition is the application of a seriation of all dentitions based on development and wear. Miles (1963) was

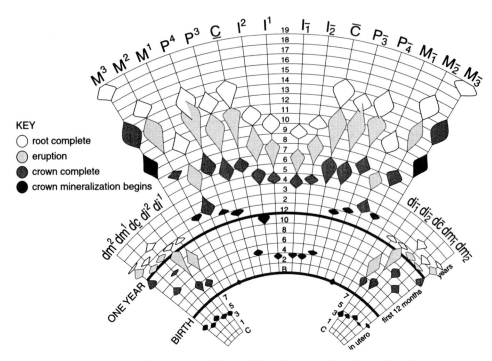

Figure 17.2 Variation in the timing of dental development, based on Gustafson and Koch (1974), with third molar data from Anderson *et al.* (1976). Range values are plus-or-minus one standard deviation for the third molars. Patterns: black, crown mineralization begins; dark gray, crown completion; light gray, eruption; white, root completion.

the first to establish a scale of attrition based on development. The basics of the technique are as utilized in the following example: A first molar accumulates about 6 years of wear before the second molar of the same individual erupts (assuming eruption at 6 and 12 years, respectively). When a similar amount of wear (6 years' worth) is found on a third molar of another individual (a molar assumed to have erupted at age 18), the age of that individual can be estimated as 18 + 6 = 24 years. Miles uses 6.0, 6.5, and 7.0 years between successive molar eruption. A variety of techniques have been applied to the quantification of tooth wear (Molnar, 1971; Scott, 1979; Brothwell, 1989; Walker *et al.*, 1991; Dreier, 1994; Mayhall and Kageyamu, 1997).

Lovejoy (1985) has concluded, for the prehistoric Libben skeletal population, a large human osteological series from the midwestern United States (see Chapter 20), that dental wear assessed by seriation procedures is an important and reliable indicator of adult age at death. He found, on the populational level, that dental wear was very regular in form and rate. As Lovejoy notes, the assessment of a single individual in a forensic setting based on dental wear allows only a gross approximation of age, but if an entire biological population is seriated, tooth wear can yield precise results. In fact, Lovejoy and colleagues (1985) concluded that dental wear is the best single indicator for determining age of death in skeletal populations. They found dental wear as an age indicator to be accurate and consistently without bias. Figure 17.3 illustrates the wear standards used by these workers.

Hillson (1986) discusses methods useful in assessing individual age based on microscopic analysis of the permanent teeth. As teeth age, for-

Table 17.1

Average Age (in years) of a Skeletal Individual Based on Assessment of Dental Development at Each Crown Position[a]

		di1	di2	dc	dm1	dm2	I1	I2	C	P3	P4	M1	M2	M3
A. MALES														
Ci	Cusp initiation	—	—	—	—	—	—	—	0.6	2.1	3.2	0.1	3.8	9.5
Cco	Cusp coalescence	—	—	—	—	—	—	—	1.0	2.6	3.9	0.4	4.3	10.0
Coc	Crown outline complete	—	—	—	—	—	—	—	1.7	3.3	4.5	0.8	4.9	10.6
Cr 1/2	Crown one half	—	—	—	—	—	—	—	2.5	4.1	5.0	1.3	5.4	11.3
Cr 3/4	Crown three-quarters	—	—	—	—	—	—	—	3.4	4.9	5.8	1.9	6.1	11.8
Crc	Crown complete	0.15	0.2	0.7	0.4	0.7	—	—	4.4	5.6	6.6	2.5	6.8	12.4
Ri	Root initiated	—	—	—	—	—	—	—	5.2	6.4	7.3	3.2	7.6	13.2
Rcl	Root cleft present	—	—	—	—	—	—	—	—	—	—	4.1	8.7	14.1
R 1/4	Root one-quarter	—	—	—	—	—	—	5.8	6.9	7.8	8.6	4.9	9.8	14.8
R 1/2	Root half	—	—	—	—	—	5.6	6.6	8.8	9.3	10.1	5.5	10.6	15.6
R 2/3	Root two-thirds	—	—	—	—	—	6.2	7.2	—	—	—	—	—	—
R 3/4	Root three-quarters	—	—	—	—	—	6.7	7.7	9.9	10.2	11.2	6.1	11.4	16.4
Rc	Root complete	1.5	1.75	3.1	2.0	3.1	7.3	8.3	11.0	11.2	12.2	7.0	12.3	17.5
A 1/2	Root apex half closed	—	—	—	—	—	7.9	8.9	12.4	12.7	13.5	8.5	13.9	19.1
		di1	di2	dc	dm1	dm2	I1	I2	C	P3	P4	M1	M2	M3
B. FEMALES														
Ci	Cusp initiation	—	—	—	—	—	—	—	0.6	2.0	3.3	0.2	3.6	9.9
Cco	Cusp coalescence	—	—	—	—	—	—	—	1.0	2.5	3.9	0.5	4.0	10.4
Coc	Crown outline complete	—	—	—	—	—	—	—	1.6	3.2	4.5	0.9	4.5	11.0
Cr 1/2	Crown one half	—	—	—	—	—	—	—	2.5	4.0	5.1	1.3	5.1	11.5
Cr 3/4	Crown three-quarters	—	—	—	—	—	—	—	3.5	4.7	5.8	1.8	5.8	12.0
Crc	Crown complete	0.15	0.2	0.7	0.3	0.7	—	—	4.3	5.4	6.5	2.4	6.6	12.6
Ri	Root initiated	—	—	—	—	—	—	—	5.0	6.1	7.2	3.1	7.3	13.2
Rcl	Root cleft present	—	—	—	—	—	—	—	—	—	—	4.0	8.4	14.1
R 1/4	Root one-quarter	—	—	—	—	—	4.8	5.0	6.2	7.4	8.2	4.8	9.5	15.2
R 1/2	Root half	—	—	—	—	—	5.4	5.6	7.7	8.7	9.4	5.4	10.3	16.2
R 2/3	Root two-thirds	—	—	—	—	—	5.9	6.2	—	—	—	—	—	—
R 3/4	Root three-quarters	—	—	—	—	—	6.4	7.0	8.6	9.6	10.3	5.8	11.0	16.9
Rc	Root complete	1.5	1.75	3.0	1.8	2.8	7.0	7.9	9.4	10.5	11.3	6.5	11.8	17.7
A 1/2	Root apex half closed	—	—	—	—	—	7.5	8.3	10.6	11.6	12.8	7.9	13.5	19.5

[a] The data are from Smith's (1991) compilation of published studies.

mation of secondary dentine reduces the coronal height of the pulp cavity. Drusini *et al.* (1997) have used this to age radiographs of adult individuals to ±5 years on 78% of teeth assessed. Other studies have shown that apical translucency of tooth roots correlates with adult age, but applications of the technique have shown it to be less useful than other methods (Kvaal *et al.*, 1994). In some forensic cases, a combination of gingival regression and root transparency may allow aging of adults over 40 and under 80 years of age with a mean error of estimation of ±10 years (Lamendin *et al.*, 1992).

17.3.3 Estimating Adult Age from Cranial Suture Closure

It has been appreciated since the 1500s that sutures between various cranial bones fuse progressively as the individual ages. In the early 1900s suture closure enjoyed widespread use in skeletal aging, but the false

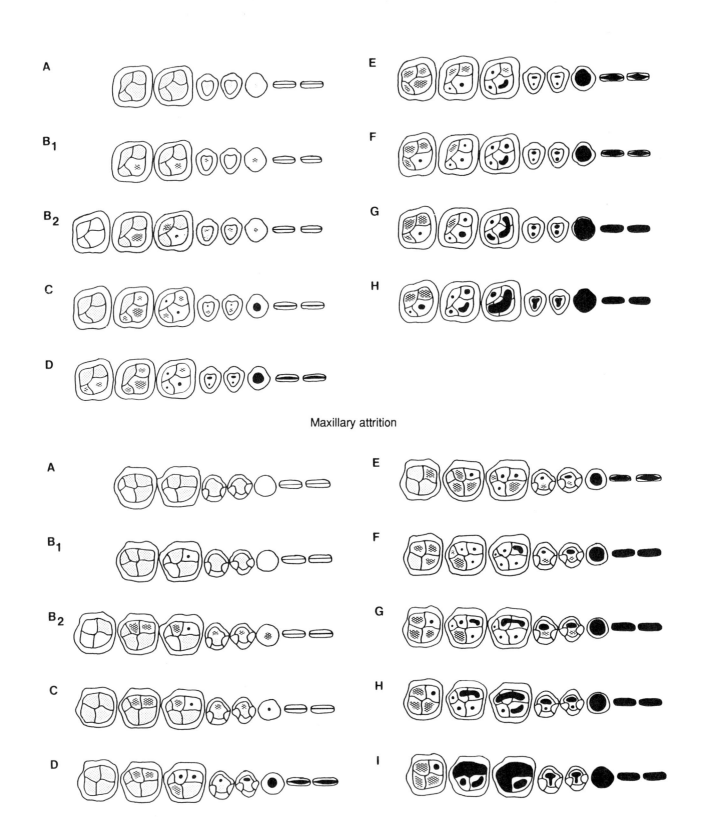

Maxillary attrition

Mandibular attrition

Figure 17.3 Modal tooth-wear patterns of a prehistoric Native American population from Libben, Ohio. Wear is divided into phases for the right maxillary (*top*) and left mandibular (*bottom*) dentitions. Exposed dentine is shown in black. Age in years for the various phases are as follows: A, 12–18; B$_1$, 16–20; B$_2$, 16–20; C, 18–22; D, 20–24; E, 24–30; F, 30–35; G, 35–40; H (maxillary), 40–50; H (mandibular), 40–45; I, 45–55. See Lovejoy (1985) for a full description.

promise of one or two accurate indicators of adult skeletal age during the 1950s (such as metamorphosis of the pubic symphysis) led to disuse of the technique. Meindl and Lovejoy (1985), however, reinvigorated the study of cranial suture closure. They chose a series of 1-cm segments of ten sutures or suture sites and scored these on a scale of 0 (open) to 3 (complete obliteration). The results erased some of the prejudice against suture closure assessment as a means of skeletal aging in the adult and stimulated more research in this area. Galera *et al.* (1998) provide a comparative analysis of different cranial suture aging methods. One cranial feature, the sphenooccipital synchondrosis, is particularly useful in aging isolated crania because at least 95% of all individuals have fusion here between 20 and 25 years of age, with a central tendency at 23 years of age (Krogman and İşcan, 1986).

Other cranial sutures show more variation in age of closure. The Ley *et al.* (1994) work on the Spitalfields population from Britain indicates that there may be sexual and interpopulational differences in the rates of suture closure. This study developed yet another technique of scoring suture closure. The Buikstra and Ubelaker Standards volume (1994) recommends that seventeen cranial suture segments each be given a numerical score. The score of 0, or **open,** is given when there is no evidence of any ectocranial closure. A score of 1 is given to suture sites with **minimal closure.** A score of 2 is given to sites with **significant closure,** and a score of 3 is given to a **completely obliterated** suture (complete fusion). Figure 17.4 shows the location of the 1-cm sutural sites.

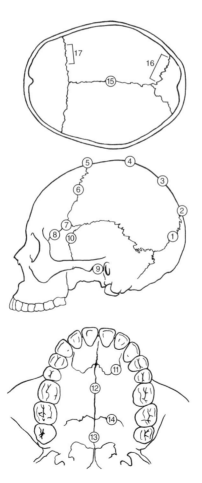

Figure 17.4 Cranial suture fusion sites, after P. Walker, in Buikstra and Ubelaker, (1994). A score of from zero (unfused) to 3 (completely obliterated) is assigned to each site. Sites are one-centimeter ectocranial segments of the sutures as shown. Endocranial segments are larger. In the Meindl and Lovejoy (1985) system, scores are independently summed for vault (#'s 1–7) and lateral-anterior (#'s 5–10) sites. Other suture sites such as the maxillary suture (Mann *et al.*, 1991) have been used to segregate individuals into even broader age categories, but its use in forensic cases has been questioned (Gruspier and Mullen, 1991). *(Continues)*

Figure 17.4 (continued)

Site Name		Description
1	Midlambdoid	Midpoint of L. lambdoid suture
2	Lambda	Intersection of sagittal and lambdoidal
3	Obelion	At obelion
4	Anterior Sagittal	One-third the distance from bregma to lambda
5	Bregma	At bregma
6	Midcoronal	Midpoint of left coronal suture
7	Pterion	Usually where parietosphenoid suture meets the frontal
8	Sphenofrontal	Midpoint of left sphenofrontal suture
9	Inferior Sphenotemporal	Intersection between left sphenotemporal suture and line between articular tubercles of the temporomandibular joint
10	Superior Sphenotemporal	On left sphenotemporal suture 2 cm below junction with parietal
11	Incisive Suture	Incisive suture separating maxilla and premaxilla
12	Anterior Median Palatine	Score entire length on paired maxillae between incisive foramen and palatine bone
13	Posterior Median Palatine	Score entire length
14	Transverse Palatine	Score entire length
15	Sagittal (endocr.)	Entire sagittal suture endocranially
16	Left Lambdoidal (endocr.)	Score indicated portion
17	Left Coronal (endocr.)	Score indicated portion

Meindl and Lovejoy (1985) "vault" sutural ages (add scores for sites 1–7).

Composite Score	Mean Age	Standard Deviation
0	—	—
1–2	30.5	9.6
3–6	34.7	7.8
7–11	39.4	9.1
12–15	45.2	12.6
16–18	48.8	10.5
19–20	51.5	12.6
21	—	—

Meindl and Lovejoy (1985) "lateral-anterior" sutural ages (add scores for sites 6–10).

Composite Score	Mean Age	Standard Deviation
0	—	—
1	32.0	8.3
2	36.2	6.2
3–5	41.1	10.0
6	43.4	10.7
7–8	45.5	8.9
9–10	51.9	12.5
11–14	56.2	8.5
15	—	—

17.3.4 Estimating Subadult Age from Long-Bone Length

In the absence of teeth and various epiphyses, subadult individual age may be estimated from long-bone length. This procedure of subadult skeletal aging is not as exact as others and should always be done with reference to the same or a closely related skeletal collection. Seriate the growth series and compare the isolated long-bone lengths to the series in order to derive ages. If series are not available, data presented in Ubelaker (1989) are useful in age assessment. For age and sex determination of fetal and neonate material, consult Weaver (1986) and Fazekas and Kosa (1978).

17.3.5 Estimating Subadult Age from Epiphyseal Closure

Fusion of a postcranial epiphysis is orderly, and an epiphysis fuses at a known age, but these ages vary by individual, sex, and population. As Stevenson (1924) points out, the intensity of epiphyseal activity is greatest between ages 15 and 23 years. Fusion of the epiphysis is progressive and is usually scored as unfused (nonunion), ¼ united, ½ united, ¾ united, or fully fused (complete union). The beginning of epiphyseal union for several elements overlaps the conclusion of tooth eruption, making these aging techniques complementary. Figures 17.5 and 17.6 illustrate the considerable interindividual variation in the chronology of epiphyseal union for several human skeletal elements, using data taken from male Korean War dead (McKern and Stewart, 1957). Much of the work on epiphyseal union has been done on long bones, but recent work on vertebrae show the utility of these elements, particularly in aging teenagers and young adults (Albert and Maples, 1995).

It should be noted that union begins earlier in females than in males, and that different individuals of the same sex can show very different times of union. The last epiphysis to fuse is usually the medial clavicle, at an age of about 21 years. Late-fusing bones such as the clavicle, however, show wide variation in age at fusion. For example, some medial clavicle epiphyses fuse before 21 years, whereas other individuals show persistent nonfusion at age 30 (for more references, consult Stevenson, 1924; Mensforth and Lovejoy, 1985; Webb and Suchey, 1985; and Krogman and İşcan, 1986). Once fusion of all epiphyses occurs, under 28 years of age for the great majority of cases, growth ends. As a result, fewer age indicators remain for the postcranial skeleton of the adult individual.

17.3.6 Estimating Adult Age from the Pubic Symphysis Surface

One of the most widely used indicators of age at death has been metamorphosis of the symphyseal surface of the pubis of the os coxae. Age-related changes on this surface continue after full adult stature has been achieved and other epiphyses of the limbs have fused. Pubic symphyses of other primates metamorphose more quickly than human ones and usually synostose with advancing age. In humans, however, changes of the symphyseal surfaces allow them to be used in generating osteologically determined age at death. The young adult human pubic symphysis has a rugged surface traversed by horizontal ridges and intervening grooves. This surface loses relief with age and is bounded by a rim by age 35. Subsequent

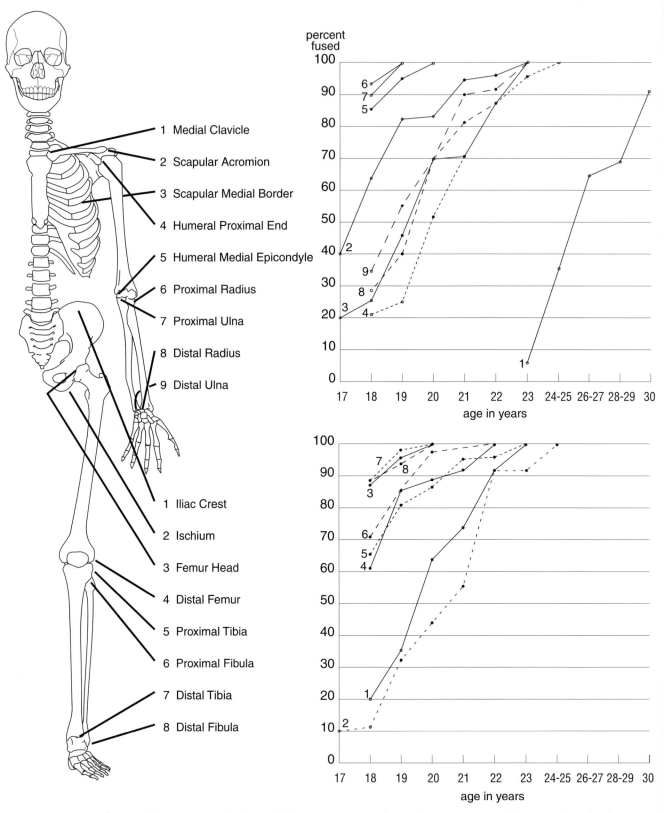

Figure 17.5 Ages of fusion for various male skeletal elements. Data on fusion from McKern and Stewart (1957). These standards, derived from U.S. military personnel who died in the Korean War, show considerable variation in fusion for any given element. For example, in the medial clavicle, Stewart and McKern found that of 10 individuals aged 17 years, none had fused epiphyses. For the clavicle, the epiphyseal cap begins to unite to the medial end of the clavicle as early as 18 years but can begin to unite at any time between 18 and 25 years. The earliest complete fusion came among some soldiers who died at 23 years, but the study showed that others lived to age 31 before fusion was complete. To use this table, choose a numbered epiphysis from above or below the waist, and find its graph to the right of the skeleton. The graph shows what percentage of adult male individuals showed full fusion of each epiphysis at any given age.

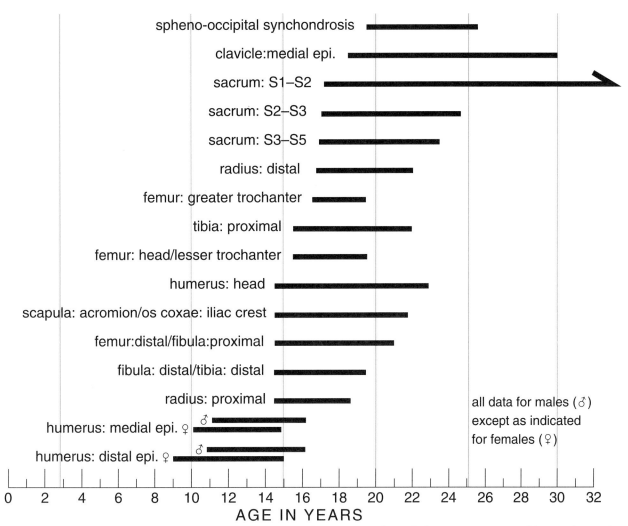

Figure 17.6 Timing of the fusion of epiphyses for various human osteological elements. Horizontal bars indicate the period during which union/fusion is occurring. Male and female data are taken from the summary Figure 20 of Buikstra and Ubelaker's Standards volume (1994).

erosion and general deterioration of the surface are progressive changes after this age. Figure 17.7 illustrates these changes.

Age-related changes at the pubic symphysis have been recognized for many years, and the first formal system for using these changes to determine age was developed by Todd (1920), based on a series of 306 males of known age at death. Todd identified four basic parts to the pubic symphysis, a surface with an irregular oval shape: (a) the ventral border (rampart); (b) the dorsal border (rampart); (c) the superior extremity; and (d) the inferior extremity. Todd noted evidence of billowing, ridging, ossific nodules, and texture on each part of the symphyseal surface. Using these observations on his sample of known ages, Todd recognized ten phases of pubic symphysis age, ranging from 18/19 years to 50+ years, and noted that these phases were more reliable age indicators between 20 and 40 years than after 40 years. He perceived three major stages in symphysis metamorphosis. His phases I–III were a "postadolescent" stage, phases IV–VI were the buildup stage, and phases VII–X represented the degenerative

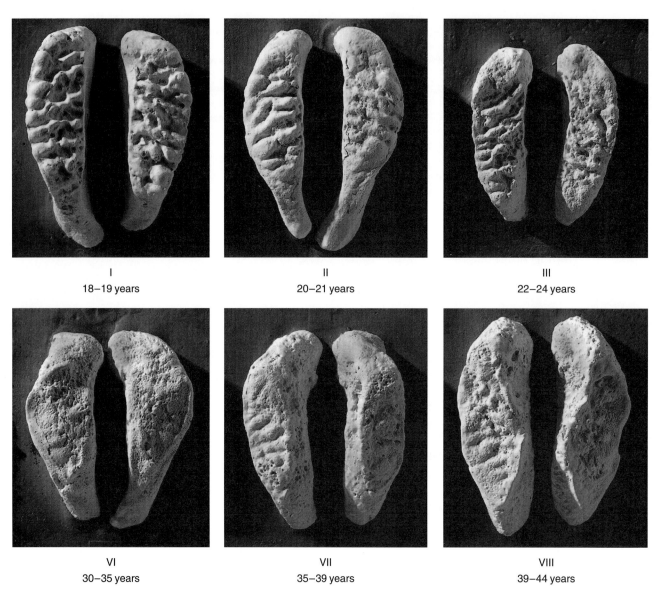

I	II	III
18–19 years	20–21 years	22–24 years

VI	VII	VIII
30–35 years	35–39 years	39–44 years

Figure 17.7 Todd's (1920) ten age phases of pubic symphysis modification in adult white males. Todd's standard specimens are shown here, natural size. The anterosuperior ends of the pubic symphyses are toward the top of the page, and the ventral margins of each symphysis pair are opposed. In the original Todd study, the phases were defined according to topography on the symphyseal surface and the nature of the margins of this surface. Note the wider age ranges for the higher stages. For more details on application of the Todd technique, consult the original 1920 publication. Many newer standards have been published since this first attempt to use changes in the topography of the pubic symphysis to age the skeleton, all relying on the bony changes that correlate with age.

stage. Figure 17.7 illustrates Todd's original standards; subsequent work on age-based changes in this anatomical region is based on this foundation.

Few tests of this method were made, although the method gained wide acceptance. Brooks (1955) found a tendency of the Todd system to overage, especially in the third and fourth decades. In 1957, McKern and Stewart used skeletal remains of 349 male Korean War dead in an effort to refine the Todd method. Their approach was to divide the symphyseal surface longitudinally into two halves, or "components"—the "dorsal demi-face"

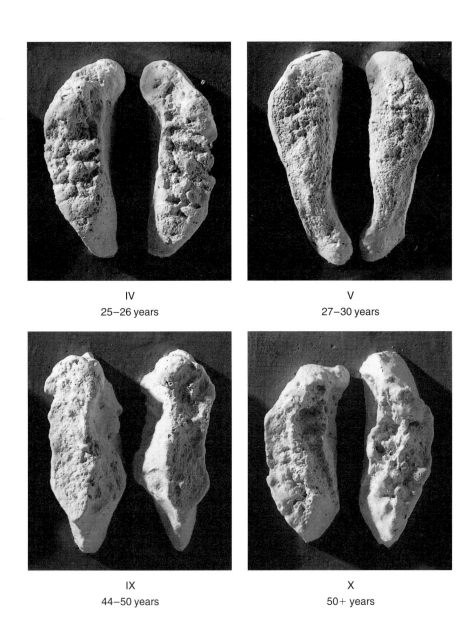

IV
25–26 years

V
27–30 years

IX
44–50 years

X
50+ years

and the "ventral demi-face." The third component of the symphysis was the "symphyseal rim." Five developmental stages were recognized for each of the three components. In using this system, the osteologist calculates a developmental stage for each component, adds these together, and derives an age of death for the specimen. This system, like that of Todd, was derived from an all-male sample of limited age range.

Gilbert and McKern (1973) used a sample of 103 females of known age to generate a component system for aging female specimens by the pubis. Since female pubic symphyses are subject to trauma during childbearing, there is a potential for premature changes in the bone surface which could lead to overaging. In 1979, Suchey tested the Gilbert and McKern method for aging the female pubic symphysis by asking twenty-three professional osteologists to age pubic faces of unrevealed age. The results showed the system to be highly unreliable, and prone to inaccurate estimates.

Meindl and colleagues (1985a) tested the accuracy of all these methods in a study of the Hamann-Todd collection. They found that the original

Todd method was more reliable than the more recent component techniques, and that all systems tended to underage. They recognize five major biological phases for the pubic symphysis and provide careful illustrative documentation of their results. They also provide a much-needed biological perspective on the metamorphosis of the pubic symphysis and assess this part of the anatomy from a comparative evolutionary background (see also Lovejoy *et al.*, 1995, 1997).

Suchey *et al.* (1986) and Katz and Suchey (1986), working on the large Los Angeles County Coroner multiracial sample, examined 739 male individuals between the ages of 14 and 92. These investigators contend that the age-at-death data for their skeletal individuals are more accurate and precise than those used to build the Todd, McKern-Stewart, and Meindl *et al.* standards. The Todd and McKern-Stewart methods were tested on the Los Angeles sample and interobserver error was assessed. As a result, modifications of the earlier techniques of pubic symphysis age estimation were suggested. Katz and Suchey (1986) suggest that Todd's methodology is excellent, but that the collection he used was inadequate. They recommend the use of a modified Todd approach with six phases defined on the entire symphyseal face. Their data clearly show a large amount of variation in ages for any one phase, particularly for older individuals. For example, their Phase V shows a mean age of 51 years, but only 95% of the sample of 241 male pubic symphyses that matched this phase was within the wide age limits of 28 to 78 years.

An assessment of "race" differences among 704 of the male pubic bones was undertaken by Katz and Suchey (1989). Following the implementation of the six-phase Suchey/Brooks system for male individuals, 273 female pubic bones were studied, and a system analogous to the male system was devised. Refined descriptions of the Suchey/Brooks method are found in Brooks and Suchey (1990) and in Figure 17.8. In a test of this and other pubic symphysis aging methods, Klepinger *et al.* (1992) concluded: "... all the aging methods based on the *os pubis* proved disappointing in regard to both accuracy and precision." They recommend that the "racial" refinement of the Suchey Brooks system be used, that any independent evidence of trauma or debilitation be considered, and that all estimated ages be reported within a standard deviation interval of plus-or-minus two.

17.3.7 Parturition Changes at the Pubic Symphysis

Childbirth, or **parturition**, may result in changes on the pubic symphysis (particularly pitting adjacent to the symphysis on the dorsal edge of the pubis), auricular surface, and preauricular area of the female os coxae. To what extent can these changes be used by the osteologist to assess whether, or how often, an individual has given birth? Studies designed to answer these questions include Kelley (1979) and Suchey *et al.* (1979). These studies show that up to 20% of nulliparous women display these or related features. Pitting acquired during pregnancy and childbirth may become obliterated with age. Medium to large pitting can occur without parturition (17 of 148 cases, Suchey *et al.*, 1979), and parturition can occur without pitting. Four of over 700 male individuals displayed medium to large pitting. In short, there is a correlation between pitting of the os coxae and pregnancy or parturition, but this correlation is not strong. The number of pregnancies cannot be predicted by the morphology of the dorsal aspect of the pubis. Tague (1988), in a study combining humans and nonhuman mammals, concludes that the severity of the resorption at the pubis is not significantly related to that of the preauricular area. Furthermore, age at death was shown to be significantly associated with resorption of the pubis. Cox and Scott (1992) note that pubic tubercle extension is also associated with childbirth, whereas preauricular sulci and dorsal pitting were not. Tague calls for further study of the link between estrogen and osteoclastic activity in these areas of the os coxae.

17.3.8 Estimating Adult Age from the Ilium's Auricular Surface

Lovejoy and colleagues (1985b) examined the auricular surface of the os coxae as a possible site of regular change corresponding to age in the Hamann-Todd collection. The use of this surface to age individual specimens has some advantages, namely, that this part of the os coxae is more likely to be preserved in forensic and archeological cases, and that the changes on the auricular surface, unlike those on the pubic symphysis, extend well beyond the age of 50 years. Lovejoy and colleagues describe age-related changes in surface granulation, microporosity, macroporosity, transverse organization, billowing, and striations that are somewhat similar to those described for the surface of the pubic symphysis. These investigators note that auricular surface aging is more difficult to master than the Todd method for the pubic symphysis, but they state that the potential rewards are worth the extra effort, as the method is independent of symphyseal aging but equally accurate.

The changes described by Lovejoy *et al.* (1985b) for the auricular surface are as follows. The young auricular surface (Figure 17.9), beginning in the first few years after postcranial epiphyseal fusion, shows a fine-grained surface texture and a pattern of regular, usually transverse surface undulations called billowing. The surface's topography is very much like the subchondral bone of an unfused epiphysis, although the billowing is not so pronounced. Beginning in adulthood, these features of the sacroiliac joint are progressively and regularly modified as age increases. Granularity of the surface becomes coarser, billowing and striae are dramatically reduced, the original transverse organization of youth is lost, and the surface begins to display perforations of its subchondral bone, a condition known as "microporosity." In the later stages of life, the surface becomes increasingly dense and disorganized. Larger subchondral defects termed **macroporosity** progressively increase with age after the fifth decade. By the sixth and seventh decades, the surface has become dense, both microporotic and macroporotic, and has lost all evidence of transverse organization. Lovejoy and colleagues have formalized a system of eight phases with which to classify this metamorphosis (Figure 17.9 and Buikstra and Ubelaker, 1994, Figure 10). The addition of the sacroiliac joint to the osteologist's arsenal of age indicators is an important one. However, both the original research and subsequent tests (Murray and Murray, 1991) on the method suggest that the method, like the pubic symphysis methods, has large estimation errors associated with it and should not be used alone in the assessment of an unknown's age-at-death.

17.3.9 Estimating Adult Age from the Sternal Rib End

İşcan and Loth (1986) have studied metamorphosis of the sternal end of the fourth rib and found that it corresponds to age but varies by sex. They

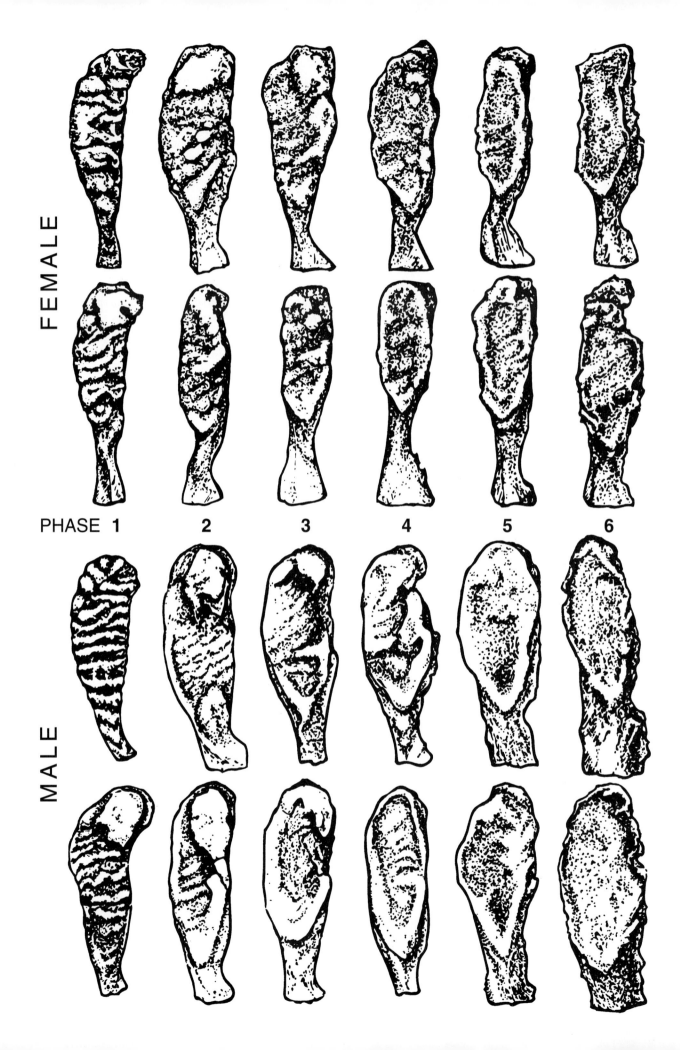

FEMALE

PHASE 1   2   3   4   5   6

MALE

examined "form, shape, texture and overall quality" of the rib end and defined a series of phases that begins with a flat or billowy end with regular and rounded edges. With age, this rim thins and becomes irregular. The surface porosity increases, and the bone becomes "ragged." As with all of the newer techniques described above, this one has not been widely

---

Figure 17.8 The Suchey/Brooks pubic symphysis scoring system. The phase descriptions below may be applied to either male or female symphysis faces, but matches of females should only be made in reference to the female phase types in the upper two rows. Phase descriptions are from Brooks and Suchey (1990, italics therein), and statistics for the Suchey/Brooks phases in females and males follow the descriptions; drawings by P. Walker in Buikstra and Ubelaker's Standards volume (1994). It is recommended that these illustrations be supplemented by casts before actual aging is attempted.

Phase 1: Symphyseal face has a billowing surface (ridges and furrows) which usually extends to include the pubic tubercle. The horizontal ridges are well-marked, and ventral beveling may be commencing. Although ossific nodules may occur on the upper extremity, *a key to the recognition of this phase is the lack of delimitation of either extremity (upper or lower).*

Phase 2: The symphyseal face may still show ridge development. *The face has commencing delimitation of lower and/or upper extremities occurring with or without ossific nodules.* The ventral rampart may be in beginning phases as an extension of the bony activity at either or both extremities.

Phase 3: Symphyseal face shows lower extremity and *ventral rampart in process of completion.* There can be a continuation of fusing ossific nodules forming the upper extremity and along the ventral border. Symphyseal face is smooth or can continue to show distinct ridges. Dorsal plateau is complete. Absence of lipping of symphyseal dorsal margin; no bony ligamentous outgrowths.

Phase 4: Symphyseal face is generally fine grained although remnants of the old ridge and furrow system may still remain. *Usually the oval outline is complete at this stage, but a hiatus can occur in upper ventral rim.* Pubic tubercle is fully separated from the symphyseal face by definition of upper extremity. The symphyseal face may have a distinct rim. Ventrally, bony ligamentous outgrowths may occur on inferior portion of pubic bone adjacent to symphyseal face. If any lipping occurs, it will be slight and located on the dorsal border.

Phase 5: *Symphyseal face is completely rimmed with some slight depression of the face itself, relative to the rim.* Moderate lipping is usually found on the dorsal border with more prominent ligamentous outgrowths on the ventral border. There is little or no rim erosion. Breakdown may occur on superior ventral border.

Phase 6: *Symphyseal face may show ongoing depression as rim erodes.* Ventral ligamentous attachments are marked. In many individuals the pubic tubercle appears as a separate bony knob. The face may be pitted or porous, giving an appearance of disfigurement with the ongoing process of erratic ossification. Crenulations may occur. The shape of the face is often irregular at this stage.

Descriptive Statistics:

Phase	Female (n = 273)			Male (n = 739)		
	Mean	Standard Dev.	95% Range	Mean	Standard Dev.	95% Range
1	19.4	2.6	15–24	18.5	2.1	15–23
2	25.0	4.9	19–40	23.4	3.6	19–34
3	30.7	8.1	21–53	28.7	6.5	21–46
4	38.2	10.9	26–70	35.2	9.4	23–57
5	48.1	14.6	25–83	45.6	10.4	27–66
6	60.0	12.4	42–87	61.2	12.2	34–86

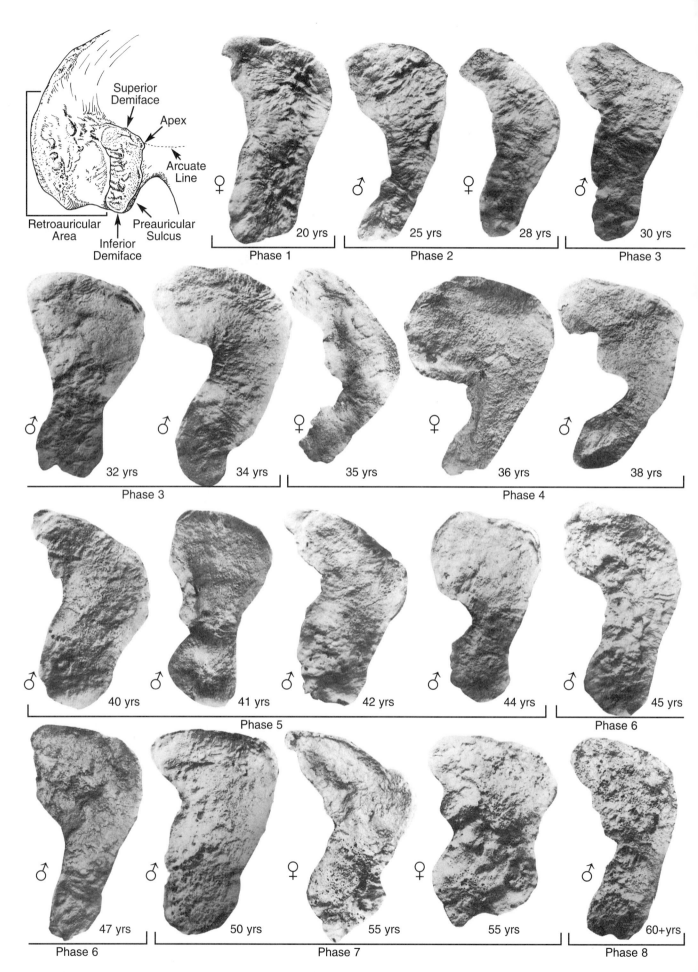

Superior Demiface

Apex

Arcuate Line

Retroauricular Area

Inferior Demiface

Preauricular Sulcus

♀ 20 yrs — Phase 1

♂ 25 yrs — Phase 2

♀ 28 yrs

♂ 30 yrs — Phase 3

♂ 32 yrs — Phase 3

♂ 34 yrs

♀ 35 yrs — Phase 4

♀ 36 yrs

♂ 38 yrs

♂ 40 yrs — Phase 5

♂ 41 yrs

♂ 42 yrs

♂ 44 yrs

♂ 45 yrs — Phase 6

♂ 47 yrs — Phase 6

♂ 50 yrs

♀ 55 yrs — Phase 7

♀ 55 yrs

♂ 60+ yrs — Phase 8

tested, and, in incomplete skeletons, identification of the fourth rib is not always possible. However, Dudar (1993) reports that rib four standards can "cautiously" be applied to other sternal rib ends. Russell *et al.* (1993) verify the utility of this technique on a sample of 100 males.

17.3.10 Estimating Adult Age by Radiographic Analysis

Changes in cancellous and cortical bone structure at the macroscopic and microscopic levels take place throughout life. Walker and Lovejoy (1985) have studied this phenomenon by assessing radiographs from the Hamann-Todd collection and the prehistoric Libben collection. Using seriation, these authors describe progressive, site-specific loss of bone with age in both the clavicle and femur. Visual seriation of radiographs showed a moderately high and significant correlation between increased age of death and decreased bone density. Macchiarelli and Bondioli (1994) show significant variation in density of the proximal femur, much of it unrelated to age. Jackes (1992) discusses problems with applications to archeological remains.

17.3.11 Estimating Adult Age by Using Bone Microstructure

The normal remodeling of bone during adult life has been proposed as a condition useful for aging skeletal material. Microscopic analysis has allowed the relationships between the number of osteons and osteon fragments and the percentages of lamellar bone and non-haversian canals to be examined. Simmons (1985), Frost (1987), and Stout (1992) provide excellent summaries of these procedures. It should be noted that these procedures are destructive to the bones under study. Many studies of histomorphometry have been undertaken on the long bones of the postcranial skeleton (see Stout, 1992, for a review). Cool *et al.* (1995) have shown that histomorphological variables in the human occipital were less reliable than those in the long bones in estimating age.

All of these methods are quantitative and depend on osteonal remodeling of bone and accumulated osteon populations (see Chapter 2). Many

---

Figure 17.9 Modal changes to the auricular surface with age. Phases described by Lovejoy *et al.* (1985b) as follows:

Phase 1: Age 20–24; billowing and very fine granularity

Phase 2: Age 25–29; reduction of billowing but retention of youthful appearance

Phase 3: Age 30–34; general loss of billowing, replacement by striae, coarsening of granularity

Phase 4: Age 35–39; uniform coarse granularity

Phase 5: Age 40–44; transition from coarse granularity to dense surface; this may take place over islands on the surface of one or both faces

Phase 6: Age 45–49; completion of densification with complete loss of granularity

Phase 7: Age 50–59; dense irregular surface of rugged topography and moderate to marked activity in periauricular areas

Phase 8: Age 60+; breakdown with marginal lipping, microporosity, increased irregularity, and marked activity in periauricular areas

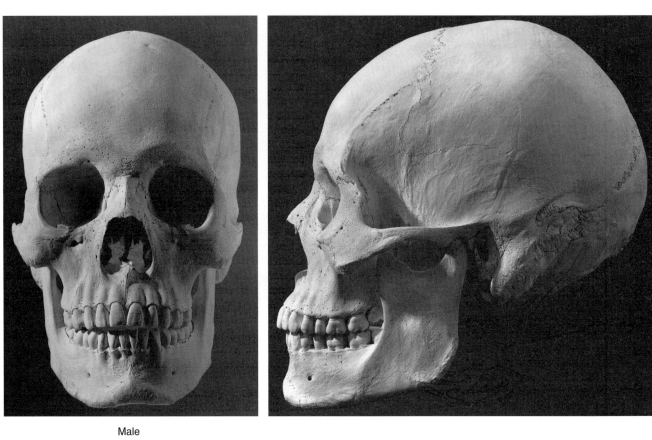

Male

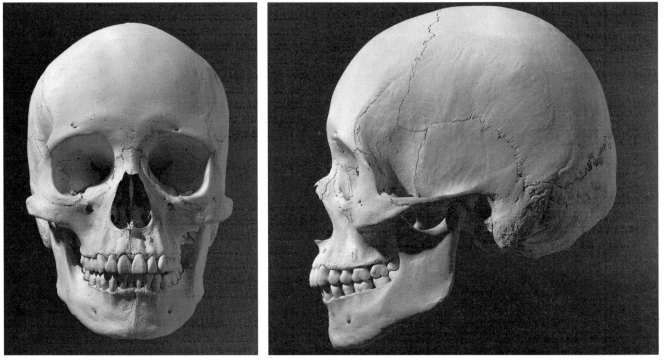

Female

Figure 17.10 Male and female adult skulls in frontal and lateral views. The female skull chosen for this illustration was taken from the "female" end of the female range. The male is the same individual used to illustrate the cranium in Chapter 4. This comparison illustrates the differences between male and female skulls discussed in the text. It should *not* be taken as a representation of the difference between *average* male and female skulls, but rather as an indication of how much sexual variation is seen in the human skull. One-half natural size.

factors can influence this process and its products, including sex, hormones, mechanical strain, and nutrition. Remodeling is the sequential removal and replacement of older lamellar bone with newer lamellar bone and takes place throughout the lifespan. Histological analysis of tissue from selected sites on the skeleton (including the ribs and clavicle; Stout and Paine, 1992; Stout *et al.*, 1994, 1996), has shown an association between age at death and the number of observable osteons per unit area in a cross section. The number of intact and fragmentary osteons per unit area is calculated for each bone (normally for at least two slices of each bone), and the result is put into regression equations that calculate the age. Stout (1992) identifies a variety of problems with the technique as applied by different investigators, and calls for more research in the forensic setting. Pfeiffer *et al.* (1995) and others have found that histological profiles vary by sample location, something that must be controlled for in application of these techniques. As Ericksen (1991) notes, it is critical that osteologists using bone microstructure to age archeological specimens be very cautious about pre-analysis exfoliation of unremodeled peripheral bone lamellae. Jackes (1992) notes other complicating factors for the use of these techniques on archeological remains. Aiello and Molleson (1993) compared pubic symphyseal aging and microscopic aging techniques and found neither to be more accurate. Wallin and colleagues (1994, p. 353) have found that their determination of age at death through microscopic bone morphometry resulted in standard deviations of over 12 years and was "considerably less precise than generally stated in the literature."

17.3.12 Multifactorial Age Estimation

Given the variety of techniques available for assessing skeletal age at death, what techniques should be used by the osteologist? In Todd's original 1920 work on the changes he had classified in the pubic symphysis, he took great pains to point out that the most accurate estimate of age can only be made after examination of the entire skeleton. However, due to the sometimes fragmentary nature of skeletal remains and the history of development of aging techniques, his advice has often been forgotten by human osteologists who followed. All osteologists use dental development, eruption, epiphyseal appearance, and fusion when aging immature skeletal material. For aging adult skeletal remains, however, osteologists are sharply divided on the question of technique. This controversy provides an important arena for the continued testing and refinement of the techniques outlined above.

Some osteologists, particularly those working in forensic contexts, favor the use of the pubic symphysis and assign other anatomical regions a lesser role in age analysis. Lovejoy *et al.* (1985a) note that this traditional forensic orientation to aging has led to problems when skeletons from large populations are aged by different observers using established methods. Furthermore, the value of skeletal age indicators has been judged on the basis of **accuracy** (differences between predicted and actual ages) without due regard to **bias** (the tendency of a given technique to over- or underage).

If more than one criterion is available for assessing skeletal age at death, all criteria should be employed. One immediate objection to this recommendation arises because of the marked differences in the reliability between different age indicators. For example, many investigators are hesitant to alter a determination of age at death based on the pubic symphysis face by additional data from cranial suture closure because of the latter's supposed unreliability (see Meindl and Lovejoy, 1985). In forensic aging of single individuals, such caution may be advisable (depending on the assessed age). However, cranial suture closure is correlated with increasing age, and in the analysis of populations the addition of data on age at death from the sequential addition of other age indicators should improve the accuracy of determination. The conclusions of their study of multifactorial age determination are very significant for all osteologists.

17.4
Determination of Sex

The terms "sex" and "gender" have increasingly become conflated in the anthropological and medical literature. They do not refer to the same thing, they are not synonyms, and they should not be used interchangeably. Gender is an aspect of a person's social identity, whereas sex refers to a person's biological identity. This distinction is important for biological anthropologists to preserve in general, and particularly important to retain in bioarcheology (Walker and Cook, 1998). In the archeological context, it is often possible to determine sex through analysis of skeletal remains, and gender roles through studies of material culture (artifacts) and context.

With a sample of fifty lowland gorilla males and fifty lowland gorilla females, even the untrained observer could sort skeletal elements by sex using size and shape. For this primate, 100% accuracy in sorting is easily obtained. The same applies to orangutans. With chimpanzees, the differences are not as marked, but when the canine teeth rather than the overall size of the cranium are examined, perfect accuracy can still be approached. Moving to a sample of fifty male and fifty female modern humans, there is far less **sexual dimorphism** in body and canine size, and the sorting accuracy is therefore reduced. Human sexual dimorphism is complex, with behavioral, physiological, and anatomical dimensions. Anatomical differences are extreme in some soft tissue areas and more limited in the skeleton. Nevertheless, skeletal differences between male and female humans do exist and can be useful to the osteologist. Buikstra and Mielke (1985) summarize in helpful tabular form the accuracy of a variety of skeletal sexing techniques.

In determining the sex of any skeletal element, the osteologist starts with 50% accuracy—random guessing will be correct half of the time. For some elements, such as the cranium, training and experience can often allow correct sorting about 80–90% of the time. It is extremely important to remember that sexual identification of human skeletal material is generally most accurate after the individual reaches maturity. Only then do the bones of different sexes become differentiated sufficiently to be useful in sexing.

In general, for all parts of the human skeleton, female elements are characterized by smaller size and lighter construction. For this reason, in a large, seriated, mixed-sex collection of elements, the largest, most robust elements with the heaviest rugosity are male. Males can average up to 20% larger in some skeletal dimensions, and in others there may be no dimorphism. The smallest, most gracile elements are female. Normal individual variation, however, always produces some small, gracile males and some large, robust females who fall toward the center of the distribution where sorting sex is difficult. In other words, the sexes overlap near the center of the distribution. For this reason, osteologists have traditionally concentrated on elements of the skull and pelvis in which sex differences in humans are the most extreme.

In addition to the complications of individual variation within the population, incorrect sex identifications are sometimes made because of variation between populations. Some populations are, on the average, composed of larger, heavier, more robust individuals of both sexes, and other populations are characterized by the opposite tendency. Because of such interpopulational differences in size and robusticity, males from one population are sometimes mistaken for females in other populations and vice versa. The osteologist should always attempt to become familiar with the skeletal sexual dimorphism of the population from which unsexed material has been drawn. As it is with aging, seriation can be a helpful approach in determining the sex of skeletal remains from a population.

17.4.1 Sexing the Skull and Dentition

Determination of sex based on parts of the skull follows the observation that males tend to be larger and more robust than females. In addition to size, tendencies such as those outlined here provide useful indications for determining the sex of isolated skulls. These characteristics are the traditional ones used by osteologists. Figure 17.10 illustrates them.

Relative to female crania, male crania are characterized by greater robusticity. Male crania typically display more prominent supraorbital ridges, a more prominent glabellar region, and heavier temporal and nuchal lines. Male frontals and parietals tend to be less bossed than female ones. Males tend to have relatively large, broad palates, squarer orbits, larger mastoid processes, larger sinuses, and larger occipital condyles than do females. Male mandibles are characterized by squarer chins, more gonial eversion, deeper mandibular rami, and more rugose muscle attachment points when compared to female mandibles.

The strength of these cranial tendencies can be summarized by the following: Where associated postcranial material is available, always use the pelvis for sex determination of the cranium. When sexing skulls, always use the entire population under study. Seriate this population according to the criteria you use, and then sort. If you are sexing only one or a few individuals, try to use comparative populations that are genetically and temporally close to the ones your sample is from.

Walker, in the Buikstra and Ubelaker Standards volume (1994), provides five aspects of skull morphology that are useful in determining sex. These are shown in Figure 17.11. In all cases, a five-point scale is used, with the more gracile, feminine features at the lower end of the range. Graw *et al.* (1999) present another scoring system focused only on the supraorbital margin.

Recent studies of the posterior border of the mandibular ramus suggested to Loth and Henneberg (1996, 1998) that this region could be used to sex unknowns with an average of about 92% accuracy. These authors claimed that adult males have a distinct angulation of the posterior border of the mandibular ramus at the level of the occlusal surface of the molars.

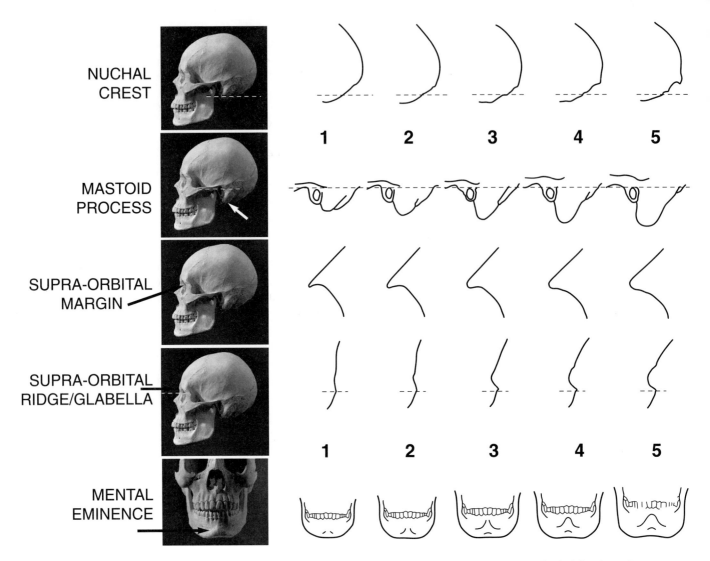

Figure 17.11 A scoring system for sexually dimorphic cranial features from Walker in Buikstra and Ubelaker (1994). In recording the features, optimal results are obtained by holding the cranium or mandible at arm's length, a few inches above the appropriate portion of this figure, oriented so that the features can be directly compared with those illustrated. Move the bone from diagram to diagram until the closest match is obtained. Score each trait independently, ignoring other features. Specific procedure for each trait is described below, with the extremes described and the intermediates illustrated in the figure.

Nuchal Crest: View the lateral profile of the occipital, and compare it with the diagrams. Feel the surface of the occipital with your hand and note any surface rugosity, ignoring the contour of the underlying bone. Focus upon the rugosity attendant to attachment of nuchal musculature. In the case of minimal expression (score = 1), the external surface of the occipital is smooth with no bony projections visible when the lateral profile is viewed. Maximal expression (score = 5) defines a massive nuchal crest that projects a considerable distance from the bone and forms a well-defined bony ledge or "hook."

Mastoid Process: Score this feature by comparing its size with that of surrounding structures such as the external auditory meatus and the zygomatic process of the temporal bone. Mastoid processes vary considerably in their proportions. The most important variable to consider in scoring this trait is the volume of the mastoid process, not its length. Minimal expression (score = 1) is a very small mastoid process that projects only a small distance below the inferior margins of the external auditory meatus and the digastric groove. A massive mastoid process with lengths and widths several times that of the external auditory meatus should be scored as 5.

Supraorbital Margin: Begin by holding your finger against the margin of the orbit at the lateral aspect of the supraorbital foramen. Then hold the edge of the orbit between your fingers to determine its thickness. Look at each of the diagrams to determine which it seems to match most closely. In an example of minimal expression (score = 1), the border should

They argued that the lack of such flexure in females allows sex to be identified. The technique was used by Donnelly *et al.* (1998) in a blind test that demonstrated a much lower accuracy (62–68%).

In an attempt to go beyond the traditional methods outlined above, Giles and Elliot in 1963 used nine standard cranial metrics to diminish the subjectivity involved in sexing the skull. However, a study by Meindl *et al.* (1985b) has shown that subjective assessment of the skull compared favorably to the discriminant functions of Giles and Elliot. In tests on Hamann-Todd crania, Meindl *et al.* found that older individuals show increasingly "masculine" morphology. Whereas 10.2% of the males in their sample of 100 were sexed incorrectly, only 4.9% of the females were misidentified. Given these facts, Meindl *et al.* suggest that overall sex ratios and age class sex ratios in prehistoric cemeteries should only be estimated from adult burials with fully preserved pelves.

Because teeth are often better preserved than other skeletal elements, there have been efforts to sex the skeleton using the teeth. The degree of sexual dimorphism in human crown sizes varies between populations. Human dental dimorphism centers on the canines, but it is not nearly as pronounced as it is in the great apes. Human lower canines show the greatest dimorphism, up to 7.3% (Hillson, 1996), followed by the upper canines. Deciduous teeth are also dimorphic, with molars and canines up to 7%. Accuracy of sexing unknowns based on dental metrics, either univariate or multivariate, varies from 60% to a claimed 90% and usually lies between 75 and 80%. When it is noted that the actual size differences between sexes in individual teeth are very small, about half a millimeter on average, it can be seen that intra- and interobserver error might play a role in sex estimation. De Vito and Saunders (1990), Bermudez de Castro *et al.* (1993), Beyer-Olsen and Alexandersen (1995), and Hillson (1996) provide reviews of the use of dental dimensions to sex human teeth.

17.4.2 Sexing the Postcranial Skeleton

As for the cranium, sexually diagnostic traits in the postcranial skeleton are difficult to identify and assess before puberty. Numerous metric studies of the postcranial skeleton have examined sexual dimorphism in size of various adult elements. Bass (1995) provides an excellent review of these. The results on the most dimorphic limb bones can be summarized by noting that single measurements, or combinations of measurements, have usually been found to correctly identify the sex of between 80 and 90% of all individuals. Incorrect identification within any population is a

---

feel extremely sharp, like the edge of a slightly dulled knife. A thick, rounded margin with a curvature approximating a pencil should be scored as 5.

Prominence of Glabella: Viewing the cranium from the side, compare the profile of the supraorbital region with the diagrams. In a minimal prominence of glabella/supraorbital ridges (score = 1), the contour of the frontal is smooth, with little or no projection at midline. Maximal expression involves a massive glabellar prominence, forming a rounded, loaf-shaped projection that is frequently associated with well-developed supraorbital ridges.

Mental Eminence: Hold the mandible between the thumbs and index fingers with thumbs on either side of the mental eminence. Move the thumbs medially until they delimit the borders of the mental eminence. In examples of minimal expression (score = 1), there is little or no projection of the mental eminence above the surrounding bone. By contrast, a massive mental eminence that occupies most of the anterior portion of the mandible is scored as 5.

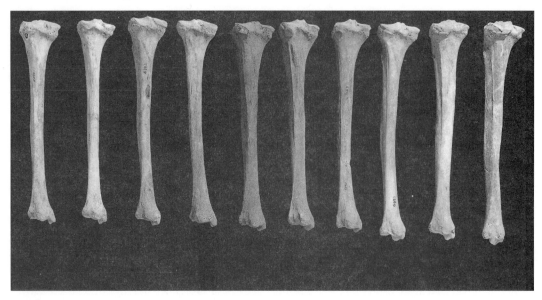

Figure 17.12 Variation in tibial size and shape among ten females (*left*) and ten males (*right*) selected at random from a single-site, sex-balanced sample of 100 prehistoric Californian skeletons. This sample, 20% of the total population, gives an indication of the normal sexual dimorphism encountered in modern human skeletal remains. One-sixth natural size.

consequence of size overlap between males and females in the center of the overall range (Figure 17.12). Many studies have been conducted on known-sex samples to derive discriminant functions capable of classifying sex accurately more than 85% of the time for a variety of elements ranging from the metacarpals (Falsetti, 1995) to the metatarsals (Robling and Ubelaker, 1997) and calcaneus (Introna *et al.*, 1997). These functions are often not tested beyond (independent of) the skeletal population on which the discriminants were based, so claims of accuracy are sometimes questionable. Rogers (1999) has claimed 92% accuracy based on four characters of the distal humerus, but testing on a wider sample will be required.

The skull has been a traditional focus of sexing studies, but a number of methods of sexing have also been applied to the pelvis. There are dramatic functional differences between male and female pelvic anatomy which extend to the bony skeleton, differences found in all modern human groups. The pelvis is of vital importance in locomotion and parturition. During human evolution selective pressures associated with these and other roles led to the sexual dimorphism seen in the modern human pelvis.

Traditional methods used to determine sex on the pelvis or its parts are based on the following tendencies: The sacrum and os coxae of females are smaller and less robust than those of males. Female pelvic inlets are relatively wider than male ones. The greater sciatic notch on female os coxae is relatively wider than the notch on male bones (Figure 17.14). Females have relatively longer pubic portions of the os coxae, including the superior pubic ramus, than males. The subpubic angle, formed between the lower edges of the two inferior pubic rami, is larger in females than in males. The preauricular sulcus is present more often in females than in males. A corollary is that the auricular surface is more elevated from the female ilium than from the male ilium even though sexual di-

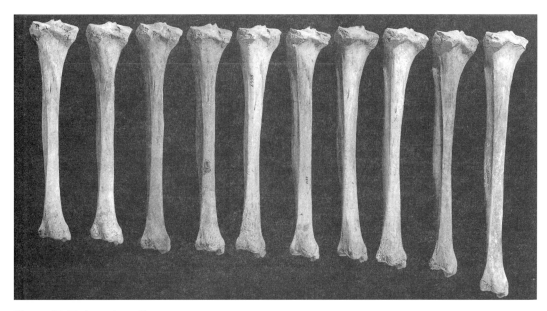

Figure 17.12 (*continued*)

morphism in the auricular surface itself is insufficient for accurate sexing (Ali and MacLaughlin, 1991). The acetabulum tends to be relatively larger in males (Figures 17.12 and 17.13).

A variety of metric techniques have been developed to express these relationships. Washburn's attempt to quantify the relative proportion of the pubic part of the os coxae is the most famous and effective of these. Washburn (1948) measured the length of the pubis relative to the length of the ischium via an index that discriminated between male and female os coxae. Rogers and Saunders (1994) provide a review of metric and morphological traits used to sex the pelvis.

In 1969, T. W. Phenice published an important new method for sexing the pelvis. This paper, "A Newly Developed Visual Method of Sexing the *Os Pubis*," describes the most accurate method yet known for determining sex of an individual from the skeleton. Until the publication of the Phenice paper, the osteologist's success at using traditional visual methods of sexing the pelvis depended, in large part, on experience—decisions were more or less subjective. The application of metric criteria was difficult because much of the material was too fragmentary for reliable measurement, and even the simplest techniques were time consuming. Phenice's method (illustrated in Figure 17.15) has changed the situation, allowing more accurate, quicker sexing on any pelvis with an intact pubic region.

In employing the Phenice method to sex an os coxae, note that not every specimen is a "perfect" male or female. When there is a criterion that does not obviously sex the specimen, discard that criterion. When there is some ambiguity concerning one or two of the criteria (most often in the medial aspect of the ischiopubic ramus, and least often in the ventral arc), there is usually one of the remaining criteria that clearly attributes the specimen to a sex. After sexing the specimen with this procedure, observe the more traditional features outlined earlier to see if they correspond to (strengthen) your diagnosis. For any case which they do not confirm, recheck your Phenice observations. Remember that female

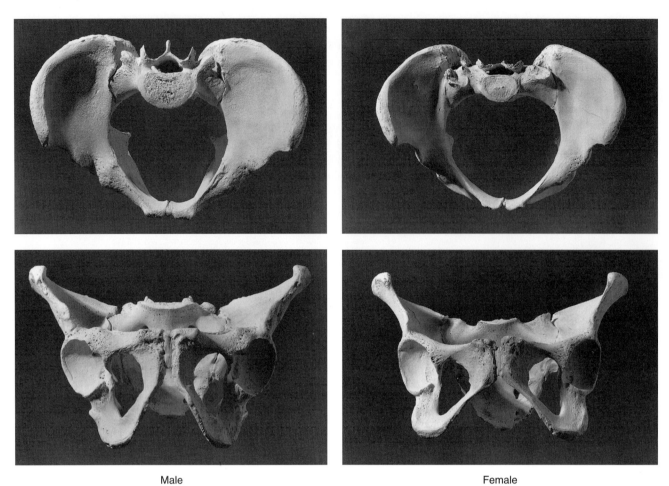

Male   Female

Figure 17.13 Sexual dimorphism in the bony pelvis, showing differences in size and shape. One-fourth natural size.

individuals are most likely to be intermediate in displaying the Phenice features. The Phenice method should only be used for fully adult material. Accuracy of sexing based on this method ranges from 96 to 100%, the highest ever achieved in the skeleton, but Lovell (1989) has suggested that accuracy might be reduced in the case of older adult specimens. In 1990 MacLaughlin and Bruce tested the Phenice characters for accuracy of sex identification on skeletal series from London, Leiden, and Scotland. They were unable to confirm the accuracy obtained by Phenice and others, achieving success on only 83% of the English, 68% of the Dutch, and 59% of the Scottish. They found the subpubic concavity to be the single most reliable indicator. Using 1284 pubic bones from the Los Angeles County coroner collection, Sutherland and Suchy (1991) reported that they achieved 96% sexing accuracy using the ventral arc alone. They note that this feature first appears at age 14 but does not become marked until age 20. The discrepancy between these two major tests of the Phenice technique remains unexplained. The best advice for sexing of the os coxae, as for aging the skeleton, is to use all of the available data.

Ventral arc. Orient the pubis so that its rough ventral surface faces you, and you are looking down along the plane of the pubic symphysis surface. The ventral arc is a slightly elevated ridge of bone that sweeps inferiorly and laterally across the ventral surface of the pubis, merging with the

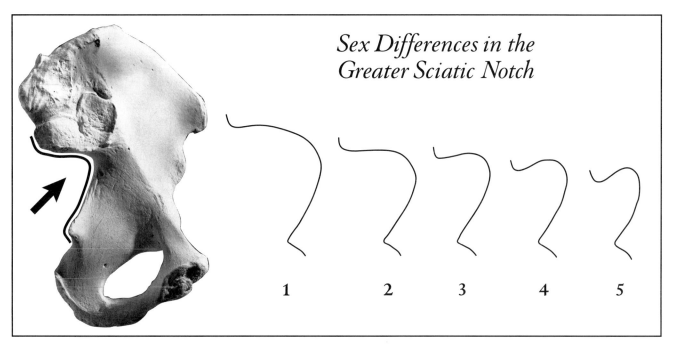

Figure 17.14 Sex differences in the greater sciatic notch, from Walker in Buikstra and Ubelaker's Standards book (1994). The greater sciatic notch tends to be broad in females and narrow in males. These shape differences are not as reliable as those in the subpubic region and should be thought of as secondary indicators. The best results for scoring are obtained by holding the os coxae above this figure so that the greater sciatic notch has the same orientation as the outlines, aligning the straight anterior portion of the notch that terminates at the ischial spine with the right side of the diagram. While holding the bone in this manner, move it to determine the closest match. Ignore any exostoses near the preauricular sulcus and the inferior posterior iliac spine. Configurations more extreme than 1 or 5 should be scored as 1 and 5, respectively. The illustration numbered 1 shows typical female morphology, whereas the higher numbers are male conformations.

medial border of the ischiopubic ramus. The ventral arc, when present, thus sets off the inferior, medial corner of the pubic bone in ventral view. It is present only in females. Male os coxae may have elevated ridges in this area, but these do not take the wide, evenly arching path of the female's ventral arc or set off the lower medial quadrant of the pubis.

Subpubic concavity. Turn the pubis over, orienting it so that its smooth, convex dorsal surface faces you and you are once again sighting along the midline. Observe the medial edge of the ischiopubic ramus in this view. Female os coxae display a subpubic concavity here; the edge of the ramus is concave in this view. Males, on the other hand, do not show the dramatic concavity here. Male edges are straight or very slightly concave.

Medial aspect of the ischiopubic ramus. Turn the pubis 90°, orienting the symphysis surface so that you are looking directly perpendicular to it. Observe the ischiopubic ramus in the region immediately inferior to the symphysis. This medial aspect of the ischiopubic ramus displays a sharp edge in females. In males the surface is fairly flat, broad, and blunt.

17.5
Estimation of Stature

Estimating individual stature from bone lengths has a long history in physical anthropology. The fact that the height (stature) of the human

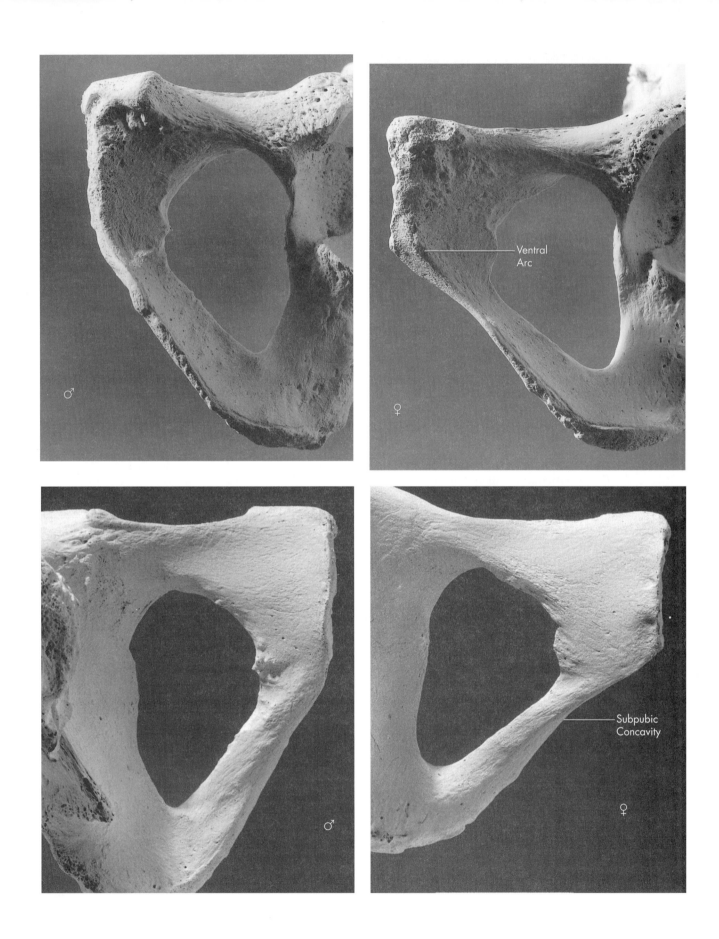

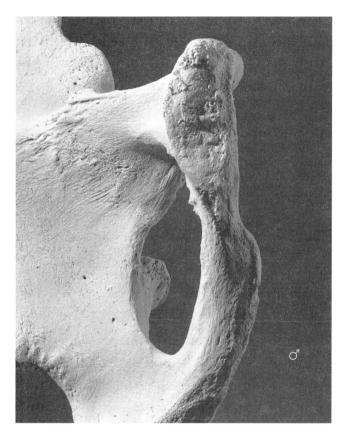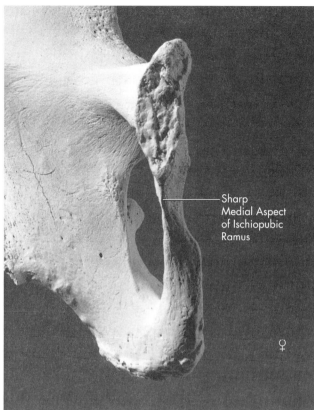

Sharp
Medial Aspect
of Ischiopubic
Ramus

♂ ♀

Figure 17.15 The Phenice (1969) technique for sexing the pubic portion of the os coxae from the left side. In each comparison, the male is on the left and the female is on the right. These os coxae are the ones illustrated in Figure 17.13. Natural size.

body correlates with limb bone length across all ages allows the osteologist to reconstruct an individual's stature from different skeletal elements. Unfortunately, the correlation is imperfect within living populations and varies between populations. Based on studies of skeletons from individuals of known stature, several investigators have derived regression equations useful in estimating stature in different human populations.

To estimate stature, based on the maximum length of a male femur from a Mesoamerican archeological site, for example, the osteologist would apply the formula derived by Genovés from modern Mexican samples and published in 1967. This formula is as follows:

$$stature \text{ (cm)} \pm 3.417 = (2.26)(femur\ length) + 66.379$$

If femur length is known, stature may be calculated with about a 68% probability that the calculated value falls within 3.417 cm of the actual stature of the individual. This formula, of course, can be validly applied only in Mexican samples, but Bass (1995) and Bennett (1993) provide useful tables for stature estimation in different human groups. Most of these formulae are based on the lengths of one or more bones.

Trotter and Gleser (1958) developed formulae for estimating stature based on the Korean War dead, extending their earlier work on World War II remains in an anatomical collection. Formulae were presented for "racial" groups. The Genovés (1967), Trotter and Gleser (1958), and Trotter

Table 17.2
Equations Used to Estimate Stature, in Centimeters, with Standard Error, from
the Long Bones of Various Groups of Individuals between 18 and 30 Years of Age[a]

White Males		**Black Males**	
3.08 Hum + 70.45	±4.05	3.26 Hum + 62.10	±4.43
3.78 Rad + 79.01	±4.32	3.42 Rad + 81.56	±4.30
3.70 Uln + 74.05	±4.32	3.26 Uln + 79.29	±4.42
2.38 Fem + 61.41	±3.27	2.11 Fem + 70.35	±3.94
2.68 Fib + 71.78	±3.29	2.19 Fib + 85.65	±4.08
White Females		**Black Females**	
3.36 Hum + 57.97	±4.45	3.08 Hum + 64.67	±4.25
4.74 Rad + 54.93	±4.24	2.75 Rad + 94.51	±5.05
4.27 Uln + 57.76	±4.30	3.31 Uln + 75.38	±4.83
2.47 Fem + 54.10	±3.72	2.28 Fem + 59.76	±3.41
2.93 Fib + 59.61	±3.57	2.49 Fib + 70.90	±3.80
East Asian Males		**Mexican Males**	
2.68 Hum + 83.19	±4.25	2.92 Hum + 73.94	±4.24
3.54 Rad + 82	±4.60	3.55 Rad + 80.71	±4.04
3.48 Uln + 77.45	±4.66	3.56 Uln + 74.56	±4.05
2.15 Fem + 72.57	±3.80	2.44 Fem + 58.67	±2.99
2.40 Fib + 80.56	±3.24	2.50 Fib + 75.44	±3.52

[a]To estimate stature of older individuals, subtract 0.06 (age in years, 30) cm; to estimate cadaver stature,
add 2.5 cm. From Trotter (1970). The tibia is not included; see text for rationale.

(1970) formulae for stature estimation are the most frequently used methods in North America. Table 17.2 is taken from the latter publication. As Feldesman and Fountain (1996) note, if the specimen's ancestry (race) is unknown, it is best to use generic equations of stature. Formicola (1993) evaluated various stature formulae on 66 archeological skeletons from seven European countries. They found that the Trotter and Gleser formulae for blacks worked better than those for whites.

There is growing debate on how to accommodate old data sets to the modern forensic world (Jantz, 1992, 1993; Giles, 1993). As Jantz (1992) notes, the most commonly used female stature formulae are derived from the Terry Collection, skeletons from people who died in the early 1900s. To what extent should formulae based on those samples be modified to reflect the secular changes in bone length and body height undergone during the last century? Giles (1991) makes further comments regarding stature loss in the elderly. Jantz and colleagues (1995) also note that the Trotter and Gleser stature formulae involving tibial length produce stature estimates averaging 2–3 cm too great when used with properly measured tibiae. They show that the original formulae involving tibiae are based on mismeasured tibiae (the malleolus was omitted in length measurements). Finally, Ousley (1995) notes that stature can be defined in several ways, ranging from forensic (for example, from a driver's license) to biological (from cadavers or living individuals). He suggests that biological stature estimations based on long bone lengths are generally less precise than many have assumed. Table 17.3 is from Ousley (1995).

Table 17.3
Regression Equations for Estimating Forensic Stature
(Estimating from Skeletal Remains)[a]

Factor	Bone measurement in mm	Constant	90% PI	N
White Males				
0.05566	Femur Max L + Tibia L	21.64	±2.5"	62
0.05552	Femur Max L + Fibula L	22.00	±2.6"	54
0.10560	Femur Max L	19.39	±2.8"	69
0.10140	Tibia L	30.38	±2.8"	67
0.15890	Ulna L	26.91	±3.1"	62
0.12740	Humerus L	26.79	±3.3"	66
0.16398	Radius L	28.35	±3.3"	59
White Females				
0.06524	Femur Max L + Fibula L	12.94	±2.3"	38
0.06163	Femur Max L + Tibia L	15.43	±2.4"	42
0.11869	Femur Max L	12.43	±2.4"	48
0.11168	Tibia L	24.65	±3.0"	43
0.11827	Humerus L	28.30	±3.1"	45
0.13353	Ulna L	31.99	±3.1"	40
0.18467	Radius L	22.42	±3.4"	38
Black Females				
0.11640	Femur Max L	11.98	±2.4"	18
Black Males				
0.16997	Ulna L	21.20	±3.3"	14
0.10521	Tibia L	26.26	±3.8"	19
0.08388	Femur Max L	28.57	±4.0"	17
0.07824	Humerus L	43.19	±4.4"	20

[a] Note that the bone measurements should be in millimeters, but that all constants were converted to predict statures and prediction intervals in inches because most North American forensic applications record stature in inches. For example, if the maximum length of the femur from a probable white male is 454 mm, the forensic stature is estimated by: 0.10560 (454) + 19.39 = 67.33 ± 2.8 inches. This person, if a white male, would have a roughly 90% chance of having a forensic stature between 64.5 and 70.1 inches (5 feet 4½ inches to 5 feet 10 inches). These figures will continue to be revised as sample sizes grow.

17.6
Estimation of Ancestry

Imagine a sample of 1000 people—a sample composed of 400 native Nigerians, 300 native Chinese, and 300 native Norwegians. If these people randomly seated themselves at a lecture, the speaker would be able to tell, simply by looking at their faces, whether their ancestry was Asian, African, or European. The sorting accomplished on the basis of soft tissue facial features would correspond perfectly to the geographic origin of the three major components of the sample.

This kind of sorting within the species *Homo sapiens* is usually termed racial sorting. However, there are no "pure" human races (A.A.P.A., 1996). By definition, all members of the same species have the potential for reproduction, and hence any subspecific classification is arbitrary. Defining the term "race" has proven difficult in the history of physical anthropology because concepts of race have often been based on composites of biological, social, and ethnic criteria used in a typological fashion. The confusion arising from these difficulties has persuaded some anthropologists to conclude that the very use of the term "race" is counterproductive. In osteological work, particularly work in forensic contexts, the determination of race, or geographic ancestry, is usually an important consideration.

The title of a recent paper by physical anthropologist Kenneth Kennedy (1995) asked, "But Professor, Why Teach Race Identification if Races Don't Exist?" Typological race concepts in physical anthropology have gone the way of the dinosaurs. But human populations are routinely divided into separate races (Blacks, Whites, Native Americans, and Hispanics) in U.S. governmentally mandated programs, the popular media, and the forensic sciences. As St. Hoyme and İşcan (1989) note, human osteologists who examine human bones must communicate with law enforcement personnel, students, and the general public. How people are categorized by others depends on law and custom. The United States government's bureaucratic approach to race is quite specific in this regard, noting that its classifications (American Indian or Alaskan native; Asian or Pacific Islander; Black; Hispanic; White; or Other) "should not be interpreted as being scientific or anthropological in nature." Most forensic applications bring the human osteologist into contact with medical examiners, law enforcement, and other government personnel who expect missing and/or found persons to be classified in terms of their bureaucratic races as defined by the government. These racial categories, of course, mix historical and social phenomena with biology. Gill (1995) provides an example of this by pointing out that in the United States, a person who is of 75% European descent, but has a black African grandparent is considered black rather than white. Today, it is common for parents of completely different geographic ancestry to have children whose anatomical configurations will defy assessment of ancestry. How does the osteologist deal with these social and biological realities?

The osteologist's role, particularly in the forensic setting, is often to individualize an unknown's osteological remains by assessing the age, sex, stature, and race of the individual. The ability to determine the geographic ancestry of a skeletal unknown is useful for narrowing the possibilities and leading to a positive identification in many cases. Yet as a biological scientist, the human osteologist knows that all variation is continuous, not discrete. As Kennedy (1995) notes, there is a paradox in the scientific rejection of "race" and its survival in medical/legal contexts. As Kennedy says (p. 798), "Forensic anthropologists are keenly aware that neither the medical examiner, the judge, the attorney client nor the sheriff would appreciate a lecture on the history of the race concept in Western thought. These professionals want to learn if the skeleton on our laboratory table is a person of Black, White, Asian or Native American ancestry. . . ." To conduct analysis of ancestral background, the osteologist may use osteological traits known to vary among different human populations in different parts of the world.

As Brace (1995, p. 172) explains, "Skeletal analysis provides no direct evidence for skin color for example, but it does allow an accurate estimate of original geographical origins. African, eastern Asian, and European ancestry can be specified with a high degree of accuracy. Africa of course entails 'black,' but 'black' does not entail African." Marks (1996) notes that the tendency for Americans to classify people in three races is an artifact of history and statistics—immigrants to North America have come mostly from ports where seafaring vessels in earlier centuries could pick them up. Hence, the American notion of "black" is actually west African, and the notion of "Asian" is actually east Asian. People from south Asia (India and Pakistan—people with darker skins and facial resemblances to Europeans) immigrated in smaller numbers and therefore didn't merit as much bureaucratic concern.

A real example of the dilemma facing human osteologists in the area of "racial" identification comes from the work of Katz and Suchey (1989) on the Los Angeles male pubic symphysis. These workers, in a paper entitled "Race Differences in Pubic Symphyseal Aging Patterns in the Male," assess the Los Angeles County Coroner sample used to generate the Suchey–Brooks system of symphyseal aging (see section 17.3.6). They segregated the symphyses into "racial" categories. They did not use the California death certificates made out by coroner investigators. As Katz and Suchey note, these examiners used nonuniform mixtures of biological, cultural, and linguistic variables in their determinations. Rather, Suchey divided the autopsied individuals into 486 Whites, 140 Blacks, and 78 Mexicans, and noted that her Mexican category is a category showing Mexican ancestry coupled with a strong American Indian racial component. Katz and Suchey found that pubic symphyseal metamorphosis was accelerated in Blacks and Mexicans, but they could not address the issue of causality.

Whereas soft tissue characteristics such as skin color, hair form, and facial features often allow unambiguous attribution of geographic ancestry among living people, the hard tissues display less reliable signatures of affinity. There are, in fact, no human skeletal markers that correspond perfectly to geographic origin.

The problems in using discrete cranial and dental features to determine ancestry are perhaps best appreciated by considering what all osteologists agree is a racial marker: the shovel-shaped incisors seen in high frequency in modern Asian populations. A review and compilation of data on incisor shoveling by Mizoguchi (1985) shows wide ranges of expressivity and incidence values in different extant human groups. Suffice it to say that incisors from Asian populations show a high incidence of shoveling, but also that the presence of shoveled incisors is hardly grounds for confident identification of a dentition as Asian.

The skull is the only part of the skeleton that is widely used in estimating geographic ancestry (but see İşcan and Cotton, 1985, for a consideration of the pelvis as a racial indicator and Baker *et al.*, 1990, for a femoral technique). Even with this element, all workers agree that racial estimations are usually more difficult, less precise, and less reliable than estimations of age, sex, or stature. Despite decades of research, much more osteological work on geographic differentiation within *Homo sapiens* remains to be done and is urgently needed. Work on modern skulls of known origin has revealed certain tendencies.

Howells (1995) notes that the human species lacks well-defined subspecies but has clear local tendencies of variation. It is simply not possible to attribute every human cranium to one or another geographically defined group on the basis of its morphology or measurements—populations of the human species are morphologically too continuous for this. Howells conducted exhaustive and long-term studies on a selected sample of 2504 human crania from around the world. He used 57 measurements on each skull and employed multivariate statistical techniques and the computer to show clearly that human variation in cranial shape, as represented by his measurements, is patterned, and that "target" skulls of unknown affinity could often be unambiguously placed in a parent "population."

Compared to populations of African or European origin, Asian populations display skulls characterized by narrow, concave nasal bones, prominent cheek bones, circular orbits, and shoveled incisors. Compared to Asians and Europeans, African crania have been characterized as showing wide interorbital distances, rectangular orbits, broad nasal apertures with poor inferior definition, gracile cranial superstructures, and pronounced total facial and alveolar prognathism. European crania have been characterized as displaying narrow nasal apertures with sharp inferior borders (sills), prominent nasal spines, heavy glabellar and supraorbital regions, receding cheek bones, and large, prominent nasal bones.

Given the limitations of using such subjective criteria for recognizing geographical ancestry, some have turned to cranial metric methods for the assessment of racial status (Giles and Elliot, 1962; Howells, 1969b). One such attempt is that of Gill (1984, 1986), which addresses the problem of sorting European from Native American crania.

Gill (1995) provides a compendium of traits useful in assessing ancestry in an American context in his article "Challenge on the Frontier: Discerning American Indians from Whites Osteologically." He notes that the Giles–Elliot discriminant function approach has been shown to be ineffective at sorting crania, particularly in the Northwestern Plains area where he works in both forensic and archeological contexts. Gill considers races to be statistical abstractions of trait complexes, not pure entities or rigidly definable types. Table 17.4 is a list of useful nonmetric traits of the dentition and cranium taken from his paper.

The attention now being given the origin of anatomically modern *Homo sapiens* in the later Pleistocene (Mellars and Stringer, 1989) should stimulate more work on skeletal differentiation within geographically separated populations of the species. Meanwhile, all of the techniques noted here, both visual and metric, should be applied only on adult remains and with comparative material. See Gill (1986), Krogman and İşcan (1986), and İşcan (1988) for further discussions of this issue.

17.7
Identifying the Individual

In paleontological and prehistoric archeological contexts, fossils are sometimes given nicknames like "Dear Boy" or "Lucy," but we will never know how members of their own species identified them. In historic archeological contexts, it is possible that skeletal remains may be identified as unique individuals, such as named Egyptian pharaohs or people buried beneath headstones in historic cemeteries. In the forensic realm, the human osteologist is often presented with unidentified skeletal remains. The positive identification of human skeletal remains—the unequivocal matching of teeth, crania, or postcranial remains with unique, named individuals—is often the most important step in the analysis. The identification of age, sex, stature, and ancestry all narrow the windows of possible identification—possible matching—to known individuals (often missing or unaccounted for). The last step in the process of identification

	"American Indians"	"Whites"	"Blacks"
Incisors	Shovel-shaped	Blade-form	Blade-form
Zygomatics	Robust, flaring	Small, retreating	
Prognathism	Moderate	Very limited	Marked alveolar and facial
Palate	Elliptic	Parabolic	Hyperbolic
Cranial sutures	Complex	Simple	Simple
Nasal spine	Medium, "tilted"	Long, large	Small
Chin	Blunt, median	Square, bilateral, projecting	Blunt, median, retreating
Ascending ramus	Wide, vertical		Narrow, oblique
Palatine suture	Straight	Jagged	Arched
Zygomatic tubercle	Present		
Incisor rotation	Present		
Nasal profile	Concavo-convex	Straight	
Sagittal arch	Low, sloping		
Wormian bones	Present		
Nasals	Low, tented	Highly arched, steeplelike	Low, flat
Nasal aperture	Medium		Wide
Zygomaticomaxillary suture	Angled	Curved	Curved
Dentition		Small, crowded	Large molars
Nasal sill		Very sharp	Very dull or absent
Nasion		Depressed	
Cranial vault		High	Low
Mandible		Cupping below incisors	
Inion hook		Present	
Postbregmatic depression			Present

From Rhine (1990) and Gill (1995).

sometimes involves matching unique features of the "unknown" skeleton with unique characters of the "known" missing.

Fingerprint analysis, of course, is a means by which forensic specialists routinely match criminals with their crimes. As the soft tissue features of the body decay or are incinerated, however, the use of fingerprints, hair, and personal items to individuate the deceased becomes impossible. Teeth, the skeletal structures most resistant to such destruction, are often used to identify people in mass disasters. Such individuation via teeth and their modifications, usually by dentists, is routinely accomplished by **forensic odontologists,** specially trained experts accomplished at such identifications. Radiographs and other dental records kept by dentists are matched against modifications on the deceased's teeth, often resulting in a positive identification (Kogon and MacLean, 1996).

Another means of establishing a positive identification on unknown skeletal remains is the comparison of those remains with medical radiographs taken when the individual was alive. Fractures, of course, can heal, but there are often trabecular and cortical points of identity through which a positive identification can be established. Postcranial skeletal characters (Owsley and Mann, 1992; Kahana *et al.*, 1998), and frontal sinus morphology in radiographs (Ubelaker, 1984) have been shown to be individually specific. Chapter 23 outlines one such case study, and the forensic literature has many more. The success of radiographic identification of unknown human remains, like that of many other techniques in human osteology, depends on the experience of the interpreter. Hogge *et al.* (1994) showed that the most accurate identifications came from cranial remains and the cervical spine and chest, whereas the least accurate identifications were made on the lower leg.

More recently, techniques which superimpose the skull of an unknown deceased individual on old photographs, motion pictures, or videotapes have been developed (İşcan and Helmer, 1993). Austin-Smith and Maples (1994) have tested the reliability of such superimposition methods and provide a good review of the techniques, limitations, and successes. When anterior dentition is recovered with the skull and a smiling photograph with the teeth in focus is available, the shapes of individual teeth and their relative positions are often distinctive enough for an identification to be made. Using only one photograph, Smith and Maples found a 9% chance of false identification, but when two photographs representing a difference of about 90 degrees in the angle of the face to the camera were used for superimposition, the chance of false identification dropped to less than 1%.

A final means of personal identification based on skeletal remains involves forensic three-dimensional facial reconstruction. A series of techniques exist for the "restoration" of the soft tissue cover of a human skull (İşcan and Helmer, 1993). The most recent review of the history of development and current status of such techniques by Tyrrell and colleagues (1997) notes that facial reconstruction still stands on the threshold between art and science. These authors conclude that current methods are useful, but insufficiently reliable to serve as evidence of positive identification in a court of law.

Suggested Further Readings

Bennett, K. A. (1993) *A Field Identification Guide for Human Skeletal Identification* (2nd Edition). Springfield, Illinois: C. C. Thomas. 113 pp.
> A compilation of tables of data useful in estimating sex, age, stature, and ancestry.

Caldwell, P. C. (1986) New questions (and some answers) on the facial reproduction techniques. In: K. J. Reichs (Ed.) *Forensic Osteology: Advances in the Identification of Human Remains.* Springfield, Illinois: C. C. Thomas. pp. 229–255.
> Discussion of the art and science of facial reproduction.

Gill, G. W., and Rhine (Eds.) (1990) *Skeletal Attribution of Race: Methods for Forensic Anthropology.* Albuquerque, New Mexico: Maxwell Museum of Anthropology. Anthropological Paper No. 4. 99 pp.
> An edited volume with a variety of articles about identifying ancestry.

Hamilton, M. E. (1982) Sexual dimorphism in skeletal samples. In: R. L. Hall (Ed.) *Sexual Dimorphism in Homo sapiens.* New York: Praeger. pp. 107–163.
> A review of the problems and prospects for the use of skeletal material in estimating sexual dimorphism.

İşcan, M. Y. (1988) Rise of forensic anthropology. *Yearbook of Physical Anthropology* 31:203–230.

> This review article traces the development of forensic anthropology and provides a wealth of citations.

İşcan, M. Y. (Ed.) (1989) *Age Markers in the Human Skeleton.* Springfield Illinois: Charles C. Thomas. 359 pp.

> This edited volume provides a variety of perspectives on the determination of age, from fetal life through adulthood, from various parts of the human skeleton and dentition. Its chapters, written by primary workers in the field of skeletal biology, summarize the limitations, advantages, and current status of various techniques used to determine skeletal age.

İşcan, M. Y., and Helmer, R. P. (Eds.) (1993) *Forensic Analysis of the Skull.* New York: Wiley-Liss. 276 pp.

> An edited volume with a variety of chapters covering aging, sexing and racing the skull, and principles and techniques of facial reconstruction.

İşcan, M. Y., and Kennedy, K. A. (Eds.) (1989) *Reconstruction of Life from the Skeleton.* New York: Alan R. Liss. 315 pp.

> This edited volume is a valuable sourcebook on a wide range of issues, including the assessment of age, sex, stature, and ancestry.

Krogman, W. M., and İşcan, M. Y. (1986) *The Human Skeleton in Forensic Medicine* (2nd Edition). Springfield, Illinois: C. C. Thomas. 551 pp.

> This updated classic is an essential, comprehensive guide to making the estimations discussed above, particularly in forensic contexts.

Lovejoy, C. O., and colleagues. (1985) Eight papers on Todd and Libben skeletal material. *American Journal of Physical Anthropology* 68:1–106.

> This collection of research papers illustrates work in both archeological and forensic contexts.

Stewart, T. D. (1979) *Essentials of Forensic Anthropology.* Springfield, Illinois: C. C. Thomas. 300 pp.

> A guide to all of the estimations introduced in this chapter.

Ubelaker, D. H. (1987) Estimating age at death from immature human skeletons: An overview. *Journal of Forensic Sciences* 32:1254–1263.

> A comprehensive review of the methods and limitations of age estimation for immature skeletal remains.

Ubelaker, D. H. (1989) *Human Skeletal Remains: Excavation, Analysis, Interpretation* (2nd Edition). Washington, D.C.: Taraxacum. 172 pp.

> A concise guide to identification of the variables covered here, including valuable tables and charts.

CHAPTER 18

Osteological and Dental Pathology

In Chapter 2, on bone biology, we introduced the concept of variation and its importance to work in human osteology. We described this variation as stemming from four main sources: age, sex, ancestry, and idiosyncrasy. Many facets of human osteological variation are illustrated by the descriptions of the various skeletal elements in Chapters 4–13, and by the analysis of age, sex, ancestry, and stature presented in Chapter 17. In the current chapter, we continue to consider variation, concentrating on variation due to pathology. Nonmetric nonpathological variations (discrete traits) are essential in reconstructing various biological dimensions of former human populations. These are reviewed in Chapter 20.

The present chapter introduces a variety of biological processes that can result in skeletal modifications before death (**premortem**). Cultural practices that take place before or around the time of death (**perimortem**) are also examined. The discussion of skeletal modifications below focuses on pathological conditions. In Chapter 19 we examine some additional perimortem modifications, but there we concentrate on changes to the bone that occur after the death of the individual (**postmortem changes**). The student should note that the most critical step in recording paleopathological conditions is the recognition of true bone abnormalities, as opposed to the normal range of variation in immature and adult healthy individuals.

Paleoepidemiology is the study of disease in a past community.

Paleopathology is the study of diseases in ancient populations as revealed by skeletal remains and preserved soft tissues (see also Chapter 21 for a description of molecular evidence). Bones and teeth can be records of events in the life of an individual, including trauma and disease. Indeed, dramatic insights into the health of individuals, and of populations, may be gained from studies in osteological pathology. As Miller and colleagues (1996) note, there are three main objectives of paleopathological research. First, the diagnosis of specific diseases in individual skeletal remains. Second, the analysis of the impact of various diseases in human populations through time and space. Third, the clarification of evolutionary interactions between humans and disease.

Pathological changes observable in osteological materials result from an imbalance in the normal equilibrium of bone resorption and formation, or growth-related disorders. This imbalance can arise as a result of many factors, including mechanical stress, changes in blood supply, inflammation of soft tissues, changes brought about by infectious diseases, hormonal, nutritional and metabolic upsets, and tumors (Mensforth *et al.*, 1978). Diagnosing the exact cause of an observed skeletal pathology is, however, not always possible. For example, an individual's growth may be interrupted by a range of factors including infectious disease, starvation, or trauma. This growth arrest may lead to the formation of **Harris lines** in the long bones, lines of increased bone density that represent the position of the growth plate at the time of insult to the organism. The lines are visible radiographically, or in cross section, and may be used to estimate the age at which the individual was stressed. Radiographic assessment is an essential component of describing and diagnosing disease in skeletal remains. Observations on different individuals may be combined to estimate the degree to which the population was stressed. From the Harris lines themselves, however, only general statements about the variety of possible stressors (including disease and diet) may be made (Maat, 1984; Hummert and Van Gerven, 1985). Furthermore, the lines may be removed by bone remodeling after they form. These Harris lines illustrate yet another problem common to comparative work in paleopathology. Macchiarelli and colleagues (1994) have shown that interobserver error may be high in radiographic interpretations, and even the scoring of the same radiographs by the same individual on different occasions (intraobserver error) can result in reported differences. Waldron and Rogers (1991) echo these concerns based on a study of interobserver variation in coding osteoarthritis in skeletal remains.

The osteologist is at a decided disadvantage compared to the forensic pathologist or clinician. Whereas a clinician can monitor progress of a disease in a patient, the paleopathologist is limited to the static appearance of the skeleton at the time of death. Furthermore, in the majority of paleopathology cases, diagnosis is necessarily based on gross appearance and radiology. In contrast, the clinician diagnosing a living patient's disease can assess patient history, soft tissue, chemistry, pathogens, and pain. It follows that paleopathological diagnoses based on skeletal lesions will rarely have the precision routinely encountered in clinical settings. Developments in molecular biology, however, are opening new doors for the paleopathologist (see Chapter 21). Unfortunately, few diseases leave signatures of any kind on the human skeleton, and those that do may cause very similar skeletal reactions. The only real advantage that the osteologist has in studying pathology is the ability to study the entire skeleton at once, without soft tissue cover. For these reasons, Steinbock (1976) suggested that the most rational approach to differential diagnosis in human skeletal pathology is to state the most likely diagnosis followed by a list of possible alternatives in order of decreasing likelihood.

As Miller *et al.* (1996) note, there are two major impediments to paleopathological work. First, there is a paucity of well-documented, clinically diagnosed skeletal samples to use as controls against which unknown skeletal samples (forensic, archeological, or paleontological) may be compared. The second problem lies with the difficulty in finding skeletal abnormalities or patterns of abnormalities that are unique to individual disease categories. Miller and colleagues suggest that diagnosis of paleopathologies be made at two levels. First, a classification of each case into one of seven categories: (1) anomaly, (2) trauma–repair, (3) inflammatory/immune, (4) circulatory (vascular), (5) metabolic, (6) neuromechanical, and (7) neoplastic (cancers). In a series of blind tests, these authors found that even trained specialists working on skeletal remains only achieved accuracies of about 43% at this broad diagnostic level, and only about 30% of more specific diagnoses were accurate. Buikstra and Ubelaker's Standards volume (1994) advocates the division of paleopathological changes into nine categories: (1) abnormalities of shape, (2) abnormalities of size, (3) bone loss, (4) abnormal bone formation, (5) fractures and dislocations, (6) porotic hyperostosis/cribra orbitalia, (7) vertebral pathology, (8) arthritis, and (9) miscellaneous conditions.

The first step in any paleopathological analysis is to establish the envelope for what is normal in size, shape, and topography for a healthy human's skeleton. When skeletal remains fall outside that envelope, pathology is one possible explanation. There are two basic steps to paleopathological work. The first is **description,** and the second is **diagnosis.** Diagnosis will rarely approach what is possible in the clinical setting because bone usually responds to insults by either resorbing or depositing, and it may therefore easily respond to different diseases in very similar ways. Description is the most important step in paleopathological work because even if the following diagnosis is incorrect, other workers can come along later and amend or modify the diagnosis. In any description, identify the nature and distribution of anomalies across the skeleton and the pattern of distribution of these anomalies across the population. These anomalies can take the form of lesions or changes in size or shape. Note the distribution of all lesions, whether single, multiple, diffuse, or concentrated. Note if bone is eaten away (lytic) or deposited (blastic). Check all bones of the skeleton for further evidence. Techniques and methods for paleopathological diagnosis may include, in addition to gross anatomy, radiographs, histology, microradiographs, scanning electron micrographs, chemical analysis, serology, and DNA techniques (see Chapter 21).

Steinbock's 1976 book, a 1981 Smithsonian volume by Ortner and Putschar, and an encyclopedia by Aufderheide and Rodriguez-Martin (1998) provide excellent guides to osteological pathology. Additional illustrations and techniques are found in Buikstra and Ubelaker (1994). The reader is urged to consult these sources when working with potentially pathological human skeletal remains. The current chapter is intended to supplement these texts and is organized accordingly. It is designed to provide the reader with an introduction to the kinds of dental and osteological pathology most commonly encountered in work with prehistoric skeletal remains. All illustrated examples of osteological pathology are drawn from the Berkeley Primate Skeletal Collection, most of them from archeological contexts. In summarizing work on this collection, Richards and Ánton (1987) note that in the five thousand individuals examined, over one-fifth showed anomalous development or the effects of pathological processes. Their data show that degenerative joint disease, periodontal disease, fracture, and osteomyelitis are the leading changes observed in this mostly prehistoric Californian collection.

18.1
Trauma

The most common pathology affecting the skeleton is degenerative change. Trauma occupies second place and affects the skeleton several

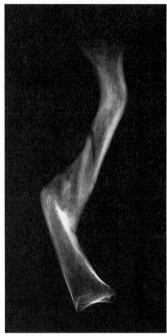

Figure 18.1 Healed fracture. This left clavicle shows a postmortem fracture (the light-colored bone around the crack visible in this inferior view) and a more medial premortem fracture that has healed. The original shaft surfaces are joined by a bony callus. The radiograph shows that this clavicle's medial end rotated counterclockwise relative to its lateral end, resulting in a dramatic misalignment of the fractured pieces during and after healing. Prehistoric, California. One-half natural size.

ways—fracturing or dislocating the bone, disrupting its blood or nerve supply, or artificially deforming it.

18.1.1 Fracture

Fracture in a bone occurs as a result of abnormal forces of tension, compression, torsion, bending, or shearing applied to the bone. Several possible types of gross fracture in a bone may result from this abnormal stress (Figures 18.1 and 18.2). A **complete fracture** is one in which broken ends of a bone become separated. An **incomplete,** or "greenstick," fracture (infraction) is one in which breakage and bending of a bone is combined. A **comminuted** fracture is one in which the bone splinters, whereas a **compound fracture** is a fracture in which the broken bone perforates the skin. Some fractures of bone are attributable to other pathological causes—**pathological fracture** occurs as a result of bones already having been weakened by other pathological or metabolic conditions such as osteoporosis.

Fractures of bones are often described by the features of the break itself: comminuted when there is shattering of bone, **compressed** when the bone is squeezed, or **depressed** when bone fragments are depressed below the adjacent surface.

Fracture healing is described in Chapter 2. Antemortem fracture may be differentiated from postmortem fracture only when a **callus,** the hard tissue formed at the site of a broken bone during the healing process, is

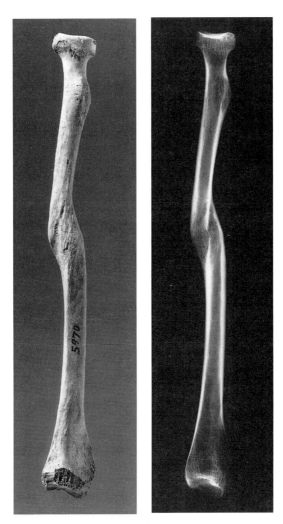

Figure 18.2 Healed fracture. This left radius has a healed midshaft fracture. The radiograph shows that the fracture offset is about equal to the width of the bone. The proximal part of the bone is offset posteriorly. Prehistoric, California. One-half natural size.

present. All other fractures that occur at or around the time of death should be diagnosed as perimortem. Full fracture healing can completely remove any gross signs of fracture, even in a radiograph. The rate of fracture repair depends on fragment alignment, the amount of movement at the fracture, and the health, age, diet, and blood supply of the individual. Some fractures never heal because of continued movement at the broken surface. Nonunions develop most frequently in the appendicular skeleton. It is possible that a new "joint," or **pseudarthrosis,** will form at the fracture site (Figure 18.3).

Trauma such as sword cuts or arrow perforations to bones constitute special kinds of fractures. Such wounds are capable of healing through the same processes described in Chapter 2. In any kind of fracture, adjacent bone is susceptible to subsequent pathological complications such as infection, tissue death, deformity, and arthritis brought on by the initial trauma.

Figure 18.3 Pseudarthrosis. This left humerus was broken just above midshaft. The fracture failed to heal (a nonunion), resulting in continuing movement, which formed the false joint. Prehistoric, California. One-half natural size.

The analysis of fractures on the populational level can be very informative in addressing questions of prehistoric behavior. For example, Lovejoy and Heiple (1981) assessed the Libben population and found that overall fracture rate was high. The low incidence of fractures in children suggested that traumatic child abuse was not practiced. The results of their analysis also suggested that fracture risk was highest in the 10–25 and

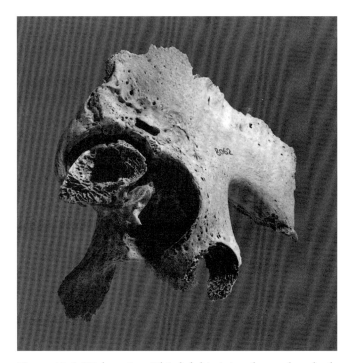

Figure 18.4 Dislocation. This left hip joint shows that the femoral head dislocated anterosuperiorly from its original place within the acetabulum. The cross section of the femoral neck is seen in the post-excavation break, which faces the viewer. Osteoarthritis secondary to the trauma is evident. Prehistoric, California. One-half natural size.

45+ age categories, and that the care of patients was enlightened and skillful among this Native American group.

18.1.2 Dislocation

In addition to causing bone fracture, trauma to the skeleton can also involve movement of joint participants out of contact and the simultaneous disruption of the joint capsule. If the bones participating in the joint remain dislocated, the result may be diagnosed osteologically (Figure 18.4). When the joint is dislocated, the articular cartilage cannot obtain nourishment from the synovial fluid, the cartilage disintegrates, and arthritic changes occur. Osteological manifestations of dislocation are usually confined to adults. The violent trauma necessary for dislocation usually separates the epiphyses in subadults, and slipped femoral epiphyses are common in juveniles. In the elderly the more brittle bone usually gives way, fracturing prior to dislocation. The two joints most often displaying osteological manifestations of dislocation are the shoulder and hip joints.

18.1.3 Artificial Deformation

Fracture and dislocation described above result from sudden trauma to the skeleton, but long-term trauma can also modify the shape of a bone. Deformities of this kind are most often induced as a result of cultural

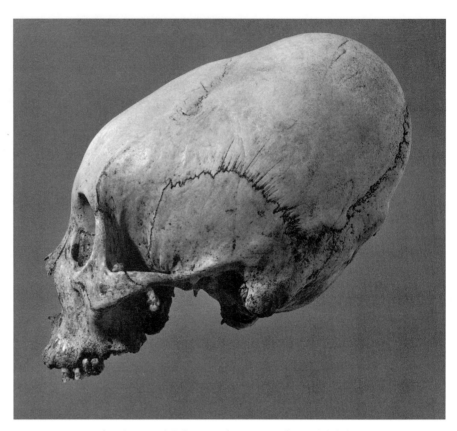

Figure 18.5 Artificial cranial deformation. Circumferential deformation is produced by wrapping the rear of the cranial vault. Prehistoric, Peru. One-half natural size.

practices such as cradleboarding, massaging, or binding the crania of infants. Another example is the foot-binding practiced by Chinese women of high status. The most common manifestations of artificial deformation of the skeleton are of the cranium. Cultures around the world have, for cosmetic reasons, altered the shape of the adult head by placing abnormal pressures on the developing skull. Ortner and Putschar (1981) describe the practice by people on every continent except Australia, but Brown's (1981) work there suggests that the practice was continentally ubiquitous. Both cultural and biological information can come from the analysis of intentionally deformed crania. For example, Ánton (1989) has used intentional anteroposterior and circumferential cranial vault deformation in a Peruvian sample to study the relationship between the cranial vault and base in the development of the craniofacial complex. Some examples of artificial cranial deformation are illustrated in Figures 18.5 and 18.6.

18.1.4 Trephination and Amputation

Ortner and Putschar (1981) describe **trephination,** or **trepanning,** as perhaps the most remarkable trauma seen by the paleopathologist (Figure 18.7). Written accounts documenting this practice extend to the ancient Greeks, but archeological work has traced it even more deeply into the past. The practice is known from Europe, the Pacific, both Americas,

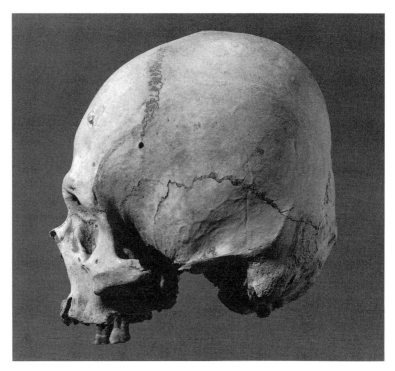

Figure 18.6 Artificial cranial deformation. Anteroposterior deformation is caused by the application of pressure from behind. Prehistoric, Peru. One-half natural size.

Africa, and Asia. Several techniques have been used to make an artificial hole in the cranial vault, including scraping a patch of bone away, cutting a bone patch out by cutting grooves through the vault, and drilling small holes around the plug of bone to be removed. The practice was undertaken to yield relief from intracranial pressure (especially from compressive fractures of the skull vault) and to relieve headaches, cure mental illness, or let out evil spirits. Success rate for this prehistoric surgery, as judged by subsequent healing around the hole, was often surprisingly high. Some postmortem trephination was done to fashion amulets for the adornment of survivors.

Evidence of **amputation** may be observed on the skeleton in the form of missing appendages or parts of appendages. Diagnosing premortem from perimortem amputation, again, depends on presence of healing or infection of the bone tissue at the point of trauma.

18.2
Infectious Diseases and Associated Manifestations

Infectious disease has long been a major cause of death in human populations. Unfortunately, dealing with osteological evidence of infectious disease can be frustrating because few infectious diseases leave any direct evidence of their existence in the skeleton. Many of the chronic infectious

Figure 18.7 Trephination. Three artificially produced holes are evident on this three-quarters view of a cranium. The hole closest to the parietal boss is nearly completely obliterated by healing. The other holes also show substantial bony healing, indicating that the individual survived the operations. Prehistoric, Peru. One-half natural size.

diseases that *do* leave osteological signs produce morphologically overlapping responses, making differential diagnosis impossible.

Osteitis is a general term for an inflammation of bone tissue caused by infection or injury and is not specific as to cause. The terms **osteomyelitis** and **periostitis** are slightly more confusing, because they serve to generally describe osteological conditions as well as to identify specific diseases; periostitis is a symptom in a disease syndrome such as syphilis, but it is also common in many other diseases (Ortner and Putschar, 1981).

18.2.1 Osteomyelitis

Osteomyelitis is bone inflammation caused by bacteria that usually initially enter the bone via a wound. This disease mainly affects the long bones (Figure 18.8) and is defined as an infection that involves the medullary cavity. However, usage in the paleopathology literature has often been imprecise. Osteomyelitis is almost always caused by pus-producing microorganisms (90% of the time by *Staphylococcus aureus*) and is thus called **suppurative,** or **pyogenic, osteomyelitis.** The microorganisms can reach the bone directly, as a result of injury at any age, or via the bloodstream (hematogenous osteomyelitis, most often found in children). Characteristic hard tissue manifestations include an **involucrum** of coarsely woven bone around the original long bone cortex and one or more open-

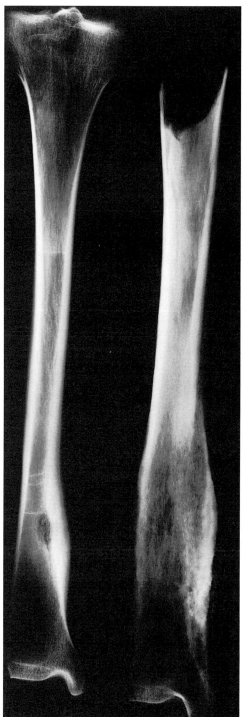

Figure 18.8 Osteomyelitis. The tibia on the left shows localized reactive bone and a cloaca; that on the right shows the result of a more extensive reaction to the infection. The radiographs of the two bones show clear involvement of the medullary cavity. Prehistoric, California. One-half natural size.

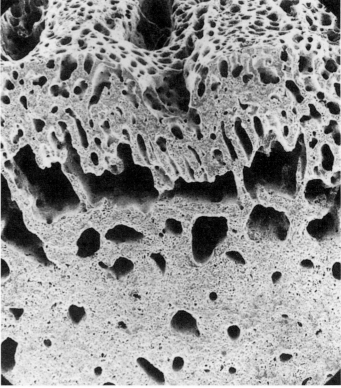

Figure 18.9 Periostitis. The right tibia and left ulna of the same individual show reactive bone that overlies the cortex. In gross aspect, the postmortem exfoliation of reactive bone has exposed the underlying cortex on the posterior ulnar edge. Prehistoric, California. One-half natural size. The scanning electron micrograph shows a cross section through the surface of a similarly affected individual. Scale: approximately 20×.

ings for pus drainage called **cloacae** (**fistulae**). The latter open through the involucrum.

18.2.2 Periostitis

Periostitis is a condition of inflammation of the periosteum caused by trauma or infection. It is not a disease. See Mensforth *et al.* (1978) for a complete review of its various etiologies. These authors demonstrate an age-specific distribution of periosteal reactions which seems to coincide with and be a response to infectious disease in infants and children in the prehistoric Libben population. Periostitis involves only the outer (cortical) bone, without involvement of the marrow cavity as in osteomyelitis. It can be acute or chronic and occurs any time that the inner surface of the periosteum reacts to insult by forming woven bone that sleeves the underlying cortical bone (Figure 18.9).

18.2.3 Tuberculosis

Tuberculosis is a chronic infectious disease that results from a bacterium, *Mycobacterium tuberculosis*. Infection is usually via the respiratory system, but other body parts, including bones, can also be affected. Bone and joint destruction can result from the infection. The presence of Precolumbian tuberculosis in the New World is assessed by Buikstra (1981b), who reviews literature on the topic that stretches back over three hundred years. She concludes that the evidence for precontact New World tuberculosis is good. Salo *et al.*'s molecular work confirmed this in 1994 (see Chapter 21).

Skeletally, a variety of bones can be affected by tuberculosis, but the vertebral column is the most common primary focus. Collapse of one or several vertebral bodies causing a sharp angle in the spine (**kyphosis**) when viewed from the side is the most common manifestation. Differential diagnoses from osteomyelitis and septic arthritis are often possible because tuberculosis shows destruction and cavitation in cancellous bone, without extensive associated reactive bone. The pattern of element involvement, with the vertebrae and os coxae as foci, marks tuberculosis. In addition, there is no evidence of sequestration, an involucrum, or fusion of the joints. Ortner and Putschar (1981) discuss and illustrate the skeletal effects of tuberculosis.

18.2.4 Treponemal Infections

Skeletally significant diseases caused by a microorganism known as a spirochaete in the genus *Treponema* are yaws, and endemic and venereal syphilis. These diseases have a worldwide distribution today, but considerable controversy exists over their origins and distribution several hundred years ago. Debate centers on whether syphilis (misdiagnosed as leprosy) was present in the Old World before Columbus returned from the New World. The controversy is well reviewed by both Steinbock (1976) and Ortner and Putschar (1981), with a more recent update by Baker and Armelagos (1988).

In syphilis, the microorganisms enter the body through skin or mucous membrane sites. Tertiary syphilitic skeletal lesions occur progressively, usually beginning between 2 and 10 years after infection. These can be complex, but there is usually an osteological focus of the disease in the frontal and parietals, the facial skeleton, and the tibia. Individual lesions may not be distinguishable from some cancers, tuberculosis, or other infectious changes. Both Steinbock (1976) and Ortner and Putschar (1981) provide further details.

18.3
Circulatory Disturbances and Hematopoietic Disorders

When blood supply to bones (see Chapter 2) is upset by trauma or other diseases, a variety of bony manifestations can occur, including death of

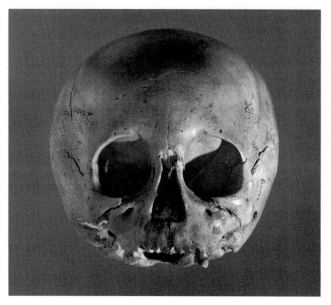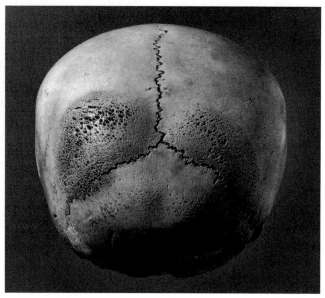

Figure 18.10 Cribra orbitalia, porotic hyperostosis. The skull of a 5-year-old child is shown in anteroinferior and posterior views to display the cribra orbitalia of the frontal (in the roofs of the orbital cavities) and spongy hyperostosis of the parietals and occipital. Prehistoric, Peru. One-half natural size.

bone tissue. Anemias can also affect the gross appearance of bones. Iron deficiencies, sickle-cell anemia, and thallasemia all result in the expansion of spaces occupied by hematopoietic marrow within the bones. The result is often visible in the skull, with a widening of the diploë leading to a thickening of the cranial vault. Zimmerman and Kelley (1982) attempt a differential diagnosis of different types of anemias.

Porotic hyperostosis, or **spongy hyperostosis,** is a condition exhibiting lesions, usually of the cranial vault. These represent a thinning and often complete destruction of the outer table of the cranial vault that results in a sievelike or "coral-like" appearance of the ectocranial surface. The lesions seem to be caused by anemia–associated hypertrophy of the diploë between the inner and outer tables. Porotic hyperostosis is most often seen in immature individuals and is recognizable as a porosity of the cranial vault. It is usually bilaterally symmetrical, focused on the parietals and the anterolateral quadrant of the orbital roofs. The orbital lesions are called **cribra orbitalia,** and a similar disturbance of the endocranial surface is called **cribra cranii** (Figure 18.10).

The causes of this bony reaction have been extensively speculated on. Its high frequency in the Southwest of North America was once thought to be because of a maize-based, iron-deficient diet. However, work on the phenomenon in a large sample of California skeletons (nonagricultural, fish-dependent) by Walker (1986) has suggested that porotic hyperostosis is sometimes due to nutrient losses associated with diarrheal diseases rather than to diet per se, a position supported by Kent (1987). Palkovich (1987), in a study of skeletal remains from the prehistoric Southwest, suggests that endemically inadequate maternal diet can combine with infection to produce very early onset of iron-deficient anemia with resultant porotic hyperostosis. Research by Stuart-Macadam (1987a,b) stresses the association between the osteological manifestations and anemia. Mensforth *et al.* (1978) demonstrate, using the prehistoric Libben sample, that a common cause of porotic hyperostosis is the normal physiological sequelae of infection. As noted above, they found a direct relationship between periostitis (infection) and porotic hyperostitis (iron deficiency) in subadults. This is consistent with clinical data that suggest that iron stores may be sequestered within the body as a defense against infection. Thus, much prehistoric porotic hyperostosis may be the secondary consequence of infectious disease, not of diet. Indeed, Stuart-Macadam (1991) posits that porotic hyperostosis is the result of an interaction between customs, diet, hygiene, parasites, and infectious diseases. She feels evidence for chronic disease in skeletal material should not be interpreted as an individual's inability to adapt to the environment but rather as evidence of the individual's fight for health against the pathogen.

18.4
Metabolic and Hormonal Imbalance

Narrowly defined, metabolic disorders of bone are disorders in which a reduction in bone mass is the result of inadequate osteoid production, or mineralization or excessive deossification of bone. Nutritional deficiencies are usually classified under metabolic disorders. Hormonal disturbances can also lead to dramatic changes in normal skeletal anatomy.

18.4.1 Scurvy

Scurvy is a metabolic disease caused by a long-term insufficient intake of vitamin C, which is essential for the production of collagen and therefore osteoid. Skeletal manifestations are most apparent in infants, usually in the form of cortical thinning and pathological fractures in rapidly growing bone areas.

18.4.2 Rickets

Rickets is most often a nutritional disease resulting from an insufficient amount of vitamin D in the diet that causes a failure of mineral deposition in the bone tissue. As a result, excessive uncalcified osteoid accumulates, and the bone tissue remains soft and flexible. The disease was described in the 1600s, but its nutritional source was not discovered until the 1920s. The osteological effects of rickets are present throughout the skeleton, but they are most pronounced in the limbs, which are usually bent and distorted. The legs are characteristically bowed outward or inward. In adults, the same dietary deficiency is called **osteomalacia**—a disease usually linked to general malnutrition, particularly deficiencies in protein, fat, calcium, and phosphorus. Its greatest effect is on bones in which remodeling is highest (ribs, sternum, vertebrae, and pelvis), which are subject to pathological fracture.

18.4.3 Osteoporosis

Osteoporosis, or **osteopenia** in the nonclinical situation, refers to the increased porosity (reduced mass) of bone that is most often part of the aging process. It is a consequence of the organism's failure to maintain the balance between bone resorption and formation. Postmenopausal women are most at risk for osteoporosis because of the cessation of estrogen production. Males are endowed initially with more bone mass than females and so do not become vulnerable to osteoporosis until later in life (in their 70s and 80s). It is estimated that the annual costs associated with osteoporosis in the United States alone involve $7 billion. Osteoporosis is associated with 1.2 million bone fractures in the elderly each year. Research into basic bone biology, particularly into the factors that activate and inactivate osteoclasts, has been spurred on by these statistics. Lenchik and Sartoris (1997) provide a review of rapidly progressing research into osteoporosis.

18.4.4 Endocrine Disturbances and Dysplasias

The growth of the skeleton is controlled, in large part, by the secretion of hormones in the pituitary and thyroid. Pathology in these glands can lead to extreme skeletal changes. **Gigantism** results from excessive production of somatotrophic hormone and consequent overstimulation of growth cartilages and gigantic proportions of the skeleton. **Acromegaly** is similarly caused by an overly productive pituitary, but after the epiphyses are fused. The most dramatic osteological manifestation of acromegaly is growth at the mandibular condyle and a resulting elongation and distortion of the lower jaw. **Dwarfism** is caused by a variety of conditions. **Achondroplasia** is a hereditary form of dwarfism with limb shortening, almost normal trunk and vault development, and a small face (Figure 18.11). It is a hereditary disease caused by the congenital disturbance of cartilage formation at the epiphyses, a skeletal dysplasia rather than an endocrine disturbance.

18.5 Tumors

Osteogenic sarcoma is less rare than bony cancer that results from metastasis. In other words, skeletal tumors usually stem from other tissue sources, but their appearance can be very dramatic. Modern classifications of skeletal tumors name over forty different tumor types in bone and associated cartilage and fibrous connective tissue. Histological specimens and biochemical data are usually needed to sort these out. Tumors are somewhat arbitrarily divided into **benign** or **malignant** and are classified according to the tissues in which they originate. **Multiple myeloma,** for example, is a rare primary malignant tumor of hematopoietic tissue. Its effect on bone tissues is a widespread pattern of lytic lesions on various skeletal elements.

The most common tumor of bone is an **osteochondroma.** These are benign tumors, usually asymptomatic. They always arise at epiphyseal

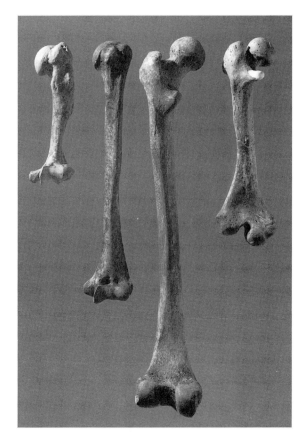

Figure 18.11 Achondroplasia. The left humerus and left femur of an achondroplastic dwarf (*far left* and *far right*) are compared with a normal human humerus and femur from a single individual (*center*). Note the disproportions and fully adult status of the achondroplastic individual. Prehistoric, California. One-fourth natural size.

lines and protrude at right angles to the long axis of a bone. They resemble ossified tendons in many cases. An **osteoma** is a mound of compact bone, usually on the ectocranial surface; these are often called "button" osteomata. They are hard, dense, and ivorylike in appearance and occur in about 1% of all people. **Ear exostoses** are "osteomata" of the region on the inner aspect of the external auditory meatus.

The most commonly formed osteogenic sarcomatas occur during the growth period. These include **osteosarcomata, chondrosarcomata,** and **Ewing's sarcoma.** Both Steinbock (1976) and Ortner and Putschar (1981) provide good information on these rare cancers.

18.6
Arthritis

Arthritis is the inflammation of a joint—a general inflammation that includes soft tissue effects. The inflammation can come as a result of

Figure 18.12 Osteoarthritis. The normal lumbar vertebra (*left*) lacks the osteophyte development seen on the anterior and lateral edges of the vertebra with degenerative arthritis (*right*). Prehistoric, California. One-half natural size.

trauma as well as of bone and joint infections (see Rogers *et al.*, 1987, for a summary and classification).

18.6.1 Osteoarthritis

Osteoarthritis, the most common form of arthritis, is characterized by the *destruction* of the articular cartilage in a joint and the *formation* of adjacent bone, in the form of bony lipping and spur formation (**osteophytes**), around the edges of the joint. A better term for this phenomenon is **degenerative joint disease** (Figures 18.12–18.14). The causes of this disease are, for the most part, mechanical. The disease occurs mostly in load-bearing joints, particularly in the spine, the hip, and the knees. Osteoarthritis is an inherent part of the aging process. For a review of current knowledge about osteoarthritis, consult Epstein (1989), and for a review of prehistoric arthritis in the Americas, see Bridges (1992).

Osteoarthritis is usually classified as either **primary,** resulting from a combination of factors that include age, sex, hormones, mechanical stress, and genetic predisposition, or **secondary,** initiated by trauma or another cause such as the invasion of the joint by bacteria (**septic,** or **pyogenic, arthritis**—often a complication of osteomyelitis). Studies of the patterning of osteoarthritic lesions of the skeleton at the individual and population level can shed light on prehistoric activity patterns. For example, Merbs (1983) was able to show that osteoarthritic changes seen in female Hudson Bay Inuit skeletons correlated with ethnographic accounts and archeological evidence for scraping and cutting animal hides and sewing, whereas lesions seen in males correlated with harpoon throwing and kayak paddling.

A phenomenon often found associated with osteoarthritis is **eburnation,** the result of subchondral bone being exposed when cartilage is destroyed. Bone affected this way takes on a polished, ivorylike appearance. The projecting spicules of bone associated with osteoarthritis are called **osteophytes.** Nearly all individuals older than 60 years exhibit these arthritic features, especially in the lower thoracic and lumbar regions. Rogers *et al.* (1987, 1989) assess classifications of osteoarthritis, and Bridges

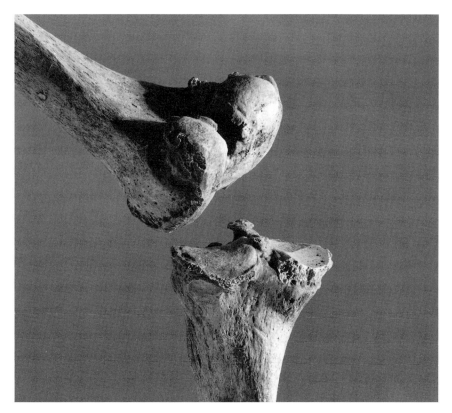

Figure 18.13 Osteoarthritis. Eburnation and marginal lipping are evident on this left knee joint, seen here in posterolateral view and lit from the lower right to show detail. The eburnation is the ivory-like, shiny patch on the medial femoral and tibial condyles. Prehistoric, California. One-half natural size.

(1993) notes that the scoring procedures used to record the disease seriously affect the results of any comparative study.

18.6.2 Rheumatoid Arthritis

Middle-age women have a predisposition for this arthritis. Its exact cause remains unknown but almost certainly varies with genetic background. In rheumatoid arthritis, the body's immune system attacks its own cartilage. Bone changes are atrophic and especially focused in the hands and feet. The lesions are usually bilaterally symmetrical. Rheumatoid arthritis is the least common arthropathy in archeological skeletal material. Woods and Rothschild (1988) argue that it is evidenced in New World skeletal remains, and Rothschild *et al.* (1990) propose recognition criteria for skeletal remains.

18.6.3 Ankylosing Spondylitis

An **ankylosis** is an abnormal immobility and fixation of a joint resulting from pathological changes in the joint. Ankylosing spondylitis is a chronic and usually progressive disease that affects the vertebral column. The

Figure 18.14 Osteoarthritis. Trauma to this left femur produced secondary osteo-arthritic changes to the joint, seen here in the form of a bony extension and defor-mation of the femoral head and a buildup of osteophytes around the perimeter of the acetabulum of the os coxae. Prehistoric, California. One-third natural size.

associated ligaments of the spine ossify, and the intervertebral joints be-come immobilized.

18.7
Dental Pathology

Because the teeth directly interact with the environment, they are suscep-tible to damage from physical and biological influences not operating on other skeletal elements. A study of dental pathology can be useful in in-vestigating the health and diets of individuals and populations. Even though tooth wear that is excessive by today's standards has characterized humans and their ancestors for millions of years, it should be noted that tooth wear and artificial tooth modification have been included in this chapter only for convenience. Lukacs (1989) divides dental diseases into four categories, infectious, degenerative, developmental, and genetic. All of these, of course, are interrelated in the dental health of the individual. Tooth wear should be thought of as a pathology only when it is so extreme that the associated bone is negatively affected. In Chapter 17 we consider how tooth wear may be used to age skeletal material, and in Chapter 20 we discuss how such wear can be used to assess prehistoric diets. Figure 18.15 shows how extreme tooth wear was under many prehistoric conditions.

Figure 18.15 Tooth wear. Heavy attrition on this adult male's dentition has eliminated all but the third molar crowns. The individual continued to chew on the stubs of his incisors, canines, premolars, and second molar after the tooth crowns had worn away. Such wear is a normal phenomenon in older individuals from aboriginal populations with grit in the diet. Prehistoric, California. Natural size.

18.7.1 Caries

Dental **caries** is a disease process characterized by the progressive decalcification of enamel or dentine. The macroscopic appearance of caries can vary from opaque spots on the crown to gaping cavities in the tooth. A prerequisite for the formation of dental caries is **dental plaque** and a diet that includes fermentable carbohydrates. Plaque is the matrix and its inhabiting community of bacteria that forms on the tooth. Carious lesions can begin anywhere that plaque accumulates, most often in the fissures on tooth crowns and in the interproximal areas. Figure 18.16 shows a large carious lesion.

Larsen (1983) observed that the prehistoric shift to agriculture on the Georgia coast led to an increase in the frequency of carious lesions, most marked in females. This finding indicated subsistence role differences between the sexes and is an example of how archeological and osteological data can be combined in insightful ways. Walker and Erlandson (1986) saw a similar but inverse shift on the California coast. Here, prehistoric people made a subsistence shift from a cariogenic diet, consisting mostly of plant foods, to an intensive exploitation of fish.

18.7.2 Dental Hypoplasia

Dental hypoplasia is a condition characterized by transverse lines, pits, and grooves on the surface of tooth crowns. These disturbances are defects

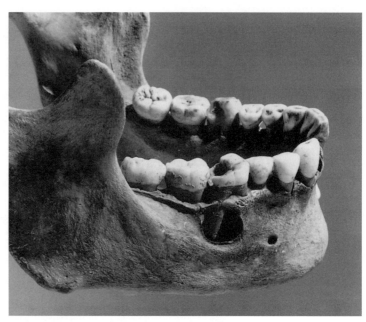

Figure 18.16 Caries. Bilateral carious lesions of the first molars are evident on this individual. A related abscess around the distal first molar root is seen posterosuperior to the right mental foramen. Prehistoric, California. Natural size.

in enamel development. Amelogenesis, enamel formation, begins at the occlusal apex of each tooth crown and proceeds rootward, ending where the crown meets the root at the cervical enamel line. During this process, stress to the organism may result in a temporary upset of ameloblastic activity, and a consequent enamel defect marking the interruption of development. These enamel hypoplastic defects can take many forms, ranging from single pits, to lines, to grooves. As Danforth *et al.* (1993) note, scoring of these features varies greatly between observers, so that interobserver and intraobserver bias must be considered in any comparison of results from different studies.

Several different factors can cause dental hypoplasias (see Goodman and Armelagos, 1985, and Skinner and Goodman, 1992, for reviews), all of them the result of metabolic insult to the organism. Like Harris lines (see above), hypoplastic bands in individuals can indicate the age at insult. Study of these developmental defects in populations can give insights into patterns of dietary and disease stress in prehistoric groups. Goodman and Rose (1991) provide a fine discussion of enamel hypoplasias as clues about the adequacy of prehistoric diet. Figure 18.17 shows heavy hypoplastic banding and pitting on a child's permanent canines.

18.7.3 Periodontal Disease

Periodontitis is the inflammation of tissues around a tooth. It can involve both soft tissues and the bone itself. Periodontal disease in skeletal remains is recognized as a result of infection of the alveolar bone and adjacent tissues. It causes recession of the alveolar bone, as either a horizontal lowering of the crest of the alveolar process or an irregular lowering of the process, with pockets or wells expanding into the cancellous bone of the

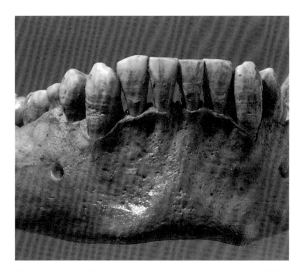

Figure 18.17 Dental hypoplasia. This 12-year-old child suffered a metabolic upset during the formation of his canines (to see when this occurred, consult Figure 17.1). This upset recorded itself in the linear hypoplasia seen bilaterally on the labial surface of the canine crowns. Calculus is also seen on the labial surface of the incisor crowns. Prehistoric, California. Natural size.

jaws. The agents of infection are microorganisms, and the disease is usually due to the combined effects of large, mixed communities of bacteria. An **abscess** is a localized collection of pus in a cavity formed by tissue disintegration. Abscesses are often found as cavities within alveolar bone near the tooth root apices. Figure 18.16 illustrates these features. Clarke and Hirsch (1991) provide a thorough account of factors influencing alveolar bone.

18.7.4 Dental Calculus

Dental calculus is mineralized plaque on a tooth surface. The fact that calculus can trap food debris has been used by Dobney and Brothwell (1986) as an approach for ascertaining aspects of prehistoric diet. Figure 18.17 shows remnants of calculus buildup.

18.7.5 Artificial Dental Modification

A variety of cultural practices (technically, forms of trauma) make impacts on the dentition. People engrave, color, and even intentionally pull out (**evulse**) teeth for cosmetic purposes. Brothwell (1981) illustrates the variety of decorations encountered on dental remains, primarily on the anterior teeth. These artificial incisions should not be confused for hypoplasias. Milner and Larsen (1991) review filing, chipping, inlays, and other alterations, referring to these as **dental mutilations,** whereas purposeful removal of usually anterior teeth is termed **ablation.**

Frayer (1991) and many others have interpreted the interproximal grooves sometimes present between adjacent teeth as grooves left by the use of toothpicks, and such features are even known from fossil hominid teeth of more than two million years in age. Fox (1992) reviews literature on cultural striations in the human dentition, noting several studies which have shown handedness based on the directionality of oblique scratch marks on the labial surfaces of upper incisors, particularly among paleolithic hominids. These marks were presumably made because food was secured by using the front teeth as a vise while pieces were cut away with a stone implement that came into contact with the teeth.

18.8
Occupation

In both forensic and particularly in archeological contexts, it is important to determine the occupational and social status of individuals now represented by skeletons. Kennedy (1989) traces the history of such studies in anthropology and medicine. Waldron's (1994) call for critical evaluation of claims is apt and enlightening. He notes that in their fervor to deduce as much as possible about the lives of ancient individuals, osteologists have made some extravagant claims about environmental stresses, parity, social status, and occupation. As he notes, some authors have been unable to resist the urge to deduce an underlying cause for every bony lesion, identifying individuals as horsemen, sling throwers, weavers, and corn grinders based on the presence of bowed limb bones, spinal deformities, and osteoarthritis at various joints.

Perhaps because information regarding occupation and status is so valuable in an archeological context, not enough attention is paid to the premises that underlie such identifications. Waldron's examination of whether patterns of osteoarthritis can be used to identify occupation is a useful departure. Proceeding from known to unknown, he notes that there is a wealth of modern clinical data on osteoarthritis and occupation. He finds no convincing epidemiological evidence of a consistent, coherent relationship between a particular occupation and a particular form of osteoarthritis. Indeed, given the fact that sex, race, weight, movement, and genetic predisposition are all known factors that influence development of the disease, this is not surprising. Waldron concludes that since we know that occupation is not the sole cause of osteoarthritis, there cannot be any likelihood of being able to deduce the former from the latter. Furthermore, even in cases in which occupationally related activity does seem to be important in determining the expression of arthritis, there are no unique features about this expression—most people who develop arthritis at the finger joints are not mill workers, even though mill workers do develop arthritis at these joints. For single skeletons, the prospects therefore seem dim.

On a populational basis, however, the prospects are better. Here, by examining the patterns of osteoarthritis in each skeleton, and among all skeletons, it might be possible to draw conclusions about activity pattern differences on a group, or population level. This is the underlying basis of the only comprehensive skeletal study of activity-induced pathology, a small populational study of Canadian Inuits in which activity patterns such as kayaking, harpoon throwing, and sewing were related to osteoarthritic patterns on a populational basis (Merbs, 1983). A study by Stirland (1992) on sailors from King Henry VIII's A.D. 1545 flagship, the *Mary Rose,* is another, smaller, and more limited sample in which an attempt to relate occupation to paleopathology is made. Even in the best of cases, however, it remains impossible to definitively conclude that any single individual had a particular occupation based on any particular arthritic joint in his or her skeleton. As Waldron (1994, p. 98) cautions, "There is a perfectly understandable drive to make the most of what little evidence survives in the skeleton and this sometimes has the effect of overwhelming the critical faculties." The osteology student should be wary of poorly supported claims about diet, disease, demography, and occupation at both the level of the individual skeleton and the level of the population sample.

Suggested Further Readings

Aufderheide, A. C., and Rodríguez-Martín, C. (1998) *The Cambridge Encyclopedia of Human Paleopathology.* Cambridge: Cambridge University Press. 496 pp.
  Descriptions and many illustrations make this a valuable reference volume.

Buikstra, J. E., and Cook, D. C. (1980) Palaeopathology: An American account. *Annual Review of Anthropology* 9:433–470.
  A comprehensive summary of work on the paleopathology of skeletal remains from the New World.

Gregg, J. B., and Gregg, P. S. (1987) *Dry Bones: Dakota Territory Reflected.* Sioux Falls, South Dakota: Sioux Printing. 236 pp.
  A book summarizing work on skeletal remains from the Upper Missouri River Basin. The book contains numerous illustrations of osteopathologies, but the book is more than an atlas, as it attempts to relate osteological health and disease to ancient culture.

Huss-Ashmore, R., Goodman, A. H., and Armelagos, G. J. (1982) Nutritional inference from paleopathology. *Advances in Archaeological Method and Theory* 5:395–474.
  A review of skeletal paleopathological indicators of nutrition.

Ortner, D. J., and Putschar, W. G. (1981) *Identification of Pathological Conditions in Human Skeletal Remains.* Smithsonian Contributions to Anthropology 28:1–479. Washington, D.C.: Smithsonian Institution Press.
  An outstanding text; essential for any work in paleopathology.

Ortner, D. J., and Aufderheide, A. C. (1991) *Human Paleopathology: Current Syntheses and Future Options.* Washington, D.C.: Smithsonian Institution Press. 311 pp.
  An edited volume with the world's leading paleopathologists contributing papers spanning the entire spectrum of skeletal research.

Rothschild, B. M. (1992) Advances in detecting disease in earlier human populations. In: S. R. Saunders and M. A. Katzenberg (Eds.) *Skeletal Biology of Past Peoples: Research Methods.* pp. 131–151. New York: Wiley-Liss.
  A review of technological and methodological advances in paleopathology.

Steinbock, R. T. (1976) *Paleopathological Diagnosis and Interpretation.* Springfield, Illinois: C. C. Thomas. 423 pp.
  A well-illustrated text on paleopathology.

Tyson, R. A., and Alcanskas, E. S. D. (Eds.) (1980) *Catalog of the Hrdlička Paleopathology Collection.* San Diego, California: San Diego Museum of Man. 359 pp.

> An atlas with hundreds of illustrations of osteological pathology and an excellent glossary.

Tyson, R. A. (Ed.) (1997) *Human Paleopathology and Related Subjects: An International Bibliography.* San Diego: San Diego Museum of Man. 716 pp.

> A modern guide to the paleopathological literature.

Ubelaker, D. H. (1982) The development of American paleopathology. In: F. Spencer (Ed.) *A History of American Physical Anthropology 1930–1980.* pp. 337–356. New York: Academic Press.

> An informative history of paleopathological studies, with an American focus.

Zimmerman, M. R., and Kelley, M. A. (1982) *Atlas of Human Paleopathology.* New York: Praeger. 220 pp.

> An atlas with a section on mummified tissue.

CHAPTER 19

Postmortem Skeletal Modification

Bones change as the individual grows. This process, **ontogeny,** continues until late in adulthood in some regions such as the pubic symphysis (see Chapter 17). Because bone is living tissue, it can respond to physical stimuli at any time in an individual's life. Individual variation in any human skeletal element, therefore, can be the result of the bone responding to genetic control, to environmental factors, or to both. The morphological changes discussed in Chapter 18 are **premortem** effects. They include changes brought about by pathology. After death, however, further morphological changes can occur in bone, brought about by biological, chemical, and physical agencies operating on it. **Postmortem modification** alters both the condition of the individual bones and the completeness of the skeleton as a whole.

The study of the processes that operate between the time of death of the organism and the time of study by the osteologist is called **taphonomy.** Taphonomy, from the Greek words for "burial" and "laws," is a word coined in the 1940s by the Russian paleontologist Efremov. Taphonomy is usually described as a subdiscipline of paleontology, but its methods and data are often applied in archeological contexts. Human skeletal remains are recovered from a variety of contexts. These include geological deposits such as cave floors, alluvial bodies, lacustrine deposits, peat bogs, and volcanic ash. Archeological contexts include house floors, wells, battlefields, megalithic structures, plague pits, cemeteries, funeral urns, refuse pits, and even hearths. Given this variety of contexts, postmortem modification of human skeletal remains can take many forms.

Taphonomy merits the attention of the human osteologist because hominid skeletal remains often bear the traces of past processes and activities useful in archeological interpretation. Bones have the potential to show modification in forensic, archeological, and paleontological contexts. Human osteologists are often called on to determine whether the distinctive patterns of damage or element representation in a skeletal assemblage or individual are the result of human behavior. An understanding of the many processes that alter bones and bone frequencies provides the basis for such interpretations. In addition to modifying the bones themselves, postmortem processes can dramatically alter the composition of skeletal populations. It is essential that these processes and their effects be understood in order to avoid the mistaken attribution of postmortem modification to premortem pathological processes (see Chapter 18). Furthermore, several pitfalls of demographic reconstruction center around the differential destruction of skeletal remains due to postmortem processes (see Chapter 20).

The forensic osteologist working in a crime scene context is a member of an investigative team searching for all available clues. However, some physical anthropologists working archeological or paleontological contexts have focused exclusively on the retrieval of bones and thereby have missed important contextual clues about past behavior. The exclusive focus on retrieval has the potential to sacrifice much critical information about the past. Archeologists often study burials to understand past cultural activities, particularly mortuary practices. Duday (1978; Duday and Masset, 1987) has consistently called attention to the need for communication between archeologist and osteologist during every phase of the recovery of skeletal remains. Nawrocki (1995) even suggests that this specialized, gap-filling field of study should be called human taphonomy. It is important to remember that bodies are usually buried, not skeletons. The disposition of the skeleton therefore provides critical information on how the body was buried. The cultural and taphonomic circumstances surrounding death and primary burial influence the skeleton recovered by the archeologist. By using all available clues exposed during excavation, the cultural arrangement of the cadaver and the infilling of the burial space may be inferred.

It is most convenient to divide taphonomic agents into two major classes, biological and physical, and to consider human-induced modifications separately. It should be realized, however, that biological agents act through both physical and chemical pathways to alter bones. In this chapter we describe and illustrate the effects of the most commonly encountered postmortem alterations of human skeletal remains. The student is referred to Lyman's 1994 volume *Vertebrate Taphonomy* as an excellent extension of much subject matter of this chapter, from an archaeological perspective. Haglund and Sorg's (1997) excellent edited volume *Forensic Taphonomy* is another valuable resource. Chapter 21 considers taphonomic changes to biomolecules.

19.1
Bone Fracture

In osteological analysis, it is critical to identify deviations of bones from the normal condition and to distinguish between deviations caused by pathological agents and those brought about by taphonomic agents. Assessing bone fracture in an archeological context is an example of the difficulties of such work. Fracture of a radius, for example, can occur a year before death, an hour before death, immediately after death, or during excavation. The causes of such fracture vary. Premortem fracture can occur as the individual falls from a tree. Postmortem fracture can be caused by the corpse being forced into a small burial crypt or by a hyena scavenging the body. Impatient archeologists also have been known to fracture bones through carelessness. Signs of bony healing around the fracture could identify the fracture as premortem. Without this healing, however, the osteologist may be forced to identify the fracture as **perimortem**

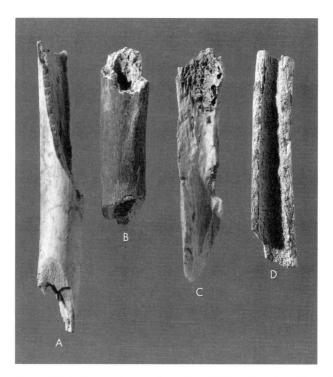

Figure 19.1 Recent (or modern, B and D) and ancient (A and C) fracture of human femoral shafts. Specimens A and C are from an archeological context in which the bone was fragmented when fresh, whereas specimens B and D were fractured during retrieval of the elements from the ground. Note the conchoidal impact scar on the internal surface of specimen C. One-half natural size.

(around the time of death), implying that it is not possible to ascertain whether the fracture occurred just before, during, or after death. Ubelaker (1992b) notes the significance of fracture of the hyoid in a forensic context. Death by strangulation sometimes involves hyoid fracture. However, dissection during autopsy can fracture the hyoid. Incomplete ankylosis of the horns to the body is sometimes mistaken for fracture. The hyoid can be fractured prior to death and not show significant remodeling at the time of death. The osteologist's contribution to a forensic investigation may be to bring all these possibilities to the attention of the investigative team.

The rate at which a bone loses its organic component and becomes "dry" as opposed to "green" or "fresh" varies widely, depending on the environment of deposition. Mineralization or "fossilization" of the bone also depends more on context than on elapsed time. Despite the variations in rate of change, it is often possible to distinguish between **ancient** perimortem fracture of bones that still retained much organic component when broken, and **recent** fractures of dry bones which occurred during excavation and transport. Discrimination is often facilitated by reference to the surface color and edge characteristics of the broken surfaces (Figure 19.1). Ancient longitudinal or spiral fractures of the shaft are usually straight, with sharp, linear edges. Since the fracture surface had already formed at the time of burial, this surface is usually the same color as the rest of the bone surface. Dry or fossilized bones that have been recently broken, on the other hand, usually have rougher, more jagged fractures, and the fracture surface is usually a different color (lighter in most unfossilized bone) than the adjacent unbroken surfaces.

As described in Chapters 4–13, individual bones perform different mechanical and physiological functions. These functions are reflected in the wide variation in size, shape, density, and internal structure of the bones. These characteristics, in turn, affect the potential for postmortem fracture of each element and even parts of elements. For example, human femoral shafts from archeological collections are far more likely to escape the ravages of biological and physical destruction than the smaller, more fragile sternum. Element representation in excavated skeletal assemblages can therefore be altered significantly from that predicted by element ratios in the intact skeleton. The absence of hand phalanges in an excavated cemetery assemblage usually does not mean that bodies were buried without fingers. Instead, such absence more often indicates that years of postburial rodent tunneling through the site displaced these small elements, or that recovery techniques were not adequate. Attribution of the disproportion to some ritual activity involving removal of the fingers would be unwarranted in such a case.

The appreciation that patterns of element disproportions in the archeological record are not all attributable to human intervention has caused archeologists to look closely at natural bone modification in the modern world. These **actualistic** studies have repeatedly shown that the structure of the bones themselves is often a major determinant of patterning in the archeological record. For example, the edge of the tibial plateau is often eroded simply because of its prominence and thin cortex, whereas the shaft of this element is rarely damaged.

19.2
Bone Modification by Physical Agents

19.2.1 Chemistry

Postmortem changes in bone range from minor alterations of bone protein to complete structural and chemical breakdown. As outlined in Chapter 2, the major constituents of bone are protein (mostly collagen) and minerals. The relationship between these constituents involves complex structural features and chemical bonds whose nature is not fully understood (Von Endt and Ortner, 1984). When the organism dies, the once dynamic bone tissue begins to disintegrate. Soil acidity (pH) and permeability, moisture, temperature, and microorganisms can all dramatically affect the rate of skeletal deterioration. Depositional environments include conditions such as the dry Egyptian or Peruvian deserts and the cold dry arctic—environments which preserve even soft tissues. Other depositional conditions ensure destruction of even the teeth. Differences in soil conditions, even within a single burial, can result in differential destruction. In general, better bone preservation is present in well-drained areas with low water tables, in soils with a neutral or slightly alkaline pH, and in temperate areas (Henderson, 1987). These generalizations are often violated, however, because preservation is so dependent on unique combinations of these variables in local depositional settings. The color and degree of fossilization are also controlled by the environment of deposi-

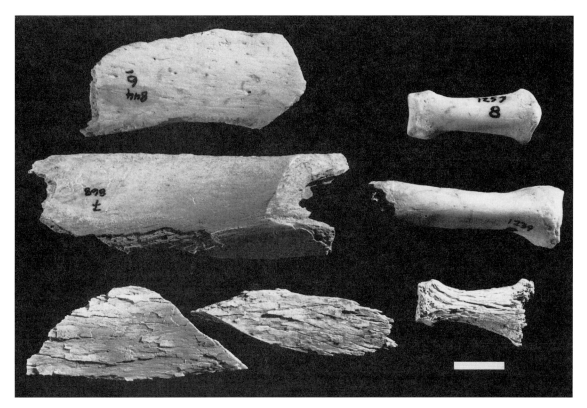

Figure 19.2 Burning and weathering of bone. The shaft fragments and phalanx in the bottom row show characteristic cracking, degreasing, and exfoliation associated with weathering. These human bones lay exposed on an unprotected surface in Colorado for about 15 years. The femur shaft fragment and phalanx in the middle row show bone deterioration, discoloration, and exfoliation indicative of burning damage. Note that the contact between the damaged and undamaged surfaces is abrupt. Soft tissue cover protected the deeper bones, whereas subcutaneous bones or bone portions are more susceptible to such damage when fleshed bones are burned, so the pattern of damage provides clues to the amount of soft tissue on the body when burning occurred. Bar equals 1 cm.

tion. Under the right conditions, a bone can become completely fossilized in a few thousand years.

When unfossilized bone is exposed to the elements, particularly rain and sun, its surface deteriorates at the same time that its organic content is lost. Weathering bones first display a network of fine, usually parallel surface cracks. These cracks progressively deepen and widen and the bone surface begins to deteriorate (Figure 19.2). The rate of weathering depends on temperature and humidity, but archeologists have attempted to use bone weathering to estimate how much time some bone assemblages took to accumulate on former land surfaces, and forensic osteologists use similar observations. Lyman and Fox (1997) discuss the pros and cons of weathering data in both realms.

19.2.2 Rock, Earth, and Ice

Bones on the surface of caves can be broken and scratched by rockfall. Buried bones can be fractured by earth movement. In colder climates, the

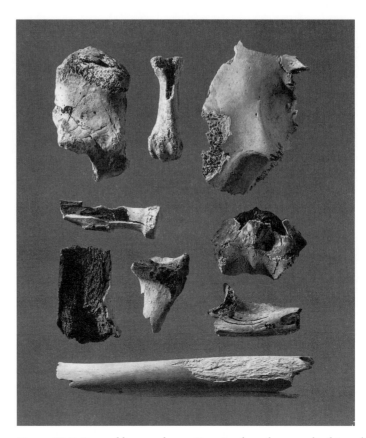

Figure 19.3 Burned human bone. Burning here has resulted in splitting, cracking, and discoloration of the specimens, exfoliation of the cranial vault bone (*upper right corner*), and destruction of the subcutaneous surface of the tibial shaft fragment (*bottom*). One-half natural size.

freeze-thaw cycle can result in damage to bones. The postmortem alteration caused by these nonhuman physical agencies may include striations and polishing that might be attributed to human intervention. In such circumstances, however, the depositional context and configuration of damage provide important clues for the accurate interpretation of the bone modifications.

19.2.3 Abrasion

Particles of grit moved in aerial or aqueous environments can abrade bones, reducing surface relief. Such sandblasting effects are commonly observed in bones exposed on the surface in desert conditions or transported in a river. Many fossil assemblages are recovered from fluviatile environments and their elements often show abrasion damage.

19.2.4 Fire

It is possible for bones to become charred by naturally occurring fires, but the effects are usually not so severe as damage caused by mortuary (cremation) or dietary (roasting) practices (Figures 19.2, 19.3). For many archeologists and osteologists, an introduction to cremated human bone is the first encounter with burned bones. This is unfortunate, because the objective of cremation as a mortuary practice is the destruction of the body. Cremated bones are typically heated to very high temperature and characteristic color changes and cracking accompanies the loss of the organic portion of the bone tissue. It is necessary to recognize, however, that bone subjected to lower temperatures for shorter periods of time is not so conspicuously altered. In fact, burning of bone tissue may so closely mimic normal bone weathering processes that microscopic or chemical analysis is necessary to distinguish the two (Taylor *et al.*, 1995). Burning (charring) of bone tissue is also sometimes confused with staining (particularly manganese staining) of bones in some depositional contexts (Shahack-Gross *et al.*, 1997). When analyzing evidence for the burning of human bones, the osteologist should always be attentive to the depth and character of the soft tissue that covered the particular osteological element at death. The molar enamel, for example, is less frequently exfoliated than incisor enamel because when the head is exposed to fire the incisors are very exposed to high temperature, whereas molars are covered by more soft tissue.

19.3
Bone Modification by Nonhuman Biological Agents

19.3.1 Nonhuman Animal

Carnivores such as hyenas, wolves, dogs, leopards, and even crocodiles can have a dramatic impact on bones and bone assemblages. These animals, particularly the canids and hyenids, are agents of bone destruction because they break bones between their teeth in an effort to retrieve the fat and marrow within. The soft, trabecular portions of bones are favored by these animals, and even a small hyena is fully capable of splintering the shaft of an adult human femur. Carnivore damage to bones is recognized by the signature of the teeth—pitting, scoring, and puncturing of the bone surface (Figure 19.4). Haglund *et al.* (1988) provide a brief account of forensic cases in which human skeletal remains were ravaged by carnivores whereas Haglund (1997) reviews canid data. Haglund (1992) contrasts rodent and carnivore damage.

Although rodents are generally smaller than carnivores, their gnawing can be just as destructive. Rodents ranging in size from mice to large porcupines chew on bones. Like carnivores, large rodents can move bones around on the landscape, often carrying them over large distances to their dens, where they accumulate and modify them by chewing. The chisel-edge of the rodent incisor is used to shave away the surface bone, producing a distinctive, fan-shaped pattern of regular, shallow, parallel, or subparallel, flat-bottomed grooves that are usually concentrated on the projecting surfaces of bones. These traces can be patterned and regular, but they should not be confused with modifications to bone made by humans (Figure 19.5).

In addition to displaying traces of chewing by mammals, bones can be scarred by the action of mammalian feet. Trampling by ungulates and polishing by constant passage of carnivores in a lair may scratch and

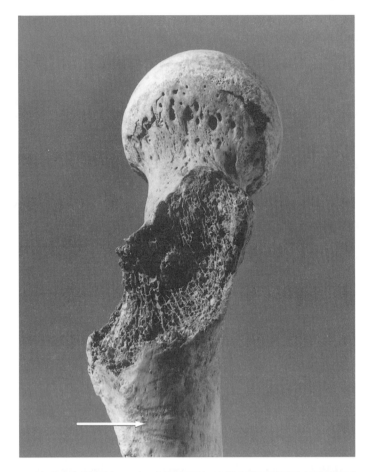

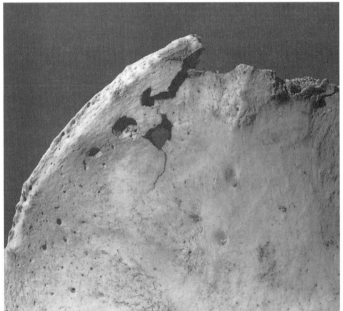

Figure 19.4 Carnivore gnawing marks on prehistoric human skeletal elements. The femur (*top left*) shows destruction of the spongy bone of the greater trochanter and associated broad horizontal grooves made by carnivore teeth just below the trochanter position. The humerus (*top right*) shows similar destruction of trabecular bone along with small punctures left by gnawing on the articular surface of the head. The os coxae (*bottom*) shows destruction of the iliac crest and adjacent perforations caused by gnawing. California. Natural size.

polish bone surfaces. The superficial striations that result from trampling might be mistaken for cutmarks until it is appreciated that these marks

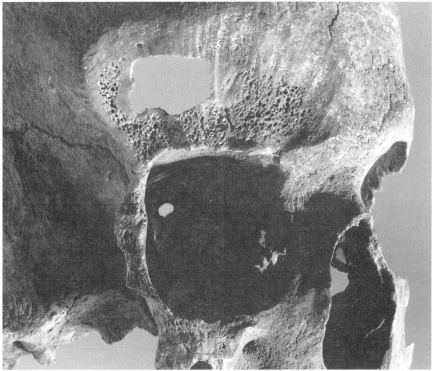

Figure 19.5 Highly patterned rodent gnawing marks on human skeletal elements. The tibia (*left*), from prehistoric California, shows gnawing by a very small rodent. The cranium (*right*), from a prehistoric African site, shows heavy gnawing, with broader gouges left by the incisors of an African porcupine. Natural size.

are usually randomly oriented and concentrated in fields of parallel striae across the most prominent parts of the bone.

19.3.2 Plant

Plants send their roots into the ground in search of water and nutrients. These roots secrete acids that can be very effective at etching the surfaces of buried bones. The pattern of root damage is usually a reticulate network of shallow grooves that should not be mistaken for the work of prehistoric engravers (Figure 19.6). This root-etched network can become so dense that the entire outer surface of a bone is etched away. Individual rootmark grooves are often whiter in color than the surrounding bone because of the decalcification brought about by the acid.

19.4
Bone Modification by Humans

Distinguishing human from nonhuman agents in the modification of skeletal remains continues to preoccupy anthropologists studying human

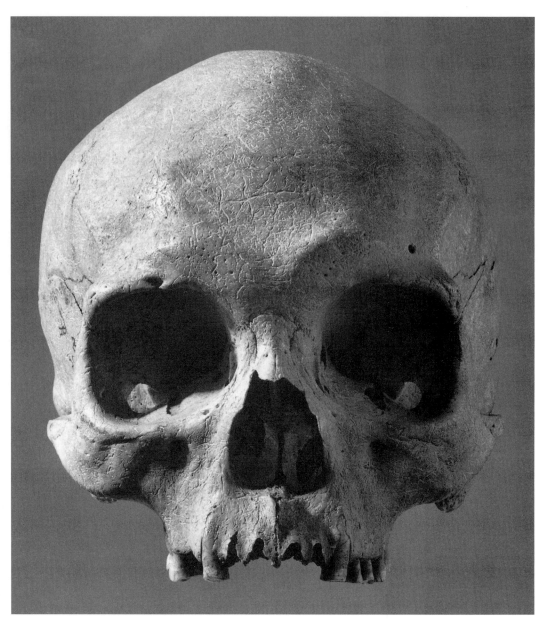

Figure 19.6 Rootmarking on a human cranium from a prehistoric California site. Such delicate, intricate etching of grooves should not be mistaken for cultural activities. Natural size.

origins in Africa, the peopling of the New World, and many other problems of prehistory. As noted previously, many actualistic studies have been conducted with the goal of discovering diagnostic attributes of human bone modification in archeological contexts. Most of these studies focus on nonhuman skeletal remains and are referred to as **zooarchaeology.** Their results also apply in cases where human remains are the objects of human modification.

Human mortuary practices may have profound effects on the disposition of a skeleton. For example, the forcing of the corpse into a small space can cause strange anatomical juxtapositions and even fractures. In second-

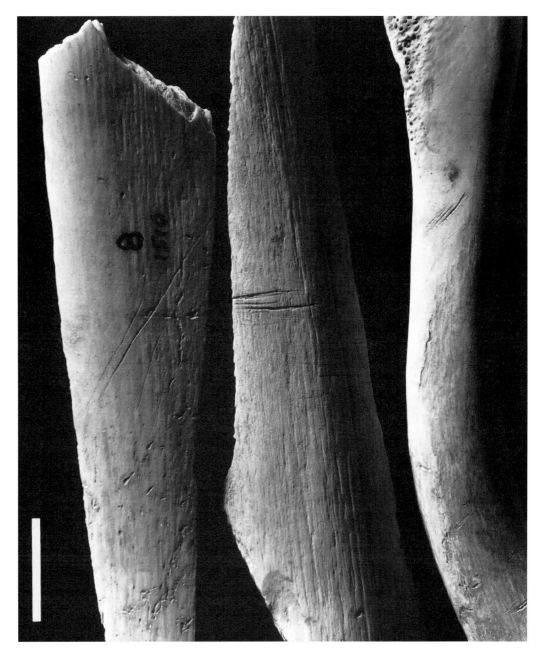

Figure 19.7 Cutmarks made by stone tools on two femur shaft fragments and a clavicle. Cutmarks are usually patterned with respect to the soft tissue that was being cut from the body—here, defleshing cutmarks to remove leg musculature and marks made in the process of decapitation. Bar equals 1 cm. From White (1992).

ary burials, there are often traces of human activity left on the bones; defleshing can leave cutmarks and scraping marks on the bones, and cremation usually causes charring (resulting in white, gray, black, and blue hues) and transverse cracking of the bones. Fortunately for the osteologist, aboriginal cremation may be inefficient, leaving identifiable fragments available for recovery and analysis.

Cannibalism in an archeological context is a topic of considerable anthropological interest. Humans can fracture long-bone shafts with ham-

Figure 19.8 Chopmarks made by stone tools on the posterior surface of a proximal tibial fragment. Bar equals 1 cm. From White (1992).

merstones and pulverize long-bone ends to extract nutritious fat. Percussion pits and anvil scars are usually seen under these conditions (see Chapter 25 and White, 1992). Cutmarks made by stone or metal tools may also appear on skeletal remains in a cannibalized assemblage. Burning of the bone associated with roasting is often concentrated on the subcutaneous surfaces of bones such as the cranium, mandible, and tibia. Study of the patterns of bone destruction through bone assemblage composition and actual physical traces left on individual bones often makes it possible to distinguish between human and nonhuman damage to bones.

As noted in Chapter 14, improper excavation, transport, and cleaning of bones in the field or laboratory can result in damage to skeletal remains. This "preparation," or cleaning, damage, is usually easily discernible from perimortem damage on the basis of its distribution and surface characteristics. Many of the world's most famous hominid fossils have been damaged in this manner. Metal instruments, including dental picks, electric drills, and even wire brushes, can all leave traces on osteological speci-

Figure 19.9 Percussion pits made by stone hammerstones. Bar equals 1 cm. From White (1992).

mens, but this damage should never be mistaken as evidence for prehistoric human behavior. The osteologist should take precautions to avoid inflicting such damage on skeletal remains. These excavation- and preparation-related defects are often easily diagnosed based on their color. An-

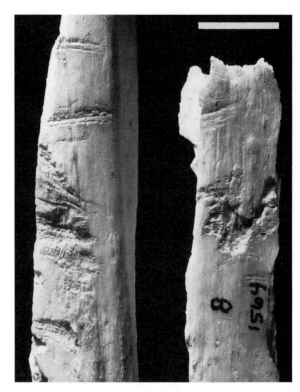

Figure 19.10 Percussion pits associated with percussion striae on long bone shaft fragments. The co-occurrence of hammerstone and anvil striae with these pits is good evidence of human involvement in the processing of these bones. Bar equals 1 cm. From White (1992).

cient surficial defects in a bone usually accumulate soil, matrix, stains, or other residues that darken them relative to the adjacent bone. Recently made defects are usually lighter in color and free of staining and microscopic and macroscopic foreign debris.

Discrimination between taphonomic agents or "actors" that anciently caused surface modifications to bones is often thought to be a more difficult task. For example, some have advocated the use of scanning electron microscopy to choose between diagnoses of carnivore chewing marks or humanly induced cutmarks. A recent study by Blumenschine *et al.* (1996) shows that blind tests of interanalyst correspondence and accuracy in identifying cutmarks, percussion marks, and carnivore tooth marks was excellent, approaching 100% for experts. Major human-made modifications to bone surfaces are outlined in the sections below and illustrated and discussed fully in White (1992).

19.4.1 Cutmarks, Chopmarks, and Scrapemarks

When the sharp, often irregular edge of a stone tool contacts a bone's surface during defleshing or disarticulation activities, a **cutmark** (Figure 19.7) is formed. These marks are usually much narrower, finer, and more V-shaped than carnivore tooth marks. Unlike the single rough furrow of a carnivore mark, or the flat-bottomed trough of a rodent incisor mark, cutmarks usually display striae within the mark and often show "shoulder

Figure 19.11 Inner conchoidal scars formed when these femoral specimens were fractured for their marrow. Bar equals 1 cm. From White (1992).

marks" or "barbs" where different parts of the tool edge contact the bone and thereby cut their own parallel or subparallel marks. Cutmarks are usually the result of slicing activities in which the blade of the tool is used perpendicular to the grain of the tissue being sliced. **Chopmarks** (Figure 19.8) are similar to cutmarks, but they result from forceful and abrupt contact between tool edge and the bone rather than from slicing activities. Chopmarks are less frequent in archeological bone assemblages modified by stone tools with fragile edges and more frequent in forensic cases where metal implements allow the chopping of tissues. **Scrapemarks,** made when the edge of the tool is scraped across the bone's surface, also show lower frequencies in archeological assemblages for the same reason. These are usually shallower than either cutmarks or chopmarks, but they cover wider areas with many parallel or subparallel striations.

19.4.2 Percussion Marks

Cutmarks and chopmarks are frequently encountered in both forensic and archeological contexts. In the forensic context, such marks can indicate antemortem trauma or postmortem attempts to disarticulate body segments. Once disarticulated and (less frequently) defleshed, the elements of the skeleton can further be reduced by direct percussion with a heavy

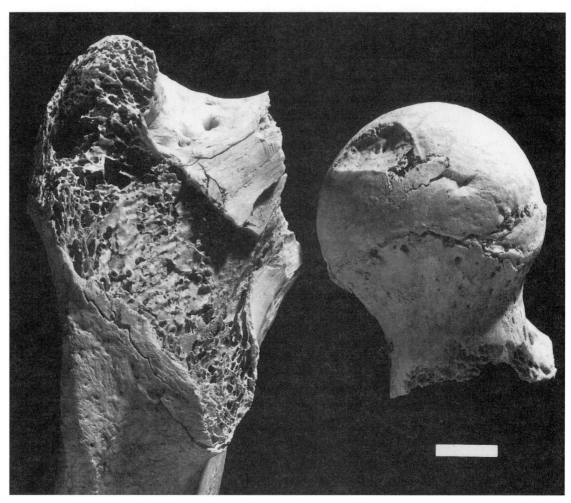

Figure 19.12 Crushing. When spongy bone is crushed by force, cortex on the adjacent areas is pushed into the crushed area, as in these femoral specimens. Note also the disarticulation cutmarks on the femoral neck. Bar equals 1 cm. From White (1992).

object. This is rarely seen in forensic contexts, but it is virtually universal in zooarcheological contexts. Fat is highly prized by many people, and bones contain fatty marrow in both their medullary cavities and their trabecular regions. To obtain the first, the shafts are cracked and pulled apart. Fat in the spongy bone can be extracted by eating the crushed trabecular portions or by boiling them to render the fat in cooking vessels. All of these cultural activities have the potential to leave traces on the bones.

When fracture of the bone is effected with a stone hammerstone, irregular, roughened **percussion pits** (Figure 19.9) that correspond to the tip of the percussor may be left on the bone (particularly if all soft tissue has been removed). If this activity is undertaken on an anvil, or if the percussor moves slightly as it impacts the bone, **percussion (or anvil) striae** (Figure 19.10) may result. This activity often produces **inner conchoidal scars** (Figure 19.11) on the medullary cavity surface of the bone shaft and **adhering flakes** on the shaft wall. When the target marrow is in the trabecular portions of the bone, the percussion to this area sometimes creates **crushing** (Figure 19.12), and when fresh bone fragments are forcefully pulled

Figure 19.13 Peeling. Peeling on a juvenile's proximal ulna and an adult's rib. Bar equals 1 cm. From White (1992).

apart, the result is sometimes **peeling,** particularly on immature bones and ribs (Figure 19.13). When the fragmented bones are boiled in ceramic vessels in an attempt to render grease, some of the shaft fragments may accumulate a peculiar form of abrasion on their tips called **pot polish** (White, 1992).

Suggested Further Readings

Behrensmeyer, A. K., and Hill, A. P. (1980) *Fossils in the Making.* Chicago: University of Chicago Press. 338 pp.
> An edited volume showing the variety of approaches that investigators have taken in taphonomic studies.

Binford, L. R. (1981) *Bones: Ancient Men and Modern Myths.* New York: Academic Press. 320 pp.
> An important book about faunal remains in archeological contexts.

Bonnichsen, R. and Sorg, M. H. (Eds.) (1989) *Bone Modification*. Orono, Maine: Center for the Study of the First Americans. 535 pp.

> This publication of the results of the First International Conference of Bone Modification is a definitive sourcebook on the subject, with papers on topics related to bone modification in a wide range of settings.

Brain, C. K. (1981) *The Hunters or the Hunted? An Introduction to African Cave Taphonomy.* Chicago: University of Chicago Press. 365 pp.

> A classic taphonomic study of bone assemblages from *Australopithecus*-bearing deposits in South Africa.

Gifford, D. P. (1981) Taphonomy and paleoecology: A critical review of archaeology's sister disciplines. *Advances in Archaeological Method and Theory* 4:365–438.

> A comprehensive review of the history and applications of taphonomy.

Haglund, W. D., and Sorg, M. H. (Eds) (1997) *Forensic Taphonomy.* Boca Raton, Florida: CRC Press.

> An edited volume that delivers the best available overview of the field.

Henderson, J. (1987) Factors determining the state of preservation of human remains. In: A. Boddington, A. N. Garland, and R. C. Janaway (Eds.) *Death, Decay and Reconstruction.* Manchester: Manchester University Press, pp. 43–54.

> A brief exposition of the complexities involved in differential preservation of human remains.

Lyman, R. L. (1994) *Vertebrate Taphonomy.* Cambridge: Cambridge University Press. 524 pp.

> The essential sourcebook in taphonomy and zooarchaeology. Comprehensive, critical, and authoritative.

White, T. D. (1992) *Prehistoric Cannibalism at Mancos SMTUMR-2346.* Princeton, New Jersey: Princeton University Press. 462 pp.

> Chapter 6, "Method and Theory: Physical Anthropology Meets Zooarchaeology," is a guide to hominid modification of bone, with many illustrations.

CHAPTER 20

The Biology of Skeletal Populations: Discrete Traits, Distance, Diet, Disease, and Demography

Sorting human from nonhuman skeletal remains and identifying the remains by element, side, age, and sex are generally the most important contributions the osteologist can make to archeological research. Such identifications, far from being trivial, are often critical in answering archeological questions. Additional information, however, can often be obtained from skeletal populations. This information can be crucial in reaching a fuller understanding of the past. The reconstruction of population biology from skeletal remains is an activity that involves potential pitfalls as well as potential benefits for the osteologist.

The aims of **paleoepidemiology** (the study of disease in ancient communities) and **paleodemography** (the study of vital statistics in ancient communities) are to make statements about past populations based on the characteristics of subsets of those populations whose skeletal remains are recovered. As Waldron (1994) notes, four extrinsic factors act on dead populations, all reducing the size of the subset available for study. These four factors are extrinsic in the sense that they are independent of the biological features of the population under study. First, only a portion of those that die are buried at the site being studied. Second, only a portion of the buried evade destruction. Third, only a portion of the undestroyed are discovered. Fourth, only a portion of the discovered are recovered for the osteologist to analyze. With any of these fractionations, the skeletal subset can be biased relative to the sample of people in the original population who actually died. Careful evaluation of such potential bias is critical to accurately reconstructing population attributes of ancient humans.

In the preceding chapters emphasis was placed on identification of skeletal parts on the individual level. The identifications of individuals and their sex, age, stature, pathology, and idiosyncratic skeletal characteristics can be critically important in forensic, archeological, or paleontological contexts. In the archeological context, however, skeletal remains allow us to take further steps in anthropological analysis. This application of human osteological information in an archeological context is sometimes called **bioarcheology.** Here, human osteology is applied in an attempt to understand biological parameters of past human populations. These parameters include relatedness, diet, disease, and demography.

20.1
Discrete Traits

One of the first and most important observations that every osteology student makes is that each human skull is different from every other human skull and can be recognized and differentiated on the basis of size, shape, and various bumps, grooves, foramina, or surface textures. Much of this variation may be partitioned according to the factors responsible for it—age, gender, and pathology. However, much of the variation is idiosyncratic and some of it is attributable to ancestry.

Minor variants of the human skeleton were noted by the ancient Greek scholar Hippocrates, who described wormian bones in human cranial sutures over 2000 years ago. Also called **discontinuous morphological traits, epigenetic variants,** or **discrete traits,** nonmetric variation is variation observed in bones and teeth in the form of differently shaped and sized cusps, roots, tubercles, processes, crests, foramina, articular facets, and other similar features. El-Najjar and McWilliams (1978) and Saunders (1989) provide reviews of work on these kinds of features in human osteology. Suffice it to say that the genetic basis for these traits, particularly the nondental ones, is unknown. Rösing (1984) provides a concise, critical review of the use of nonmetric skeletal traits, concluding that standards for their determination are poor or lacking.

Although all of the labels applied to nonmetric variation imply that it is discrete, this is not necessarily the case. As Mizoguchi (1985) points out, it is rare that nonmetric "traits" are really discontinuous and discrete, even though they are usually scored by osteologists in a nonmetric fashion (either as scored values or by presence/absence). The expression of many of the traits can, indeed, be quantified. These dental and skeletal nonmetric traits can be useful in gauging affinity of extinct human populations. The kinds of nonmetric variation sometimes described in human osteology and used in assessing population affinity are introduced here (for a more complete review of human skeletal nonmetric traits, see El-Najjar and McWilliams, 1978, and Saunders, 1989). Saunders (1989) divides discrete traits into eight categories as follows: Hyperostotic, hypostotic, foramina/canals/grooves, supernumerary vault sutures, craniobasal structures, spinal structures, prominent bony processes, and facet variations. The Buikstra and Ubelaker Standards volume (1994) recognizes two basic categories of skeletal nonmetric traits presented in Table 20.1. Some nonmetric skeletal traits, such as the oval window in the middle ear, have proven valuable to forensic osteologists in establishing "racial" affinities of unknown crania (Napoli and Birkby, 1990).

20.1.1 Dental Nonmetric Variation

Because teeth are often the most abundant elements in archeological skeletal series, and because tooth size and morphology are often more directly

Table 20.1
Skeletal Nonmetric Traits[a]

PRIMARY TRAITS	SUPPLEMENTAL TRAITS, CRANIAL
Metopic suture	Frontal grooves
Supraorbital notch	Ethmoidal foramina
Supraorbital foramen	Supratrochlear notch or foramen
Infraorbital suture	Trochlear spine
Infraorbital foramen	Double occipital condylar facet
Zygomatico-facial foramina	Paracondylar process
Parietal foramen	Jugular foramen bridging
Sutural bones	Pharyngeal tubercle
epipteric bone	Clinoid bridges or spurs
coronal ossicle	Accessory lesser palatine foramina
bregmatic bone	Palatine torus
sagittal ossicle	Maxillary torus
apical bone	Rocker mandible
lambdoid ossicle	Suprameatal pit or spine
asterionic bone	Divided parietal bone
occipitomastoid ossicle	Os Japonicum
parietal notch bone	Marginal tubercle
Inca bone	
Condylar canal	SUPPLEMENTAL TRAITS, POSTCRANIAL
Divided hypoglosal canal	Retroarticular bridge
Flexure of superior sagittal sulcus	Accessory transverse foramen (C3–6)
Foramen ovale incomplete	Vertebral number shift
Foramen spinosum incomplete	Accessory sacroillac articulation
Pterygo-spinous bridge	Suprascapular foramen or notch
Pterygo-alar bridge	Accessory acromial articular facet
Tympanic dihiscence	Unfused acromial epiphysis
Auditory exostosis	Glenoid fossa extension
Mastoid foramen	Circumflex sulcus
Mental foramen	Sternal foramen
Mandibular torus	Supratrochlear spur
Mylohyoid bridge	Trochlear notch form
Atlas bridging	Allen's fossa
Accessory transverse foramina	Porier's facet or extension
Sepal aperture	Third trochanter
	Vastus notch
	Squatting facets, distal tibia
	Squatting facets, talus
	Talar articular surface (calcaneus)

[a] Buikstra and Ubelaker (1994) divide traits into "primary" (for which they provide scoring standards) and "supplemental." Their Standards volume provides illustrations and a standard recording form recommended for the compilation of data on the "primary" skeletal discrete traits, as well as references to all of these characters.

tied to underlying genetics than are other osteological features, teeth have been examined in detail and used widely in osteological analysis. Dental anthropologists use nonmetric variation of tooth crowns to assess biological affinity. Supernumerary teeth, crown fissure patterns, cusp numbers, accessory crown features, and root number, size, and shape combine with a variety of other traits under the heading of nonmetric dental variation. Dahlberg's casts of dental nonmetric traits, available in many human osteology laboratories, provided a standard for work in this area. The Arizona State University Dental Anthropology System is the current, most

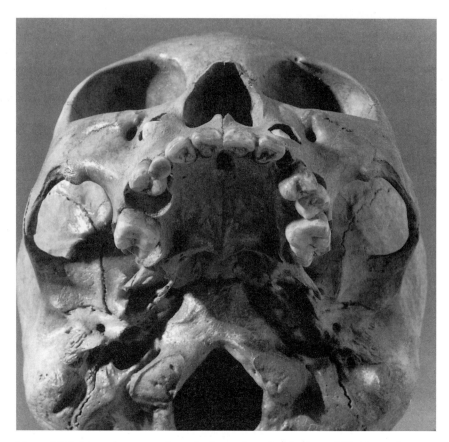

Figure 20.1 Dental nonmetric variations: shoveled incisors. Misdirected, unerupted (heterotopic) upper canines have here resulted in the retention of the deciduous canines. The anomaly is bilateral, with the unerupted canine crowns visible through holes in the anterior surface of the maxillae. Prehistoric, California. Natural size.

widely employed set of standards in dental anthropology (Turner *et al.* 1991; Scott and Turner, 1997). Figure 20.1 illustrates two nonmetric variants possessed by the same individual. Occasionally there is insufficient space in the jaw for tooth eruption, and **crowding** and **impaction** of teeth are the consequences. These, in turn, can result in pathology of associated soft and hard tissues.

20.1.2 Cranial Nonmetric Variation

A wide variety of nonmetric variants in and between the bones of the skull have been used to differentiate skulls and groups of skulls. El-Najjar and McWilliams (1978) describe forty-four such nonmetric traits, whereas Hauser and De Stefano (1989) characterize eighty-four. A few examples suffice to show the nature of the characters in question: presence or absence of a metopic suture (Figure 20.2), parietal foramina, extra bones at pterion, wormian bones, multiple mental foramina, and mylohyoid bridges have all been used in nonmetric analysis. Developmental anomalies of the skull such as scaphocephaly (long narrow skulls caused by pre-

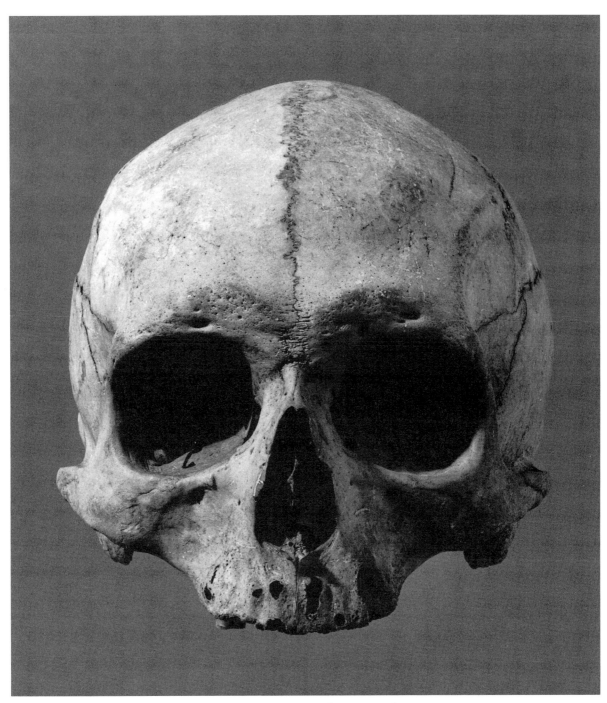

Figure 20.2 Cranial nonmetric variation. A metopic suture has persisted into adulthood in this male individual. Prehistoric, North America. One-half natural size.

mature closure of the sagittal suture) have also been considered by some to represent nonmetric variants. Others simply consider them to be developmental anomalies. A skeletal anomaly is usually considered to be pathological if it is disadvantageous to the individual.

20.1.3 Postcranial Nonmetric Variation

Finnegan (1978) provides a description of thirty nonmetric traits observed in the postcranial skeleton. He emphasizes postcranial traits because they show bilateral expression and because the appropriate skeletal elements are often preserved in the archeological context. Finnegan lists features such as the talar facets, femoral third trochanters, and bipartite cervical transverse foramina. Many of the listed features relate to the presence, shape, size, and orientation of articular surfaces. Many of the traits identified by Finnegan may be highly susceptible to environmental influence, in contrast to the dental nonmetric traits described in section 20.1.1. For this reason, dental nonmetric traits are the most widely and successfully used skeletal indicators in gauging population affinity.

20.2
Measuring Biological Distance

The assessment of affinity (biological relationship) based on skeletal form has a long history in physical anthropology (Armelagos *et al.*, 1982). Larsen (1997) provides a good review of such studies in bioarchaeology, and Chapter 21 considers the application of molecular methods to these problems. Osteologists who study the skeletal remains of anatomically modern humans often work on microevolutionary problems. To estimate the degree of genetic relatedness between populations, the osteologist works under an important assumption: populations that display the most similarity are the most closely related. The degree to which this assumption is met in practice depends on two major factors: adequacy of **sampling** and choice of **characters** (osteological traits) for comparison (Ubelaker, 1989).

Osteologists observe only **samples** of biological populations. The populations themselves are no longer available for study. Therefore, the strength of osteologically based conclusions about affinity depends on the degree to which the samples accurately reflect the real populations that once existed. The conclusions about relationship can be weakened if the sample is too small or if its composition has been altered in some systematic way. Ubelaker (1989) recommends unbiased adult samples of 100 individuals for each group being compared in biodistance studies.

Osteological assessment of the biological relationships among past populations must be made on the basis of anatomical traits. These traits should ideally be directly and exclusively controlled by genes. The more susceptible to environmental (including cultural) influence a skeletal trait is, the less valuable it is in establishing affinity. For example, flattening of the occipital by the cultural practice of cradleboarding can be observed in distantly related people, but to conclude that two populations manifesting cradleboard-induced occipital flattening were biologically closely related would be misleading. Unfortunately, no single skeletal trait is free from environmental influence.

An example of how a skeletal nonmetric trait might be influenced by activity pattern comes with auditory exostoses. These hyperplastic bony growths sometimes form in the medial tract of the external auditory canal and have long been known to have a close relationship between incidence and exposure to cold water. The interesting study of Manzi *et al.* (1991) on the incidence of auditory exostoses contrasted urban and rural communities during Roman Imperial times (A.D. 1st to 3rd century). They found that this acquired discrete trait differentiated the two skeletal samples they studied. Auditory exostoses were found at higher incidence in the middle class population of Portus who habitually used thermal baths than in the lower class slaves and farm laborers represented in a nearby necropolis. Obviously, the presence and absence of discrete characters with low heritability and high environmental influence can tell us about behavior but not much about genetic relatedness. For this reason, the advent of molecular methods to address the latter questions promises to revolutionize biodistance studies of skeletal and dental material (see Chapter 21).

Osteologists have traditionally used both metric and nonmetric traits in their assessments of biological distance between skeletal populations. Multivariate statistics have been employed, mostly with cranial measures, to gauge relatedness. Nonmetric characters have also been used, alone or in combination.

The dentition has been most effectively used to assess relationships between modern and ancient populations for several reasons. Teeth exhibit a variety of anatomical details that have been demonstrated to be stable through time, to have a high genetic component to their formation, and to differentiate living human populations. Teeth are often better preserved than bone. In addition, the effects of environment, gender, and age have less influence on tooth morphology than on most bony anatomy. For these reasons, teeth have figured prominently in reconstructing the biological history of various human populations. Standardization of traits and the methods for scoring them (for example, Carabelli's cusp, fissure patterns, number of cusps, and incisor shoveling) have considerably facilitated and enhanced the accuracy of dental nonmetric analysis (Turner *et al.*, 1991; Scott and Turner, 1997).

As an introduction to the use of discrete dental traits in gauging biological distance, we begin with a classic study by Scott and Dahlberg (1982). These investigators scored the presence or absence of thirteen dental morphological traits on dental impressions taken from a sample of 1251 modern Native Americans from the southwestern United States. Several tribes from four linguistic families were represented by the sample. Table 20.2 illustrates some of the data obtained by this study. When Scott and Dahlberg calculated mean measures of divergence based on their data, they found that the distance between Southwest Native Americans and American whites was eight times greater than the mean measure of divergence among the seven Native American groups. Thus, the major geographical races differ markedly in their crown trait profiles.

Turner's (1987) work on the late Pleistocene and Holocene population history of East Asia demonstrates the power of nonmetric dental traits in assessing affinities of prehistoric skeletal remains. He used twenty-eight dental traits to demonstrate two major regional groups in East Asia, the Southeast Asians plus Jomonese ("Sunadont"), and a more northern cluster ("Sinodont"). Turner concludes that Southeast Asia was a "geogenetic hub" from which all of the Pacific Basin and Rim populations (including Native Americans) radiated.

One example that illustrates how Turner's discrete analysis of dental nonmetric traits and multivariate analysis of cranial metrics have led to useful insights on biological affinities of skeletal and modern populations

Table 20.2
Crown Trait Frequency Variation among American White and
Selected Southwest Native American Groups

	Hopi	Navajo	Zuni	Apache	Mojave	White
I^1 shoveling	.448	.537	.474	.613	.646	.000
I^1 winging	.314	.239	.200	.172	.327	.041
C^1 Tubercle	.737	.656	.909	.737	.684	.720
M^1 hypocone	.843	.735	.705	.837	.896	.908
M^1 carabelli's trait	.803	.613	.745	.583	.723	.795
M^1 cusp 5	.189	.212	.029	.154	.068	.104
C_1 distal accessory ridge	.627	.446	.790	.500	.650	.219
P_4 lingual cusp number	.154	.235	.304	.174	.308	.509
M_1 deflecting wrinkle	.378	.397	.260	.667	.486	.018
M_2 hypoconulid	.763	.714	.571	.632	.533	.131
M_1 protostylid	.344	.357	.575	.292	.250	.048
M_1 cusp 6	.498	.445	.452	.562	.098	.061
M_1 cusp 7	.246	.184	.222	.082	.268	.245

Data from Scott and Dahlberg, 1982.

is the case of the Ainu of Japan. The Ainu are a minority in modern Japan, separated from most Japanese by linguistic, cultural, and physical characteristics. The evolutionary relationship between Ainu and other Japanese has long been a subject of debate, with some workers advocating a "caucasoid" origin for the Ainu. Turner (1976, 1987) worked on dental nonmetric traits from prehistoric Chinese (An-yang) and prehistoric Japanese (Jomon) burials. He compared the patterns characterizing these skeletal series with dental data from modern Ainu and other modern Japanese. Howells (1986) worked on the same problem using multivariate analysis of cranial metric data. These authors have independently shown considerable morphological distance between modern Japanese and Ainu, while at the same time linking the prehistoric Jomon and Ainu. They conclude that non-Ainu Japanese have closer microevolutionary ties to the Chinese.

Another example of how skeletal data have been used in attempts to elucidate population history comes from Africa. In the late 1960s a group of investigators began to analyze skeletal collections from the Wadi Halfa area along the Nile in Nubia. Over 1000 prehistoric burials from Sudanese Nubian sites were studied. The remains spanned the time period from the Mesolithic (10,000 B.C.) to the Christian (A.D. 550–1400) periods, and a subsistence regime from hunting and gathering to intensive agriculture. Prior to work on these skeletal remains, most investigators argued that this area had seen dramatic population change brought about by invading people. Major migration and hybridization of "negroid" and "caucasoid" peoples were posited to explain the archeological record. Work on discrete dental and cranial trait variation led Greene and others to argue that this model of the area's history was simply incorrect. Rather, their skeletal analysis demonstrated that there was sound evidence for genetic continuity in the area through several thousand years of Nubian occupation. Good evidence for craniofacial change through time was found and related to changing biocultural adaptations in the area (Carlson and Van Gerven, 1977; Armelagos et al., 1981; Greene, 1982; Van Gerven et al., 1995). However, Turner and Markowitz (1990) hypothesized that the ancestry of recent Nubians was not derived from local late Pleistocene populations, and that a population replacement event occurred during the Holocene. Irish and Turner (1990) addressed the issue with comparisons of dental traits. As Larsen (1997) notes, additional data are needed, both biological and cultural, before the discontinuity model may be accepted.

Finally, we consider a case in which the use of skeletal nonmetric traits has been applied to small-scale studies of affinity. Howell and Kintigh (1996) examined the cemeteries at Hawikku, in the American Southwest. This site was occupied for about 340 years and was the place of first contact between Spanish and Native American (Puebloan) people. Burials at Hawikku were distributed in spatially discrete clusters, or cemeteries. The research question centered on whether these cemeteries represented kin groups. Distributions of dental nonmetric traits were compiled for the cemeteries, and a nonrandom patterning of genetic markers, as expressed by the dental traits, was found. This identification of kin-group cemetery specificity based on biological indicators suggests to these authors that studies utilizing spatial information and biological measures of genetic distance hold great potential for elucidating many aspects of social structure. The arbitrary divisions between archeology and physical anthropology that have become ingrained in academic departments of anthropology must clearly be overcome in any comprehensive understanding of the past. Larsen (1997) examines many other bioarcheological case studies in his chapter on skeletal biodistance studies. It is increasingly apparent that the molecular revolution in osteology holds the potential to become the definitive arbiter of much biodistance work with human skeletal remains (see Chapter 21).

20.3 Diet

One of the primary goals of archeological research is the reconstruction of subsistence patterns in past human populations. A multidisciplinary approach is usually taken in this endeavor, with specialists analyzing floral, faunal, and fecal material recovered in habitation sites, and still other archeologists examining the remains of technology used to exploit different food resources. Such an approach uses information from many disciplines to elucidate the past. In an example of biomechanical analysis of more recent skeletal populations, Ruff *et al.* (1984), building on previous work by Lovejoy *et al.* (1976) and Lovejoy and Burstein (1977), were able to show a decline in bone strength of the femur at a time when agriculture was adopted, suggesting a decline in physical demands with the change in subsistence. The osteologist can make contributions to understanding the diet of prehistoric people by examining skeletal pathologies, analyzing dental wear, and, more recently, through the analysis of trace and major elements extracted from the skeletal remains themselves (Chapter 21).

The interaction between nutrition and skeletal pathology is a complex, difficult subject area and the focus of a great deal of current research. For a comprehensive review of the topic, see Martin *et al.* (1985) and Larsen

(1997). Indicators such as Harris lines, dental hypoplasia, and dental asymmetry (see Chapter 18) may be used on a populational basis to determine nutritional adequacy in prehistory. Unfortunately, however, stress markers in bone are nonspecific, and only patterns and trends of nutritional stress on the populational level can be ascertained. On the opposite end of dietary reconstruction, the focus can be individual and the results very specific, for example, when colon contents can reveal a meal (Shafer *et al.*, 1989).

Dental wear and dental caries have long been used in attempts to characterize prehistoric diet. Teeth interface directly with foodstuffs, and the physical and chemical composition of food has a direct influence on wear and decay. Prehistoric people who incorporated large amounts of grit into their diet through food preparation techniques such as grinding food between stones exhibit pronounced dental wear. The limitations of using dental wear to assess diet are easily understood by considering two imaginary prehistoric populations eating exactly the same diet. If one population used sandstone grinders to prepare the bulk of its diet, while the other group used wooden mortars, the rate and nature of dental wear would be very different, even though the nongrit content of the diet was identical.

Microscopic examination of worn teeth reveals pits and striations in the enamel and dentine. Use of the scanning electron microscope has made it possible to carefully study microscopic wear on teeth (Walker, 1981; Teaford, 1991, 1994). As with paleopathology, study of dental wear must be done on a populational basis to yield reliable dietary reconstructions. This is particularly true with microwear analysis because the microscopic signature of the individual's last meal or set of meals may not be indicative of what the average diet was over the life of the individual. Much additional research in this area is required. It is clear, however, that dental microwear study is an important compliment to macroscopic wear, bone chemistry, and pathological assessment in dietary analysis.

Unlike dental wear, dental caries are a pathological condition (see Chapter 18) whose incidence is under the influence of many factors, including diet. The incidence of caries, for example, has been shown to be generally higher in agricultural than in hunting and gathering economies. Smith (1984) and Schmucker (1985) summarize studies aimed at elucidating prehistoric diet through the analysis of tooth wear, and they both find that hunters and gatherers can be distinguished from farmers on the basis of macroscopic tooth wear. Powell (1985) reviews the use of dental wear and caries in reconstructing prehistoric diet.

Dental caries is the disease process characterized by demineralization of dental hard tissues by organic acids produced when bacteria ferment dietary carbohydrates (especially sugars). Because carious lesions are readily apparent on teeth, there is a very large literature associated with them, even for prehistoric populations. Osteologists have been studying temporal trends in caries since the 1800s. Changes in processing technology and food had important implications for the oral health of past human populations. Cariogenic foods lead to a higher prevalence of caries in a population. Within a population, sex and status differences in the amount of cariogenic food eaten may play important roles in determining the frequencies of caries (Walker and Hewlett, 1990). Larsen (1997) reviews the use of caries frequencies in studies of modern and archeological skeletal samples.

20.4
Disease and Demography

The study of populations, **demography,** is concerned with the vital statistics of populations—life expectancy, mortality rates, birth rates, and population growth, size, and density. Demographers interested in modern people use data collected by census takers who census the living. **Paleodemography** is the study of the demography of prehistoric populations. The vital statistics of these populations can be reconstituted by use of their skeletons. The osteologist can reconstruct these populations by censusing the dead. A major assumption used here by the osteologist is that the rates of growth and aging established for modern humans can be appropriately applied to individuals who lived in the distant past. The more ancient the populations under study, the less valid this assumption is likely to be.

The reliability of demographic reconstructions built on skeletal material depends on the accuracy of individual age and sex estimations for the skeletons. In addition, reliability depends on how accurately the sample of skeletons represents what was once the living population. Van Gerven and Armelagos (1983), Greene *et al.* (1986), Boddington (1987), Bocquet-Appel and Masset (1982), Wood *et al.* (1992), Jackes (1992), and Konigsberg and Frankenberg (1994) provide good reviews of the assumptions and limitations of archeological data in demographic reconstructions. Figure 20.3 is an illustration of a large cemetery excavated during the 1960s in the midwestern United States. Here, 1327 articulated skeletons were recovered, ranging in age from *in utero* individuals to elderly adults. The excavators estimate that this represents a 300-year occupation (Lovejoy *et al.,* 1985a).

To better understand the constraints and limitations of demographic reconstructions based on skeletal remains, imagine an ancient population in which all of the dead were buried in a single cemetery over the span of one hundred years. In this imaginary case, no people died away from home or were cremated or eaten by carnivores. None of the skeletons were disturbed after burial by biological or physical agents. Furthermore, imagine that the entire cemetery was preserved intact through the centuries. Finally, imagine that all of it was excavated, and that all of the individuals, including very young infants, were recovered. Provided that recovery was complete, record keeping was good, none of the skeletal material was lost subsequent to excavation, and the osteologist could accurately age and sex all of the individuals, these data might be used directly to reconstruct demographic attributes of the population.

As Ubelaker (1989) notes, demographic reconstruction from skeletal remains is not difficult mathematically. For example, consider survival through time, beginning with live births. At birth, survival would be 100%. By age 5, with high infant mortality, perhaps only 60% of the original population would have survived. This would mean that 40% of the cemetery population would have been made up of children in the 0- to 5-year-old age range. By plotting the age estimate for these, and the other burials, in 5-year intervals through time, one could reconstruct a **survivorship curve** for the population (Figure 20.4). One might examine survivorship by sex or make deductions about life expectancy in the population. In short, it would be possible to examine the paleodemography of the population under study.

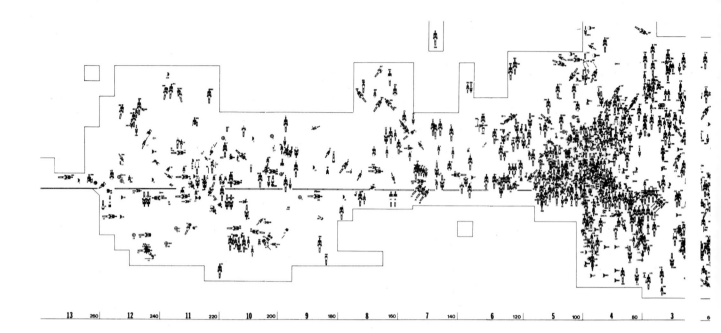

Figure 20.3 Plan of the prehistoric Libben site, Ottawa County, Ohio. Studies of skeletal populations such as this one can lead to insights into demographic aspects of early human populations. The area in bold outline is a blowup of part of the plot.

It is important to derive demographic facts for past human groups. One must note, however, that accurate and reliable demographic reconstruction was only achieved under the specified conditions in the imaginary case outlined above. Most archeologically derived skeletal samples do not meet these conditions, and survivorship curves that they generate are

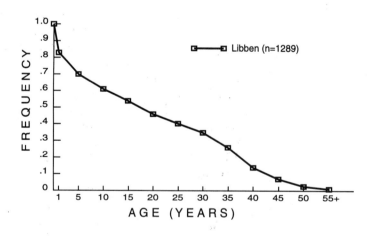

Figure 20.4 Survivorship curve based on the prehistoric Libben skeletal population. Data from Lovejoy *et al.* (1977).

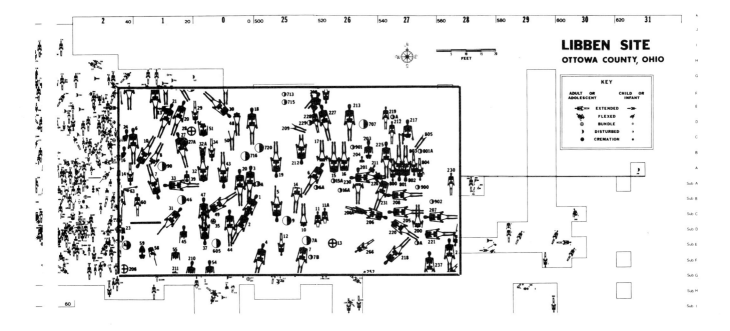

prone to systematic error as a result. For example, many human groups differentially dispose of the dead. If there is bias in the burial practices, the demographic profile of that population cannot be accurately determined. Many cemeteries show differential preservation that favors young adult individuals over children or elderly adults because bones of the former are stronger and less prone to destruction by taphonomic agents (evidence for bias due to preservation can be found by careful analysis of the sample; see Chapter 19 and Walker *et al.*, 1988). Many cemeteries are nonrandomly excavated or incompletely sampled. Only intact specimens are saved subsequent to collection in many archeological excavations. Many skeletal samples are poorly curated, with the loss of much material. In short, most archeologically derived skeletal populations are inadequate to provide accurate paleodemographic reconstructions. If an understanding of paleodemographic aspects of ancient populations is the goal of a research project, it is imperative that the osteologist work closely with the archeologist to see that sampling strategy does not bias the ultimate results.

Waldron (1991) expresses the linkages and pitfalls of studying demography and disease in skeletal populations as follows (p. 24):

The underlying assumption that is inherent in any attempt to use a death assemblage to predict something about the living is that the dead population is representative—or at least typical—of the live population. Given all the nonrandom events that surround death and burial, not to mention preservation and recovery, this is at best an approximation, and at worst the two (the live and the dead) bear no epidemiological relation to each other whatsoever. However, it is clearly important to know where on this spectrum a particular group, or set of groups, lies, especially if the data derived from their study are to be used to construct life tables, to make inferences about changing patterns in disease or dietary habits, or to draw any of the other demographic conclusions that are so commonly bandied about.

The relationship between the adoption/intensification of agriculture and population size and health has been an object of anthropological inquiry for many decades. Paleodemographic and paleopathological data have been brought to bear on this subject for many years, and as of the beginning of this decade, the idea that agriculture brought with it a decreased quality of life and increased mortality rates was widely accepted. In a sobering and influential contribution, Wood and colleagues (1992) reminded anthropologists that the study of prehistoric populations and their health is a complex undertaking, never straightforward or simple. These authors question a basic assumption made by many osteologists regarding lesions on skeletal remains. They argue that rather than reflecting declining community health, such lesions indicate that the affected individuals survived some disease, that such survival might actually indicate an improvement in health, and that individuals who lived long enough to manifest pathological lesions on their skeletons were advantaged relative to people who succumbed to disease before their skeletons were affected. Furthermore, these authors note that large numbers of immature skeletons may indicate more about fertility than mortality. These observations are in sharp contrast to the received wisdom in paleopathology and paleodemography. How, then, are osteologists to interpret paleopathology on a populational basis? Wood *et al.* (1992) say that considerably more critical research is required, but that in the meantime, the skeletal evidence pertaining to the transition from hunting and gathering to settled agriculture is equally consistent with an improvement or a deterioration of health (see also Cohen, 1994).

Applying human skeletal data from historic and prehistoric contexts to important questions about culture and biology is an important avenue of anthropological investigation. Decades of such application have brought an increased understanding of the complexities involved in such studies. It is clear that anthropologists will continue to use skeletal populations in efforts to better understand the past and will do so with increasingly sophisticated techniques and heightened cautions built upon a better appreciation of the fragmentary and biased nature of the records that they study.

Suggested Further Readings

Cohen, M. N., and Armelagos, G. J. (Eds.) (1985) *Paleopathology at the Origins of Agriculture.* Orlando, Florida: Academic Press. 615 pp.
> An edited volume featuring investigators working with archeological and osteological remains from around the globe.

Greenberg, J. H., Turner, C. G., and Zegura, S. L. (1986) The settlement of the Americas: A comparison of the linguistic, dental and genetic evidence. *Current Anthropology* 27:477–497.
> This paper illustrates how osteology can contribute to a multidisciplinary approach to understanding population history. Greenberg (a linguist), Turner

(an osteologist), and Zegura (a geneticist) consider the origin of Native American peoples.

Hauser, G., and DeStefano, G. F. (1989) *Epigenetic Variants of the Human Skull.* Stuttgart: E. Schweizerbartsche Verlagsbuchhandlung. 301 pp.
> An excellent, comprehensive treatment of nomenclature, gross anatomy, function, embryology, development, genetics, medical relevance, scoring methods, and variation of 84 cranial nonmetric traits.

Howells, W. W. (1989) Skull shapes and the map: Craniometric analyses in the dispersion of modern *Homo. Papers of the Peabody Museum of Archaeology and Ethnology* 79:1–189.
> This monograph by the dean of craniometric analyses assesses the evolutionary divergence in cranial shape among different geographic areas.

Huss-Ashmore, R., Goodman, A. H., and Armelagos, G. J. (1982) Nutritional inference from paleopathology. *Advances in Archaeological Method and Theory* 5:395–474.
> A review of skeletal paleopathological indicators of nutrition.

Larsen C. S. (1997) *Bioarchaeology: Interpreting Behavior from the Human Skeleton.* Cambridge: Cambridge University Press. 461 pp.
> A comprehensive summary of all aspects of bioarchaeology: the standard volume in the field.

Scott, G. R., and Turner, C. G. (1997) *The Anthropology of Modern Human Teeth: Dental Morphology and Its Variation in Recent Human Populations.* New York: Cambridge University Press. 382 pp.
> A comprehensive look at how teeth can be used to assess population biology.

Teaford, M. F. (1991) Dental Microwear: what can it tell us about diet and dental function? In: M. A. Kelley, and C. S. Larsen (Eds.) *Advances in Dental Anthropology.* pp. 341–356. New York: Wiley-Liss.
> This paper summarizes work on the microwear of mammalian teeth, providing a good summary of the accomplishments and goals of using dental microwear to establish diet in extant and skeletal populations.

Ubelaker, D. H. (1989) *Human Skeletal Remains: Excavation, Analysis, Interpretation* (2nd Edition). Washington, D.C.: Taraxacum. 172 pp.
> Chapter 5 provides a concise review on prehistoric population dynamics.

Verano, J. W., and Ubelaker, D. H. (Eds.) (1992) *Disease and Demography in the Americas.* Washington, D.C.: Smithsonian Institution Press. 294 pp.
> An edited volume to commemorate the Columbus Quincentenary by examining the effects of Europeans contacting New World populations. The contributions of skeletal studies to this field are summarized by leading experts for all regions of North and South America.

Waldron, T. (1994) *Counting the Dead: The Epidemiology of Skeletal Populations.* West Sussex: John Wiley. 109 pp.
> An excellent introduction and critical evaluation, useful in paleodemography and paleopathology.

CHAPTER **21**

Molecular Osteology

Howells (1995, p. 2) poses the question: "Is a given unknown skull, with a malc assignment [based on morphological characters identified in Chapter 17], indeed a male rather than a somewhat extreme female? There is no way of telling, and that is the problem." This is no longer the case.

Traditional methods for analyzing skeletal remains are based on examining the sizes and shapes of bones. Recent advances in other fields have given rise to a new set of methods that allow the osteologist to analyze the molecular constituents of bone. Under certain conditions, skeletal remains retain sufficient densities of deoxyribonucleic acid (DNA), amino acids, and various isotopes to permit their recovery and analysis. The techniques used are often complex and require specially equipped laboratory facilities. However, the accuracy and precision of the results obtained may be superior to those of traditional methods, depending on the nature of the question. DNA analysis, in particular, allows the bioarcheologist or forensic osteologist to address questions that are beyond the range of morphological methods. Molecular techniques are thus becoming the method of choice for various types of osteological analysis. Even so, these techniques are best applied as part of an overall osteological analysis (Chapter 15), rather than in isolation.

21.1
Sampling

The use of new high-tech molecular methods does not mean that the fundamentals of proper recovery, preparation, and documentation of skeletal remains can be ignored. Indeed, mistakes made prior to or during the sampling of material for molecular analysis can seriously compromise results. In addition to the standard set of recommendations for recovery and documentation of skeletal remains (Chapter 14), investigators wishing to obtain good samples for molecular analysis should consider the following:

- If molecular analysis is to be employed on newly recovered skeletal material, recovery procedures should be modified to minimize contamination of the remains with modern compounds. The specific procedures will depend on the nature of the site, the available time and resources, and the type of analysis planned. For DNA analysis, these would ideally include using disposable latex gloves and hair nets, and sterilizing excavation tools. If this is not practical for the entire sample, a reasonable compromise may be to employ such procedures on the specific elements destined for DNA analysis.

- The exposure of skeletal remains substantially alters their environment. This may in turn lead to further decay of their molecular constituents. Currently, little is known about the effects of preservatives and changes in temperature, humidity, moisture, and air circulation on the preservation of various biomolecules. Until more information is available, the wisest course is to minimize the magnitude of such changes as much as is practical. Consultation with the specialist who will be conducting the molecular analysis is advisable.

- The evaluation of the accuracy of molecular analyses depends in part upon proper documentation of the provenience of the samples. If molecular analysis is to be employed on newly recovered skeletal material, particular attention should be paid to properly documenting both the context of the remains and the excavation methods employed. Depending on the type of analysis planned, it may be advisable to obtain soil samples from the area surrounding the elements to be analyzed. Here again, consultation with the appropriate specialist is crucial.

- Whether the molecular methods are to be applied to newly recovered skeletal remains or to museum collections, the choice of which specimens to sample must be made carefully. Since molecular techniques are typically destructive, specimens for sampling should be chosen to minimize the morphological information lost while maximizing the potential information gained in the molecular analysis (DeGusta and White, 1996). In order to adequately weigh these often competing goals, it is imperative that the skeletal remains first be examined for signs of pathology, bone modification, and other morphological variations (see Chapters 18–20). Only then can an accurate assessment be made of the "morphological value" of the skeletal specimens and the various portions of individual specimens. It is often useful for molecular analysis to be first attempted on nonhuman remains or on human specimens of dubious provenience to establish the feasibility of the method prior to the destructive sampling of more valuable specimens.

- Removal of skeletal tissue for molecular analysis destroys information about the morphology of the bone, but this loss can be greatly reduced by proper documentation prior to sampling. The exact methods employed to record the morphology will depend on the anticipated degree of destruction and the importance of the specimen. Minimally, high-quality photographs (with a scale bar) and radiographs should be taken (Chapter 15) and developed prior to sampling. Molding and casting of specimens (Chapter 14) provides a three-dimensional record of the morphology. This, in combination with photographs, minimizes the loss of information. Documentation only preserves morphological information if it remains permanently accessible, so casts and photographs of the relevant specimens (along with copies of the results of the molecular analysis) should be properly and promptly deposited in the appropriate archives.

21.2
DNA

Deoxyribonucleic acid (DNA) is the molecule of heredity. The genetic code in DNA is based on four chemical building blocks called nucleotides: adenine, cytosine, thymine, and guanine. These nucleotides can be thought of as forming a four letter alphabet which spells out the assembly instructions for all the proteins that make up an organism. Almost every cell in a person's body has a complete copy of their DNA, and all these copies are essentially identical.

Because an organism's DNA contains all the necessary information to assemble a complete organism, it is potentially the most biologically informative molecule. There is a tremendous amount of research aimed at extracting various types of information from DNA, in part because many diseases are thought to have a genetic component. Specific diseases—such as Huntington's disease and sickle cell anemia—can be diagnosed based on DNA alone. The sex and general ancestry of an unknown individual can be determined from DNA. Comparisons of DNA samples can be used to establish identity and paternity.

The ability to determine sex, ancestry, disease status, and identity from DNA has obvious applications in osteology, forensics, archeology, and even paleontology. These applications, though, all depend on the ability to obtain DNA from organic remains of various antiquity. After an organism dies, however, the highly organized molecules of DNA degrade rapidly. The key conceptual breakthrough that lead to the field of "ancient DNA" was the recognition that despite this decay, fragments of DNA are sometimes present in remains of great antiquity and that, with only a few modifications, existing techniques for isolating modern DNA could also be used to retrieve this DNA (Higuchi et al., 1984; Pääbo, 1985). These techniques originally required a large amount of preserved DNA, a criteria met only in cases of exceptional preservation (e.g., mummies and ice-embedded animals).

21.2.1 PCR and Methodology

The development of the polymerase chain reaction (PCR) made it possible to retrieve exponentially smaller amounts of DNA (Mullis and Faloona, 1987). In order to analyze DNA, it is necessary to have a sufficient quantity of it. Using conventional techniques, even a dozen molecules of DNA are effectively invisible—they are too small to detect. The PCR acts as a "molecular photocopy machine" by making literally millions of copies of a section of DNA. This is referred to as amplification, and PCR can amplify a section of DNA starting from only a few original molecules. The large amount of DNA which results can then be analyzed quite easily using a variety of standard techniques. Using PCR, researchers were able to retrieve DNA from ancient skeletal remains (Hagelberg et al., 1989).

This advance made ancient DNA methodology applicable to a broad range of questions in osteology.

The general methodology for obtaining and analyzing DNA from a bone or tooth involves three general stages. First, the DNA must be extracted and isolated. This involves reducing about a gram of bone or tooth to powder. The powder is then treated chemically to remove proteins and other compounds and to concentrate the DNA. Second, a predetermined section of the DNA is amplified using the PCR. Finally, the resulting DNA sample is analyzed, typically by determining the nucleotide sequence.

21.2.2 Contamination

Despite the exciting potential of ancient DNA, the applications are still limited by methodological difficulties (Handt *et al.*, 1994; Kelman and Kelman, 1999). The major problem is contamination by exogenous DNA (DNA not from the targeted individual). Living organisms are constantly shedding DNA-bearing tissues in the form of skin cells, hair, saliva, and other secretions. Archeological skeletal remains, for example, can be contaminated by the DNA of organisms in the soil, microorganisms growing in the bones themselves, excavators, curators, or the DNA analysts. The problem of contamination is worsened by the nature of the PCR. Polymerase chain reaction preferentially amplifies well-preserved DNA molecules, which are more likely to be modern contaminants than truly ancient DNA. Since PCR produces large amounts of highly concentrated DNA, laboratories often encounter problems with the products of previous PCR reactions contaminating current work. Contamination is of extreme concern when attempting to retrieve DNA from ancient human remains, as humans are also the main source of exogenous DNA, making contamination more difficult to detect. Several published DNA sequences from very ancient remains are now widely held to be inauthentic (Lindahl, 1997). Since the materials used for the extraction of DNA are often unique, and the analysis is time consuming, independent replication of results is not always carried out. A number of techniques have been developed to reduce the chances of contamination occurring and to increase the likelihood of contamination being recognized. The independent verification of ancient DNA results is also becoming more common. So, whereas contamination will likely continue to be of concern, the body of reliable ancient DNA work will also continue to expand.

21.2.3 Taphonomy of DNA

Beyond a certain time period, perhaps 100,000 years (Poinar *et al.*, 1996), no DNA is likely to be preserved in skeletal remains. Within that time range, the factors leading to preservation are not understood, though recent work on amino acids may help remedy this situation (see below). For example, some skeletal remains have yielded DNA, whereas other remains, of similar antiquity or even younger, do not preserve any DNA. The specifics of death, burial, and **diagenesis** (a change in the chemical, physical, or biological composition of bone subsequent to death) that result in the preservation of DNA in some cases, but not in others, are unknown. The preservation of DNA does seem to be primarily influenced by environmental conditions rather than by time, at least for remains younger than about 10,000 years (Parsons and Weedn, 1997). Bones and teeth that are macroscopically and microscopically well-preserved seem more likely to yield DNA. Empirical evidence also suggests that remains from colder regions may preserve DNA better than remains from warmer areas (Poinar *et al.*, 1996). It is also clear that mitochondrial DNA (mtDNA)—the small portion of the genome that is inherited only from the mother—is easier to retrieve than nuclear DNA (Parsons and Weedn, 1997). This is likely due to the greater number of copies of mtDNA and perhaps its smaller, circular structure.

21.2.4 Applications

As molecular biologists develop ways to extract more information from modern DNA, the types of questions that can be addressed with ancient DNA will expand as well. Currently, there are five major questions about a deceased individual that DNA analysis of skeletal remains can potentially address:

- What sex was this individual? Methods of sexing skeletal remains based on morphology depend on the preservation of sexually dimorphic elements and have a nontrivial error rate even for adult remains (Chapter 17). If DNA can be obtained, the sex of any individual (regardless of age) can be determined with extremely high precision from even very fragmentary skeletal remains (Stone *et al.*, 1996).

- What diseases did this individual have? A number of diseases are genetic in nature and could potentially be screened for in past populations using DNA analysis. Disease processes characterized by long-term infection by substantial densities of viral or bacterial pathogens might also be detected through recovery of the pathogen's DNA. As yet, very few studies of paleopathology have utilized DNA methods. The most notable application to date was the amplification of the *Mycobacterium tuberculosis* DNA from a Peruvian mummy to verify a pre-Columbian occurrence of tuberculosis in the New World (Salo *et al.*, 1994).

- What ancestral population was the individual a member of? A number of morphological techniques have been developed to assess the geographic affinity of skeletal remains, but they are of limited accuracy and require relatively complete remains (Chapter 20). Reliable methods to assess the populational affinities of skeletal remains are sorely needed in archeology (to assess the relationships between past populations) and forensics (to provide information on an isolated skeleton that may lead to identification). DNA typing of skeletal remains has the potential to provide the best available information regarding the populational affinity of the individual. Mitochondrial DNA (mtDNA) is a small portion of the human genome that is inherited only from the mother. Several regions of mtDNA are highly variable within modern humans, and the sequencing of these regions can permit the estimation of the ancestral maternal population (Connor and Stoneking, 1994). Archeologically, these techniques have been used to assess the relationships of prehistoric New World populations with various modern Native American groups (Stone and Stoneking, 1993; Hauswirth *et al.*, 1994; Kaestle, 1995). Forensically, analysis of mtDNA has been used to help identify the remains of U.S. military personnel in Vietnam (Holland *et al.*, 1993).

- Who were the familial relatives of this individual? A relatively common problem in forensic osteology is to establish the identity of a skeletonized individual. In some cases, a possible identity will be established based on other clues (e.g., a recently missing person of similar age and sex), but confirmation is needed. DNA analysis is the best method for testing hypotheses about the identity of skeletal remains. The general approach is to compare the DNA from the skeleton with the DNA of the presumed relatives. For a number of variable regions of the DNA, the odds of a match between unrelated individuals is extremely low. Exactly how low is a matter of debate for cases involving blood samples from living individuals (Devlin *et al.*, 1994), but in osteological contexts this is rarely if ever a concern. DNA typing has been used to identify skeletonized individuals in contexts involving mass deaths (the Branch Davidian incident in Waco, Texas; Houck *et al.*, 1996), mass graves (Guatemala and former Yugoslavia; Boles *et al.*, 1995; Primorac *et al.*, 1996), remains of military personnel (Vietnam; Holland *et al.*, 1993), war criminals (Josef Mengele; Jeffreys *et al.*, 1992), historic figures (the Romanov family; Ivanov *et al.*, 1996), and numerous forensic cases involving murder victims (e.g., Hagelberg *et al.*, 1991; Sweet and Sweet, 1995). Even though the determination of familial relationships is most applicable in forensic or historical contexts, archeological analysis of mortuary rituals and burial practices can often be advanced if the general relationships of the interred individuals can be established (Stone and Stoneking, 1993). Establishing the familial relationships between individuals in the same prehistoric population requires more detailed analysis than is usually attempted.

- To what species did this individual belong? The sequence of certain portions of DNA varies between species, and thus in theory DNA typing can be used to evaluate the species to which a bone belongs. However, the determination of whether fragmentary skeletal remains are human or nonhuman can be done quickly, cheaply, and accurately by visual inspection of the morphology, rendering DNA analysis unnecessary for this question in most osteological contexts. In paleontological settings, though, morphological methods for determining the species affiliation of hominid fossils often produce ambiguous results. In the case of Neanderthals, for example, paleoanthropologists disagree as to whether or not there was gene flow between Neanderthals and contemporaneous early modern humans. A comparison of Neanderthal DNA with the DNA of early modern humans would provide a crucial test of hypotheses about their relations. The first step in this has been taken by Krings *et al.* (1997), who successfully retrieved a portion of the mtDNA from a Neanderthal humerus. This result was replicated in another laboratory, and the sequence obtained is quite distinct from any known modern human sequence. Even though the processes of fossilization and DNA degradation preclude the application of DNA techniques to most of the fossil record, questions about species and evolutionary relationships within the last few hundred thousand years may be addressed using this method.

21.3
Amino Acids

Amino acids are the chemical building blocks of proteins in all living organisms. Each type of amino acid comes in two mirror image forms known as antiomers: the D form and the L form. All amino acids incorporated into a protein are in the L form, but over time they gradually convert into the D form, a process known as racemization. Attempts to use the ratio of D to L forms as an absolute dating method have generally failed, as this ratio is significantly affected by diagenesis, but the technique has found two applications in modern osteology.

First, it has been suggested that the D/L ratio of aspartic acid in teeth is indicative of age-at-death (Ohtani and Yamamoto, 1991, 1992; Ohtani, 1995; Carolan *et al.*, 1997). However, the error of the estimate is figured at about ± 15 years, and since the ratio is affected by diagenesis, the method is only applicable in modern contexts. Due to these limitations, the application of amino acid racemization for establishing age-at-death is of little use.

More recently, Poinar *et al.* (1996) have shown that the degree of racemization of aspartic acid in skeletal remains is correlated with the preservation of DNA. Beyond a certain degree of racemization, no DNA was able to be amplified. This technique takes advantage of the sensitivity of the racemization process to environmental factors, since it is just those factors which speed the decay of DNA. The main use of amino acid racemization in ancient DNA work is to assess the prospects of retrieving DNA and to help confirm the authenticity of the DNA obtained.

21.4
Isotopic Osteology

In Chapter 20 we looked at the various ways that anthropologists have traditionally assessed the dietary preferences of individuals and populations in the past. Developments in the chemical analysis of osteological remains have opened new windows on the past. Analysis of trace elements and isotopes in human osteology has played an increasingly important role in dietary reconstruction during the last two decades (for reviews, see Price, 1989; Sandford, 1993; Schoeninger, 1995; and Larsen, 1997).

Traditionally, the issues of diet and affinity have been approached in human osteology via morphological assessment. The application of chemical methods of analysis has been mostly directed toward osteological remains from archeological contexts. As Larsen (1997) notes, documentation of diet in the past provides the context for studies of growth, stress, disease, and subsistence activities. Conventional approaches to diet utilized archeological materials, particularly plant and animal remains. The archeological record has long been known to be biased in its preservation of food remains, however, with plant remains often approaching invisibility due to difficulties in preservation. The prospect of an independent and objective means to generate consumption profiles of different foods eaten in the past is therefore an enticing one.

Organisms comprise common elements such as hydrogen (H), carbon (C), oxygen (O), nitrogen (N), calcium (Ca), and less common (trace) elements such as strontium (Sr). Many elements come in different isotopes. These isotopes differ in how many neutrons they possess. Some isotopes are unstable and radioactive, and by measuring their decay geochronologists are able to determine the age of many materials. Heavier isotopes have more neutrons in their nuclei. Lighter isotopes (^{3}C relative to ^{4}C, for example) break and form chemical bonds more rapidly than heavier isotopes. These facts of chemistry mean that ratios of the stable isotopes can be examined in efforts to deduce aspects of paleoecology and human behavior, including diet. Carbon and nitrogen are the elements that have received most attention in studies of human osteological chemistry.

There are two stable isotopes of carbon, ^{12}C and ^{13}C. Ratios of these isotopes in mammalian bones reflect diet (for example, plant tissues consumed during life). Plants use two photosynthetic pathways. The so-called C_3 plants discriminate against the heavier isotope of carbon, and their tissues are enriched in ^{12}C. Organisms eating more of these plants will therefore show higher ^{12}C/^{13}C ratios in their bones. Maize and other C_4 plants do not discriminate as much and have more of the heavier isotope, decreasing this ratio of carbon isotopes in the bone collagen of consuming organisms. Similarly, the heavier ^{15}N isotope of nitrogen concentrates as it travels up through the food chain. Marine plants have higher concentrations of this isotope than land plants, and animals higher up in the marine food chain have, as a consequence, higher ^{15}N/^{14}N ratios in their bones. People feeding on marine mammals are thus expected to have higher ratios than those subsisting on terrestrial food sources. Thus, the isotopic compositions of the bone tissue may be indicators of diet. By taking a tiny sample of bone tissue, the osteologist can convert the sample into a gas and measure the isotope values with a mass spectrometer. The isotopic ratio values can be compared between different skeletons. With these tools, in theory, bioarcheologists are positioned to evaluate subsistence changes through time by direct reference to the chemical composition of skeletal remains.

The initial 1970s chemical studies of bone were received with great enthusiasm because the techniques appeared to provide direct, quantitative means for reconstructing diet (Sandford, 1993). By the 1980s, trace element and stable isotope research was heralded as a breakthrough, and more researchers began to conduct these studies. As Sandford (1993) notes, however, the early optimism was soon curbed by studies that demonstrated that elemental concentrations are influenced by many complex and often interrelated processes. By the late 1980s authors were referring to the "abuse of bone analyses for archeological dietary studies" (Hancock et al., 1989) and proclaiming that there were "no more easy answers" (Sillen et al., 1989).

During life, elemental deposition in the skeleton is governed by more than just the abundance of elements in the diet. After burial, bones can be subjected to diagenesis. Concerns over how these variables influence bone chemical composition has generated a great deal of additional research during the 1990s. As Radosevich's "The Six Deadly Sins of Trace Element Analysis: A Case of Wishful Thinking in Science" (1993, p. 318) states: "It is possible that a viable field of trace element analysis of bone in this field can still be constructed, but examinations of basic geochemistry and taphonomy of soil-buried bone must be carried out first, not as an afterthought." As Ambrose (1993) has pointed out, studies of stable isotopes have been developed mostly in geochemistry and plant physiology rather than in anthropology. The most substantial insights into prehistoric diet and land use have been achieved in the areas involving the introduction of C_4 plants like maize, the demonstration that diets of high-status individuals were different from lower-status individuals in some societies, and the assessment of marine versus terrestrial foodwebs. Larsen (1997) reviews the results of these applications. As with the use of biomolecules in human osteology, stable isotope and trace element analyses require thorough grounding in chemistry, biochemistry, physiology, physics, and other laboratory sciences.

Suggested Further Readings

Brown, T. A., and Brown, K. A. (1992) Ancient DNA and the archaeologist. *Antiquity* 66:10–23.
> This review article discusses the biological importance of DNA, the techniques available for studying it, and the use of ancient DNA in archeology.

Herrmann, B., and Hummel, S. (Eds.) (1994) *Ancient DNA.* New York: Springer-Verlag. 263 pp.
> This edited volume provides an overview of the field, with an emphasis on techniques rather than applications.

Kelman, L. M., and Kelman, Z. (1999) The use of ancient DNA in paleontological studies. *Journal of Vertebrate Paleontology* 19:8–20.
> An excellent introduction to the subject with good cautions regarding contamination.

Krings, M., Stone, A., Schmitz, R. W., Krainitzki, H., Stoneking, M., and Pääbo, S. (1997) Neandertal DNA sequences and the origin of modern humans. *Cell* 90:19–30.
> The first successful application of DNA methods to the human fossil record.

Li, W.-H., and Graur, D. (1991) *Fundamentals of Molecular Evolution.* Sunderland, Massachussetts: Sinauer. 294 pp.
> A textbook that presents both the essential information on DNA as well as the use of DNA in studies of evolution.

Parsons, T. J., and Weedn, V. W. (1997) Preservation and recovery of DNA in postmortem specimens and trace samples. In: W. D. Haglund and M. H. Sorg (Eds.) *Forensic Taphonomy: The Postmortem Fate of Human Remains.* pp. 109–138. Boca Raton: CRC Press.
> A review of the preservation of DNA in postmortem samples, including both forensic and ancient skeletal remains, and how it can be retrieved and analyzed.

Sandford, M. K. (Ed.) (1993) *Investigations of Ancient Human Tissue: Chemical Analyses in Anthropology.* Langhorne, Pennslyvania: Gordon and Breach Science Publishers, SA. 431 pp.
> An edited volume with chapters by practitioners of isotopic work on human skeletal remains, whose combined message is caution.

Schoeninger, M. J. (1995) Stable isotope studies in human evolution. *Evolutionary Anthropology* 3:83–98.

> A good introduction to the use of isotopes in dietary studies of archeological human remains.

Stone, A. C., and Stoneking, M. (1993) Ancient DNA from a pre-Columbian Amerindian population. *American Journal of Physical Anthropology* 92:463–471.

> One of the few actual analyses of a prehistoric skeletal sample using DNA methods.

Stoneking, M. (1995) Ancient DNA: How do you know when you have it and what can you do with it? *American Journal of Human Genetics* 57:1259–1262.

> Addresses the standards needed in ancient DNA work and identifies productive areas of future research.

CHAPTER 22

Forensic Case Study
Homicide: "We Have the Witnesses but No Body"

The previous 21 chapters have outlined the basics of human osteology. Our focus has been on the use of skeletal remains in forensic, archeological, and paleontological contexts. The remainder of the book is devoted to presenting case studies selected to show the great excitement and breadth of studies that share the foundation of human osteology.

Human osteologists routinely assist law-enforcement agencies, coroners, and medical examiners by identifying skeletal remains. The remains themselves are recovered from a variety of contexts, including aircraft crash sites, makeshift graves, and open fields. When skeletal material is found, the primary forensic concerns usually initially involve the age, sex, stature, and ancestry of the individuals in question and often require positive identifications. This case study provides an example of a difficult but successful investigation in forensic osteology—an investigation that led to the arrest, confession, and conviction of a murderer.

Many investigations in forensic human osteology involve fairly straightforward matches between the unknown remains and missing individuals. Where the evidence is complete, positive identification is usually easily obtained, even by investigators not trained in osteology. In the case of more fragmentary remains, however, this work of identification becomes more difficult, and osteologists may be in the unique position of performing the identification. Owsley *et al.* (1993) discuss their forensic work in the notorious Jeffrey Dahmer case, in which fragmentation of the victim's remains was intentional. As Rougé and colleagues note (1993), radiographic identification of human remains may often be accomplished by focusing on deformities and anomalies of the postcranial skeleton. Owsley (1993) provides an example of the kind of situation that a forensic osteologist may face in conducting investigations in the developing world by discussing the identification of the remains of two U.S. journalists seven years after their disappearance in Guatemala. Like these, the case study presented below is not typical of forensic osteological investigations because of the extreme circumstances surrounding disposal and identification of the body. However, it illustrates the importance of basic detective work, teamwork, and basic osteological identification—the fundamentals of osteology in any forensic context.

The case presented here, number 191613, is documented by materials on file at the Cuyahoga County Coroner's Office, Cleveland, Ohio. Official reports, print media reports, and photographs in the case file were used to write this chapter. The dates and names of the victim, witness, assailant, and scenes used in this chapter have been changed to protect these individuals.

22.1
A Disappearance in Cleveland

Katie Jones telephoned police to report the disappearance of her older brother Harry in July of 1980. She told police of an argument between her brother and Mr. Charles Cook, aka Chuckie, the owner of Chuckie's Corner, a nightclub in Cleveland, Ohio. Jones and Cook had been feuding for two years, and on this particular Saturday night the argument was over an alleged assault of a woman. Cook insisted that Jones leave his establishment. When Jones left, Cook followed him up Ashland Avenue.

Witnesses last saw Harry Jones being pursued east and north, the pursuer firing several shots at Jones. Because of the argument in the nightclub, Charles Cook became a suspect in the disappearance of Harry Jones. When questioned, suspect Cook admitted to owning a gun but insisted that his gun had been stolen on the night of the disappearance. He denied killing Jones.

22.2
Investigation

Because of the possibility that Jones had been shot, Cleveland homicide detectives were assigned to the case. They quickly learned that Cook had boasted to another witness that he had, indeed, killed Jones, but that no body would be found: "I burned him up and police won't find any evidence." Homicide detective Jon T. Qualey noted: "This is a new one for me. Usually we have a victim and no witnesses. This time we have the witnesses but no body."

On further investigation, the detectives learned that Cook was an assistant supervisor at the Animal Resource Center at the Case Western Reserve University School of Medicine, where his duties included the disposal of research animal carcasses in an incinerator. Following this realization, in the words of detective Qualey, "We put one and one together and we came up with two." Detectives contacted the Cuyahoga County Coroner's office. Members of its staff joined the Cleveland police in a preliminary examination of the contents of the incinerator where animal remains were disposed. Sifting through the incinerator debris, authorities identified what appeared to be human bone fragments. A melted piece of lead about the mass and size of a .38-caliber bullet was found in the debris.

22.3
Inventory

Cleveland Homicide Unit detectives organized a full investigation of the incinerator's remains. The investigation team included the deputy Cuyahoga County Coroner, Dr. Elizabeth Balraj, and Barbara Campbell of the Trace Evidence Department of the coroner's office. The Cuyahoga County Coroner's office retains a number of consultants who contribute to its investigations as necessary. In this case, consulting anthropologist Dr. C. Owen Lovejoy of Kent State University, consulting dentists Drs. Elizabeth Robinson and James Simmelink, and consulting radiologist Dr. Benjamin Kaufman joined the incinerator investigation team.

The contents of the incinerator were emptied into twenty-five labeled metal bins. The incinerator was divided into an upper and a lower section, and the contents of these sections were kept separate. The contents removed from the top section of the incinerator had a total capacity of 40 gallons and weighed 60 kg. The bottom section had a 55-gallon capacity of contents weighing 75 kg. The contents comprised fragments of cremated skeletal remains, ash, masses of synthetic material, and metal and wire mesh (Figure 22.1).

Contents of each of the twenty-five metal bins were sifted through a fine wire screen. Sorting the human from nonhuman remains was the task of the consultant in anthropology, Dr. Lovejoy. All skeletal remains that were diagnostically human were set aside for further analysis. All skeletal remains that were diagnostically nonhuman were separated, photographed, and stored. Amalgam, ash, and all metal pieces were separated from the debris.

The human remains were extremely fragmentary, all of the specimens being brittle, grayish-white, and showing excessive shrinkage and exfoliation consistent with their exposure to high temperature. The incinerator was normally heated to between 1400 and 2000° F, hotter than normal incinerators but not as hot as a crematorium, which runs at 2300°. The skeleton had suffered greatly due to the incineration, but fragments of many body parts remained and were available for further analysis.

Figure 22.1 Bags of bone fragments from the incinerator at the Animal Resource Center at Case Western Reserve University.

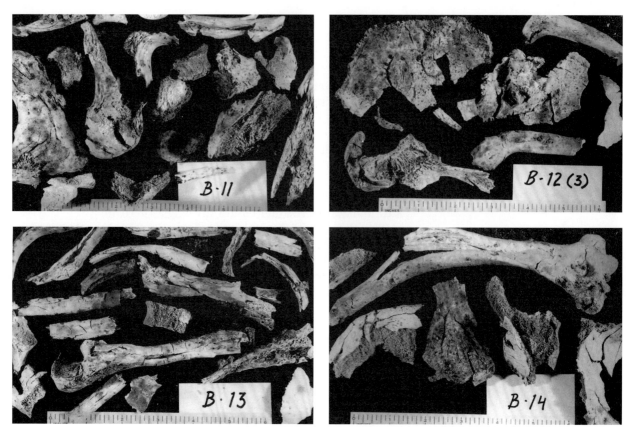

Figure 22.2 Some of the human bone fragments from bins B-11, B-12, B-13, and B-14. Scales in inches.

Figure 22.2 illustrates the condition of the human skeletal parts recovered from the incinerator.

22.4 Identification

Once the 163 diagnostically human bone fragments were separated from the animal bone in the incinerator, the first question for the investigators was how many human individuals were represented by the remains. The second question was about the identity of the individual(s) whose bones were present.

Portions of elements from the entire skeleton were recovered. Mandible ramus, clavicle, scapula, ulna, os coxae, and femur were represented by portions on both right and left sides. Careful comparison of those pieces that were present bilaterally showed antimeric correspondence. Furthermore, no cases of mismatched right and left sides were found among the human remains. Finally, despite the large number of human fragments, no skeletal element was duplicated in the collection. All of these observations made it highly probable that a single human individual was cremated in the incinerator, and that the fragmentary skeletal remains of this individual were subsequently mixed in the incinerator debris with animal remains.

The investigative team then turned to the question of identity. Individuation in this case proceeded along two complementary lines. The first set of questions centered on the standard issues of age, sex, and ancestry of the individual in question. The second major question was whether the remains were those of the missing Harry Jones.

Determination of age, sex, and ancestry was difficult due to the extremely fragmentary nature of the remains. Fortunately, the recovery of a few important areas allowed Dr. Lovejoy to go beyond the determination of "adult." Several sexually dimorphic portions of the skeleton were available for analysis. These included the femoral head, the right supraorbital region including the sinus system, the external occipital protuberance of the occipital, and the mastoid of the temporal bone. All of these features indicated to Dr. Lovejoy that the remains were those of a male individual.

A portion of the pubic symphyseal face was preserved. Here, the rampart was obviously complete with no remodeling scars. The surface was granular, with early rim formation. From these traits, Lovejoy concluded that the specimen was in the fourth decade of life. Because two complete auricular surfaces of the bony pelvis were recovered, Dr. Lovejoy was able to make an additional age assessment from this part of the skeleton. The surfaces had uniformly coarse granularity, with no significant macroporosity or microporosity. There were no islands of density and apical activity was moderate. Based on this, an age estimate of 36 ($\pm$5) years was estimated.

Having established that the individual was a male in his late thirties at the time of death, the next step was to see whether the remains belonged to Harry Jones. Harry Jones was 37 years of age when he disappeared, so Lovejoy's age estimates were suggestive but not conclusive.

In many forensic cases, physical anthropologists can work from intact skeletons, or intact parts of skeletons. In such cases, the dentition or the intact cranium can often establish a definite tie with photographs or dental records taken before death. In the analysis of the incinerated remains, however, investigators were faced with the task of somehow matching the small fragments of skeleton they had with knowledge about Harry Jones. No teeth were recovered for the analysis. The challenging task of individuation was accomplished because of some skillful detective work and radiographic analysis.

Because investigators strongly suspected that the remains from the incinerator were those of Harry Jones, they searched for radiographs that had been taken of Jones during his life. Radiographs of Harry Jones were taken in 1972, 1974, 1976, and 1978 and were available from St. Luke's Hospital, University Hospital, and Cleveland Metropolitan General Hospital.

Drs. Simmelink and Robinson, forensic odontologists, focused on one of the human bone fragments recovered from the ashes of bin B-12. This fragment was most of a right human mandible, including corpus and ramus (Figure 22.3). The 1977 films of Jones showed that he retained some upper and lower teeth at that time, but the 1981 films showed an edentulous Harry Jones. Radiographs of the mandibular specimen recovered from the incinerator were compared to the 1977 and 1981 films of Harry Jones. In addition to the lack of teeth or radiolucent sockets, the investigators discovered a 5-mm diameter "calcified density" in the body of the mandible that matched the premortem films. Furthermore, several other areas

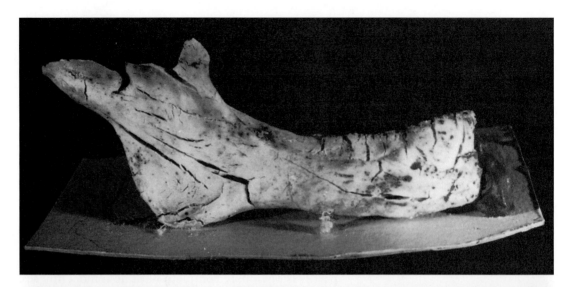

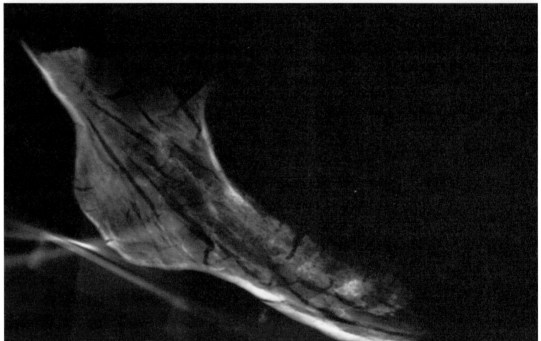

Figure 22.3 Right mandible from bin B-12. Natural size.

of trabecular bone along the mandibular canal and inferior border of the mandible also matched. Finally, the outline of the mandible from the incinerator matched perfectly with that of Harry Jones's radiographs (Figure 22.3 and 22.4). In summary, there were no radiographic inconsistencies between the 1981 films of Harry Jones and the 1984 films of the unknown mandible fragment from the incinerator bin. In the opinion of the consulting forensic odontologists, "this detailed comparison of the right jaw bone indicates positively that (the unknown) right mandibular bone piece is from the skeleton of Harry Jones."

Dr. Kaufman, the consulting radiologist on the case, concurred with the analysis of the odontologists, also noting the identity of the conden-

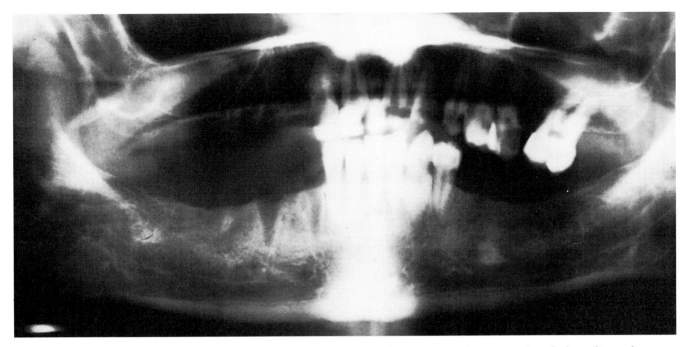

Figure 22.4 A comparison of the radiograph of the mandible found in the incinerator (Figure 22.3) with the radiograph taken of Harry Jones before his death (shown here) revealed a correspondence in the lack of teeth, or radiolucent sockets, in the approximately 5-mm diameter "calcified density" in the body of the mandible, in trabecular bone patterns along the mandibular canal and inferior border of the mandible, and in the outline of the mandible. Aside from the loss of teeth, there are no radiographic inconsistencies between the 1981 films of Harry Jones and the 1984 films of the unknown mandible fragment from the incinerator bin. This provided a positive identification of the deceased.

sation ("sclerosis") of bone in the mandible body, of the bony trabeculations, and of the position and shape of the mental foramen. Furthermore, Dr. Kaufman noted no inconsistency in the appearance of the temporal bone and the vertebrae between the incinerator specimens and Harry Jones's radiographs. Finally, antemortem radiographic views of Harry Jones's hands taken in 1979 matched radiographs of the hand bones recovered from the incinerator. Thus, Dr. Kaufman confirmed the finding of the dentists, concluding that the skeletal fragments from the incinerator were from the skeleton of the recently deceased Harry Jones.

22.5 Conclusion

Deputy coroner Balraj filed her report on the incinerator investigation, concluding that the human skeletal remains found among the cremated animal remains within the incinerator of the Animal Resource Department of the Case Western Reserve University School of Medicine were those of Mr. Harry Jones. A check of the closed-circuit television system that monitors and tapes activity around the clock in the medical school building produced pictures of the suspect, Charles Cook, backing a university truck up to the loading dock and entering the building at 3:10 A.M. on Sunday morning, about six hours after the quarrel with Harry Jones. The receiving dock is located about 20 feet from the incinerator. Records at the university showed that the suspect, who was scheduled to report to work at 6:00 that morning, had clocked in at 3:15, just after he appeared on the television monitor. He did not, however, sign in at the security guard's desk, as required of anyone entering the building before 6:00 A.M.

Secure in the belief that the body would never be found, Charles Cook at first denied the murder of Harry Jones. When the evidence against him mounted, however, he reversed his position and pleaded guilty to the murder of Harry Jones. He was sentenced to fifteen years to life in prison.

This case study was chosen because it effectively illustrates the unique contributions that the osteologist can make in the forensic arena. A knowledge of the basic principles of element identification and siding, and of individuation in human osteology, was the key in the analysis. In some ways, the forensic osteology of the Harry Jones homicide was unusual. The trail to the suspect was a short one, and witnesses were able to assist the detectives in locating the suspect as well as the deceased. The skeletal remains were very fragmentary, but an excellent radiographic history of the victim was available for comparative work. More often, the unknown skeletal remains are more complete, the suspect is not identified, and the possible victims are many. In any forensic situation, however, the keys to success are competent identification of the remains, careful, critical observation of the available clues, and close collaboration with other authorities on the investigation team.

CHAPTER 23

Forensic Case Study
Child Abuse, the Skeletal Perspective

In forensic human osteology a key concern is often the identification of individuals based on the analysis of skeletal remains. The case study documented in Chapter 22 presented one example of how even the most fragmentary skeletal remains can be recovered, analyzed, and identified in a criminal investigation. Forensic work with skeletal remains often involves the documentation of events as well as identities. Within the last decade, for example, forensic osteologists have worked with authorities in Haiti, El Salvador, and Bosnia in efforts to reveal secrets of the very recent past. The current chapter presents a case study that illustrates the contribution that a careful analysis of skeletal remains can make in the realm of forensic science.

Child abuse is widespread in modern society. The magnitude of nonaccidental pediatric injuries is staggering. The National Council for the Prevention of Child Abuse estimates more than 1.1 million cases of child abuse in the United States during 1994–1995. Approximately 15% of these cases involved serious injuries resulting in more than 1200 nonaccidental deaths in infants and children (Lancon *et al.*, 1998). Before the 1990s, very little attention had been paid to the problems of identifying child abuse in the skeletal remains of children. Phil Walker and colleagues have changed this, with an important paper documenting their work on five case studies. One of these is detailed in the account below, an account that draws exclusively from Walker *et al.* (1997).

23.1
Child Abuse and the Skeleton

Forensic cases involving the skeletal remains of chronically abused children are common. When such a child is killed, the abusers may attempt to dispose of the body surreptitiously and claim that a kidnap occurred.

Under such circumstances, time may pass before the body is discovered, and a fragmentary, partial skeleton may be the only evidence remaining.

Such cases are very difficult for the forensic pathologist or radiologist who typically lack experience in dealing with defleshed skeletal remains. The patterns of scars, bruises, and soft tissue trauma seen by the medical examiner or forensic pathologist are no longer available as evidence under these conditions. Even the picture of the battered child syndrome seen radiographically is very different from the one studied by the forensic osteologist who is directly examining the bones themselves. In the case presented below, it was the expertise of the forensic osteologists that led to the documentation of evidence not apparent to pathologists or radiologists—evidence crucial to the demonstration that child abuse had occurred.

23.2
A Missing Child Found

Police investigating a report of a boy who had been missing for five years discovered the partially skeletonized remains of a three-year-old child in the trunk of the family car. His parents first told law enforcement officials that the boy had died after slipping and hitting his head while taking a bath. Although at first they said that they had buried him, the discovery of his skeleton made it clear that instead they had carried the dead child in the trunk of their car for five years.

When the remains of the child were autopsied, the cause of death was not determined. The parents were charged with illegal disposal of the body. The remains were then sent to forensic osteologist Phil Walker of the University of California at Santa Barbara. An expert in both the forensic and bioarchaeological areas, Walker was well qualified to take a second look at the child's bones.

23.3
The Analysis

Dental development was used to provide a precise age at death for the child using the techniques discussed in Chapter 17. Combined with long bone measurements, these data indicated an age of 3 to 4 years at death. More detailed histological work on the teeth, focussing on Retzius line and cross striation counts in histological sections of the child's teeth were consistent with an age of 3 years, 7 months at death. Furthermore, they indicated that the child had suffered disruption of dental development, the last occurring about two months before his death.

Although the remains of the child were left in the car trunk for five years, considerable soft tissue covered the bones. When the desiccated tissue was carefully cleaned away, a linear fracture was seen to cross the left occipital bone, extending from the foramen magnum to the lambdoid suture. A 3.5-cm^2 area of subperiosteal new bone formation was observed

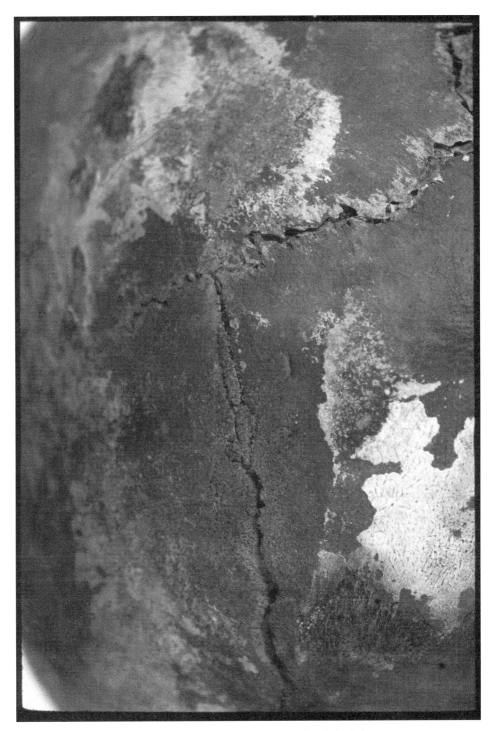

Figure 23.1 Posterioinferior view of the cranium. The lambdoidal suture traverses the vault from left to right. The vertical fissure is a partially healed fracture of the occipital. Close examination of this fracture revealed at least two stages of healing, showing that the child incurred the injury at least a month before his death. Photo courtesy of Phil Walker.

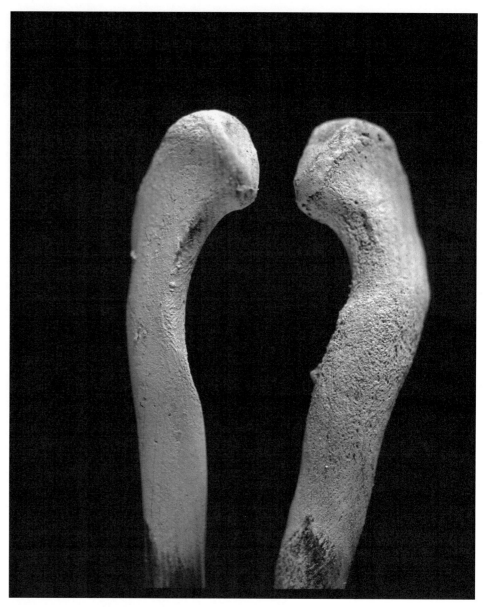

Figure 23.2 Healed fracture of the clavicle shaft (right) compared to the normal opposite side. Photo courtesy of Phil Walker.

below the lambdoid suture, confined to the occipital, and extending to the fracture line (Figure 23.1).

Gross and histological analyses showed that the area of bone formation recorded at least two stages of healing. Most of the affected area lacked large porosities and was comparatively dense, reintegrating with the external vault table. To Walker's practiced eye, this indicated a month or more between the injury that produced the fracture and the death of the child. But along the borders of the fracture, the well-healed bone was overlain by a second more recent episode of bone formation. Some of the fracture edge was starting to heal, and this newer, more porous bone would have taken more than a week to form.

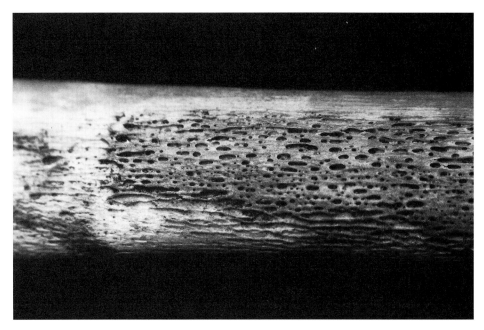

Figure 23.3 From a similar case, an area of subperiosteal bone formation on the fibula. This shows porosities and sharp margins indicative of recent healing. Some of the bone at the end of the lesion has been lost through postmortem flaking damage. Photo from Walker *et al.* (1997), courtesy of Phil Walker.

Disruption of the healing process is commonly seen in child abuse cases. Multiple traumatic episodes lead to these osteological patterns. Parents involved in the chronic, repeated beating of their children usually avoid seeking medical treatment for the child for fear of detection of their abusive behavior. Untreated, the bone begins to heal, but the fracture can be reopened with further trauma.

Turning to the teeth, Walker noticed that an upper and a lower incisor had antemortem fractures. He could tell that they occurred before death because their occlusal surfaces were both worn. Such fractures, of course, could occur without any abusive parental behavior, but such injuries are found at high frequency among abused children, reinforcing the idea that this child had suffered repeated injury.

The rest of the skeleton held more evidence. The clavicle showed a healed fracture (Figure 23.2). The left radius and ulna showed areas of subperiosteal new bone formation. These lesions are thin layers of new bone that form beneath the periosteum in response to trauma and subperiosteal bleeding (Figure 23.3). They are often asymmetrically distributed and can result from beating or stripping of the periosteum from the bone when the limbs are forcefully traumatized. In this case, the forearm had been traumatized in this manner with an area of subperiosteal new bone on the distal half of the ulna that nearly encircled the shaft (Figure 23.4).

No other long bones showed evidence of subperiosteal formation. Walker's work on other child abuse cases has shown that asymmetrical distribution of such subperiosteal lesions in vulnerable areas where bones are subcutaneous is common. Here, the borders of the lesions on the radius and ulna were beginning to integrate into the adjacent cortical bone. This healing indicated that the trauma that had caused them occurred a month prior to the death of the child. None of these subperiosteal lesions were

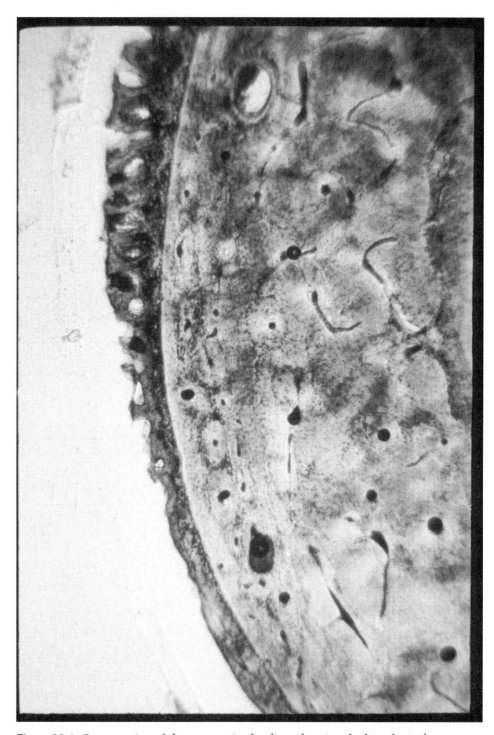

Figure 23.4 Cross section of the traumatized radius, showing the histological appearance of subperiosteal new bone formation in response to trauma. Normal, dense subperiosteal bone is seen at the margin of the lesion, with thickened, vascularized bone near the lesion's center. Photo from Walker *et al.* (1997), courtesy of Phil Walker.

visible on the high-resolution radiographs. Indeed, they are usually less than a half of a millimeter thick, but readily apparent to the osteologist. Here, the bare-bones osteologist had the advantage of seeing what was invisible to the radiologist and forensic pathologist.

Radiographs of the child's long bones showed that there were many Harris lines (see Chapter 18). These were bilaterally symmetrical in the distal radius, with 15 lines in the distal 18 mm of the bone. Some Harris lines are normally present in children of this age, but fewer than 5% of children between the ages of 2.5 and 4 years have as many Harris lines as this child.

23.4
The Result

Added up, the skeleton and teeth of the child whose remains had been recovered from the trunk of his parent's car showed that the months before his death had been punctuated with trauma. It was a pattern consistent with child abuse. Multiple injuries in different stages of healing are consistent with abuse. An accidental explanation for such injuries becomes increasingly unlikely as the number of traumatic episodes increases. The frequency of fractures produced by severe physical abuse decreases with advancing age, probably because smaller children are more easily held by their arms and legs and beaten, whereas such abuse is more difficult to inflict on older children because of their size and ability to resist. In this case, armed with the osteological evidence of severe physical abuse over a prolonged period, prosecutors charged the parents with second-degree murder, to which they eventually pled guilty.

CHAPTER 24

Archeological Case Study
The Bioarcheology of the Stillwater Marsh, Nevada

The previous two case studies dealt with osteological remains in the very recent past. Such forensic osteology cases represent an application of knowledge about the human skeleton to specific questions regarding individual identification and reconstructions of very recent human behavior. The next four case studies show how knowledge of human osteology can be applied in attempts to understand the more distant past. The first two case studies involve the relatively recent archeological past, whereas the final two involve human paleontology.

The term **bioarcheology** refers to the study of the human biological component of the archeological record. Bioarcheology is therefore a new name for an old subfield of human osteology. Just like other subfields such as forensic osteology or hominid paleontology, bioarcheology is multidisciplinary and uses the latest techniques to reveal as much as possible from skeletal remains. For example, the work of Phil Walker and colleagues (for example, Lambert and Walker, 1991) on native populations inhabiting the area adjacent to the Santa Barbara Channel in California has employed data from deep sea cores and tree rings, artifact assemblage change, archeological evidence of trade, spatial patterning and density of habitation sites through time, and even early mission records. These data have been integrated with osteological evidence of paleopathology, demography, and isotopic composition in an effort to understand the evolution of social complexity. The present case study is another bioarcheological analysis, the case of the Stillwater Marsh in Nevada. The account below is drawn directly from Larsen and Kelly (1995) and Larsen *et al.* (1996).

24.1
Background

The lives of hunting and gathering people were thought of by early anthropologists as short and difficult, but ethnographic studies changed that

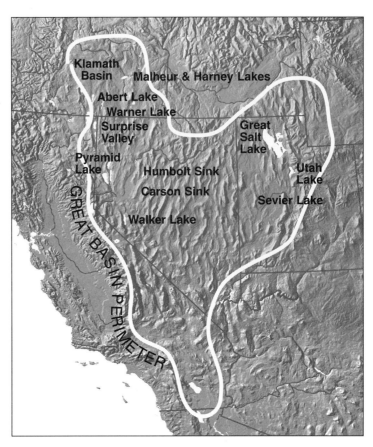

Figure 24.1 The Great Basin. Stillwater Marsh is on the eastern edge of the Carson sink. From Larsen and Kelly (1995).

view by the 1960s. Even though all hominid subsistence was based on foraging economies until the relatively recent advent of agriculture, by the time anthropologists could scientifically study hunting and gathering societies these foragers had already been forced into mostly marginal habitats by surrounding agriculturalists. Their lifeways had been substantially disturbed through contact. Because of this, the archeological record provides a unique window on the past. Because skeletal and dental indicators provide a cumulative biological history of an individual's lifetime and can record stress, nutrition, disease history, and physical activities, osteological remains are paramount in the study of human adaptation in the past. The conditions in the Great Basin of western North America (Figure 24.1) have preserved a remarkable archeological record of occupation, a record that is now being enhanced more than ever before through the detailed, multidisciplinary analysis of human skeletal remains.

24.2
Geography of the Carson Sink

The Carson Desert is an extensive area of sand dunes, alkali flats, and slightly alkaline marshes covering an area of about 2800 km² at a distance

Figure 24.2 Overview of the Stillwater Marsh area. The circle represents a probable house feature exposed by flooding. From Larsen and Kelly (1995).

of about 100 km east of the present city of Reno, Nevada. This area, one of the lowest in the Great Basin, is the landlocked drainage terminus for the Carson River, which feeds the Stillwater Marsh, an ecologically rich area inhabited for many years by pre-Columbian aboriginal populations who subsisted on the area's bounty (Figure 24.2). For the most part, these people buried their dead throughout the marsh rather than at cemeteries, a mortuary pattern likely to be characteristic of many prehistoric foraging groups.

24.3
Exposure and Recovery

Between 1982 and 1986, record winter precipitation resulted in massive flooding of several Great Basin wetlands, including the Stillwater Marsh. As the floodwaters withdrew in 1985 and 1986, many archeological sites were exposed. Hundreds of burials were uncovered, as were trash and cache pits, artifacts, and pithouse sites. The alarming nature of the exposed human remains led a local resident and amateur archeologist to alert state and federal authorities to the exposure of these valuable resources. The Nevada State Museum salvaged the disturbed remains, including 416 burials or individuals. Wind and wave erosion continued to disturb burials in the region, and in 1987 archeological crews under the

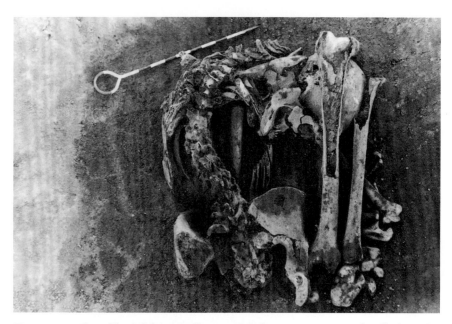

Figure 24.3 Flexed burial from Stillwater Marsh. From Larsen and Kelly (1995).

direction of Clark Larsen and Robert Kelly surveyed the most heavily impacted marsh shoreline. Additional remains were recovered by this effort. The combined skeletal sample is the largest of any reported to date from the Great Basin. In turn, many of the analytical techniques employed in the assessment of these remains represent the first of their type to be conducted on prehistoric Great Basin skeletal remains.

Six of the burials in the Stillwater series were radioisotopically dated, using radiocarbon. The dates range from ca. 2300 to 300 B.P. The predominant period represented by projectile points found in the sometimes associated archeological sites was the Undertown phase (1250–650 B.P.). The skeletal series was treated, for the purpose of the analysis, as a single population. However, it is obvious that it is merely a tiny sample of the total human population inhabiting the Stillwater area over thousands of years. It is important to note that the human remains from Stillwater Marsh were not from cemeteries but rather from isolated graves scattered throughout the marsh region, indicating a probable lack of formal disposal areas (Figures 24.3 and 24.4). The remains represent a series of small samples drawn at unknown intervals from a larger population over a span of centuries, if not millennia, rendering paleodemographic analysis problematic.

24.4
Analysis

The biocultural analysis therefore necessitated a different approach from the ones usually employed on cemetery assemblages of skeletons. What were the problems encountered by the analytical team? First, there was an increased probability that the remains were not representative of the

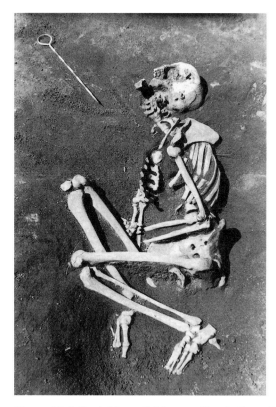

Figure 24.4 Burial from Stillwater Marsh. From Larsen and Kelly (1995).

population from which they were drawn. Second, relative chronological placement was very difficult, particularly because most burials did not contain directly associated diagnostic artifacts.

The individual remains were all analyzed. Age, sex, and stature were all determined according to standard procedures (Chapter 17). Computed tomographic scanning of selected humeri and femora was undertaken at the Veterans Administration Hospital in Reno. Bone fragments were sampled for stable isotope and genetic analyses. Discrete trait analysis showed homogeneity among the remains, as did skeletal measurements (Brooks *et al.*, 1988), so they were treated as a single sample for the purposes of the analysis. The five goals of the project were as follows: (1) to provide a description of the remains, (2) to assess the quality of life, (3) to improve documentation and understanding of population history, (4) to characterize diet, and (5) to identify physical activity patterns.

24.5 Affinity

There are long-standing debates about population movements in the Great Basin. Linguistic data suggest that Numic-speaking people arrived between 1000 and 700 B.P. Study of serum albumin derived from the skeletal remains revealed similarity with modern Numic speakers, but because the skeletal remains with the shared allele were undated, these results did not bear on the issue of entry of these people into the area. The analysis of mitochondrial DNA showed a very low frequency of the 9-basepair deletion, a deletion observed in some prehistoric and extant Native American populations. This suggested that the Stillwater population was probably not ancestral to any group with a high frequency of this deletion (e.g., California Penutian, Zuni, Yuman, Washo, or Southern-Uto-Aztecan language groups). The molecular analyses left several possibilities open regarding the identification of ancestral-descendant relationships in the Great Basin and beyond, but this study constitutes an important step in the ongoing work in this direction.

24.6 Osteoarthritis

Over three quarters of the individuals in the skeletal series were affected by osteoarthritis, and all individuals over 30 years of age showed this pathology. The highest frequency was in the lumbar vertebrae, but cervical vertebrae and elbows had frequencies of more than 50%. Controlling for age, there were several differences between males and females. The Stillwater series exhibited a pattern of sexual dimorphism in osteoarthritis prevalence, with the males more affected than females, with the exception of the lumbar vertebrae (Figure 24.5). This suggests that women frequently carried heavy loads (perhaps children, firewood, water, and/or food). Males had significantly higher frequencies of osteoarthritis in the hip and ankle.

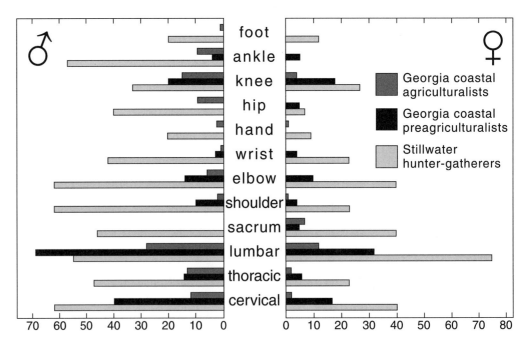

Figure 24.5 Prevalence of osteoarthritis among different Native American Groups. Stillwater rates are shown as the lighter columns. From Larsen and Kelly (1995).

The rates of osteoarthritis prevalence values are very high relative to other skeletal series. This was interpreted to mean that these foragers engaged in physically demanding activities, particularly high levels of mechanical loading of the spine. Several individuals had vertebral compression fractures (Figure 24.6). The investigators conclude that frequent foot transport of heavy loads might be implicated in these pathologies. They make reference to ethnographic accounts suggesting that the recent inhabitants of this region routinely engaged in physically demanding activities.

24.7
Limb Shaft Cross-Sectional Anatomy

The Stillwater series showed a consistent pattern of elevated bone strength relative to comparative samples, which the investigators interpreted to reflect high bending and torsional loading modes. The humeral values were low relative to the femoral values. The low bone mass (small total amount of cortical bone) in both elements relative to other skeletal series, when controlled for age, was interpreted to reflect episodic undernutrition. Comparison of males and females showed high dimorphism, suggesting that females were less mobile and more "tethered" to the marsh setting.

24.8
Physiological Stress

Dental hypoplasias are nonspecific growth arrest markers in teeth that give evidence of the periodicity and intensity of stress (Chapter 18). Two thirds of the individuals studied showed at least one hypoplasia, concentrated at 3–4 years of age. Overall, the Stillwater series showed relatively low levels of hypoplasia prevalence compared to other populations. Only 16% of the individuals showed periosteal inflammation, mostly of the tibia. This was also low relative to most other North American forager series. Although stress was not severe, it was present at appreciable levels. The investigators suggest that the narrow hypoplasias and the bone mass data suggest that the Stillwater populations suffered from episodic nutritional stress.

24.9
Dietary Reconstruction

Stable carbon and nitrogen isotopic analysis of human bone samples showed that a variety of foods were consumed. The juvenile samples analyzed fell completely within the adult range, suggesting no age-related differences in diet. There were no differences between adult males and

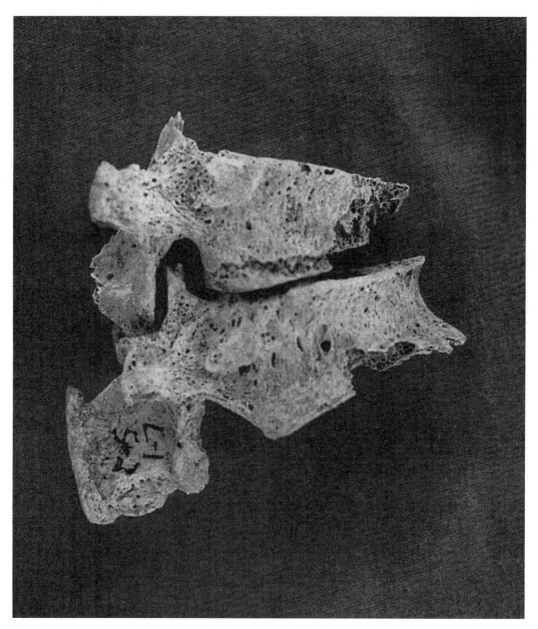

Figure 24.6 Compression fractures of lumbar vertebral bodies. From Larsen and Kelly (1995).

females and no differences between burials from different sites. The wide dispersion of isotopic values in the series indicated that some individuals consumed almost only C_3 plants, whereas others ate significant amounts of food with the C_4 isotopic signature. The relatively positive ^{15}N values were interpreted to indicate that the diet did not include pinyon pine nuts as a major dietary component, but rather the cattail and desert-blite.

The frequency of dental caries in the series is very low. The high rate of occlusal surface wear, largely attributable to the introduction of grit into the diet via seed grinding on stone metates and via a sandy environment, may have contributed to this low frequency, but most available foods were noncariogenic in the first place. There was a low frequency of cribra orbitalia and porotic hyperostosis, suggesting low rates of iron deficiency anemia.

24.10
The Future

The investigation concluded that the populations inhabiting the Stillwater Marsh comprised individuals who were physically robust, ate varied diets, were in relatively good health, and were not sedentary, with heavy workloads and considerable mobility. These conclusions were reached on the basis of a small sample of the population who lived and died at one place in the Great Basin over a long time period. Even though several of the conclusions could be challenged on the basis of the small and dispersed samples (a restriction that the investigators could not avoid), their bioarchaeological study shows the way for more comprehensive and definitive studies that will surely follow.

All of the Stillwater Marsh burial recoveries and excavations and all of the subsequent skeletal analyses (including the destructive ones on already broken specimens) were conducted with permission from the Fallon Paiute-Shoshone Tribe under a Memorandum of Understanding between the U.S. Fish and Wildlife Service and the Tribe. All remains discussed here and in the monograph have been reinterred in a subterranean crypt on U.S Fish and Wildlife Service land. With appropriate permissions from the Tribe and the U.S. Fish and Wildlife Service, these remains could be made accessible to other researchers in years to come, as new research questions arise and new, more sophisticated and precise techniques are developed by skeletal biologists in the future.

CHAPTER 25

Archeological Case Study
Anasazi Remains from Cottonwood Canyon

In the following two case studies, the fundamentals of osteology outlined in Chapters 1–20 of this book are applied to fossils. Plio-Pleistocene hominid bones and teeth are rare, usually fragmentary, and almost always fossilized. A case study from an entirely different context is presented here. The skeletal remains are from anatomically modern humans found during an archeological excavation.

In the post-Pleistocene archeological record, many human skeletons recovered in and around habitation sites consist of single burials. When intact, these burials are relatively easy to analyze. Any one of hundreds of case studies involving such remains could have been chosen for this chapter, but most would have added little information to that provided in previous chapters. In contrast to the assessment of primary burials, bundle burials and ossuaries are progressively more difficult for the osteologist to deal with because the skeletal remains in them are usually more mixed and fragmentary. Even bigger challenges for the osteologist involve analysis of mixed cremations and other cases in which the skeletal remains have been deliberately damaged. In this chapter we present one such challenging case.

Characteristics of skeletal remains from some archeological sites in the American Southwest have led investigators to conclude that sporadic cannibalism was practiced by the Anasazi, a prehistoric Native American group responsible for, among other things, constructing impressive cliff dwellings such as those at Mesa Verde in Colorado. In this chapter we describe the discovery, recovery, and analysis of one such assemblage from the state of Utah. This case study demonstrates the importance of being able to identify fragmentary osteological remains. Furthermore, it provides an illustration of how the study of osteological remains in an archeological context can make significant contributions to the understanding of past events and behaviors.

25.1
Cannibalism and Archeology

Cannibalism is a subject that holds considerable interest for the anthropologist. Workers in all three subdisciplines of anthropology—ethnology, archeology, and physical anthropology—have become involved with the study of cannibalism over the past century. Textbooks in anthropology typically report on cannibalism in both the recent and the deep past. Students learn about cannibalism at Choukoutien, among the Aztec, and in highland New Guinea.

In 1979 William Arens wrote *The Man-Eating Myth,* in which he investigated some of the most popular and best-documented cases of cannibalism in the ethnohistorical record. Arens concluded that, aside from survival conditions, there was inadequate documentation of cannibalism as a custom in any form in any society. This conclusion sparked considerable controversy, but most critics agreed that, if cannibalism were as widespread as anthropologists had traditionally maintained, better documentation would be required to demonstrate it. Ethnologists have, however, run out of time to provide the documentation. Even in the remote corners of the world where cannibalism was widely reported in the 1800s, the practice no longer exists. The documentation, if it is to be forthcoming, will therefore have to come from the archeological record.

Because the early historical and ethnographic accounts of cannibalism are riddled with doubts and because ethnographic observation is no longer possible, archeology is the only remaining tool for investigating the existence and extent of cannibalism. But how is cannibalism recognized in the archeological record? A long history of work on faunal remains from archeological contexts provides the answer. The faunal analyst studies the context of the nonhuman bone assemblages from archeological sites and the composition and modifications to these assemblages (cutmarks, hammerstone percussion for marrow removal, and other trauma). The butchery and consumption of animals can be understood from these observations of faunal remains. When human remains from an archeological site are consistent with a nutritionally motivated breakdown—when patterns of burning, cutmarks, percussion, crushing, and other fracture on human remains match what is seen on faunal remains—the assemblage is usually interpreted as evidence of cannibalism.

Over the past twenty years there has been an accumulation of evidence in the American Southwest that indicates the occurrence of cannibalism among the Anasazi. Anasazi burials are typically primary burials and are often accompanied by grave goods. During excavation of several sites in the Four Corners area of Colorado, Arizona, New Mexico, and Utah, however, human skeletal remains obviously not in primary contexts have been encountered. These remains are extremely fragmentary, with obvious cutmarks, intentional fracture, and signs of burning. Turner (1983b) and White (1992) summarized the evidence and concluded that cannibalism was practiced at these localities. Turner and Turner (1999) have extended these studies. The practice was very uncommon, however, as the number of recorded instances of cannibalism is very small when compared to the thousands of Anasazi sites that have yielded evidence of intentional, primary, considerate burial.

25.2
Cottonwood Canyon Site 42SA12209

Site 42SA12209 is an Anasazi Pueblo I habitation site located in Utah's Cottonwood Canyon, near the town of Blanding (Fetterman *et al.*, 1988). The site was damaged by a uranium mining road in the 1950s or 1960s and not recorded until a 1971 archeological survey covered the area. Because the site was seasonally impacted by road maintenance, the Forest Service arranged for testing in 1986 and salvage excavation by Woods Canyon Archaeological Consultants, Inc., in July of 1987.

The site is located about 2000 m above sea level in the semiarid Upper Sonoran life zone. The site was part of a large Cottonwood Canyon Pueblo I community. More than one hundred sites have already been located in the area. Based on dendrochronological analysis and ceramic seriation, occupation dates to the latter half of the ninth century (A.D. 880–910). The material culture, pollen, and macrobotanical remains from the site all indicate that the prehistoric inhabitants were agriculturalists who relied heavily on the cultivation of corn for their subsistence.

25.3
Discovery

The salvage fieldwork was aimed at excavating the portion of the site in danger of disturbance by road maintenance. This part of the site consisted of a plaza and seven surface rooms. Associated with the fill of one of the structures (Structure 3) was a feature consisting of hundreds of human bone fragments. The main concentration of bones was in an area 70 × 60 cm, vertically concentrated in a section of room fill 20 cm deep (Figures 25.1 and 25.2). The fill in this area was charcoal-stained and ashy, with much refuse. Among the hundreds of bone fragments, only one articulated hand was found (Figure 25.3). The "interment" was therefore not primary, and it obviously consisted of more than one individual. The remains were interred sometime after the abandonment of Structure 3 because the pit was dug into the roof-fall of the structure. To ascertain as much as possible about the biology of the individuals whose skeletal remains had been discovered in Structure 3, Jerry Fetterman of Woods Canyon Archaeological Consultants engaged human osteologists at the University of California at Berkeley.

25.4
Analysis

25.4.1 Patterns of Damage

Approximately 700 specimens, mostly fragments of bone, arrived in the Berkeley laboratories in autumn 1987. Unlike the remains from several other localities in the Southwest which yielded this kind of material, the

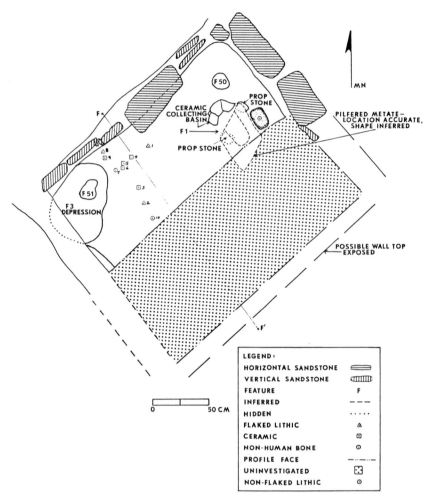

Figure 25.1 Plan map of Structure 3 showing the Feature 3 (F3) depression from which the human bones were recovered. This shallow pit extended below the surface in the west corner of the room and was filled with charcoal-stained and ashy material, with a good deal of refuse (from Fetterman *et al.*, 1988).

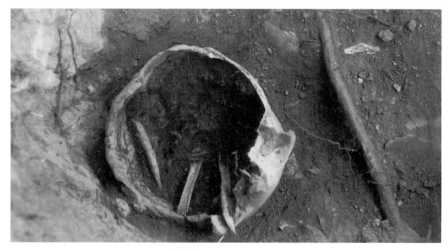

Figure 25.2 Inverted human cranium lacking the base, with fragments of human limb bone shaft inside, *in situ* in Feature 3. Note the nearby plant roots.

Figure 25.3 Articulated human hand and wrist *in situ* in Feature 3. Note the nearby roots and the limb bone shaft splinters projecting from the excavation wall.

Cottonwood assemblage was not particularly well preserved. Root activity had etched the bone surfaces and weakened the fragments, making the collection difficult to handle and analyze. It was immediately apparent, however, that the patterns of damage on the assemblage were similar to those already reported by Nickens (1975) and Turner (1983b) at other sites (i.e., possible cannibalism).

25.4.2 Sorting and Refitting

Procedures involving the identification of specimens in an assemblage like the one from 42SA12209 differ considerably from standard bioarchaeological practice. In normal osteological work, all broken elements are glued together, and all elements are then identified by direct comparison with complete skeletal material. The methodology developed to deal with assemblages like the one from Cottonwood Wash proceeds through several different analytical steps to ensure that the results can be compared directly to results reported by faunal analysts who rarely practice **refitting (conjoining,** the fitting together of broken pieces). Data on the Cottonwood remains were therefore collected for use by both faunal analysts and physical anthropologists (White, 1992).

The Cottonwood Wash specimens received in Berkeley were unwrapped and sorted into cranial and postcranial elements (Figure 25.4). Because of the delicate nature of the bones, care was taken in handling the material. All specimens were identified and sorted by element category, side, and age. Primary identifications were made very conservatively because guessing about element identity at this stage can have negative effects on refitting broken pieces in later phases of the analysis. For example, if a femur fragment is first misidentified as a tibia fragment, it will not be checked against femur fragments later in the analysis, and potential joins will thus be missed. Because of this potential problem, a series of four indeterminate postcranial element categories were used during the element sort.

Figure 25.4 Analysis of the Cottonwood osteological sample. The specimens were identified and sorted by element, and each fragment was examined for possible refitting.

All recognizably nonhuman elements were separated at this stage of the analysis. None of these were from the feature in Structure 3. The 691 remaining elements were all either recognizably human or indeterminate, with no way to eliminate the possibility that they were human. The number of identifiable human pieces and the lack of identifiable nonhuman pieces in the collection suggested that most of the unidentifiable fragments were human.

Once all of the specimens were sorted into element categories, data on specimen number, siding, fragmentation, element identity, age, identity in standard faunal analysis, and trauma were entered into a computerized database file. Specimens were examined for signs of perimortem trauma by eye and hand lens under strong directional lighting. Cutmarks, burning, hammerstone impact scars, adhering flakes (flakes that adhere to the specimen on the edges of impact points), anvil damage, and crushing were all scored as present or absent.

All specimens were systematically checked against each other within each element category to see whether or not they joined. A systematic refitting exercise is time-consuming, but it is necessary for maximum restoration of the bones. Each cranial piece, for example, was checked against every other cranial piece in the collection for which a join was possible. For each pair of bones, all broken edges of the first piece were systematically checked against all broken edges of the second piece.

Figure 25.5 Refitting of the Cottonwood sample in progress. Within element categories, each specimen is checked against all other specimens for possible joins in an effort to restore skeletal elements.

Of course, a join between two intact right temporals is impossible and need not be checked. Similarly, there is no reason to check a broken temporal edge against a frontal, because these bones do not articulate. Thus many pieces can be ruled out as impossible joins before any checking of broken edges is done. For pieces of vault or limb bone shaft, however, whenever the cortical thickness makes a join a possibility, that possibility must be completely explored. This means checking every appropriate broken edge of a given fragment against every other appropriate broken edge in the collection. This systematic refitting involves physically passing one broken edge against another, looking for a join. When the bone locks into place, the analyst should check for matches in the cortical thickness as well as in anatomical structures that cross the break. Color and edge length are not factors to consider in such analysis, because postdepositional factors can differentially stain bones in the ground. Furthermore, several fragments can join along the broken edge of a single specimen. Figure 25.5 shows systematic refitting in progress.

The systematic refitting exercise produced many joins across the anciently fractured surfaces of the bone collection from Cottonwood. Nearly 140 pieces were found to join within thirty-three sets. Some sets were made up of nearly thirty pieces. Joins found during the exercise were temporarily taped together for analysis and photography. Gluing together specimens in an assemblage in which breakage was ancient is not recommended, because it gives future investigators an inaccurate portrait of the assemblage and makes comparison with nonhuman faunal remains more difficult. A few of the fractures were obviously of recent origin because of color and surface texture. These were glued together and notes in the catalog were made to this effect.

25.4.3 Minimum Number, Age, and Sex of the Individuals

The archeologists wanted to know exactly how many human individuals were represented by the collection of nearly 700 osteological specimens. It was possible, but extremely improbable, that each unrefitted, nonantimeric (not mirror images from opposite sides) fragment was from a different individual. A more appropriate method for estimating how many individuals were involved in the collection was the determination of the minimum number of individuals (MNI, see Chapter 14) that must have been represented to account for the remains recovered. A true minimum number count for a bone assemblage takes into account element, side, age, sex, occlusion, articulation, and antimeric partners.

The original on-site feature report for 42SA12209 noted the presence of at least two individuals and called for further analysis. For this site, as for many ravaged paleontological and archeological assemblages, the MNI was determined on the basis of the cranial and dental evidence. The MNI of four individuals was calculated independently from both the dental and the cranial evidence (Figure 25.6). There were two immature frontal bones that probably belonged to the two individuals identified on the basis of teeth. There was an intact frontal belonging to a major vault portion of an adult specimen that included mostly intact parietals. A second adult individual was indicated by a large conjoined set from a partial vault with parietal areas duplicating those of the first adult.

These MNI results were achieved after refitting. Identical results were obtained by looking at the dental evidence prior to refitting. Using all of the available evidence, the four individuals based on craniodental evidence were defined as follows:

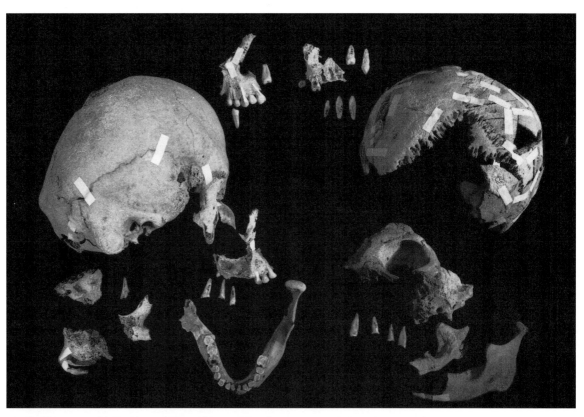

Figure 25.6 The minimum number of four individuals in the Cottonwood sample was estimated on the basis of cranial parts. Specimens are shown here after refitting. White tape holds temporary joins. One-fourth natural size.

- **Individual 1.** A fragmentary maxilla with associated teeth. An age of 12 years is indicated by barely open canine and premolar root apices.
- **Individual 2.** Another maxilla fragment with associated teeth. This individual is slightly advanced in root fusion over individual 1, with a probable age of 12.5 years.
- **Individual 3.** Conjoining cranial specimens indicate a robust individual. Probably associated, on the basis of robusticity and size, is another craniofacial conjoining set, a mandibular conjoining set, and isolated teeth. This individual is an old adult.
- **Individual 4.** This individual is represented by most of a cranial vault and probably associated mandible. This individual is also an old adult.

In summary, the assemblage contained a minimum of two immature individuals at about age 12 years and two old adults. Nothing in the postcranial sample was at odds with this assessment.

Sexing was not possible for the immature individuals. One of the adults was obviously male because of size and robusticity. The smaller adult vault was probably of a female individual. Strong artificially produced flattening of the posterior parietal and occipital areas, a cradle-boarding effect, was evident on both adult crania. The adult male had thickened cranial vault bones, suggesting a response to anemia. The presumed mandible of this individual showed extensive premortem loss of the posterior teeth and subsequent alveolar resorption.

The presumed adult female individual showed heavy mandibular tooth wear and a large carious lesion on one first molar with an associated abscess in the mandibular alveolar region. A small adult metatarsal head showed minor arthritic lipping.

25.5
What Happened? The Osteological Contribution

The large number of joins within the assemblage, particularly in the cranial vaults and the postcranial conjoining sets, suggests that the bones were interred in Structure 3 soon after they were fractured (Figures 25.7 and 25.8). There is no doubt that the fractures crossed by the conjoins represent perimortem damage.

Beyond element preservation and representation, evidence for trauma was preserved in the form of surface modification of the bones in the assemblage. The preservation of the Cottonwood Wash assemblage, with much of the surface detail lost to root etching and erosion, made accurate observation of surface trauma to the bones extremely difficult. This poor preservation rendered quantification of cutmarks, hammerstone and anvil damage, burning, and other modification meaningless and misleading; most of the evidence of perimortem surface modification had been erased by postmortem depositional modification.

The fracture on the cranial vaults of all four individuals was ancient, as indicated by the percussion scars, adhering flakes, crushing, anvil damage, and internal and external vault release (Figures 25.9 and 25.10). This pattern of fracture has been seen in other assemblages in the Southwest, particularly the one from Mancos Canyon (Nickens, 1975; White, 1992).

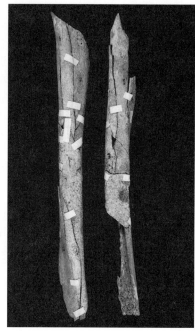

Figure 25.7 Adult femora from the Cottonwood sample: *left,* individual fragments of shaft prior to refitting; *right,* two femoral shafts restored in the refitting exercise. Note that the bone ends are still missing. Masking tape marks and binds the temporary joins. One-fourth natural size.

Fracture of the postcranial elements followed the pattern seen at Mancos and elsewhere, with shaft splinters dominating the assemblage. Fracture appears to have occurred while the bone was fresh. Indeterminate cranial fragments were the next most frequently encountered items. Estimates showed that fewer than sixty specimens in the assemblage, less than 10% of the sample, would have been considered identifiable by a faunal analyst.

The portions of each skeletal element represented in the refitted assemblage were also revealing. There was an absence of parts composed of spongy bone; for example, proximal humeri and vertebral bodies were absent. This pattern of element representation and element preservation has also been described for other assemblages in the Southwest. The intentional crushing of spongy bone portions may be involved with extraction of nutritive value from bone as described in ethnographic situations; it is suggestive of cannibalism.

Crushing of the bone was most evident on cranial pieces, and stone-on-bone impact seems to have been responsible. A few specimens showed clear scars of hammerstone impacts that did not result in fracture at the impact point. Several specimens showed flakes of bone still attached to the region immediately adjacent to the hammerstone impact point. These flakes were particularly evident on the long-bone shaft fragments and on the cranial vault pieces. There was some evidence of anvil scratching. The cutmark evidence from the collection was particularly poor, these superficial marks being susceptible to erasure by root action and bone exfoliation. There were, however, examples of cutmarks preserved (Figure 25.11).

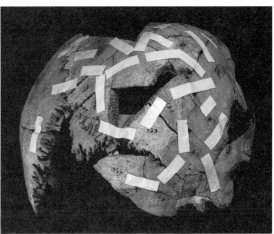

Figure 25.8 Adult cranium from the Cottonwood sample: individual fragments of cranium prior to refitting, and the partial cranium restored in the refitting exercise. Masking tape marks and binds the temporary joins. View is three-quarters, from the right rear. One-half natural size.

Unlike the better preserved Mancos Canyon collection, the Cottonwood Wash material was difficult to assess with respect to burning because of the preservation problems described above. It was evident that many of the fragments were burned, some of them extremely calcined. Little can be said about the burning of cranial elements prior to fracture of the vault or burning of the postcranial elements. After fracture, however, some of the fragments were intensely burned. This postfracture burning included both vault and postcranial pieces in which conjoining showed fresh unburned bone on one side of a fracture edge and heavily burned bone on the opposite side, implying that some bone fragments were discarded in a fire after fracture.

The data from conjoining combine with the evidence of trauma detailed above to show that fragmentation of the assemblage resulted from percussion blows directed at cranial vaults and limb bones. Refitting results also show that many parts were missing from the assemblage.

Figure 25.9 Cranial fragments from the Cottonwood sample that show evidence of hammerstone percussion. Note also the extensive root etching of the outer bone surface. Magnification 2×.

Several important conclusions were drawn from the analysis of the assemblage from Cottonwood Wash. The assemblage from the feature in Structure 3 included no recognizable nonhuman elements. It included artificially fragmented bones and teeth representing the cranial and post-cranial remains of at least four individuals. Two of these individuals were

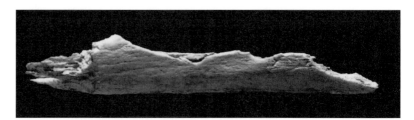

Figure 25.10 Limb bone shaft fragment from the Cottonwood sample shows evidence of hammerstone percussion. Magnification 1.5×.

Figure 25.11 Cutmarks made by a stone tool on a long bone shaft fragment. Magnification 4×.

old adults, one a male and one probably a female. Two were young individuals of approximately 12 years of age.

The fragmentation in the assemblage and the data from conjoining studies indicate that there was extensive perimortem human involvement with the bone material. This involvement included skinning, flesh removal, or disarticulation activities indicated by cutmarks. Burning was difficult to assess, but some bone fragments were heated to very high temperature after fracture. Fracture of the vault and limb bone shafts was accomplished by percussion with a hammerstone, resulting in anvil scars, adhering flakes, and crushing. The lack of spongy bone parts of elements suggests that these portions of the bone were also crushed by hammerstone impact.

The composition and characteristics of the fragmented human bone assemblage from Cottonwood Wash site 42SA12209 are similar to what is seen in a variety of sites across the Southwest. These assemblages have been interpreted as evidence of cannibalism (White, 1992). It will be necessary to address patterns within and variation between these assemblages of human bone from the prehistoric Southwest before we are in a position to understand their full behavioral significance. This endeavor will necessarily involve the combined skills of the archeologist and the human osteologist if progress in understanding cannibalism is to be made.

CHAPTER 26

Paleontological Case Study
The Pit of the Bones

Research on human origins is often called **paleoanthropology.** Once conducted by a few people searching for hominid fossil bones, investigations into human origins and evolution now involve large, multidisciplinary, often international teams of field and laboratory specialists who seek to reveal the past in a detail thought impossible only two decades ago. In today's paleoanthropological projects, paleontologists specializing in various plant and animal groups ranging from pollen to elephants are integrated with physical anthropologists, archeologists, geologists, geochronologists, and remote sensing specialists. The present case study presents the most significant hominid discovery ever made in Europe, a discovery which is rewriting the textbooks of human evolution.

In 1856 a partial skeleton was recovered from a cave in the Neander valley near Dusseldorf in Germany. Recognized by their peculiar skeletal and cranial features, many more Neanderthals have since been found in Europe and the Middle East. *Homo neanderthalensis* is now a widely accepted side branch in the human evolutionary tree. But from where did this form come? When, where, and how did it evolve its anatomical peculiarities? For years, paleoanthropologists posed these and many other questions. Answers came slowly, fragment by fragment, as bits of crania, mandible, and postcrania were revealed at sites such as La Chapelle, Le Moustier, Heidelberg, and Swanscombe. Discoveries at a series of sites in northern Spain, collectively known as Atapuerca, have recently shaken European paleoanthropology. This research is ongoing, but the spectacular finds made by teams of paleoanthropologists working to reveal the Pleistocene Europeans are revolutionizing our views of these ancestors and relatives. This case study is about the most impressive set of Atapuerca discoveries, that of the skeletal remains of over thirty individuals found deep in a cave system, in a small cavity known as the Sima de los Huesos, the Pit of the Bones (Arsuaga *et al.,* 1997). This work involves building knowledge of the past through applying the principles of skeletal identification and analysis stressed in previous chapters. Like most osteological work, it involves many steps, illustrated in Figure 26.1.

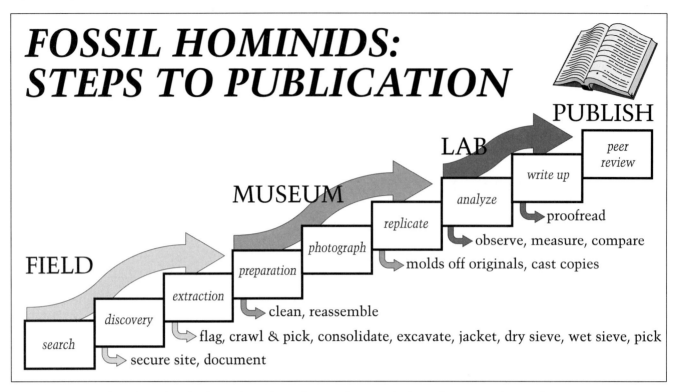

FOSSIL HOMINIDS: STEPS TO PUBLICATION

PUBLISH

LAB

MUSEUM

FIELD

search

discovery

extraction

preparation

photograph

replicate

analyze

write up

peer review

secure site, document

flag, crawl & pick, consolidate, excavate, jacket, dry sieve, wet sieve, pick

clean, reassemble

molds off originals, cast copies

observe, measure, compare

proofread

Figure 26.1 Bioarchaeological and paleoanthropological research proceeds through a series of steps that intervene between project conception and publication. Depending on the nature of the fieldwork and laboratory analysis, such work can take large teams of specialists decades to complete.

26.1
Atapuerca

The Sierra de Atapuerca is 12 km northeast of the historic city of Burgos, in the northern part of Spain. The hills here feature a variety of karst systems, including extensive cave systems, sinkholes, and collapsed caves in upper Cretaceous marine rocks. The cutting of a deep and narrow trench for a former railway exposed some of these caves and their fillings. Excavations in the Gran Dolina, one of the cave fillings exposed in the wall of the railway cut at Atapuerca, recently yielded the earliest dated hominid remains from Europe, ca. 800,000-year-old remains that show evidence of cannibalism. The younger Sima de los Huesos is found close by, across the railroad trench, opposite the Gran Dolina.

To get to the Sima today, one enters the Cueva Major-Cueva del Silo system about 500 m away. To reach the Pit of the Bones one must descend into the cave system and traverse a labyrinth of angular, tortuous, maze-like passages. Sometimes stooping, sometimes walking crablike, and often crawling and sliding through tiny openings, one passes 500 m through the dark, silent, and always muddy passages and galleries of this cold subterranean complex. Getting to the Sima and back out is physically exhausting. Near the end of the approach, the cave system opens up into the colossal Cyclops Gallery (Figure 26.2), with a narrow passageway lead-

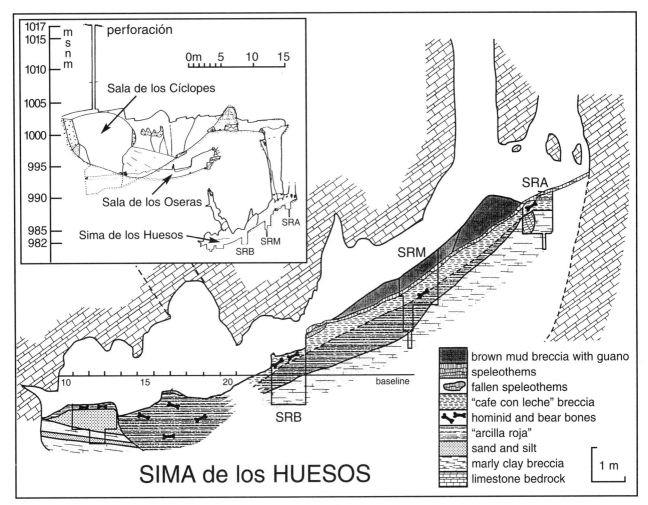

baseline

■ brown mud breccia with guano
▓ speleothems
⌐ fallen speleothems
▒ "cafe con leche" breccia
🦴 hominid and bear bones
▤ "arcilla roja"
▦ sand and silt
▥ marly clay breccia
▨ limestone bedrock

1 m

SRB

SIMA de los HUESOS

Figure 26.2 Profiles of the Cyclops Gallery and the Sima de los Huesos. From Arsuaga *et al.* (1997).

ing the Sala de las Oseras (the bear nest chamber), where bear hibernation nests and claw marks are found in the clay adhering to the chamber's walls. Excavations there revealed bears who died during hibernation. In the opinion of the researchers, there was probably a small entrance to the cave system near the Cyclops Gallery that allowed bears to get into this part of the cave system to hibernate. Humans apparently only came here once or a few times because no archaeological remains are present, and this entrance to the cave system was blocked by a cave collapse in the Middle Pleistocene and sealed.

Today, after traversing the Cyclops Gallery, one passes down a steeply sloping, 2-m-wide passageway leading to a vertical shaft which connects to the Sima below. The only entrance to the Sima is through this vertical shaft. With climbing safety ropes attached, one descends slowly down a narrow ladder into the cavity below (Figure 26.3), reaching an inclined surface known as La Rampa, the ramp. The Sima de los Huesos is the small cavern at the foot of the ramp.

Figure 26.3 Team leader Arsuaga climbs down the 13-m chimney into the Sima de los Huesos. Photo by and courtesy of Javier Trueba.

26.2
Discovery

The first hominid fossils from the Sima were found in 1976 during a sampling of its bones by a graduate student of Professor E. Aguirre of Madrid. The sediments of the Sima had been badly disturbed by amateur collectors attracted by the abundant bear bones. A small sample of the disturbed sediment was collected during a brief 1983 visit by Aguirre's team. It yielded three additional hominid teeth among the bear bones. In 1984 the systematic removal of the Sima's disturbed sediment began and the first fossils were found *in situ*. In those days the Sima was choked with disturbed sediment, bones, and fallen limestone blocks.

26.3
Recovery

In order to reach the *in situ* remains, the overlying mass of debris and disturbed sediments had to be cleared. Large blocks were broken by hammer and chisel. Everything was removed, pulled up through the shaft. Only a few workers could work in the Sima at one time, and oxygen was rapidly exhausted because there was no ventilation. All the disturbed sediment, several tons worth, was removed in the backpacks of workers who climbed out of the pit and snaked their way back to the surface with their precious cargo (Figure 26.4). The sediment was taken to a nearby river where it was wet-sieved to recover the fragmented remains of hundreds of bears and the few hominid remains.

Figure 26.4 The Sima de los Huesos team at the entrance of the Cueva Major. During the field season these workers crawl through 500 m of narrow passageways and cave galleries each day as they descend into the cold, dark, Pit of the Bones. The team leader, Juan Luis Arsuaga, is centered in the back row. Photo by and courtesy of Javier Trueba, from Arsuaga *et al.* (1997).

Most of the bones were broken, some of them into many dozens of pieces. The resulting mixture of mostly bear and hominid remains was sorted piece by piece, first by taxon and then element. The slow and tedious job of restoration proceeded simultaneously. It took over five years to remove the uppermost disturbed sediments from the cave, but the hominid count climbed. In 1987 a suspended scaffolding was installed in the Sima to allow the paleontologists to work without stepping on the newly exposed, *in situ* sediment (Figure 26.5). A shaft was drilled through the roof of the Cyclops Gallery to allow a more direct removal of sediment.

As the *in situ* deposit was excavated, it was found to contain a bone-bearing breccia with clay matrix, mainly composed of *Ursus deningeri*. This Middle Pleistocene bear was the ancestor of the larger, later Pleistocene cave bear. By the end of the 1995 season, a minimum number of 166 bear individuals was calculated from the thousands of bear bones recovered from the excavations. A total of 1685 hominid pieces had been found to represent a minimum number of 32 hominid individuals. Every part of the hominid skeleton was represented, often by many individuals. A few other carnivores and micromammals were found, but no ungulates and no archaeological remains. None of the hominid bones were found in articulation, but there are some significant associations. For instance, Cranium 5 and its mandible were found together, as were two hip bones and sacrum of the complete pelvis and many bones of the same hand. Almost all of the bones in the deposit were broken, and restorations and individual associations were difficult to perform.

Excavation of the *in situ* remains had to proceed very slowly because of the logistical conditions and the fragility of the specimens. The bones

Figure 26.5 The Sima de los Huesos team at work, deep in the pit. They are working on a suspended scaffold which protects the unexcavated, fragile fossils beneath them. Photo by and courtesy of Javier Trueba.

Figure 26.6 A portion of the hominid fossil breccia during the 1992 excavation. All of the pieces are hominid, except for a bear's rib fragment at the top center. Multiple crania, a mandible, and postcranial elements are visible. The bones are very soft and delicate at this stage of extraction, and can only be extracted, handled, and studied after preservative is applied. Anatomical detail is excellent. Photo by and courtesy of Javier Trueba, from Arsuaga *et al.* (1997).

were extremely soft and fragile, and wooden implements were used to slowly remove the wet clay in which they were embedded. Careful application of preservative was required for each piece. In some parts of the deposit, there was more bone than matrix (Figure 26.6). Only a small part of the Sima deposit has so far been excavated, and it is certain that many more hominid fossils will be recovered. Dating efforts are continuing, but it is clear from biochronologic and radioisotopic considerations that the site is probably about 300,000 years old. Already, however, this assemblage is staggering in its size and importance—Atapuerca's Sima de los Huesos is already the largest known repository of fossil hominids from the Middle Pleistocene and a tremendous source of knowledge about the skeletal biology of a hominid population from the deep past.

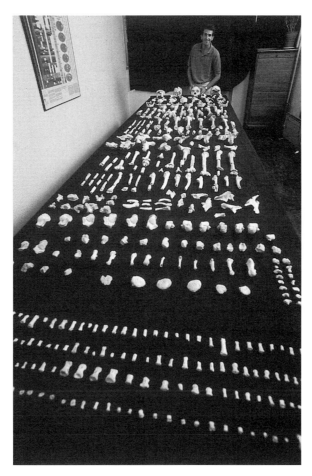

Figure 26.7 Arsuaga stands behind the tabletop of hominid bones recovered from the Sima. Analysis of this large sample could only begin after a meticulous excavation, careful cleaning and preservation, and a full sorting, identification, and refitting. This is only a fraction of the hominid sample that will eventually come from the Pit of the Bones, but it is already the largest and most significant fossil hominid assemblage ever found. Photo by and courtesy of Javier Trueba.

26.4
Paleodemography

After the tens of thousands of bear bones had been segregated from the c. 1600 hominid pieces, and after a massive refitting exercise joined as many of the bones from the ever-growing sample as was possible (Figure 26.7), it was necessary to estimate the minimum number of hominid individuals that had been recovered so far. This was done by Bermúdez de Castro and colleagues, working on the mandibular, maxillary and dental remains. A minimum of 32 individuals was represented, and of these, a balanced sex ratio was calculated (Figure 26.8). Age at death was estimated by applying the Miles method described in Chapter 17. The resulting survivorship curve showed low representation of infants and children, and a high representation of adolescents and prime age adults.

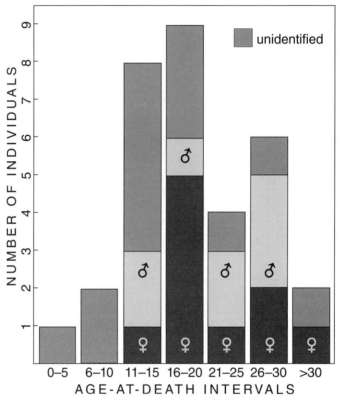

Figure 26.8 Age-at-death distribution of the 32 individuals identified in the Atapuerca Sima de los Huesos fossil hominid assemblage. Sexes are differentiated here where determination was possible. From Arsuaga *et al.* (1997).

26.5 Paleopathology

Analysis of the hominid sample showed that it was characterized by a high incidence of temporomandibular joint disease. Signs of this pathology were found in 70% of individuals, well above values seen in historic populations. One skull showed an extensive maxillary osteitis associated with a dental apical abscess, and another apical abscess in its mandible. There were no fractures or clear traumatic lesions among 1200 postcranial elements, but one immature individual showed a severe traumatic lesion on its left browridge. Enamel hypoplasias were present, and most commonly emplaced between birth and seven years. There were significantly fewer hypoplasias than are found in either Neanderthals or most modern human populations.

26.6
Functional and Phylogenetic Assessment

The skeletal remains from the Sima, because of their excellent preservation and completeness, and because of the large number of individuals, allow for considerations not often available to paleoanthropologists. For example, it is evident that the sexual dimorphism in these hominids was no greater than that seen among modern human populations. It is evident that not all of the peculiar morphological characters defining later Neanderthals had yet evolved in the Sima people. However, a set of traits uniting these Middle and Late Pleistocene forms was detected. For example, the Sima hominids, and the Neanderthals that followed them, share a more laterally oriented scapular glenoid cavity than any other hominid, a transversely oval humeral head, and a host of characters of the occipital, frontal, and facial skeleton. These novel traits, characters shared exclusively between the two hominid forms, suggest to the Atapuerca team that they have discovered the exclusive ancestors of Neanderthals. This interpretation would mean that European populations had a distinct evolutionary role for hundreds of thousands of years. Such findings lead the Atapuerca team to predict that the ancestry of anatomically modern humans will not be found in European fossils but perhaps in Asian or African contemporaries of the Sima people.

26.7
Continuing Mysteries

The fossil hominids from the Pit of the Bones, and the still larger sample that lies entombed within the Pit, will keep paleoanthropologists busy for decades, if not centuries. All these remains will come under the scrutiny of new workers and new techniques. How did this unique collection of fossils come to be deposited in this deep chamber in the first place? This is the central mystery of the Sima de los Huesos. Some facts surrounding the assemblage illustrate the magnitude of solving this puzzle. The demographic profile suggests part of the age pyramid of a living population. Deposition was far from a cave entrance. There are a few rodents, no herbivores, no food refuse, no stone tools, and no human modifications on the bones of either the bears or the hominids (there are some carnivore tooth marks on both). Bears, by far, are the most abundantly represented mammals in the deposit. There is no clear size sorting or alignment of the bones. All elements of the body are represented. The main biases are against sternae, vertebrae, and ribs. No bones are articulated.

Arsuaga and colleagues believe that the Sima was a natural trap for bears in the Middle Pleistocene, and that animals falling into the pit could not climb up the chimney from the pit. They suggest that mortuary practices were responsible for the accumulation of men, women, and children in the Sima. As bears fell into the pit, and desperately struggled in vain to dig their way out, they disarticulated, trampled, unintentionally fractured during digging, and occasionally gnawed other bones already in the Sima. They, in turn, died—only to later become disarticulated, fractured skeletal remains themselves. These remains stayed trapped in the wet, cold mud of the Sima until their scientific rescue began late in the 20th century. It is important to remember that only a small volume of sediments has so far been excavated. It is too early for meaningful studies of spatial distribution to be made, and it is virtually certain that additional element associations will be found. As excavations by this remarkable team of Spanish investigators continue into the next century, more clues to the central mystery of the Sima de los Huesos will continue to bring us closer to a solution—and to a better understanding of the deep past.

CHAPTER 27

Paleontological Case Study
Australopithecus Mandible from Maka, Ethiopia

This final case study illustrates the application of human osteology in a deeper paleontological context. Like the previous case studies, this one shows that research in human osteology is a team effort and must be built on the basic skills and techniques of identification, recovery, and analysis.

27.1
Historical Background

Three of the world's great rift valleys intersect in the Afar Triangle of northeastern Ethiopia. This large, depressed region is a hot, dry desert inhabited by the Afar people, who are nomadic pastoralists. Flowing through this desert is the Awash River which drains the Ethiopian highlands. Paleoanthropological research came relatively late to this region, after the discoveries of the Leakeys at Olduvai and Koobi Fora and work in the lower Omo valley by Clark Howell and colleagues. Maurice Taieb, a young French geologist, was the first to realize the paleoanthropological potential of the Afar as he performed geological mapping in this region in the 1960s.

Since Taieb's first explorations, the Afar has become the most important place on earth for the study of human origins and evolution. The region is now perhaps best known for a relatively complete and well-preserved fossilized skeleton nicknamed "Lucy" by anthropologist Donald Johanson shortly after her discovery in 1974. Other discoveries at Hadar—the "Lucy" site—led to the recognition of a species of human ancestor known as *Australopithecus afarensis*. When this species was named in the late 1970s, its remains came from two sites, Hadar in the Afar and Laetoli in Tanzania, where Mary Leakey's team had excavated fossilized bones and teeth as well as footprints preserved in hardened volcanic ash.

The 3- to 4-million-year time range had been a virtual blank for human paleontologists before the recovery of the Hadar and Laetoli fossils in the

1970s. The *A. afarensis* fossils brought new insights and fueled ongoing debates. Controversy surrounded the interpretation that these fossils represented a single species. More controversy involved interpretations of the creature's locomotion. Some contended that it was a bipedal terrestrial primate, and others posited more arboreality in its behavior. In human paleontology, particularly for ancestors as remote as *A. afarensis*, the fossil record is often highly incomplete. Even the relatively intact, uniquely preserved skeleton of "Lucy" is only a single individual. There are very few places in Africa, let alone the rest of the world, where a dead organism has a good chance of becoming fossilized. There are even fewer places where these fossilized remains are accessible to the scientists who seek them.

One such place lies about 75 km south of the "Lucy" site in Ethiopia. This study area is called the Middle Awash. Here, beginning in the 1970s, fossil hominid remains from multiple time horizons have been recovered. In 1981 a team led by J. Desmond Clark and I visited a set of exposed Pliocene sediments on the eastern side of the Awash river where a fragmentary proximal femur of a subadult *Australopithecus afarensis* was found. That discovery was followed by further survey in 1990. The case study below is an account of that survey and its results (see White *et al.*, 1993).

27.2
The Geography and Geology of Maka

Today the Awash River supports a wide, green ribbon of vegetation that winds through an otherwise parched landscape. Each of the small drainages that flow west into the modern Awash River bears an Afar name. The Middle Awash project uses these names to identify paleontological and archeological collection areas. Today the Maka drainage is usually dry. The Maka catchment is cut into sediments that date to the Pliocene. The ancient beds exposed here by recent and ongoing erosion were originally deposited in a layer-cake fashion. Coarser-grained beds, the Maka sands, represent fluviatile deposition that characterized this region when the huge lakes that preceded them disappeared due to climatic and tectonic changes. These fluviatile Maka deposits embedded the remains of many mammals such as hippopotami, elephants, giraffes, pigs, kudus, crocodiles, various carnivores, and, rarely, human ancestors. It must be stressed that the desolate badlands of the Maka area today (Figure 27.1) bear no resemblance to the swampy, riverside habitats available to the Pliocene creatures who roamed this place millions of years ago. Indeed, paleoanthropologists have come to understand that it is sometimes best to ignore the modern lakes, rivers, streams, and mountains that they see today in their efforts to reassemble the past. Most of these modern features of today's geography had not even formed at the time of deposition. Not only were the ancient landscapes very different from those that we see today, but the deposition on these landscapes took place at higher elevations, long before the Afar floor had dropped to its present low elevation by massive tectonic movements associated with rift formation. The paleoanthropological challenge at Maka was to recreate this vanished world of the Pliocene and to understand the human ancestors who lived there.

Figure 27.1 An overview of the Maka catchment in the Middle Awash study area of Ethiopia. Arrow indicates an individual standing at the location of the hominid mandible described in the text.

Determining the age of the Maka sediments was a challenge in 1990. The associated animal remains were very similar to those found at Hadar, to the north, dated to between 3 and 4 million years. Fortunately, project scientists were able to locate and sample a lens of volcanic ash that was interbedded in the Maka Sands. Although this ash could not itself be dated by argon/argon radioisotopic techniques, its chemical composition allowed the team to match it with other, already-dated outcrops in the region and beyond. The volcanic lens at Maka was only a small remnant of a blanket of debris representing a volcanic eruption that had broadcast ash widely over Eastern Africa, from the Turkana Basin (where it was known as the Tulu Bor tuff) to the Omo (where it had been labeled Tuff B) to Hadar (where it had been called the Sidi Hakoma tuff) and even into the Gulf of Aden (where it had been found in Deep Sea Drilling Project cores). This volcanic horizon is well dated at 3.4 million years. Thus, most of the fossils from Maka could be directly related to the dated horizon by stratigraphic work—and most were therefore 3.4 million years old.

The coarse, yellow-gray Maka sands and gravels are poorly consolidated. As a result, they erode rapidly when infrequent torrential rain showers cross the Middle Awash. Each rain brings more erosion, exposing more of the buried fossils to the paleontologist. In large collection fields such as Maka, fossils reaching the surface should be recovered as soon as possible because they are very brittle. Organic components of these remains have long since disappeared. The bones usually fragment below the surface before the paleoanthropologist can even see them. Once exposed, they scatter more and more widely in successive rainstorms as they weather out of the ancient sediment. Most of the important fossil remains from sites such as Hadar, Olduvai, Laetoli, Aramis, and Koobi Fora are found as fragments of surface specimens by survey teams walking and crawling these eroding outcrops of ancient sediments.

Given the geological facts and geomorphological settings of these open-air fossil fields, *in situ* fossil hominids are rarely found, and excavations are usually limited to recovering already disturbed and scattered remains of fossils that are found by paleontologists surveying the eroding surface of the sediments. The human osteologist working under such conditions must therefore be able to identify small, scattered, and often highly fragmentary skeletal and dental elements. Indeed, were it not for the osteological skills of the surveyors, many fossils such as the famous *Homo erectus* Turkana Boy KNM WT-15,000 specimen from northern Kenya would remain unidentified and unrecovered. If the survey osteologist dismisses a small hominid fragment as an unidentifiable scrap of bone in the field, it may be many years before anyone returns to the area.

It must be emphasized that it is extremely rare for fossil hominid crania or other body parts to be found intact at such open air paleontological sites. Indeed, the term "site" is not an apt description of these occurrences, which are better referred to as "fossil fields." Compounding the problems associated with erosion, hominids were never common components of the total biomass. They had lifetimes much longer than the average mammal's. As a consequence, relatively few hominid carcasses were available for burial and fossilization. Because of their intelligence and lifestyles, they rarely died in swampy or lakeside conditions favoring fossilization. Hominids are therefore rare members of fossil assemblages. The Maka team faced all of these problems in 1990. The area had last been surveyed in 1981, and the intervening nine years of erosion had scattered and destroyed some fossils already on the surface, while eroding the surface to expose others. The team's goal was to find as many significant fossils as possible in an effort to shed some light on the very deep past.

27.3
Search and Discovery

The Middle Awash team members were familiar with the challenge of finding and recognizing the few, usually broken hominid bones and teeth scattered on the Maka outcrops. The Maka outcrops are unusually rich in vertebrate fossils, with abundant bones and teeth representing dozens of other ancient species. The concept of a **search image** is important in this kind of surface survey. It is obvious that a paleontologist wishing to recover shrew remains spends a great deal of time on his or her knees, closely scrutinizing the surface of the ground for tiny bones and teeth. The paleontologist searching for fossil elephant remains, in contrast, can walk quickly across the outcrops, scanning the surface for much larger remains. The paleontologist seeking hominid remains should carry an intermediate search image. Isolated hominid teeth are usually discernible

Figure 27.2 Photograph of the first fragment of the mandible, as it was found by the author.

only from a bending or kneeling stance on the outcrop, whereas a distal femur can be spotted from standing position.

Late on the afternoon of Friday, October 12, 1990, I came across a slight rise high in the Maka catchment. I had passed this location nine years earlier, but much erosion had occurred in the meantime. The sun was

Figure 27.3 View of the hominid discovery site. The man is pointing to the position of the first mandible fragment. Subsequent surface crawling established additional fragments both above and below this position. The mandible had weathered from the top of the Maka sands, about 1 m above the man's head.

already very low, casting shadows across the outcrops, and it was time to return to camp. Other collectors were having their last looks at the outcrops and heading back to the vehicles. The Maka sands in this area are very rich in fossils, and shattered bits of bone and teeth were scattered everywhere. During the day each survey team member had seen hundreds of thousands of such fragments. Most of them were unidentifiable pieces of shattered ungulate remains, broken tortoise carapaces, crocodile scutes, and the like. But as I peered into a tiny rivulet, I saw the side of a tooth crown just exposed by the last rainstorm before the dry season (Figure 27.2).

As I bent down to have a closer look, I saw a very exciting fossil. It was the side of a hominid molar and the tooth was embedded in a fragment of mandible. The slope was steep and rocky, and I gently retreated in my own footprints to avoid crushing any other potential fragments. I immediately established a perimeter around the fossil so that others would not tread on the specimen or other pieces that I suspected lay nearby. It is very important to control the passage of people across a new discovery site because even well-meaning efforts have the potential to push fragile fossils into the soft surface, or break them underfoot (see Chapter 14). After

Figure 27.4 The lower left canine in place. Fossils often become part of a surface lag such as this one when the finer sediments are winnowed away. The lag may be many centimeters thick, and all of it needs to be screened to recover small fragments.

carefully photographing the specimen in the dying light and mapping the discovery on an aerial photograph, we built a cairn on a prominent nearby hillside (the project did not yet have a GPS unit). Darkness was falling, and there was no chance that the fossil would be disturbed during the night because the area was uninhabited and no rain was coming. It was safer to leave it than to attempt to remove it in the dark. I carefully placed a cap stone above the specimen to protect it from the unlikely passage of a wild animal, and we left the site in darkness.

27.4 Collection

When the team returned to the site, the first order of business was to photograph and video document the original find. It was clear from the position of the fossil that it was not *in situ* (Figure 27.3). A close examination of its broken edges revealed that the fracture was very recent,

Figure 27.5 The left mandibular corpus very close to its original position. The premolar crowns and mental foramen are visible.

suggesting the strong possibility that other pieces lay nearby. In a situation like this, it is very important to establish the distribution of all pieces of the hominid fossil. We do this with pin flags set at each piece. This procedure is important because it provides critical clues to the position in which the fossil was originally buried, and provides a guide to the recovery operation that must follow.

The most opportune time to find a fossil is just after the first part of it has been exposed by erosion. Unfortunately, the paleontologist usually arrives too late. As discussed above, many fossils have weathered out of their original deposit, but by carefully assessing the geomorphology and lithostratigraphy of the site, the surface scatter, and matrix on the specimen, it is often possible to identify the original stratum that contained the fossil, and the spot from which it most likely came.

In this case, we immediately found additional teeth and mandibular fragments in the rivulet that the original piece occupied. We circumscribed the local catchment area with a surveyor's twine perimeter around what we considered to be the most likely geographic area and stratigraphic horizon from which the piece had weathered. We then began at the bottom of the slope and crawled slowly and systematically up slope on hands and knees. Each fossil fragment, no matter how small and whether immediately identifiable or not, was collected—even nonhominid elements. Each identifiable piece of the hominid was pin flagged. No artifacts were found during the crawling pickup operation (nor have any ever been found in sediments of this age). The operation quickly revealed how the combination of gravity, water, wind, and animal trampling as well as vegetation had scattered the pieces.

Pinpointing the *in situ* resting place of a fossil can be much like placer mining—locating the spot of burial by looking at the concentration and patterning of fossil fragments and taking local topography into account.

Figure 27.6 While one excavator collects the surface scree, other workers sieve buckets of sediment in an attempt to retrieve small and buried fragments of tooth and bone that might belong to the shattered mandible.

Careful excavation of the displaced, original mandible fragment revealed that it was from the right corpus and had a freshly broken fourth premolar root and the first two molar crowns. Our hopes built as we crawled up slope. We found the left canine (Figure 27.4) and more fragments of mandibular body. Near the top of the slope we found most of the rest of the specimen—the left corpus with all the premolars and molars. This large fragment had just weathered out of the sandy deposit (Figure 27.5). After a day of careful pickup we had established that the specimen had eroded out of the Maka sands near the top of the slope, fragmented, and tumbled and

Figure 27.7 At camp, accumulating fragments of mandible suggest that the specimen is fairly complete.

washed down to the base of the slope and even farther down the gully at the slope's base.

From this point forward, recovery of the remainder of the specimen was the challenge. Many fossils have lain on land surfaces for years, suffering destruction by weathering, or being washed away by successive rains. In this case we had been lucky. The fossil had probably weathered out in the last two years, so we had a chance to recover most of the broken pieces— if we were careful. After the surface crawl had stripped all of the fossil bone from the surface, we removed the large clasts of sandstone mantling the surface and organized a sieving operation. Some team members swept the loose surface debris into buckets. Others sifted the debris through a 1-mm sieve (Figure 27.6). Each screen was carefully picked by at least three workers. Each fossil piece was kept, even those that did not belong to the hominid. Over 8000 pieces were recovered from the hillside. All of these were washed and sorted.

Sorting the recovered pieces was very conservative. Many pieces obviously belonged to the specimen and were immediately segregated (Figure 27.7). Each piece was checked and rechecked. All of the recovered bones were sorted into hominid, nonhominid, and indeterminate pieces. Bones were placed in the nonhominid category only when two senior project hominid paleontologists agreed to this designation. This conservative

Figure 27.8 A total of 109 fragments recovered during the surface crawl and sieving operation comprise the adult mandible.

measure is recommended because misidentification at this primary sorting stage of the process has the potential to eliminate hominid pieces prematurely.

27.5 Restoration

Finally, at the end of several days, roughly 300 hominid or potentially hominid fossil pieces were left in consideration. These were taken from the field to the Paleoanthropology Laboratory at the National Museum in Ethiopia's capital, Addis Ababa. Here the restoration process began. Work with the readily identifiable hominid pieces (Figure 27.8) facilitated restoration. Color, texture, and dimensions of broken surfaces were clues to

Figure 27.9 Work at the Paleoanthropology Laboratory at the National Museum of Ethiopia by Dr. Gen Suwa (left), Dr. Berhane Asfaw, and the author led to a restoration of the Maka mandible.

joining tiny pieces of bone. Gradually, the mandible took shape over the course of two weeks of work. This restoration time was unusually brief because the fossil fragments had no adhering matrix. Fitting pieces were found, cleaned, and glued together in a sandbox (Figure 27.9). The Maka mandible was quickly becoming the most complete lower jaw of this species ever found. A total of 109 of the recovered pieces eventually went into the restored mandible (Figure 27.10).

27.6
Photography and Casting

To make permanent records of the restored specimen, it was photographed in black-and-white and color, and a video record was made. Silicon rubber molds were then prepared (see Chapters 14 and 15 for details of these procedures). The first "perfect" casts (dental plaster for dimensional stability) produced from these molds were designated as record casts and

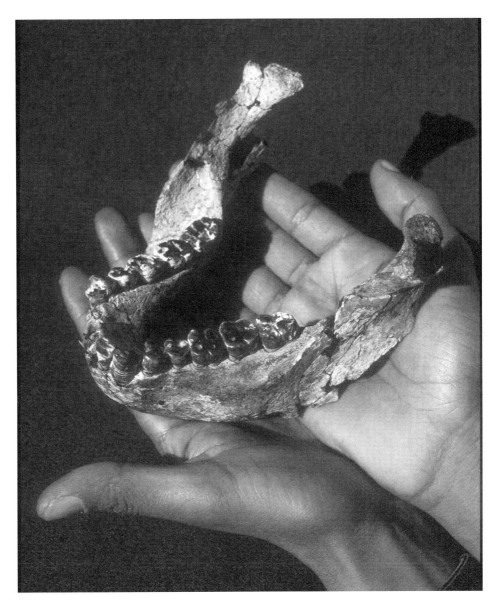

Figure 27.10 The restored mandible from Maka. This is the most complete mandible of *Australopithecus afarensis.*

locked in a protected location in case the original fossils were ever lost or destroyed. Finally, after months of field and laboratory work—after screening all the sediments judged to possibly contain fragments of the specimen, checking each recovered fragment against all other ones for possible joins, gluing the fragments together, strengthening the joins and surfaces with preservatives, photographing and molding the fossil—the new Maka specimen was ready to be studied and interpreted. Meanwhile the geologists were performing laboratory work to establish the antiquity of the associated volcanic ash. Project paleontologists were working at restoring other mammalian fossils found in the same horizon, from which biochronological and paleoenvironmental results could be derived. The other hominid fossils found at Maka in 1990, including a massive humerus, were also being restored and analyzed.

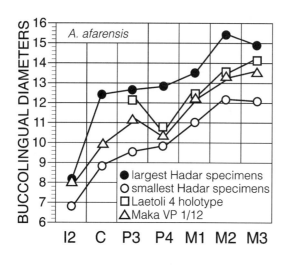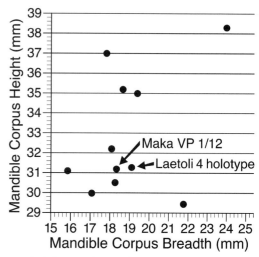

Figure 27.11 Dental and mandibular measurements compiled during the analytical phase of the research demonstrate the position of the new mandible relative to the "Lucy" mandible (A.L. 288-1) and the type specimen from Laetoli (L.H.-4). From White *et al.* (1993).

27.7 Analysis

When *Australopithecus afarensis* was created, the Laetoli mandible L.H.-4 was designated as its type specimen. Some argued that the geographic distance between Laetoli (in Tanzania) and Hadar (in Ethiopia) was too great for a single hominid species. Of course, *Homo erectus* occupied a range far greater, so that argument was dismissed by most workers. However, the Maka mandible represented an important new addition to the debate about this aspect of the hominid fossil record.

In analyzing the new adult Maka mandible, Berhane Asfaw, Gen Suwa, and I compared it to all of the known early hominid fossils. In addition, we used the large Hamman–Todd collection of modern African apes. Our comparisons revealed very close resemblances between Maka and the Laetoli holotype, particularly in the dentition. The Maka specimen was centered in the *A. afarensis* range of dental variation, and its mandibular dimensions were very close to those of the Laetoli holotype (Figure 27.11). Functionally, the new specimen provided an important look at incisor and premolar function, as well as the biomechanical attributes of symphyseal shape and the form of the ramus. Finally, it displayed a paleopathology—the earliest evidence of temporomandibular joint disease in the hominid fossil record.

27.8 Publication

By the summer of 1993 the Middle Awash research team had finished its preliminary work on the geology, geochronology, and paleontology of the

1990 Maka discoveries. Because of the general importance of the results, the team chose to publish an announcement of the finds in the international journal of science, *Nature* (published in London; White *et al.*, 1993). As with any modern, multidisciplinary research into human origins, our project included personnel working on many different aspects of the Maka discoveries. Nine of these scientists contributed significantly to the paper's paleontological, geochronological, and geological results. All nine were coauthors, representing eight different institutions in four countries. The Maka mandible was thus announced to the scientific community as the earliest, most complete mandible of an early hominid, three years and one month after its discovery.

Appendix: Photographic Methods and Provenance

A.1
Methods

This book focuses primarily on the topical features of human bone. To adequately convey this information in photographs, we endeavored to minimize the confusing and unnecessary aspects of natural bone material, such as stains and the translucence of some bones and teeth, and to concentrate on the important surface features. Preparation of the bone, camera equipment, lighting and staging, and film were each significant variables in meeting these goals.

A.1.1 Preparation of the Bone

Osteological material was coated in such a way that stains and glare were reduced without loss of surface details. After the bone material was degreased, a finely ground, opaque titanium pigment suspended in a solution of ethelene glycol and water was applied through the nozzle of a 0.2-mm double-action airbrush. Air pressure was adjusted between 24 and 40 psi during application, depending on a variety of circumstances including the nature of the bone surface and humidity. One coat was generally sufficient to obscure all stains and create an opaque nonreflective surface.

Other methods of coating, including ammonium carbonate smoking and direct brushing with a variety of paints, proved unacceptable. The titanium coating used was easily removed with water and/or alcohol.

A.1.2 Equipment

Cameras used to record the images included a Sinar F2 4 × 5 view camera, a Hasselblad 500CM (medium format), and a Nikon F3T (35-mm format). The use of three formats allowed similar proportional enlargements of the negatives regardless of subject size, because subject-to-image ratios could be kept somewhat the same. This resulted in a fairly consistent grain character for each figure. The primary 4×5 lens was a Schneider Symmar HM 240-mm f/5.6 with a Copal #3 shutter. A 120-mm f/2.8 Zeiss S-Planar with a B55 extension tube provided the images in the medium format, and a 55-mm f/2.8 Micro-Nikkor with a PK-13 extension tube worked on the tooth shots in the 35-mm format.

A.1.3 Lighting and Staging

Lighting was exclusively artificial (daylight balanced). Two Speedotron electronic strobe packs at full power were used to drive one quad head (4800 Ws) main light and one single head (2400 Ws) background. Each strobe was raw (without reflector or diffusion).

Staging began with a 4×8-ft neutral gray sheet of formica on the floor. An 18×24-inch sheet of nonglare glass was suspended about 16 inches over the background on two C-stands. Several sheets of white foam core board formed a tunnel through which the background strobe flashed. An additional board acted as a bounce reflector at the end of the tunnel. The only outlet for background light was through the glass stage from below. To maximize surface detail, the main light flashed at a very low angle to the subject. This light bounced off of a white card on the opposite side of the subject to fill in shadows and reduce contrast.

Each bone was placed on the glass and oriented on a variety of pliable and tacky substances hidden from view. In most cases, the angle and power of the flash created a slight reflection of the subject on the glass. To absorb excess light and minimize this effect, a black mask was cut for most subjects and placed out of view on the underside of the bone.

The direction of light and subject orientation follow traditional scientific illustration convention. In general, the light falls on the subject from the upper left relative to the viewer. Within this reference a hierarchy is followed for the preferred orientation: dorsal top, ventral bottom; proximal up, distal down; anterior left, posterior right; superior up, inferior down; and any special view required to convey specific information. Lateral views are left lateral where bilateral symmetry exists. All skull part figures are oriented in approximate Frankfurt Horizontal with the aforementioned conventions in mind. Any deviation from these conventions is mentioned in the figure legends.

A.1.4 Film and Development

Ilford FP4 black-and-white film (rated ISO 125) was chosen for its moderate contrast, wide latitude, availability in all formats, and fine-grain characteristics. Preliminary tests indicated that an ASA of 80 worked best for the high-key subjects of the project. Polaroid instant films type 55 and 554 were employed to check contrast, lighting, depth of field, and orientation before most final exposures.

Because extension tubes and long bellows extensions were used in most shots, depth of field (focus) became a concern. Thus a near minimum lens aperture was used whenever possible to balance sharpness with depth of field. A slower film speed would have required a very close and difficult-to-control flash-to-subject distance. Also, the additional fine grain available from the slower films, though desirable, was superfluous because of the limitations of photolithography and the printing process.

All film was developed normally. Most images were printed on Ilford Polycontrast resin-coated paper with #4 and #4.5 filters. Some were printed on Ilfobrom fiberbase paper, contrast grade 3. Certain areas of each image were burned with no filter to bring out highlight detail; other areas were dodged to open shadows. The prints were slightly overexposed to avoid loss of texture in the highlights and to carry all the detail evident in the negative. This artistic compromise maximized the educational content.

Main light-to-subject distance remained largely constant. The camera-to-stage distance varied in some cases to maintain a consistent film plane-to-subject focus distance as the size of the subject changed. Certain deep subjects (e.g., proximal views of limb bones) dictated adjustments that resulted in great camera-to-stage distances. These greater camera-to-stage distances led to unavoidable variations in lighting ratios between the main and background lights. Since the background is superfluous to the intent, we printed the figures to maximize the information on the bone and allowed the neutral dark gray background to vary as much as one-and-a-half gray scale zones. The reader will notice some variation in background throughout the book for this reason.

A.2 Provenance

The osteological specimens illustrated in this text are listed here by figure number and source.

Lowie: Lowie (now Hearst) Museum, The University of California, Berkeley, California

Atkinson: Atkinson Collection, A. W. Ward Museum of Dentistry, The University of the Pacific School of Dentistry, San Francisco, California

Hamann-Todd: Hamann-Todd Collection, The Cleveland Museum of Natural History, Cleveland, Ohio

UCB: Berkeley Teaching Collection (without provenance), The University of California, Berkeley, California

A.2.1 Chapter 2

Figure 2.1 *(top)*	Left to Right: Lowie 12-8796, -9326, -10289, -10236, -8864, -10323.
Figure 2.1 *(middle)*	Left to Right: Lowie 12-5289, -9292, -5388, -8770, -8865, -5375.
Figure 2.1 *(bottom)*	Left to Right: Lowie 12-3793, -110, -8851, -8296, -8771a, -10444.
Figure 2.2	Left to Right: Lowie 12-10259, -10257, -8796, -10215, -10311, -8847, -10191, -8778, -5388, -5391, -8833.
Figure 2.3	Left to Right: Lowie 12-8869, -5394, -5378, -10352, -10354.
Figure 2.4	Left to Right: Lowie 12-10237, -10353, -10249 (reversed), -8951 (reversed), -8856, -9516, -10245A (reversed), -10291.
Figure 2.6	UCB.
Figure 2.8	Left to Right: Lowie 12-3793, -110, -8851, -8296, -8771a, -10444.

A.2.2 Chapter 4

Figure 4.1–4.5	Atkinson B-139.
Figure 4.6	Atkinson D-79.
Figure 4.7	Atkinson A-24.
Figure 4.7	Atkinson A-65.
Figure 4.7	Atkinson A-96.
Figure 4.7	Atkinson B-139.
Figure 4.7	Atkinson C-276.
Figure 4.9–4.13	Atkinson B-139.
Figure 4.14–4.40	UCB.
Figure 4.41–4.43	Atkinson B-139.

A.2.3 Chapter 5

Figure 5.1–5.20	Atkinson D-37 (deciduous) and D-79 (permanent).

A.2.4 Chapter 6–13

All Figures	Hamman–Todd 857 except for Growth Sections: 1-year-old = Hamman–Todd 2370; 6-year-old = Hamman–Todd 2144 and 624.

A.2.5 Chapter 17

Figure 17.9a	Atkinson B-139.
Figure 17.9b	Atkinson C-209.
Figure 17.11a	Females: Left to Right: Lowie 12-5391, -5384, -8943, -9723, -10242, -10343, -8848, -5290, -8793, -10322.
Figure 17.11b	Males: Left to Right: Lowie 12-9326, -9327, -8822, -8941, -8567, -8865, -10350, -10327, -8792, -8771.
Figure 17.12	Male: Lowie 12-9043; Female: Lowie 12-4639.

A.2.6 Chapter 18

Figure 18.1	Lowie 12-5571.
Figure 18.2	Lowie 12-5970.
Figure 18.3	Lowie 12-6430.
Figure 18.4	Lowie 12-8062.
Figure 18.5	Lowie 12-3293.
Figure 18.6	Lowie 12-1916.
Figure 18.7	Lowie 12-87.
Figure 18.8	Lowie 12-281, -3650.
Figure 18.9	Lowie 12-5696.
Figure 18.10	Lowie 12-2009.
Figure 18.11	Lowie 12-6670, -3454c.
Figure 18.12	UCB.
Figure 18.13	Lowie 12-6418.
Figure 18.14	Lowie 12-7932.
Figure 18.15	Lowie 12-3454K.
Figure 18.16	Lowie 12-6971b.
Figure 18.17	Lowie 12-7265.

A.2.7 Chapter 19

Figure 19.1   On the left: Mancos; UCB.
Figure 19.2   Mancos (see White, 1992).
Figure 19.3   Lowie 12-432 and Mancos.
Figure 19.4   Lowie 12-9376.
Figure 19.5   Lowie 12-10099, -11054.
Figure 18.6   Lowie 12-309.
Figures 19.7–19.13 Mancos.

A.2.8 Chapter 20

Figure 20.1   Lowie 12-7265.
Figure 20.2   Lowie 12-317.

Glossary

Ablation The removal of part of the body; usually used when referring to the removal of teeth.

Abscess A localized collection of pus in a cavity formed by tissue disintegration; often found as cavities within alveolar bone near the tooth root apices.

Achondroplasia A hereditary form of dwarfism with limb shortening, almost normal trunk and vault development, and a small face; caused by congenital disturbance of cartilage formation at the epiphyses.

Acromegaly A condition caused by an overly productive pituitary, but after the epiphyses are fused; the most dramatic osteological manifestation of acromegaly is growth at the mandibular condyle and a resulting elongation and distortion of the lower jaw.

Agenesis The lack of tooth formation at a given position.

Alveolar resorption The removal of alveolar bone.

Alveolus A tooth socket.

Ameloblasts The cells that form enamel through a process known as amelogenesis.

Ankylosing spondylitis A condition in which the ligaments of the spine ossify, immobilizing the adjacent vertebrae.

Ankylosis An abnormal complete immobility or fixation of a joint resulting from pathological changes in the joint.

Anlage The aggregation of cells indicating the first trace of an organ during embryogenesis.

Annulus fibrosus The peripheral part of an intervertebral disk.

Antimere One half of a bilaterally symmetrical structure.

Apophysis An outgrowth or small bony projection.

Appendicular skeleton Bones of the limbs, including the shoulder and pelvic girdles (but not the sacrum).

Arthritis Inflammation of a joint.

Arthropathy Any disease affecting the joints.

Articulation A place where two adjacent bones contact.

Atrophy Wasting away and reduction in size, particularly after an organ has matured.

Attrition Tooth wear.

Auditory exostosis A bony growth on the walls of the external auditory meatus.

Avulsion The forcible tearing away of part of a structure.

Axial skeleton Bones of the trunk, including the vertebrae, ribs, and sternum.

Ball and socket joint A spheroidal joint such as the hip joint with the hemispherical femur head fitting into the acetabulum; allows for movement in many directions.

Basicranium Bones of the cranial base.

Bioarcheology The study of human remains from archeological contexts.

Buccal Pertaining to the cheek.

Calculus Tartar, a deposit of calcified dental plaque on teeth.

Callus Hard tissue formed in the osteogenic layer of the periosteum as a fracture repair tissue; normally replaced.

Calotte A calvaria without the base.

Calvaria The cranium not including the face.

Cancellous bone A spongy, porous lightweight bone found under protuberances where tendons attach, in the vertebral bodies, in the ends of long bones, in short bones, and sandwiched within flat bones; also called trabecular bone.

Caries A disease characterized by the progressive decalcification of enamel or dentine; the hole or cavity left by such decay.

Cartilage A form of connective tissue consisting of cells embedded in a matrix.

Cartilaginous joint A joint in which the articulating bones are united by means of cartilage and very little movement is allowed.

Cementum A bonelike tissue that covers the external surface of tooth roots, surrounding the dentin of the root and neck of a tooth.

Cervicoenamel line or junction (CEJ) The line encircling the crown of the tooth which is at the most rootward extent of the enamel.

Collagen A fibrous structural protein constituting about 90% of bone's organic content.

Commingled Bone assemblages containing remains of several individuals, often incomplete and fragmentary.

Comminuted fracture A fracture in which the bone splinters.

Compact bone The solid, dense bone that is found in the walls of bone shafts and on external bone surfaces, also called cortical bone.

Complete fracture A fracture in which broken ends of a bone become separated.

Compound fracture A fracture in which the broken bone perforates the skin.

Congenital Acquired during development in the uterus and not through heredity.

Coronal suture The suture that lies between the frontal and parietals.

Cranial sutures Fibrous joints of the skull.

Craniosynostosis Same as craniostosis; premature fusion of cranial sutures resulting in abnormal skull shape.

Cranium Bones of the skull except for the mandible.

Cremation A mortuary practice involving the intentional burning of the body.

Cribra cranii Lesions on the endocranial surface, usually of the frontal region.

Cribra orbitalia Usually lesions in the form of bilateral pitting of the orbital part of the frontal bone.

CT (or CAT) Computed (axial) tomography; a radiological technique using computer processing to generate scans through an object.

Deciduous teeth The first teeth to form, erupt, and function in the first years of life.

Dehiscence A developmental abnormality that results in a perforation of the tympanic plate of the temporal bone; Sometimes called foramen of huschke.

Demography The study of population statistics.

Dental hypoplasia A condition characterized by transverse lines, pits, and grooves on the surface of tooth crowns; such disturbances are defects in dental development.

Dental plaque The matrix and its inhabiting community of bacteria that forms on the tooth.

Dentin A special type of calcified but slightly resilient connective tissue; primary dentin develops during growth, whereas secondary dentin forms after the root formation is complete.

Dentinoenamel junction The boundary between the enamel cap and the underlying dentin.

Deoxyribonucleic acid (DNA) The molecule of heredity that contains the genetic code.

Diachronic Changes or events considered through time.

Diagenesis Chemical, physical, and biological changes undergone by a bone through time.

Diaphysis The shaft of a long bone.

Discontinuous morphological traits Also called epigenetic variants, discrete traits, or nonmetric variation; variation observed in bones and teeth and in the form of differently shaped and sized cusps, roots, tubercles, processes, crests, foramina, articular facets and similar features.

Discrete variables Variables composed of a finite number of values, such as nonmetric traits which are usually scored as present or absent.

Dysplasia An abnormal development of bone tissue.

Eburnation Worn, polished, ivorylike appearance of bone resulting from exposure and wear of adjacent subchondral bone.

Edentulous Without teeth.

Enamel A layer of extremely hard, brittle material that covers the crown of a tooth.

Endochondral ossification The process of bone growth preceded by cartilage precursors called cartilage models.

Endosteum An ill-defined and largely cellular membrane that lines the inner surface of bones.

Epiphysis The cap at the end of a long bone that develops from a secondary ossification center.

External auditory meatus The ear hole.

Exostosis A bony growth from a bone surface, often involving ossification of muscular or ligamentous attachments.

Focal Localized.

Fontanelles Soft spots of cartilaginous membrane in the newborn's skull that eventually harden and turn into bone.

Foramen magnum The large oval hole in the base of the skull.

Forensic osteology Osteological work aimed at identification of the relatively recently deceased, usually done in a legal context.

Frankfurt Horizontal A plane defined by three osteometric points—the right and left porion points and the left orbitale.

Gigantism A condition resulting from excessive production of somatotrophic hormone and consequent overstimulation of growth cartilages and gigantic proportions of the skeleton.

Gomphosis The joint between the roots of the teeth and the bone of the jaws.

Hallux The first or big toe.

Harris lines Lines of increased bone density that represent the position of the growth plate at the time of insult to the organism and formed on long bones due to growth arrest.

Haversian canals Also known as secondary osteons; free anastomosing canals in compact bone that contain blood and lymph vessels and nerves and marrow.

Hemopoietic Related to the production and development of red blood cells.

Hinge joint A joint allowing movement limited mostly to one plane such as the knee and elbow joints.

Homeobox A family of highly conserved 180-bp DNA sequences that encode small proteins that activate specific genes.

Hyaline cartilage Smooth cartilage that covers the articular surfaces of bones.

Hydroxyapatite A dense inorganic filling; the second component of bone.

Hypercementosis A condition in which an excess of cementum forms on the root of the tooth.

Hyperostosis Abnormal growth of bone tissue.

Hyperplasia Overgrowth.

Hypoplasia Undergrowth; for teeth, dental enamel hypoplasia is a disturbance of enamel formation which often manifests itself in transverse lines, pits, or other irregularities on the enamel surface.

Hypertrophy Increase in volume of a tissue or organ.

Idiosyncratic Pertaining to the individual.

Inca bone Extrasutural bone occasionally found in the rear of human crania.

Incomplete fracture or "greenstick" fracture A fracture in which breakage and bending of a bone is combined.

Inhumation Burial or interment.

Intramembranous ossification Process of bone development in which bones ossify by apposition on tissue within an embryonic connective tissue membrane.

Interment Burial.

Interobserver error Error from variation in precision among data recorded by different observers analyzing the same things.

Interproximal Between adjacent surfaces.

Intraobserver error Error from variation in precision among data recorded by the same observer analyzing the same things at different times.

Joint Any connection between different skeletal elements.

Kyphosis The collapse of one or several vertebral bodies causing a sharp angle in the spine.

Labial Pertaining to the lips.

Lambdoidal suture The suture that passes between the two parietals and the occipital bone.

Lamellar bone Bone whose microscopic structure is characterized by collagen fibers arranged in layers or sheets around Haversian canals.

Lesion An injury or wound; an area of pathologically altered tissue.

Lingual Pertaining to the tongue.

Lipping Bone projecting beyond the margin of the affected articular surface, usually in osteoarthritis.

Malocclusion The condition in which upper and lower teeth do not occlude, or meet, properly.

Masseter muscle Muscle originating on the inferior surface of the zygomatic arch and inserting on the lateral surface of the mandibular ramus and gonial angle of the mandible.

Medial pterygoid muscle Muscle originating on the medial surface of the lateral pterygoid plate of the sphenoid and inserting on the medial surface of the mandibular gonial angle.

Medullary cavity The canal inside the shaft of a long bone.

Metaphyses The expanded, flared ends of the shaft.

Metopic suture The suture that passes between unfused frontal halves and only rarely persists into adulthood.

Minimum number of individuals (MNI) The minimum number of individuals necessary to account for all of the elements in the assemblage.

Mitochondrial DNA (mtDNA) A small genome that is inherited only maternally, often used in ancient DNA analysis because it is easier to retrieve than nuclear DNA.

Morphogen Molecule that influences morphogenesis.

Morphogenesis The generation of form.

Morphology The form and structure of an object.

Multiple interment A burial in which more than one individual is present.

Necrosis Physiological death of a cell or group of cells.

Nonmetric trait Dichotomous, discontinuous, discrete, or epigenetic traits; non-pathological variations of tissues difficult to quantify by measurement.

Occlusal surface The chewing surface of the tooth.

Odontoblasts Cells that form dentin through a process known as odontogenesis.

Ontogeny The development or course of development of an individual; growth.

Ossuary A communal grave made up of secondary remains of individuals initially stored somewhere else.

Osteitis A general term for inflammation of bone tissue caused by infection or injury but not specific as to cause.

Osteoarthritis Also called degenerative joint disease; the most common form of arthritis, characterized by the destruction of the articular cartilage in a joint and the formation of adjacent bone, in the form of bony lipping and spur formation.

Osteoblasts The bone-forming cells responsible for synthesizing and depositing bone material.

Osteochondroma The most common tumor of bone; benign tumors, usually asymptomatic, always arise at epiphyseal lines and protrude at right angles to the long axis of a bone.

Osteoclasts Cells responsible for the resorption of bone tissue.

Osteocyte Living bone cell developed from an osteoblast.

Osteogenesis Bone formation and development.

Osteoma often called "button" osteomata; a mound of compact bone, usually on the ectocranial surface.

Osteomalacia A disease that causes softening of the bones, usually linked to general malnutrition, particularly deficiencies in protein, fat, calcium, and phosphorous.

Osteometric Relating to the measurement of bones.

Osteomyelitis A bone inflammation caused by bacteria that usually initially enter the bone via a wound.

Osteon A Haversian system, the structural unit of compact bone composed of a central vascular (Haversian) canal and the concentric lamellae surrounding it; a primary osteon is composed of a vascular canal without a cement line, whereas the cement line and lamellar bone organized around the central canal characterize a secondary osteon.

Osteophyte A small abnormal bony outgrowth often found at the margins of articular surfaces as a feature of osteoarthritis.

Osteoporosis Increased porosity of bone.

Osteosarcoma A malignant tumor of the bone cells.

Paleodemography The study of the demography of prehistoric populations.

Paleopathology The study of diseases in ancient populations as revealed by skeletal remains and preserved soft tissues.

Parasagittal section Any planar slice which parallels the sagittal plane.

Pathological fracture A fracture that occurs as a result of bones already having been weakened by other pathological or metabolic conditions such as osteoporosis.

Pattern formation The spatiotemporal ordering of molecules, cells, or tissues to form a pattern, which can then develop at different scales.

Perikymata Transverse ridges and grooves on the surface of tooth enamel.

Perimortem Before or around the time of death.

Periodontitis The inflammation of tissues around a tooth; can involve both soft tissues and the bone itself.

Periosteum Thin tissue covering the outer surface of bones except areas of articulation.

Periostitis A condition of inflammation of the periosteum caused by trauma or infection; can be acute or chronic.

Periostosis Abnormal bone formation on the periosteal surface of a bone.

Piriform aperture or anterior nasal aperture The hole below and between the orbits; the nose hole.

Planar joint A joint that allows two bones to slide across one another.

Pollex The thumb or first digit of the hand.

Polymerase chain reaction (PCR) An important technique that acts as a "molecular photocopy machine" to produce many copies of a section of DNA starting from just a few molecules of DNA; this allows DNA to be retrieved from some skeletal remains.

Porosity A condition in which many small openings pass through a surface.

Porotic hyperostosis or spongy hyperostosis A condition exhibiting lesions, usually of the cranial vault, respresenting a thinning and often complete destruction of the outer table of the cranial vault.

Post-cranial skeleton All bones except the cranium and mandible.

Postmortem changes Changes to the bone that occur after the death of an individual.

Premortem changes Changes to the bone that occur before the death of an individual.

Primary bony callus Woven bone formed when a callus is subsequently mineralized.

Primary interment A burial in which all the bones are in anatomically "natural" arrangement; an articulated skeleton buried in the flesh.

Primary ossification center The first site where bone begins to form during growth of the shaft of a long bone or body of other bones.

Provenience The stratigraphic and spatial position of the specimen; provenance.

Pseudarthrosis A new or false joint arising between parts of a fractured bone that do not heal.

Radiogram An image produced on photographic film when exposed by x-rays passing through an object.

Reactive bone Bone in the process of being formed or lost, often in response to a pathological stimulus.

Remodeling A cyclical process of bone resorption and deposition at one site.

Resorption The process of bone destruction by osteoclasts.

Rheumatoid arthritis Inflammation and degeneration of the joints, particularly those of the hands and feet; usually chronic and accompanied by deformities.

Rickets A form of osteomalacia resulting from vitamin D deficiency.

Saddle-shaped or sellar joint A joint that is saddle-shaped such as the thumb joint.

Sagittal plane A plane that divides the body into symmetrical right and left halves.

Sagittal suture The suture that passes down the midline between the parietal bones.

Sarcoma A malignant tumor of connective tissue.

Secondary center of ossification Center of bone formation that appears after the primary center, as the epiphyses of long bones.

Secondary interment A burial in which the bones of a skeleton are disarticulated.

SEM Scanning electron microscope; a device which produces the image of the surface of a metal coated specimen by the reflection of electrons.

Serial homology Correspondence of parts in sequential bones, as in the vertebrae.

Sesamoid bones Small bones that lie within tendons near a joint.

Sexual dimorphism Differences between males and females.

Shovel-shaped incisors Incisors with strongly developed mesial and distal lingual marginal ridges, imparting a "shovel" appearance to the tooth.

Sinuses Void chambers in the cranial bones that enlarge with the growth of the face; four basic sets, one each, include maxillae, frontal, ethmoid, and sphenoid.

Skull The bones of the head, including cranium and mandible.

Sphenooccipital or basilar suture Actually a synchondrosis; lies between the sphenoid and the occipital.

Splanchnocranium The facial skeleton.

Squamosal suture An unusual, scalelike, beveled suture between the temporal and parietal bones.

Standard Anatomical Position Standing with feet together and pointing forward, looking forward, with none of the long bones crossed from viewer's perspective and palms facing forward.

Subchondral bone The compact bone that is covered by cartilage at joints.

Supernumerary teeth Teeth that exceed the expected number of teeth in any given category.

Sutures Articulations of the skull bones along joints with interlocking, sawtooth, or ziperlike articulations.

Symphysis A variety of cartilaginous joint in which the fibrocartilage between the bone surfaces is covered by a thin layer of hyaline cartilage.

Syndesmoses Tight, inflexible fibrous joints between bones that are united by bands of dense fibrous tissue in the form of membranes or ligaments.

Synostosis The result when any two bony elements fuse together.

Synovial joints Freely moving joints such as the hip, elbow, knee, and thumb that are coated with a thin layer of hyaline cartilage, lubricated by synovial fluid within a fibrous joint capsule.

Talon(id) The distal (posterior) portion of a primate molar, added to the modified original triangle of cusps; use suffix -*id* for lower molars.

Taphonomy The study of processes that affect skeletal remains between death and curation.

Taurodontism The condition in which the pulp chamber is inflated relative to the normal condition.

Temporalis muscle The muscle originating on the side of the cranial vault inferior to the superior temporal line and inserting on the sides, apex, and anterior surface of the coronoid process of the mandible.

TMJ The temporomandibular joint.

Trephination (trepanation) A practice in which an artificial hole is made in the cranial vault.

Trigon(id) The mesial (anterior) portion of a primate molar, comprising the modified original triangle of cusps; use suffix -*id* for lower molars.

Typology The practice of choosing one individual to characterize a species.

Wolff's law States that bone is laid down where needed and resorbed where not needed.

Wormian bone Small, irregular bones along the cranial sutures.

Bibliography

Adams, R. E. W. (1984) Resolution. *American Antiquity* 49:215–216.

Aiello, L. C., and Molleson, T. (1993) Are microscopic ageing techniques more accurate than macroscopic ageing techniques? *Journal of Archaeological Science* 20:689–704.

Ambrose, S. C. (1993) Isotopic analysis of paleodiets: Methodological and interpretive considerations. In: M. K. Sandford (Ed.) *Investigations of Ancient Human Tissue: Chemical Analyses in Anthropology.* pp. 59–130. Langhorne, Pennslyvania: Gordon and Breach Science Publishers, SA.

American Association of Physical Anthropologists (1996) Statement on biological aspects of race. *American Journal of Physical Anthropology* 101:569–570.

Albert, A. M., and Maples, W. R. (1995) Stages of epiphyseal union for thoracic and lumbar vertebral centra as a method of age determination for teenage and young adult skeletons. *Journal of Forensic Sciences* 40:623–633.

Ali, R. S., and MacLaughlin, S. M. (1991) Sex identification from the auricular surface of the adult human ilium. *International Journal of Osteoarchaeology* 1:57–61.

Alper, J. (1994) Boning up: Newly isolated proteins heal bad breaks. *Science* 263:324–325.

Anderson, D. L., Thompson, G. W., and Popovich, F. (1976) Age of attainment of mineralization stages of the permanent dentition. *Journal of Forensic Sciences* 21:191–200.

Anderson, J. E. (1962) *The Human Skeleton: A Manual for Archaeologists.* Ottawa: Dept. of Northern Affairs and National Resources.

Anton, S. C. (1989) Intentional cranial vault deformation and induced changes of the cranial base and face. *American Journal of Physical Anthropology* 79:253–267.

Arens, W. (1979) *The Man-Eating Myth.* Oxford: Oxford University Press.

Armelagos, G. J., Jacobs, K. H., and Martin, D. L. (1981) Death and demography in prehistoric Sudanese Nubia. In: S. C. Humphries and H. King (Eds.) *Mortality and Immortality: The Archaeology of Death.* pp. 33–57. Orlando, Florida: Academic Press.

Armelagos, G. J., Carlson, D. S., and Van Gerven, D. P. (1982) The theoretical foundations and development of skeletal biology. In: F. Spencer (Ed.) *A History* of American Physical Anthropology 1930–1980. pp. 305–328. New York: Academic Press.

Arsuaga, J. L., Bermúdez de Castro, J. M., and Carbonell, E. (Eds.) (1997) Special Issue: The Sima de los Huesos Hominid Site. *Journal of Human Evolution* 33: 105–421.

Aufderheide, A. C., and Rodríguez-Martín, C. (1998) *The Cambridge Encyclopedia of Human Paleopathology*. Cambridge: Cambridge University Press.

Austin-Smith, D., and Maples, W. R. (1994) The reliability of skull/photograph superimposition in individual identification. *Journal of Forensic Sciences* 39: 446–455.

Avery, J. K. (Ed.) (1987) *Oral Development and Histology*. Baltimore, Maryland: Williams and Wilkins.

Baker, B. J., and Armelagos, G. J. (1988) The origin and antiquity of syphilis. *Current Anthropology* 29:703–737.

Baker, S. J., Gill, G. W., and Kieffer, D. A. (1990) Race and sex determination from the intercondylar notch of the distal femur. Anthropological Papers Number 4: 91–95. In: G. W. Gill and S. Rhine (Eds.) *Skeletal Attribution of Race: Methods for Forensic Anthropology*. Albuquerque, New Mexico: Maxwell Museum of Anthropology.

Bass, W. M. (1995) *Human Osteology: A Laboratory and Field Manual* (3rd Edition). Columbia, Missouri: Missouri Archaeological Society.

Behrensmeyer, A. K., and Hill, A. P. (1980) *Fossils in the Making*. Chicago: University of Chicago Press.

Bennett, K. A. (1993) *A Field Identification Guide for Human Skeletal Identification* (2nd Edition). Springfield, Illinois: C. C. Thomas.

Bermudez de Castro, J. M., Durand, A. I., and Ipiña, S. L. (1993) Sexual dimorphism in the human dental sample from the SH site (Sierra de Atapuerca, Spain): A statistical approach. *Journal of Human Evolution* 24:43–56.

Berryman, H. E., Bass, W. M., Symes, S. A., and Smith, O. C. (1991) Recognition of cemetery remains in the forensic setting. *Journal of Forensic Sciences* 36:230–237.

Beyer-Olsen, E. M. S., and Alexandersen, V. (1995) Sex assessment of medieval Norwegian skeletons based on permanent tooth crown size. *International Journal of Osteoarchaeology* 5:274–281.

Binford, L. R. (1981) *Bones: Ancient Men and Modern Myths*. New York: Academic Press.

Black, S., and Scheuer, L. (1997) The ontogenetic development of the cervical rib. *International Journal of Osteoarchaeology* 7:2–10.

Bloom, W., and Fawcett, D. W. (1994) *A Textbook of Histology* (12th Edition). New York: Chapman and Hall.

Blumenschine, R. J., Marean, C. W., and Capaldo, S. D. (1996) Blind tests of inter-analyst correspondence and accuracy in the identification of cut marks, percussion marks, and carnivore tooth marks on bone surfaces. *Journal of Archaeological Science* 23:493–507.

Boddington, A. (1987) From bones to population: The problem of numbers. In: A. Boddington, A. N. Garland, and R. C. Janaway (Eds.) *Death, Decay and Reconstruction*. pp. 180–197. Manchester: Manchester University Press.

Bonnichsen, R., and Sorg, M. H. (Eds.) (1989) *Bone Modification*. Orono, Maine: Center for the Study of the First Americans.

Bocquet-Appel, J. P., and Masset, C. (1982) Farewell to paleodemography. *Journal of Human Evolution* 11:321–333.

Boles, T. C., Snow, C. C., and Stover, E. (1995) Forensic DNA testing on skeletal remains from mass graves: A pilot project in Guatemala. *Journal of Forensic Sciences* 40:349–355.

Brace, C. L. (1995) Region does not mean "race"—Reality versus convention in forensic anthropology. *Journal of Forensic Sciences* 40:171–175.

Brain, C. K. (1981) *The Hunters or the Hunted? An Introduction to African Cave Taphonomy.* Chicago: University of Chicago Press.

Breathnach, A. S. (Ed.) (1965) *Frazer's Anatomy of the Human Skeleton* (6th Edition). London: J. and A. Churchill.

Bridges, P. S. (1992) Prehistoric arthritis in the Americas. *Annual Review of Anthropology* 21:67–91.

Bridges, P. S. (1993) The effect of variation in methodology on the outcome of osteoarthritic studies. *International Journal of Osteoarchaeology* 3:289–295.

Brooks, S., and Suchey, J. M. (1990) Skeletal age determination based on the os pubis: A comparison of the Acsádi-Nemeskéri and Suchey-Brooks methods. *Human Evolution* 5:227–238.

Brooks, S. T., Haldeman, M. B., and Brooks, R. H. (1988) Osteological analyses of the Stillwater skeletal series. *U.S. Fish and Wildlife Service, Region 1, Cultural Resource Series* 2.

Brooks, S. T. (1955) Skeletal age at death: Reliability of cranial and pubic age indicators. *American Journal of Physical Anthropology* 13:567–597.

Brothwell, D. R. (1981) *Digging Up Bones* (3rd Edition). Ithaca, New York: Cornell University Press.

Brothwell, D. R. (1989) The relationship of tooth wear to aging. In: M. Y. İşcan, (Ed.) *Age Markers in the Human Skeleton.* pp. 303–316. Springfield, Illinois, C. C. Thomas.

Brown, P. (1981) Artificial cranial deformation: A component of the variation in Pleistocene Australian aboriginal crania. *Archaeology in Oceania* 16:156–167.

Brown, T. A., and Brown, K. A. (1992) Ancient DNA and the archaeologist. *Antiquity* 66:10–23.

Buikstra, J. E. (1981a) A specialist in ancient cemetery studies looks at the reburial issue. *Early Man* 3(3):26–27.

Buikstra, J. E. (Ed.) (1981b) *Prehistoric Tuberculosis in the Americas.* Evanston, Illinois: Northwestern University Archaeological Program.

Buikstra, J. E. (1983) Reburial: How we all lose. *Society for California Archaeology Newsletter* 17:1.

Buikstra, J. E., and Cook, D. C. (1980) Paleopathology: An american account. *Annual Review of Anthropology* 9:433–470.

Buikstra, J. E., and Ubelaker, D. H. (1994) *Standards for Data Collection from Human Skeletal Remains.* Fayetteville, Arkansas: Arkansas Archaeological Survey Report Number 44.

Buikstra, J. E., and Mielke, J. H. (1985) Demography, diet and health. In: R. I. Gilbert and J. H. Mielke (Eds.) *The Analysis of Prehistoric Diets.* pp. 359–422. Orlando, Florida. Academic Press.

Burr, D. B. (1980) The relationships among physical, geometrical and mechanical properties of bone, with a note on the properties of nonhuman primate bone. *Yearbook of Physical Anthropology* 23:109–146.

Caldwell, P. C. (1986) New questions (and some answers) on the facial reproduction techniques. In: K. J. Reichs (Ed.) *Forensic Osteology: Advances in the Identification of Human Remains.* pp. 229–255. Springfield, Illinois: C. C. Thomas.

Carlson, D. S., and Van Gerven, D. P. (1977) Masticatory function and post-Pleistocene evolution in Nubia. *American Journal of Physical Anthropology* 46: 495–506.

Carolan, V. A., Gardner, M. L. G., Lucy, D., and Pollard, A. M. (1997) Some considerations regarding the use of amino acid racemization in human dentine as an indicator of age at death. *Journal of Forensic Sciences* 42:10–16.

Carter, D. R. (1980) SI: The international system of units. In: V. H. Frankel and M. Nordin (Eds.) *Basic Biomechanics of the Skeletal System.* pp. 1–11. Philadelphia: Lea and Febiger.

Cartmill, M., Hylander, W., and Shafland, J. (1987) *Human Structure.* Cambridge, Massachussetts: Harvard University Press.

Clarke, N. G., and Hirsch, R. S. (1991) Physiological, pulpal, and periodontal factors influencing alveolar bone. In: M. A. Kelley and C. S. Larsen (Eds.) *Advances in Dental Anthropology.* pp. 241–266. New York: Wiley-Liss.

Clarke, R. J., and Howell, F. C. (1972) Affinities of the Swartkrans 847 Hominid Cranium. *American Journal of Physical Anthropology* 37:319–335.

Cohen, M. N. (1994) The osteological paradox reconsidered. *Current Anthropology* 35:629–637.

Cohen, M. N., and Armelagos, G. J. (Eds.) (1985) *Paleopathology at the Origins of Agriculture.* Orlando, Florida: Academic Press.

Connor, A., and Stoneking, M. (1994) Assessing ethnicity from human mitochondrial DNA types determined by hybridization with sequence-specific oligonucleotides. *Journal of Forensic Sciences* 39:1360–1371.

Cool, S. M., Hendrikz, J. K., and Wood, W. B. (1995) Microscopic age changes in the human occipital bone. *Journal of Forensic Sciences* 40:789–796.

Cox, M., and Scott, A. (1992) Evaluation of the obstetric significance of some pelvic characters in an 18th century British sample of known parity. *American Journal of Physical Anthropology* 89:431–440.

Corruccini, R. S. (1978) Morphometric analysis: Uses and abuses. *Yearbook of Physical Anthropology* 21:134–150.

Corruccini, R. S. (1987) Shape in morphometrics: Comparative analyses. *American Journal of Physical Anthropology* 73:289–303.

Currey, J. (1984) *The Mechanical Adaptations of Bones.* Princeton, New Jersey: Princeton University Press.

Danforth, M. E., Herndon, K. S., and Propst, K. B. (1993) A preliminary study of patterns of replication in scoring linear enamel hypoplasias. *International Journal of Osteoarchaeology* 3:297–302.

DeGusta, D., and White, T. D. (1996) On the use of skeletal collections for DNA analysis. *Ancient Biomolecules* 1:89–92.

De Vito, C., and Saunders, S. R. (1990) A discriminant function analysis of deciduous teeth to determine sex. *Journal of Forensic Sciences* 35:845–858.

Devlin, B., Risch, N., and Roeder, K. (1994) Comments on the statistical aspects of the NRC's report on DNA typing. *Journal of Forensic Sciences* 39:28–40.

Di Maio, D. and Di Maio, J. M. (1989) *Forensic Pathology.* New York: Elsevier.

Dirkmaat, D. C., and Adovasio, J. M. (1997) The role of archaeology in the recovery and interpretation of human remains from an outdoor forensic setting. In: W. D. Haglund and M. H. Sorg (Eds.) *Forensic Taphonomy.* pp. 39–64. Boca Raton, Florida: CRC Press.

Dobney, K., and Brothwell, D. (1986) Dental calculus: Its relevance to ancient diet and oral ecology. *British Archaeological Reports: International Series* 291: 55–81.

Donnelly, S. M., Hens, S. M., Rogers, N. L., and Schneider, K. L. (1998) Technical note: A blind test of mandibular ramus flexure as a morphologic indicator of sexual dimorphism in the human skeleton. *American Journal of Physical Anthropology* 107:363–366.

Dreier, F. G. (1994) Age at death estimates for the protohistoric Arikara using molar attrition rates: A new quantification method. *International Journal of Osteoarchaeology* 4:137–147.

Drusini, A. G., Toso, O., and Ranzato, C. (1997) The coronal pulp cavity index: A biomarker for age determination in human adults. *American Journal of Physical Anthropology* 103:353–363.

Dudar, J. C. (1993) Identification of rib number and assessment of intercostal variation at the sternal end. *Journal of Forensic Sciences* 38:788–797.

Duday, H. (1978) Archaeologie funeraire et anthropologie. *Cahiers d'Anthropologie* 1:55–101.

Duday, H., and Masset, C. (Eds.) (1987) *Anthropologie Physique et Archaeologie.* Paris: C.N.R.S.

El-Najjar, M. Y., and McWilliams, K. R. (1978) *Forensic Anthropology.* Springfield, Illinois: C. C. Thomas.

Epstein, F. H. (1989) The biology of osteoarthritis. *New England Journal of Medicine* 320:1322–1330.

Ericksen, M. F. (1991) Histologic estimation of age at death using the anterior cortex of the femur. *American Journal of Physical Anthropology* 84:171–179.

Falsetti, A. B. (1995) Sex assessment from metacarpals of the human hand. *Journal of Forensic Sciences* 40:774–776.

Fazekas, G., and Kosa, F. (1978) *Forensic Fetal Osteology.* Budapest: Akadémiai Kiadó.

Feder, H. A. (1991) *Succeeding as an Expert Witness.* New York: Van Nostrand Reinhold.

Feldesman, M. R., and Fountain, R. L. (1996) "Race" specificity and the femur/stature ratio. *American Journal of Physical Anthropology* 100:207–224.

Feldmann, R. M., Chapman, R. E., and Hannibal, J. T. (Eds.) (1989) *Paleotechniques.* Knoxville, Tennessee: Paleontological Society Special Publication Number 4.

Fetterman, J., Honeycutt, L., and Kuckelman, K. (1988) Salvage Excavations of 42SA 12209, A Pueblo I habitation site in Cottonwood Canyon, Mati-Lasal National Forest, Southeastern Utah. Report to USDA Forest Service, Mati-Lasal National Forest, P.O. Box 820, Monticello, Utah 84535; in partial fulfillment of Contract # 53-8462-7-10024.

Finnegan, M. (1978) Non-metric variation of the infracranial skeleton. *Journal of Anatomy* 125:23–37.

Foley, R., and Cruwys, E. (1986) Dental anthropology: Problems and perspectives. *British Archaeological Reports: International Series* 291:1–20.

Formicola, V. (1993) Stature reconstruction from long bones in ancient population samples—an approach to the problem of its reliability. *American Journal of Physical Anthropology* 90:351–358.

Fox, C. L. (1992) Information obtained from the microscopic examination of cultural striations in human dentition. *International Journal of Osteoarchaeology* 2:155–169.

France, D. L., Griffin, T. J., Swanburg, J. G., Lindemann, J. W., Davenport, G. C., Tramell, V., Armbrust, C. T., Kondratieff, B., Nelson, A., Castellano, K., and Hopkins, D. (1992) A multidisciplinary approach to the detection of clandestine graves. *Journal of Forensic Sciences* 37:1445–1458.

Frayer, D. W. (1985) Multivariate morphometrics. *Reviews in Anthropology* 12: 289–298.

Frayer, D. W. (1991) On the etiology of interproximal grooves. *American Journal of Physical Anthropology* 85:299–304.

Frost, H. M. (1987) Secondary osteon populations: An algorithm for determining mean bone tissue age. *Yearbook of Physical Anthropology* 30:221–238.

Galera, V., Ubelaker, D. H., and Hayek, L. C. (1998) Comparison of macroscopic cranial methods of age estimation applied to skeletons of the Terry Collection. *Journal of Forensic Sciences* 43:933–939.

Galloway, A., and Simmons, T. L. (1997) Education in forensic anthropology: Appraisal and outlook. *Journal of Forensic Sciences* 42:796–801.

Galloway, A., and Snodgrass, J. J. (1998) Biological and chemical hazards of forensic skeletal analysis. *Journal of Forensic Sciences* 43:1144–1147.

Garn, S. M. (1972) The course of bone gain and the phases of bone loss. *Orthopedic Clinics of North America* 3:503–520.

Genovés, S. (1967) Proportionality of long bones and their relation to stature among Mesoamericans. *American Journal of Physical Anthropology* 26:67–78.

Getlin, J. (1986) Hearts and bones. *Los Angeles Times Magazine.* October 12, 1986, pp. 11–27.

Gifford, D. P. (1981) Taphonomy and paleoecology: A critical review of archaeology's sister disciplines. *Advances in Archaeological Method and Theory* 4: 365–438.

Gilbert, B. M., and McKern, T. W. (1973) A method for aging the female *Os pubis. American Journal of Physical Anthropology* 38:31–38.

Giles, E. (1993) Modifying stature estimation from the femur and tibia. *Journal of Forensic Sciences* 38:758–760.

Giles, E. (1991) Corrections for age in estimating older adult's stature from long bones. *Journal of Forensic Sciences* 36:898–901.

Giles, E., and Elliot, O. (1962) Race identification from cranial measurements. *Journal of Forensic Sciences* 7:147–147.

Giles, E., and Elliot, O. (1963) Sex determination by discriminant function analysis of crania. *American Journal of Physical Anthropology* 21:53–68.

Gill, G. W. (1984) A forensic test case for a new method of geographical race determination. In: T. A. Rathburn and J. E. Buikstra (Eds.) *Human Identification: Case Studies in Forensic Anthropology.* pp. 329–339. Springfield, Illinois: C. C. Thomas.

Gill, G. W. (1986) Craniofacial criteria in forensic race identification. In: K. J. Reichs (Ed.) *Forensic Osteology: Advances in the Identification of Human Remains.* pp. 143–159. Springfield, Illinois: C. C. Thomas.

Gill, G. W. (1995) Challenge on the frontier: Discerning American Indians from whites osteologically. *Journal of Forensic Sciences* 40:783–788.

Gill, G. W., Fisher, J. W., and Zeimens, G. M. (1984) A pioneer burial near the historic Bordeaux trading post. *Plains Anthropologist* 29:229–238.

Gill, G. W., and Rhine, S. (Eds.) (1990) *Skeletal Attribution of Race: Methods for Forensic Anthropology.* Albuquerque, New Mexico: Maxwell Museum of Anthropology. Anthropological Papers Number 4.

Goldberg, K. E. (1985) *The Skeleton: Fantastic Framework.* New York: Torstar Books.

Goldstein, L. (1995) Politics, law, pragmatics, and human burial excavations: An example from northern California. In: A. L. Grauer (Ed.) *Bodies of Evidence:*

Reconstructing History Through Skeletal Analysis. pp. 3–17. New York: Wiley-Liss.

Goldstein, L., and Kintigh, K. (1990) Ethics and the reburial controversy. *American Antiquity* 55:585–591.

Goodman, A. H., and Armelagos, G. J. (1985) Factors affecting the distribution of enamel hypoplasias within the human permanent dentition. *American Journal of Physical Anthropology* 68:479–493.

Goodman, A. H., and Rose, J. C. (1991) Dental enamel hypoplasias as indicators of nutritional status. In: M. A. Kelley and C. S. Larsen (Eds.) *Advances in Dental Anthropology.* pp. 279–293. New York: Wiley-Liss.

Grant, J. C. B. (1972) *An Atlas of Anatomy* (6th Edition). Baltimore, Maryland: Williams and Wilkins.

Grauer, A. L. (Ed.) (1995) *Bodies of Evidence: Reconstructing History Through Skeletal Analysis.* New York: Wiley-Liss.

Graw, M., CzarNetzki, A., and Haffner, H.-T. (1999) The form of the supraorbital margin as a criterion in identification of sex from the skull: Investigations based on modern human skulls. *American Journal of Physical Anthropology* 108: 91–96.

Greenberg, J. H., Turner, C. G., and Zegura, S. L. (1986) The settlement of the Americas: A comparison of the linguistic, dental and genetic evidence. *Current Anthropology* 27:477–497.

Greene, D. L. (1982) Discrete dental variations and biological distances of Nubian populations. *American Journal of Physical Anthropology* 58:75–79.

Greene, D. L., Van Gerven, D. P., and Armelagos, G. J. (1986) Life and death in ancient populations: Bones of contention in paleodemography. *Human Evolution* 1:193–207.

Gregg, J. B., and Gregg, P. S. (1987) *Dry Bones: Dakota Territory Reflected.* Sioux Falls, South Dakota: Sioux Printing.

Gruspier, K. L., and Mullen, G. J. (1991) Maxillary suture obliteration: A test of the Mann method. *Journal of Forensic Sciences* 36:512–519.

Gustafson, G., and Koch, G. (1974) Age estimation up to 16 years of age based on dental development. *Odontologisk Revy* 25:297–306.

Hagelberg, E., Sykes, B., and Hedges, R. (1989) Ancient bone DNA amplified. *Nature* 342:485.

Hagelberg, E., Gray, I. C., and Jeffreys, A. J. (1991) Identification of the skeletal remains of a murder victim by DNA analysis. *Nature* 352:427–429.

Haglund, W. D. (1992) Contributions of rodents to postmortem artifacts of bone and soft tissue. *Journal of Forensic Sciences* 37:1459–1465.

Haglund, W. D. (1997) Dogs and coyotes: Postmortem involvement with human remains. In: W. D. Haglund and M. H. Sorg (Eds.) *Forensic Taphonomy.* pp. 367–381. Boca Raton, Florida: CRC Press.

Haglund, W. D., Reay, D. T., and Swindler, D. R. (1988) Tooth mark artifacts and survival of bones in animal scavenged human skeletons. *Journal of Forensic Sciences* 33:985–997.

Haglund, W. D., and Sorg, M. H. (Eds.) (1997) *Forensic Taphonomy.* Boca Raton, Florida: CRC Press.

Hamilton, M. E. (1982) Sexual dimorphism in skeletal samples. In: R. L. Hall (Ed.) *Sexual Dimorphism in Homo sapiens.* pp. 107–163. New York: Praeger.

Hancock, R. G. V., Grynpas, M. D., and Pritzker, K. P. H. (1989) The abuse of bone analyses for archaeological dietary studies. *Archaeometry* 31:169–179.

Handt, O., Höss, M., Krings, M., and Pääbo, S. (1994) Ancient DNA: Methodological challenges. *Experientia* 50:524–529.

Hansen, M. (1993) Believe it or Not. *American Bar Association Journal*, June, 1993, pp. 64–67.

Hauser, G., and DeStefano, G. F. (1989) *Epigenetic Variants of the Human Skull.* Stuttgart: E. Schweizerbart'sche Verlagsbuchhandlung.

Hauswirth, W. W., Dickel, C. D., Rowold, D. J., and Hauswirth, M. A. (1994) Inter- and intrapopulation studies of ancient humans. *Experientia* 50:585–591.

Heathcote, G. M. (1981) The magnitude and consequences of measurement error in human craniometry. *Canadian Review of Physical Anthropology* 3:18–40.

Henderson, J. (1987) Factors determining the state of preservation of human remains. In: A. Boddington, A. N. Garland, and R. C. Janaway (Eds.) *Death, Decay and Reconstruction.* pp. 43–54. Manchester: Manchester University Press.

Herrmann, B., and Hummel, S. (Eds.) (1994) *Ancient DNA.* New York: Springer-Verlag.

Higuchi, R., Bowman, B., Freiberger, M., Ryder, O. A., and Wilson, A. C. (1984) DNA sequences from the quagga, an extinct member of the horse family. *Nature* 312:282–284.

Hildebrand, M. (1968) *Anatomical Preparations.* Berkeley, California: University of California Press.

Hillson, S. (1986) *Teeth.* Cambridge: Cambridge University Press.

Hillson, S. W. (1992) Impression and replica methods for studying hypoplasia and perikymata on human tooth crown surfaces from archaeological sites. *International Journal of Osteoarchaeology* 2:65–78.

Hillson, S. (1996) *Dental Anthropology.* Cambridge: Cambridge University Press.

Hogge, J. P., Messmer, J. M., and Doan, Q. N. (1994) Radiographic identification of unknown human remains and interpreter experience level. *Journal of Forensic Sciences* 39:373–377.

Holland, M. M., Fisher, D. L., Mitchell, L. G., Rodriquez, W. C., Canik, J. J., Merril, C. R., and Weedn, V. W. (1993) Mitochondrial DNA sequence analysis of human skeletal remains: Identification of remains from the Vietnam War. *Journal of Forensic Sciences* 38:542–553.

Hollien, H. (1990) The expert witness: Ethics and responsibilities. *Journal of Forensic Sciences* 35:1414–1423.

Hoppa, R., and Saunders, S. (1998) Two quantitative methods for rib seriation in human skeletal remains. *Journal of Forensic Sciences* 43:174–177.

Hoshower, L. M. (1998) Forensic archeology and the need for flexible excavation strategies: A case study. *Journal of Forensic Sciences* 43:53–56.

Houck, M. M., Ubelaker, D., Owsley, D., Craig, E., Grant, W., Fram, R., Woltanski, T., and Sandness, K. (1996) The role of forensic anthropology in the recovery and analysis of Branch Davidian compound victims: Assessing the accuracy of age estimations. *Journal of Forensic Sciences* 41:796–801.

Howell, T. L., and Kintigh, K. W. (1996) Archaeological identification of kin groups using mortuary and biological data: An example from the American southwest. *American Antiquity* 61:537–554.

Howells, W. W. (1989) Skull shapes and the map: Craniometric analyses in the dispersion of modern *Homo. Papers of the Peabody Museum of Archaeology and Ethnology* 79:1–189.

Howells, W. W. (1995) Who's who in skulls: Ethnic identification of crania from measurements. *Papers of the Peabody Museum of Archaeology and Ethnology, Harvard University* 82:1–108.

Howells, W. W. (1969a) Criteria for selection of osteometric dimensions. *American Journal of Physical Anthropology* 30:451–458.

Howells, W. W. (1969b) The use of multivariate techniques in the study of skeletal populations. *American Journal of Physical Anthropology* 31:311–314.

Howells, W. W. (1973) *Cranial Variation in Man.* Cambridge, Massachusetts: Peabody Museum of Archaeology and Ethnology Papers.

Howells, W. W. (1983) Origins of Chinese people: Interpretations of the recent evidence. In: D. N. Keightly (Ed.) *The Origins of Chinese Civilization.* pp. 297–319. Berkeley: University of California Press.

Howells, W. W. (1986) Physical anthropology of the prehistoric Japanese. In: R. J. Pearson (Ed.) *Windows on the Japanese Past.* pp. 85–99. Ann Arbor, Michigan: Center for Japanese Studies.

Hubert, J. (1989) A proper place for the dead: A critical review of the "reburial" issue. In: R. Layton (Ed.) *Conflict in the Archaeology of Living Traditions.* pp. 131–166. London: Unwin Hyman.

Hummert, J. R., and Van Gerven, D. P. (1985) Observations on the formation and persistence of radiopaque transverse lines. *American Journal of Physical Anthropology* 66:297–306.

Huss-Ashmore, R., Goodman, A. H., and Armelagos, G. J. (1982) Nutritional inference from paleopathology. *Advances in Archaeological Method and Theory* 5: 395–474.

Introna, F., Jr., Di Vella, G., Campobasso, C. P., and Dragone, M. (1997) Sex determination by discriminant analysis of calcanei measurments. *Journal of Forensic Sciences* 42:725–728.

Irish, J. D., and Turner, C. G. (1987) West African dental affinity of Late Pleistocene Nubians: Peopling of the Eurafrican-South Asian triangle II. *Homo* 41:42–53.

İşcan, M. Y. (1988) Rise of forensic anthropology. *Yearbook of Physical Anthropology* 31:203–230.

İşcan, M. Y. (Ed.) (1989) *Age Markers in the Human Skeleton.* Springfield, Illinois: Charles C. Thomas.

İşcan, M. Y., and Cotton, T. S. (1985) The effect of age on the determination of race from the pelvis. *Journal of Human Evolution* 14:275–282.

İşcan, M. Y. and Kennedy, K. A. R. (Eds.) (1989) *Reconstruction of Life from the Skeleton.* New York: Alan R. Liss.

İşcan, M. Y., and Loth, S. R. (1986) Estimation of age and determination of sex from the sternal rib. In: K. J. Reichs (Ed.) *Forensic Osteology: Advances in the Identification of Human Remains.* pp. 68–89. Springfield, Illinois: C. C. Thomas.

İşcan, M. Y., and Helmer, R. P. (1993) *Forensic Analysis of the Skull.* New York: Wiley-Liss.

Ivanov, P. L., Wadhams, M. J., Roby, R. K., Holland, M. M., Weedn, V. W., and Parsons, T. J. (1996) Mitochondrial DNA sequence heteroplasmy in the Grand Duke of Russia Georgij Romanov establishes the authenticity of the remains of Tsar Nicholas II. *Nature Genetics* 12:417–420.

Jackes, M. (1992) Paleodemography: Problems and techniques. In: S. R. Saunders and M. A. Katzenberg (Eds.) *Skeletal Biology of Past Peoples: Research Methods.* pp. 189–224. New York: Wiley Liss.

Jantz, R. L. (1992) Modification of the Trotter and Gleser female stature estimation formulae. *Journal of Forensic Sciences* 37:1230–1235.

Jantz, R. L. (1993) Modifying stature estimation from the femur and tibia: Author's response. *Journal of Forensic Sciences* 38:760–763.

Jantz, R. L., Hunt, D. R., and Meadows, L. (1995) The measure and mismeasure of the tibia—implications for stature estimation. *Journal of Forensic Sciences* 40: 758–761.

Jarvik, E. (1980) *Basic Structure and Evolution of Vertebrates,* Vol. 2. New York: Academic Press.

Jeffreys, A. J., Allen, M. J., Hagelberg, E., and Sonnberg, A. (1992) Identification of the skeletal remains of Josef Mengele by DNA analysis. *Forensic Science International* 56:65–76.

Johanson, D. C. (1979) A consideration of the *"Dryopithecus* pattern." *Ossa* 6: 125–138.

Jones, D. G., and Harris, R. J. (1998) Archeological human remains. *Current Anthropology* 39:253–264.

Joukowsky, M. (1980) *A Complete Manual of Field Archaeology: Tools and Techniques of Field Work for Archaeologists.* Englewood Cliffs, New Jersey: Prentice-Hall.

Kaestle, F. A. (1995) Chapter 5. Mitochondrial DNA evidence for the identity of descendants of the prehistoric Stillwater Marsh population. In: C. S. Larsen and R. L. Kelley (Eds.) *Bioarchaeology of the Stillwater Marsh: Prehistoric Human Adaptation in the Western Great Basin. American Museum of Natural History Anthropological Papers* 77:73–80.

Katz, D., and Suchey, J. M. (1985) Age determination of the male *os pubis. American Journal of Physical Anthropology* 69:427–435.

Katz, D., and Suchey, J. M. (1989) Race differences in pubic symphyseal aging patterns in the male. *American Journal of Physical Anthropology* 80:167–172.

Kelley, M. A. (1979) Parturition and pelvic changes. *American Journal of Physical Anthropology* 51:541–546.

Kelley, M. A., and Larsen, C. S. (Eds.) (1991) *Advances in Dental Anthropology.* New York: Wiley-Liss.

Kelman, L. M., and Kelman, Z. (1999) The use of ancient DNA in paleontological studies. *Journal of Vertebrate Paleontology* 19:8–20.

Kennedy, K. A. R. (1989) Skeletal markers of occupational stress. In: M. Y. İşcan and K. A. R. Kennedy (Eds.) *Reconstruction of Life from the Skeleton.* New York: Alan R. Liss. pp. 129–160.

Kennedy, K. A. R. (1995) But professor, why teach race identification if races don't exist? *Journal of Forensic Sciences* 40:797–800.

Kent, S. (1987) The influence of sedentism and aggregation on porotic hyperostosis and anaemia: A case study. *Man* 21:605–636.

Kieser, J. A. (1990) *Human Adult Odontometrics.* Cambridge: Cambridge University Press.

Killiam, E. W. (1990) *The Detection of Human Remains.* Springfield, Illinois: C. C. Thomas.

King, C. A., İşcan, M. Y., and Loth, S. R. (1998) Metric and comparative analysis of sexual dimorphism in the Thai femur. *Journal of Forensic Sciences* 43:954–958.

Klesert, A. L., and Powell, S. (1993) A perspective on ethics and the reburial controversy. *American Antiquity* 58:348–354.

Kogan, J. D. (1978) On being a good expert witness in a criminal case. *Journal of Forensic Science* 23:190–200.

Kogon, S. L., and MacLean, D. F. (1996) Long-term validation study of bitewing dental radiographs for forensic identification. *Journal of Forensic Sciences* 41: 230–232.

Konigsberg, L. W., and Frankenberg, S. R. (1994) Paleodemography: "Not quite dead." *Evolutionary Anthropology* 2:92–105.

Krings, M., Stone, A., Schmitz, R. W., Krainitzki, H., Stoneking, M., and Pääbo, S. (1997) Neandertal DNA sequences and the origin of modern humans. *Cell* 90: 19–30.

Krogman, W. M., and İşcan, M. Y. (1986) *The Human Skeleton in Forensic Medicine* (2nd Edition). Springfield, Illinois: C. C. Thomas.

Kvaal, S. I., Sellevold, B. J., and Solheim, T. (1994) A comparison of different non-destructive methods of age estimation in skeletal material. *International Journal of Osteoarchaeology* 4:363–370.

Lambert, P. M., and Walker, P. L. (1991) Physical anthropological evidence for the evolution of social complexity in coastal Southern California. *Antiquity* 65: 963–973.

Lamendin, H., Baccino, E., Humbert, J. F., Tavernier, J. C., Nossintchouk, R. M., and Zerilli, A. (1992) A simple technique for age estimation in adult corpses: The two criteria dental method. *Journal of Forensic Sciences* 37:1373–1379.

Lampl, M., Veldhuis, J. D., and Johnson, M. L. (1992) Saltation and stasis: A model of human growth. *Science* 258:801–803.

Lampl, M., and Johnston, F. E. (1996) Problems in the aging of skeletal juveniles: Perspectives from maturation assessments of living children. *American Journal of Physical Anthropology* 101:345–355.

Lancon, J. A., Haines, D. E., and Parent, A. D. (1998) Anatomy of the shaken baby syndrome. *The New Anatomist* (*The Anatomical Record*) 253:13–18.

Larsen, C. S. (1983) Behavioral implications of temporal change in cariogenesis. *Journal of Archaeological Science* 10:1–8.

Larsen, C. S. (1997) *Bioarchaeology: Interpreting Behavior from the Human Skeleton.* Cambridge: Cambridge University Press.

Larsen, C. S., and Kelly, R. L. (Eds.) (1995) Bioarchaeology of the Stillwater Marsh: Prehistoric human adaptation in the Western Great Basin. *American Museum of Natural History Anthropological Papers* 77:1–170.

Larsen, C. S., Kelly, R. L., Ruff, C. B., Schoeninger, M. J., and Hutchinson, D. L. (1996) Biobehavioral adaptations in the western Great Basin. In: E. J. Reitz, L. A. Newson, and S. J. Scudder (Eds.) *Case Studies in Environmental Archaeology.* pp. 149–174. New York: Plenum Press.

Layton, R. (Ed.) (1989) *Conflict in the Archaeology of Living Traditions.* London: Unwin Hyman.

Leiggi, P., and May, P. J. (Eds.) (1994) *Vertebrate Paleontological Techniques.* Cambridge: Cambridge University Press.

Lenchik, L., and Sartoris, D. J. (1997) Current concepts in osteoporosis. *American Journal of Roetgenology* 168:905–911.

Ley, C. A., Aiello, L. C., and Molleson, T. (1994) Cranial suture closure and its implications for age estimation. *International Journal of Osteoarchaeology* 4: 193–207.

Li, W.-H., and Graur, D. (1991) *Fundamentals of Molecular Evolution.* Sunderland, Massachussetts: Sinauer.

Lindahl, T. (1997) Facts and artifacts of ancient DNA. *Cell* 90:1–3.

Liversidge, H. M. (1994) Accuracy of age estimation from developing teeth of a population of known age (0–5.4 years). *International Journal of Osteoarchaeology* 4:37–45.

Loth, S. R., and Henneberg, M. (1996) Mandibular ramus flexure: A new morphologic indicator of sexual dimorphism in the human skeleton. *American Journal of Physical Anthropology* 99:473–485.

Loth, S. R., and Henneberg, M. (1998) Mandibular ramus flexure *is* a good indicator of sexual dimorphism. *American Journal of Physical Anthropology* 105:91–92.

Lovejoy, C. O. (1978) A biomechanical review of the locomotor diversity of early hominids. In: C. J. Jolly (Ed.) *Early Hominids of Africa.* pp. 403–429. London: Duckworth.

Lovejoy, C. O. (1985) Dental wear in the Libben population: Its functional pattern and role in the determination of adult skeletal age at death. *American Journal of Physical Anthropology* 68:47–56.

Lovejoy, C. O. (1988) Evolution of human walking. *Scientific American* 256(11): 118–125.

Lovejoy, C. O., and Burstein, A. H. (1977) Geometrical properties of bone sections determined by laminography and physical section. *Journal of Biomechanics* 10: 527–528.

Lovejoy, C. O., and Heiple, K. G. (1981) The analysis of fractures in skeletal populations with an example from the Libben Site, Ottawa County, Ohio. *American Journal of Physical Anthropology* 55:529–541.

Lovejoy, C. O., Burstein, A. H., and Heiple, K. G. (1976) The biomechanical analysis of bone strength: A method and its application to platycnemia. *American Journal of Physical Anthropology* 44:489–506.

Lovejoy, C. O., and colleagues. (1985) Eight papers on Todd and Libben skeletal material. *American Journal of Physical Anthropology* 68:1–106.

Lovejoy, C. O., Meindl, R. S., Tague, R. G., and Latimer, B. (1997) The comparative senescent biology of the hominoid pelvis and its implications for the use of age-at-death indicators in the human skeleton. In: R. Paine (Ed.), *Integrating Archaeological Demography: Multidisciplinary Approaches to Prehistoric Population.* pp. 43–63. Carbondale, Illinois: Center for Archaeological Investigations, Southern Illinois University, Occasional Paper 24.

Lovejoy, C. O., Meindl, R. S., Pryzbeck, T. R., Barton, T. S., Heiple, K. G., and Kotting, D. (1977) The palaeodemography of the Libben Site, Ottowa County, Ohio. *Science* 198:291–293.

Lovejoy, C. O., Meindl, R. S., Mensforth, R. P., and Barton, T. J. (1985a) Multifactorial determination of skeletal age at death: A method and blind tests of its accuracy. *American Journal of Physical Anthropology* 68:1–14.

Lovejoy, C. O., Meindl, R. S., Pryzbeck, T. R., and Mensforth, R. P. (1985b) Chronological metamorphosis of the auricular surface of the ilium: A new method for the determination of adult skeletal age at death. *American Journal of Physical Anthropology* 68:15–28.

Lovejoy, C. O., Meindl, R. S., Tague, R. G., and Latimer, B. (1995) The senescent biology of the hominoid pelvis. *Rivista di Antropologia (Roma)* 73:31–49.

Lovejoy, C. O., Cohn, J., and White, T. D. (1999) The evolution of mammalian morphology: A developmental perspective. In: P O'Higgins and M. Cohn (Eds.) *Development, Growth, and Evolution: Implications for the Study of the Hominid Skeleton.* London: Academic Press.

Lovell, N. C. (1989) Test of Phenice's technique for determining sex from the Os Pubis. *American Journal of Physical Anthropology* 79:117–120.

Lukacs, J. R. (1989) Dental pathology: Methods for reconstructing dietary patterns. In: M. Y. İşcan and K. A. R. Kennedy (Eds.) *Reconstruction of Life from the Skeleton.* pp. 261–286. New York: Alan R. Liss.

Lyman, R. L. (1994) *Vertebrate Taphonomy.* Cambridge: Cambridge University Press.

Lyman, R. L., and Fox, G. L. (1997) A critical evaluation of bone weathering as an indication of bone assemblage formation. In: W. D. Haglund and M. H. Sorg (Eds.) *Forensic Taphonomy.* pp. 223–247. Boca Raton, Florida: CRC Press.

Lynnerup, N., Hjalgrim, H., Nielsen, L. R., Gregersen, H., and Thuesen, I. (1997) Non-invasive archaeology of skeletal material by CT scanning and three-dimensional reconstruction. *International Journal of Osteoarchaeology* 7:91–94.

Maat, G. J. R. (1981) Human remains at the Dutch whaling stations on Spitsbergen. In: A. G. F. van Holk, H. K. s'Jacob, and A. A. Temming *Early European Exploitation of the Northern Atlantic, 800–1700.* pp. 153–201. Groningen, The Netherlands: Arctic Centre, University of Groningen.

Maat, G. J. R. (1984) Dating and rating of Harris's lines. *American Journal of Physical Anthropology* 63:291–299.

Maat, G. J. R. (1987) Osteology of human remains from Amsterdamøya and Ytre Norskøya. *Norsk Polarinstituut Rapportserie* 38:35–53.

Macchiarelli, R., Bondioli, L., Censi, L., Hernaez, M. K., Salvadei, L., and Sperduti, A. (1994) Intra- and interobserver concordance in scoring Harris lines: A test on bone sections and radiographs. *American Journal of Physical Anthropology* 95: 77–83.

MacLaughlin, S. M., and Bruce, M. F. (1990) The accuracy of sex identification in European skeletal remains using the Phenice characters. *Journal of Forensic Sciences* 35:1384–1392.

Mann, A., and Monge, J. (1987) Reproducing our ancestors. *Expedition: The University of Pennsylvania University Museum Magazine of Archaeology* 29(1): 2–9.

Mann, R. W., Jantz, R. L., Bass, W. M., and Willey, P. S. (1991) Maxillary suture obliteration: A visual method for estimating skeletal age. *Journal of Forensic Sciences* 36:781–791.

Mann, R. W. (1993) Technical note: A method for siding and sequencing human ribs. *Journal of Forensic Sciences* 38:151–155.

Manzi, G., Sperduti, A., and Passarello, P. (1991) Behavior-induced auditory exostoses in Imperial Roman society: Evidence from coeval urban and rural communities near Rome. *American Journal of Physical Anthropology* 85:253–260.

Maples, W. R. (1986) Trauma analysis by the forensic anthropologist. In: K. J. Reichs (Ed.) *Forensic Osteology: Advances in the Identification of Human Remains.* pp. 218–228. Springfield, Illinois: C. C. Thomas.

Maples, W. R., and Browning, M. (1994) *Dead Men Do Tell Tales.* New York: Doubleday.

Marks, J. (1996) Science and race. *American Behavioral Scientist* 40:123–133.

Martin, D. L., Goodman, A. H., and Armelagos, G. J. (1985) Skeletal pathologies as indicators of quality and quantity of diet. In: R. I. Gilbert and J. H. Meikle (Eds.) *The Analysis of Prehistoric Diets.* pp. 227–279. Orlando, Florida: Academic Press.

Martin, R., and Saller, K. (1957) *Lehrbuch der Anthropologie,* Vol. 1. Stuttgart: Gustav Fischer.

Mayhall, J. T., and Kageyamu, I. (1997) A new, three-dimensional method for determining tooth wear. *American Journal of Physical Anthropology* 103: 463–469.

McGuire, R. (1989) The sanctity of the grave: White concepts and American Indian burials. In: R. Layton (Ed.) *Conflict in the Archaeology of Living Traditions.* pp. 167–184. London: Unwin Hyman.

McKee, J. K., and Molnar, S. (1988) Measurements of tooth wear among Australian Aborigines: II. Intrapopulational variation in patterns of dental attrition. *American Journal of Physical Anthropology* 76:125–136.

McKern, T. W., and Stewart, T. D. (1957) Skeletal age changes in young American males. Natick, Massachusetts: Quartermaster Research and Development Command Technical Report EP-45.

McMinn, R. M. H., and Hutchings, R. T. (1977) *Color Atlas of Human Anatomy.* Chicago: Year Book Medical Publishers.

McMinn, R. M. H., Hutchings, R. T., and Logan, B. M. (1987) *The Human Skeleton: A Photographic Manual.* Chicago: Year Book Medical Publishers.

Mead, E. M. and Meeks, S. (1989) Photography of archaeological and paleontological bone specimens. In: R. Bonnichsen and M. H. Sorg, (Eds.) (1989) *Bone Modification.* pp. 267–281. Orono, Maine: Center for the Study of the First Americans.

Meighan, C. W. (1992) Some scholar's views on reburial. *American Antiquity* 57: 704–710.

Meighan, C. W. (1994) Burying American archaeology. *Archaeology* Nov/Dec. 1994: 65–68.

Meindl, R. S., and Lovejoy, C. O. (1985) Ectocranial suture closure: A revised method for the determination of skeletal age at death based on the lateral-anterior sutures. *American Journal of Physical Anthropology* 68:57–66.

Meindl, R. S., Lovejoy, C. O., and Mensforth, R. P. (1985a) A revised method of age determination using the *Os pubis,* with a review and tests of accuracy of other current methods of pubic symphyseal aging. *American Journal of Physical Anthropology* 68:29–45.

Meindl, R. S., Lovejoy, C. O., Mensforth, R. P., and Carlos, L. D. (1985b) Accuracy and direction of error in sexing of the skeleton: Implications for paleodemography. *American Journal of Physical Anthropology* 68:79–85.

Melby, J., and Jimenez, S. B. (1997) Chain of custody from the field to the courtroom. In: W. D. Haglund and M. H. Sorg (Eds.) *Forensic Taphonomy.* pp. 65–75. Boca Raton, Florida: CRC Press.

Mellars, P. and Stringer, C. (Eds.) (1989) *The Human Revolution.* Edinburgh: Edinburgh University Press.

Mensforth, R. P., and Lovejoy, C. O. (1985) Anatomical, physiological, and epidemiological correlates of the aging process: A confirmation of multifactorial age determination in the Libben skeletal population. *American Journal of Physical Anthropology* 68:87–106.

Mensforth, R. P., Lovejoy, C. O., Lallo, J. W., and Armelagos, G. J. (1978) The role of constitutional factors, diet, and infectious disease in the etiology of porotic hyperostosis and periosteal reactions in prehistoric infants and children. *Medical Anthropology* 2(1):1–59.

Merbs, C. F. (1983) Patterns of activity-induced pathology in a Canadian Inuit population. *Archaeological Survey of Canada, Paper No. 119.* National Museum of Man, Mercury series. 1–200.

Metress, J. F., and Conway, T. (1974) A guide to the literature on the dental anthropology of Post Pleistocene Man. *Toledo Area Aboriginal Research Club Bulletin* Supplementary Monograph No. 1, pp. 1–142.

Micozzi, M. S. (1991) *Postmortem Change in Human and Animal Remains.* Springfield, Illinois: C. C. Thomas.

Miles, A. E. W. (1963) Dentition in the estimation of age. *Journal of Dental Research* 42:255–263.

Miller, E., Ragsdale, B. D., and Ortner, D. J. (1996) Accuracy in dry bone diagnosis: A comment on palaeopathological methods. *International Journal of Osteoarchaeology* 6:221–229.

Milner, G. R., and Larsen, C. S. (1991) Teeth as artifacts of human behavior: Intentional modification and accidental modification. In: M. A. Kelley and C. S. Larsen (Eds.) *Advances in Dental Anthropology.* pp. 357–378. New York: Wiley-Liss.

Mincer, H. H., Berryman, H. E., Murray, G. A., and Dickens, R. L. (1990) Technical note: Methods for physical stabilization of ashed teeth in incinerated remains. *Journal of Forensic Sciences* 35:971–974.

Mincer, H. H., Harris, E. F., and Berryman, H. E. (1993) The A.B.F.O. study of third molar development and its use as an estimator of chronological age. *Journal of Forensic Sciences* 38:379–390.

Mizoguchi, Y. (1985) Shovelling: A statistical analysis of its morphology. *Bulletin, University of Tokyo Museum* 26:1–176.

Molnar, S. (1971) Human tooth wear, tooth function, and cultural variability. *American Journal of Physical Anthropology* 34:175–190.

Moore-Jansen, P. H., and Jantz, R. L. (1989) *Data Collection Procedures for Forensic Skeletal Material. Report of Investigations No. 48.* Knoxville: University of Tennessee.

Morell, V. (1998) Kennewick Man's trials continue. *Science* 280:190–192.

Mori, J. L. (1970) Procedures for establishing a faunal collection to aid in archaeological analysis. *American Antiquity* 35:387–389.

Morton, R. A. (Ed.) (1984) *Photography for the Scientist* (2nd Edition). London: Academic Press.

Müller, W. A. (1997) *Developmental Biology.* New York: Springer.

Mullis, K. B., and Faloona, F. (1987) Specific synthesis of DNA in vitro via a polymerase-catalyzed chain reaction. *Methods in Enzymology* 155:335–350.

Murray, K. A., and Murray, T. (1991) A test of the auricular surface aging technique. *Journal of Forensic Sciences* 36:1162–1169.

Napoli, M. L., and Birkby, W. H. (1990) Racial differences in the visibility of the oval window in the middle ear. In: G. W. Rhine and S. Rhine (Eds.) *Skeletal Attribution of Race: Methods for Forensic Anthropology.* pp. 27–32. Albuquerque, New Mexico: Maxwell Museum of Anthropology, Anthropological Papers Number 4.

Nawrocki, S. P. (1995) Taphonomic processes in historic cemeteries. In: A. L. Grauer (Ed.) *Bodies of Evidence: Reconstructing History Through Skeletal Analysis.* pp. 49–66. New York: Wiley-Liss.

Nickens, P. R. (1975) Prehistoric cannibalism in the Mancos Canyon, southwestern Colorado. *The Kiva* 40:283–293.

Nordby, J. J. (1992) Can we believe what we see, if we see what we believe?—Expert disagreement. *Journal of Forensic Sciences* 37:1115–1124.

Odwak, H., and Schulting, R. J. (1996) A simple technique for aiding the interpretation and enhancement of radiographs. *International Journal of Osteoarchaeology* 6:502–505.

Ogden, J. A. (1990) Histogenesis of the musculoskeletal system. In: D. J. Simmons (Ed.) *Nutrition and Bone Development.* pp. 3–36. New York: Oxford University Press.

Ohtani, S. (1995) Studies on age estimation using racemization of aspartic acid in cementum. *Journal of Forensic Sciences* 40:805–807.

Ohtani, S., and Yamamoto, K. (1991) Age estimation using the racemization of amino acid in human dentin. *Journal of Forensic Sciences* 36:792–800.

Ohtani, S., and Yamamoto, K. (1992) Estimation of age from a tooth by means of racemization of an amino acid, especially aspartic acid—Comparison of enamel and dentin. *Journal of Forensic Sciences* 37:1061–1067.

O'Rahilly, R. (1989) Anatomical terminology, then and now. *Acta Anatomica* 134: 291–300.

Ortner, D. J., and Aufderheide, A. C. (1991) *Human Paleopathology: Current Syntheses and Future Options.* Washington, D.C.: Smithsonian Institution Press.

Ortner, D. J., and Putschar, W. G. (1981) *Identification of Pathological Conditions in Human Skeletal Remains.* Smithsonian Contributions to Anthropology 28: 1–479. Washington, D.C.: Smithsonian Institution Press.

Ousley, S. (1995) Should we estimate biological or forensic stature? *Journal of Forensic Sciences* 40:768–773.

Owsley, D. W. (1993) Identification of the fragmentary, burned remains of two U.S. journalists seven years after their disappearance in Guatemala. *Journal of Forensic Sciences* 38:1372–1382.

Owsley, D. W., and Mann, R. W. (1992) Positive personal identity of skeletonized remains using abdominal and pelvic radiographs. *Journal of Forensic Sciences* 37:332–336.

Owsley, D. W., Mann, R. W., Chapman, R. E., Moore, E., and Cox, W. (1993) Positive identification in a case of intentional extreme fragmentation. *Journal of Forensic Sciences* 38:985–996.

Owsley, D. W., Ubelaker, D. H., Houck, M. M., Sandness, K. L., Grant, W. E., Craig, E. A., Woltanski, T. J., and Peerwani, N. (1995) The role of forensic anthropology in the recovery and analysis of Branch Davidian compound victims: Techniques of analysis. *Journal of Forensic Sciences* 40:341–348.

Pääbo, S. (1985) Molecular cloning of ancient Egyptian mummy DNA. *Nature* 314: 644–645.

Paine, R. R., and Godfrey, L. R. (1997) The scaling of skeletal microanatomy in non-human primates. *Journal of Zoology, London* 241:803–821.

Palkovich, A. M. (1987) Endemic disease patterns in paleopathology: Porotic hyperostosis. *American Journal of Physical Anthropology* 74:527–537.

Parsons, T. J., and Weedn, V. W. (1997) Preservation and recovery of DNA in postmortem specimens and trace samples. In: W. D. Haglund and M. H. Sorg (Eds.) *Forensic Taphonomy: The Postmortem Fate of Human Remains.* pp. 109–138. Boca Raton: CRC Press.

Pfau, R. O., and Sciulli, P. W. (1994) A method for establishing the age of subadults. *Journal of Forensic Sciences* 39:165–176.

Pfeiffer, S., Lazenby, R., and Chiang, J. (1995) Brief communication: Cortical remodeling data are affected by sampling location. *American Journal of Physical Anthropology* 96:89–92.

Phenice, T. W. (1969) A newly developed visual method of sexing in the *Os pubis. American Journal of Physical Anthropology* 30:297–301.

Poinar, H. N., Höss, M., Bada, J. L., and Pääbo, S. (1996) Amino acid racemization and the preservation of ancient DNA. *Science* 272:864–866.

Pollanen, M. S., and Ubelaker, D. H. (1997) Forensic significance of the polymorphism of hyoid bone shape. *Journal of Forensic Sciences* 42:890–892.

Powell, M. L. (1985) The analysis of dental wear and caries for dietary reconstruction. In: R. I. Gilbert and J. H. Meikle (Eds.) *The Analysis of Prehistoric Diets.* pp. 307–338. Orlando, Florida: Academic Press.

Preston, D. (1997) The Lost Man. *New Yorker* June 16, 1997: 70–81.

Price, T. D. (1989) *The Chemistry of Human Bone.* Cambridge: Cambridge University Press.

Primorac, D., Andelinovic, S., Definis-Gojanovic, M., Drmic, I., Rezic, B., Baden, M. M., Kennedy, M. A., Schanfield, M. S., Skakel, S. B., and Lee, H. C. (1996) Identification of war victims from mass graves in Croatia, Bosnia, and Herzegovina by the use of standard forensic methods and DNA typing. *Journal of Forensic Sciences* 41:891–894.

Radosevich, S. C. (1993) The six deadly sins of trace element analysis: A case of wishful thinking in science. In: M. K. Sandford (Ed.) *Investigations of Ancient Human Tissue: Chemical Analyses in Anthropology.* Langhorne, Pennslyvania: Gordon and Breach Science Publishers, SA. pp. 269–332.

Rathburn, T. A., and Buikstra, J. E. (Eds.) (1984) *Human Identification: Case Studies in Forensic Anthropology.* Springfield, Illinois: C. C. Thomas.

Reader, J. (1988) *Missing Links: The Hunt for Earliest Man* (2nd Edition). Middlesex, England: Penguin Books.

Reichs, K. J. (Ed.) (1986) *Forensic Osteology: Advances in the Identification of Human Remains.* Springfield, Illinois: C. C. Thomas.

Reyment, R. A., Blackith, R. E., and Campbell, N. A. (1984) *Multivariate Morphometrics* (2nd Edition). London: Academic Press.

Rhine, S. (1990) Non-metric skull racing. In: G. W. Gill and S. Rhine (Eds.) *Skeletal Attribution of Race: Methods for Forensic Anthropology.* pp. 9–20. Albuquerque, New Mexico: Maxwell Museum of Anthropology, Anthropological Papers Number 4.

Richards, G. D., and Anton, S. C. (1987) The Berkeley Primate Skeletal Collection: A new resource for studies in paleopathology. *Paleopathology Newsletter* 57:8–11.

Richards, L. C., and Miller, S. L. J. (1991) Relationships between age and dental attrition in Australian aboriginals. *American Journal of Physical Anthropology* 84:159–164.

Richtsmeier, J. T., Cheverud, J. M., and Lele, S. (1992) Advances in anthropological morphometrics. *Annual Reviews in Anthropology* 21:283–305.

Robling, A. G., and Ubelaker, D. H. (1997) Sex estimation from the metatarsals. *Journal of Forensic Sciences* 42:1062–1069.

Rogers, J., Waldron, T., Dieppe, P., and Watt, I. (1987) Arthropathies in palaeopathology: The basis of classification according to most probable cause. *Journal of Archaeological Science* 14:179–193.

Rogers, S. L. (1986) *The Personal Identification of Living Individuals.* Springfield, Illinois: C. C. Thomas.

Rogers, T., and Saunders, S. (1994) Accuracy of sex determination using morphological traits of the human pelvis. *Journal of Forensic Sciences* 39:1047–1056.

Rogers, J., Waldron, T., Dieppe, P. and Watt, I. (1987) Arthropathies in paleopathology: The basis of classification into most probable cause. *Journal of Archaeological Science* 14:179–193.

Rogers, J., and Waldron, T. (1989) Infections in paleopathology: The basis of classification into most probable cause. *Journal of Archaeological Science* 16:611–625.

Rogers, T. L. (1999) A visual method of determining the sex of skeletal remains using the distal humerus. *Journal of Forensic Sciences* 44:57–60.

Rohen, J. W., and Yokochi, C. (1988) *Color Atlas of Anatomy* (2nd Edition). New York: Igaku-Shoin.

Rose, J. C., Green, T. J., and Green, V. D. (1996) NAGPRA is forever: Osteology and the repatriation of skeletons. *Annual Review of Anthropology* 25:81–103.

Rösing, W. (1984) Discreta of the human skeleton: A critical review. *Journal of Human Evolution* 13:319–323.

Rothschild, B. M. (1992) Advances in detecting disease in earlier human populations. In: S. R. Saunders and M. A. Katzenberg (Eds.) *Skeletal Biology of Past Peoples: Research Methods.* pp. 131–151. New York, Wiley Liss.

Rothschild, B. M., Woods, R. J., and Ortel, W. (1990) Rheumatoid arthritis "in the buff": Erosive arthritis in defleshed bones. *American Journal of Physical Anthropology* 82:441–449.

Rougé, D., Telmon, N., Arrue, P., Larrouy, G., and Arbus, L. (1993) Radiographic identification of human remains through deformities and anomalies of postcranial bones: A report of two cases. *Journal of Forensic Sciences* 38:997–1007.

Ruff, C. B., Larsen, C. S., and Hayes, W. C. (1984) Structural changes in the femur with the transition to agriculture on the Georgia coast. *American Journal of Physical Anthropology* 64:125–136.

Russell, K. F., Simpson, S. W., Genovese, J., Kinkel, M. D., Meindl, R. S., and Lovejoy, C. O. (1993) Independent test of the fourth rib aging technique. *American Journal of Physical Anthropology* 92:53–62.

Russell, M. D., and LeMort, F. (1986) Cutmarks on the Engis 2 calvaria? *American Journal of Physical Anthropology* 69:317–323.

Salo, W. L., Aufderheide, A. C., Buikstra, J., and Holcomb, T. A. (1994) Identification of *Mycobacterium tuberculosis* DNA in a pre-Columbian Peruvian mummy. *Proceedings of the National Academy of Sciences of the United States of America* 91:2091–2094.

Sandford, M. K. (Ed.) (1993) *Investigations of Ancient Human Tissue: Chemical Analyses in Anthropology.* Langhorne, Pennslyvania: Gordon and Breach Science Publishers, SA.

Saunders, S. R. (1989) Nonmetric skeletal variation. In: M. Y. İşcan and K. A. R. Kennedy, (Eds.) *Reconstruction of Life from the Skeleton.* pp. 95–108. New York: Alan R. Liss.

Saunders, S. R., and Herring, A. (Eds.) (1995) *Grave Reflections: Portraying the Past Through Cemetery Studies.* Toronto: Canadian Scholars' Press.

Saunders, S. R., and Katzenberg, M. A. (Eds.) (1992) *Skeletal Biology of Past Peoples: Research Methods.* New York: Wiley Liss.

Schmucker, B. J. (1985) Dental attrition: A correlative study of dietary and subsistence patterns in California and New Mexico Indians. In: C. F. Merbs and R. J. Miller (Eds.) *Health and Disease in the Prehistoric Southwest.* Tempe, Arizona: Arizona State University Anthropological Research Papers 34:275–323.

Schoeninger, M. J. (1995) Stable isotope studies in human evolution. *Evolutionary Anthropology* 3:83–98.

Schultz, M. (1997a) Microscopic structure of bone. In: W. D. Haglund and M. H. Sorg (Eds.) *Forensic Taphonomy.* pp. 187–199. Boca Raton, Florida: CRC Press.

Schultz, M. (1997b) Microscopic investigation of excavated skeletal remains: A contribution to paleopathology and forensic medicine. In: W. D. Haglund and M. H. Sorg (Eds.) *Forensic Taphonomy.* pp. 201–222. Boca Raton, Florida: CRC Press.

Schwartz, J. H. (1995) *Skeleton Keys.* New York: Oxford University Press.

Scott, E. C. (1979) Dental wear scoring technique. *American Journal of Physical Anthropology* 51:213–218.

Scott, G. R., and Dahlberg, A. A. (1982) Microdifferentiation in tooth crown morphology among Indians of the American Southwest. In: B. Kurtén (Ed.) *Teeth: Form, Function, and Evolution.* pp. 259–291. New York: Columbia University Press.

Scott, G. R., and Turner, C. G. (1988) Dental anthropology. *Annual Review of Anthropology* 17:99–126.

Scott, G. R., and Turner, C. G. (1997) *The Anthropology of Modern Human Teeth: Dental Morphology and Its Variation in Recent Human Populations.* New York: Cambridge University Press.

Shafer, H. J., Marek, M., and Reinhard, K. J. (1989) A Mimbres burial with associated colon remains from the NAN Ranch Ruin, New Mexico. *Journal of Field Archaeology* 16:17–30.

Shafritz, A. B., Shore, E. M., Gannon, F. H., Zasloff, M. A., Taub, R., Muenke, M., and Kaplan, S. (1996) Overexpression of an osteogenic morphogen in fibrodysplasia ossificans progressiva. *New England Journal of Medicine* 335:555–561.

Shahack-Gross, R., Bar-Yosef, O., and Weiner, S. (1997) Black-coloured bones in Hayonim Cave, Israel: Differentiating between burning and oxide staining. *Journal of Archaeological Science* 24:439–446.

Shipman, P., Walker, A., and Bichell, D. (1985) *The Human Skeleton.* Cambridge, Massachusetts: Harvard University Press.

Sillen, A., Sealey, J. C., and van der Merwe, N. J. (1989) Chemistry and paleodietary research: No more easy answers. *American Antiquity* 54:504–512.

Skinner, M., and Goodman, A. H. (1992) Anthropological uses of developmental defects of enamel. In: S. R. Saunders and M. A. Katzenberg (Eds.) *Skeletal Biology of Past Peoples: Research Methods.* pp. 153–174. New York: Wiley Liss.

Skinner, S. A., Saunders, C., Poirier, D. A., Krofina, D. L., and Wheat, P. (1998) Be prepared: The archeology merit badge is here. *Common Ground* 3:1:38–43.

Smith, B. H. (1991) Standards of human tooth formation and dental age assessment. In: M. A. Kelley and C. S. Larsen (Eds.) *Advances in Dental Anthropology.* pp. 143–168. New York: Wiley-Liss.

Smith, B. H., and Garn, S. M. (1987) Polymorphisms in eruption sequence of permanent teeth in American children. *American Journal of Physical Anthropology* 74:289–303.

Smith, H. B. (1984) Patterns of molar wear in hunter-gatherers and agriculturalists. *American Journal of Physical Anthropology* 63:39–56.

Smith, J., and Latimer, B. (1989) A method for making three-dimensional reproductions of bones and fossils. *Kirtlandia* (Cleveland Museum of Natural History) 44:3–16.

Snow, C. C. (1982) Forensic anthropology. *Annual Review of Anthropology* 11:97–131.

Spoor, C. F., Zonneveld, F. W., and Macho, G. A. (1993) Linear measurements of cortical bone and dental enamel by computed tomography: Applications and problems. *American Journal of Physical Anthropology* 91:469–484.

Steele, D. G., and Bramblett, C. A. (1988) *The Anatomy and Biology of the Human Skeleton.* College Station, Texas: Texas A&M University Press.

Steinbock, R. T. (1976) *Paleopathological Diagnosis and Interpretation.* Springfield, Illinois: C. C. Thomas.

Stevenson, P. H. (1924) Age order of epiphyseal union in man. *American Journal of Physical Anthropology* 7:53–93.

Stewart, T. D. (1979) *Essentials of Forensic Anthropology.* Springfield, Illinois: C. C. Thomas.

Stirland, A. (1992) Diagnosis of occupationally related paleopathology: Can it be done? In: D. J. Ortner and A. C. Aufderheide (Eds.) *Human Paleopathology: Current Syntheses and Future Options.* pp. 40–47. Washington, D.C.: Smithsonian Institution Press.

Stirland, A. J. (1994) The angle of femoral torsion: An impossible measurement? *International Journal of Osteoarchaeology* 4:31–35.

St. Hoyme, L. E., and İşcan, M. Y. (1989) Determination of sex and race: Accuracy and assumptions. In: M. Y. İşcan and K. A. R. Kennedy (Eds.) *Reconstruction of Life from the Skeleton.* pp. 53–93. New York: Alan R. Liss.

Stone, A. C., Milner, G. R., Pääbo, S., and Stoneking, M. (1996) Sex determination of ancient human skeletons using DNA. *American Journal of Physical Anthropology* 99:231–238.

Stone, A. C., and Stoneking, M. (1993) Ancient DNA from a pre-Columbian Amerindian population. *American Journal of Physical Anthropology* 92:463–471.

Stoneking, M. (1995) Ancient DNA: How do you know when you have it and what can you do with it? *American Journal of Human Genetics* 57:1259–1262.

Stout, S. D. (1992) Methods of determining age at death using bone microstructure. In: S. R. Saunders and M. A. Katzenberg (Eds.) *Skeletal Biology of Past Peoples: Research Methods.* pp. 21–35. New York: Wiley Liss.

Stout, S. D., Porro, M. A., and Perotti, B. (1996) Brief communication: A test and correction of the clavicle method for histological age determination of skeletal remains. *American Journal of Physical Anthropology* 100:139–142.

Stout, S. D., Dietze, W. H., İşcan, M. Y., and Loth, S. R. (1994) Estimation of age at death using cortical histomorphometry of the sternal end of the fourth rib. *Journal of Forensic Sciences* 39:778–784.

Stout, S. D., and Paine, R. R. (1992) Histological age estimation using rib and clavicle. *American Journal of Physical Anthropology* 87:111–116.

Stuart-Macadam, P. (1987a) A radiographic study of porotic hyperostosis. *American Journal of Physical Anthropology* 74:511–520.

Stuart-Macadam, P. (1987b) Porotic hyperostosis: New evidence to support the anemia theory. *American Journal of Physical Anthropology* 74:521–526.

Stuart-Macadam, P. (1991) Porotic hyperostosis: Changing interpretations. In: D. J. Ortner and A. C. Aufderheide (Eds.) *Human Paleopathology: Current Syntheses and Future Options.* pp. 36–39. Washington, D.C.: Smithsonian Institution Press.

Suchey, J. M., Wiseley, D. V., Green, R. F., and Noguchi, T. T. (1979) Analysis of dorsal pitting in the *Os pubis* in an extensive sample of modern American females. *American Journal of Physical Anthropology* 51:517–540.

Suchey, J. M., Wiseley, D. V., and Katz, D. (1986) Evaluation of the Todd and McKern-Stewart methods for aging the male *Os-Pubis*. In: K. J. Reichs (Ed.) *Forensic Osteology: Advances in the Identification of Human Remains.* pp. 33–67. Springfield, Illinois: C. C. Thomas.

Sundick, R. I. (1984) Ashes to ashes, dust to dust or where did the skeleton go? In: T. A. Rathburn and J. E. Buikstra (Eds.) *Human Identification: Case Studies in Forensic Anthropology.* pp. 412–423. Springfield, Illinois: C. C. Thomas.

Sutherland, L. D., and Suchey, J. M. (1991) Use of the ventral arc in pubic sex determination. *Journal of Forensic Sciences* 36:501–511.

Sweet, D. J., and Sweet, H. W. (1995) DNA analysis of dental pulp to link incinerated remains of homicide victim to crime scene. *Journal of Forensic Sciences* 40:310–314.

Tague, R. G. (1988) Bone resorption of the pubis and preauricular area in humans and nonhuman mammals. *American Journal of Physical Anthropology* 76:251–267.

Taylor, R. E., Hare, P. E., and White, T. D. (1995) Geochemical criteria for thermal alteration of bone. *Journal of Archaeological Science* 22:115–119.

Taylor, R. M. S. (1978) *Variation in Morphology of Teeth: Anthropologic and Forensic Aspects.* Springfield, Illinois: C. C. Thomas.

Teaford, M. F. (1991) Dental microwear: What can it tell us about diet and dental function? In: M. A. Kelley and C. S. Larsen (Eds.) *Advances in Dental Anthropology.* pp. 341–356. New York: Wiley Liss.

Teaford, M. F. (1994) Dental microwear and dental function. *Evolutionary Anthropology* 3:17–30.

Tekiner, R. (1971) *Human Skeletal Morphology: A Laboratory Manual for Anthropology.* New York: Continental House.

Thomas, D. H. (1986) *Refiguring Anthropology: First Principles of Probability and Statistics.* Prospect Heights, Illinois: Waveland Press.

Tobias, P. V. (1991) On the scientific, medical, dental, and educational value of collections of human skeletons. *International Journal of Anthropology* 6: 277–280.

Todd, T. W. (1920) Age changes in the pubic bone: I. The white male pubis. *American Journal of Physical Anthropology* 3:467–470.

Trodden, B. J. (1982) A radiographic study of the calcification and eruption of the permanent teeth in Inuit and Indian children. *Archaeological Survey of Canada Papers* 112:1–136.

Trotter, M. (1970) Estimation of stature from intact long bones. In: T. D. Stewart (Ed.) *Personal Identification in Mass Disasters.* pp. 71–83. Washington, D.C.: Smithsonian Institution Press.

Trotter, M., and Gleser, G. C. (1958) A re-evaluation of estimation of stature based on measurements of stature taken during life and of long bones after death. *American Journal of Physical Anthropology* 16:79–123.

Turner, C. G. (1976) Dental evidence on the origins of the Ainu and Japanese. *Science* 193:911–913.

Turner, C. G. (1983b) Taphonomic reconstructions of human violence and cannibalism based on mass burials in the American Southwest. In: G. M. LeMoine and A. S. MacEachern (Eds.) *A Question of Bone Technology.* pp. 219–240. Calgary: University of Calgary Archaeological Association.

Turner, C. G. (1986) What is lost with skeletal reburial? I. Adaptation. *Quarterly Review of Archaeology* 7(1):1–3.

Turner, C. G. (1987) Late Pleistocene and Holocene population history of East Asia based on dental variation. *American Journal of Physical Anthropology* 73: 305–321.

Turner, C. G., and Markowitz, M. A. (1990) Dental discontinuity between Late Pleistocene and recent Nubians: Peopling of the Eurafrican–South Asian triangle I. *Homo* 41:32–41.

Turner, C. G., and Turner, J. A. (1999) *Man Corn: Cannibalism and Violence in the American Southwest.* Salt Lake City: University of Utah Press.

Turner, C. G., Nichol, C. R., and Scott, G. R. (1991) Scoring procedures for key morphological traits of the permanent dentition. In: M. A. Kelley and C. S. Larsen (Eds.) *Advances in Dental Anthropology.* pp. 13–31. New York: Wiley-Liss.

Twelfth International Congress of Anatomists. (1989) *International Anatomical Nomenclature Committee: Nomina Anatomica* (6th Edition). New York: Churchill-Livingstone.

Tyrrell, A. J., Evison, M. P., Chamberlain, A. T., and Green, M. A. (1997) Forensic three-dimensional facial reconstruction: Historical review and contemporary developments. *Journal of Forensic Sciences* 42:653–661.

Tyson, R. A. (Ed.) (1997) *Human Paleopathology and Related Subjects: An International Bibliography.* San Diego: San Diego Museum of Man.

Tyson, R. A., and Alcanskas, E. S. D. (Eds.) (1980) *Catalog of the Hrdlička Paleopathology Collection.* San Diego: San Diego Museum of Man.

Ubelaker, D. H. (1982) The development of American paleopathology. In: F. Spencer (Ed.) *A History of American Physical Anthropology 1930–1980.* pp. 337–356. New York: Academic Press.

Ubelaker, D. H. (1984) Positive identification from the radiographic comparison of frontal sinus patterns. In: T. Rathburn and J. Buikstra (Eds.) *Human Identification.* pp. 399–411. Springfield, Illinois: C. C. Thomas.

Ubelaker, D. H. (1987) Estimating age at death from immature human skeletons: An overview. *Journal of Forensic Sciences* 32:1254–1263.

Ubelaker, D. H. (1990) Positive identification of American Indian skeletal remains from radiograph comparison. *Journal of Forensic Sciences* 35:466–472.

Ubelaker, D. H. (1989) *Human Skeletal Remains: Excavation, Analysis, Interpretation* (2nd Edition). Washington, D.C.: Taraxacum.

Ubelaker, D. H. (1992a) North American Indian population size. In: J. W. Verano and D. H. Ubelaker (Eds.) *Disease and Demography in the Americas.* pp. 169–176. Washington, D.C.: Smithsonian Institution Press.

Ubelaker, D. H. (1992b) Hyoid fracture and strangulation. *Journal of Forensic Sciences* 37:1216–1222.

Ubelaker, D. H. (1996) Skeletons testify: Anthropology in forensic science (AAPA Luncheon Address, April 12, 1996). *Yearbook of Physical Anthropology* 39: 229–244.

Ubelaker, D. H., Berryman, H. E., Sutton, T. P., and Ray, C. E. (1991) Differentiation of hydrocephalic calf and human calvariae. *Journal of Forensic Sciences* 36: 801–812.

Ubelaker, D. H. and Grant, L. G. (1989) Human skeletal remains: Preservation or reburial. *Yearbook of Physical Anthropology* 32:260–287.

Van Gerven, D. P., and Armelagos, G. J. (1983) "Farewell to Paleodemography?" Rumors of its death have been greatly exaggerated. *Journal of Human Evolution* 12:353–360.

Van Gerven, D. P., Sheridan, S. G., and Adams, W. Y. (1995) The health and nutrition of a Midieval Nubian population: The impact of political and economic change. *American Anthropologist* 97:468–480.

Verano, J. W., and Ubelaker, D. H. (Eds.) (1992) *Disease and Demography in the Americas.* Washington, D.C.: Smithsonian Institution Press.

Vitelli, K. D. (Ed.) (1996) *Archaeological Ethics.* Walnut Creek, California: Altamira Press.

Von Endt, D. W., and Ortner, D. J. (1984) Experimental effects of bone size and temperature on bone diagenesis. *Journal of Archaeological Science* 11:247–253.

Waldron, T. (1991) Rates for the job. Measures of disease frequency in palaeopathology. *International Journal of Osteoarchaeology* 1:17–25.

Waldron, T. (1994) *Counting the Dead: The Epidemiology of Skeletal Populations.* West Sussex, England: John Wiley.

Waldron, T., and Rogers, J. (1991) Inter-observer variation in coding osteoarthritis in human skeletal remains. *International Journal of Osteoarchaeology* 1:49–56.

Walker, A. C. (1981) Dietary hypotheses and human evolution. *Philosophical Transactions of the Royal Society of London,* Series B, 292:57–63.

Walker, P. L. (1986) Porotic hyperostosis in a marine-dependent California Indian population. *American Journal of Physical Anthropology.* 69:345–354.

Walker, P. L., and Erlandson, J. M. (1986) Dental evidence for prehistoric dietary change on the northern Channel Islands, California. *American Antiquity* 51: 375–383.

Walker, P. L., Cook, D. C., and Lambert, P. M. (1997) Skeletal evidence for child abuse: A physical anthropological perspective. *Journal of Forensic Sciences* 42: 196–207.

Walker, P. L., and Cook, D. C. (1998) Gender and sex: Vive la difference. *American Journal of Physical Anthropology* 106:255–259.

Walker, P. L., and Hewlett, B. S. (1990) Dental health diet and social status among Central African foragers and farmers. *American Anthropologist* 92:383–398.

Walker, P. L., Dean, G., and Shapiro, P. (1991) Estimating age from tooth wear in archaeological populations. In: M. A. Kelley and C. S. Larsen (Eds.) *Advances in Dental Anthropology*. pp. 169–178. New York: Wiley-Liss.

Walker, P. L., Johnson, J. R., and Lambert, P. M. (1988) Age and sex biases in the preservation of human skeletal remains. *American Journal of Physical Anthropology* 76:183–188.

Walker, R. (1985) *A Guide to Postcranial Bones of East African Mammals*. Palo Alto, California: Hylochoerus Press, Blackwell Scientific.

Walker, R. A., and Lovejoy, C. O. (1985) Radiographic changes in the clavicle and proximal femur and their use in the determination of skeletal age at death. *American Journal of Physical Anthropology* 68:67–78.

Wallin, J. A., Tkocz, I., and Kristensen, G. (1994) Microscopic age determination of human skeletons including an unknown but calculable variable. *International Journal of Osteoarchaeology* 4:353–362.

Warwick, R., and Williams, P. L. (Eds.) (1973) *Gray's Anatomy* (35th British Edition). Philadelphia: W.B. Saunders.

Washburn, S. L. (1948) Sex differences in the pubic bone. *American Journal of Physical Anthropology* 6:199–207.

Weaver, D. S. (1986) Forensic aspects of fetal and neonatal specimens. In: K. J. Reichs (Ed.) *Forensic Osteology: Advances in the Identification of Human Remains*. pp. 90–100. Springfield, Illinois: C. C. Thomas.

Webb, P. A., and Suchey, J. M. (1985) Epiphyseal union of the anterior iliac crest and medial clavicle in a modern multiracial sample of American males and females. *American Journal of Physical Anthropology* 68:457–466.

Webb, S. (1987) Reburying Australian skeletons. *Antiquity* 61:292–296.

Webster, W. P., Murray, W. K., Brinkhous, W., and Hudson, P. (1986) Identification of human remains using photographic reconstruction. In: K. J. Reichs (Ed.) *Forensic Osteology: Advances in the Identification of Human Remains*. pp. 256–289. Springfield, Illinois: C. C. Thomas.

White, T. D. (1992) *Prehistoric Cannibalism at Mancos 5MTUMR-2346*. Princeton, New Jersey: Princeton University Press.

White, T. D., and Folkens, P. A. (1991) *Human Osteology* (First Edition). San Diego, California: Academic Press.

White, T. D., Suwa, G., Hart, W. K., Walter, R. C., Wolde-Gabriel, G., deHeinzelin, J., Clark, J. D., Asfaw, B., and Vrba, E. (1993) New discoveries of *Australopithecus* at Maka in Ethiopia. *Nature* 366:261–265.

White, T. D., and Toth, N. (1989) Engis: Preparation damage, not ancient cutmarks. *American Journal of Physical Anthropology* 78:361–367.

Willey, P. (1981) Another view by one of the Crow Creek researchers. *Early Man* 3(3):26.

Willey, P., and Emerson, T. E. (1993) The osteology and archaeology of the Crow Creek massacre. *Plains Anthropologist* 38:227–269.

Woelfel, J. B., and Schied, R. C. (1997) *Dental Anatomy: Its Relevance to Dentistry.* Baltimore, Maryland: Williams and Wilkins.

Wolf, D. J. (1986) Forensic anthropology scene investigations. In: K. J. Reichs (Ed.) *Forensic Osteology: Advances in the Identification of Human Remains.* pp. 3–23. Springfield, Illinois: C. C. Thomas.

Wolff, J. (1869) Über die Bedeutung der Architektur der spongiösen Substanz. *Zentralblatt für die medizinische Wissenschaft* 6: Yahrgang, pp. 223–234.

Wood, J. W., Milner, G. R., Harpending, H. C., and Weiss, K. M. (1992) The osteological paradox: Problems of inferring prehistoric health from skeletal samples. *Current Anthropology* 33:343–370.

Woods, R. J., and Rothschild, B. M. (1988) Population analysis of symmetrical erosive arthritis in Ohio Woodland Indians (1200 years ago). *Journal of Rheumatology* 15:1258–1263.

Zimmerman, L. J. (1981) How the Crow Creek archeologists view the question of reburial. *Early Man* 3(3):25–26.

Zimmerman, L. J. (1987a) Webb on reburial: A North American perspective. *Antiquity* 61:462–463.

Zimmerman, L. J. (1987b) The impact of the concepts of time and past on the concept of archaeology: Some lessons from the reburial issue. *Archaeological Review From Cambridge* 6:42–50.

Zimmerman, L. J. (1989) Made radical by my own: An archaeologist learns to accept reburial. In: R. Layton (Ed.) *Conflict in the Archaeology of Living Traditions.* pp. 60–67. London: Unwin Hyman.

Zimmerman, L. J. (1994) Sharing control of the past. *Archaeology* Nov/Dec. 1994: 65–68.

Zimmerman, L. J. (1997) Anthropology and responses to the reburial issue. In: T. Biolsi and L. J. Zimmerman, (Eds.) *Indians and Anthropologists: Vine Deloria Jr., and the Critique of Anthropology.* pp. 92–112. Tucson, Arizona: University of Arizona Press.

Zimmerman, M. R., and Kelley, M. A. (1982) *Atlas of Human Paleopathology.* New York: Praeger.

Zollikofer, C. P. E., Ponce De León, M. S., and Martin, R. D. (1998) Computer assisted paleoanthropology. *Evolutionary Anthropology* 6:41–54.

Index

Suture
   cranial suture closure, adult age
       estimation, 345, 347
   definition, 42, 56, 530
   types in skull, 56–57
Symphysis, definition, 21, 530
Syndesmoses, definition, 21, 530
Synostosis, definition, 21, 530
Synovial joint, definition and
       classification, 21, 530

Talus, tarsal features, 261–263
Taphonomy, see Postmortem skeletal
       modification
Tarsal bones
   calcaneus, 263–264
   confusion with other bones,
       268–269
   cuboid, 265
   growth, 268
   intermediate cuneiform, 267–268
   lateral cuneiform, 268
   medial cuneiform, 266–267
   navicular, 265–266
   talus, 261–263
Taurodontism, definition, 530
Teaching, osteology, 6–7, 12
Teeth, see also specific teeth
   ablation, 523
   age estimation
       adult age, 343–345
       subadult age, 342–343
   anatomy, 112–115
   attrition, 109, 343–345, 524
   biological distance measurement,
       431–433
   child abuse, 463
   development, 115–116, 342
   diet determination, 433–434
   directional terms, 38–39, 111
   discrete traits, 426–428
   evolution, 109, 116
   identification of individuals, 377
   identification of teeth
       arch localization, 124–129
       class of tooth, 117, 120
       deciduous versus permanent
           teeth, 111, 120
       overview, 116–117
       right versus left side, 129–132,
           135
       upper versus lower, 120–124
   impression molding, 298
   junctions, 112
   pathology

artificial modifications, 403–404
   calculus, 403
   caries, 401
   classification of diseases, 400
   hypoplasia, 115, 401, 402
   periodontitis, 402–403
sexing, 365
shorthand terminology, 111
supernumarary teeth, 115, 529
types
   canine, 110–111, 117
   incisor, 110, 117
   molar, 111, 120
   premolar, 111, 117, 120
Temporal bones
   anatomy, 72–76
   confusion with other bones, 77
   growth, 76–77
   siding, 77
Temporalis muscle, mastication, 106,
       530
Tendon, definition, 21
Thoracic vertebrae
   identification, 147–149
   overview, 146
   siding, 149
Thorax, see Ribs; Sternum
Tibia
   anatomy, 242, 245
   confusion with other bones, 248
   growth, 245
   siding, 248–249
Torus, definition, 42
Trabecular bone, definition, 23, 524
Transport, skeletal remains
   bagging and boxing, 291
   damage minimization, 290
   jacketing, 290
   packaging, 291
Transverse plane, definition, 37
Trapezium, carpal features, 205–206
Trapezius muscle, function, 177
Trapezoid, carpal features, 206
Trepanning, see Trephination
Trephination, history, 388–389, 530
Treponemal infection, periostitis, 393
Triceps brachii, function, 197–198
Triquetral, carpal features, 204
Trochanter, definition, 41
Tubercle, definition, 41
Tuberculosis, periostitis, 393
Tuberosity, definition, 41
Tumor
   ear exostoses, 397
   osteochondroma, 396–397
   osteoma, 397

sarcomatas, 397
   types, 396
Typology
   definition, 530
   limitations, 19–20

Ulna
   anatomy, 192–194
   confusion with other bones, 195,
       197
   growth, 195
   siding, 197
Ulnar, definition, 200

Variation
   definition, 15, 381
   modern versus ancient species,
       16–17, 19
   sources, 16, 381
   typology limitations, 19–20
Vertebrae, see also Cervical vertebrae;
       Lumbar vertebrae; Thoracic
       vertebrae
   anatomy, 139, 142–143
   functional aspects, 156
   growth, 143
Vertex, cranial osteometric point, 61
Volkmann's canal, function, 26
Vomer
   anatomy, 87–88
   confusion with other bones, 88
   growth, 88
   siding, 88

Wolff's law, 20, 530
World Wide Web, resources in
       osteology, 7–11
Wormian bone, definition, 530
Wrist, functional aspects, 197–198

Xiphoid process, sternum, 161

Zygion, cranial osteometric point, 62
Zygomatics
   anatomy, 93–94
   confusion with other bones, 95
   growth, 95
   siding, 95
Zygomaxillare, cranial osteometric
       point, 62
Zygoorbitale, cranial osteometric
       point, 63